Annual Conference of the International Deep Drawing Research Group (IDDRG 2016)

Challenges in Forming High-Strength Sheets

IOP Conference Series: Materials Science and Engineering
Volume 159

Linz, Austria
12 – 15 June 2016

Editors:

Edwin Till
Melanie Baumgartner
Benjamin Hackl
Bruno Hribernik
Christian Walch

ISBN: 978-1-5108-3381-4
ISSN: 1757-8981

Printed from e-media with permission by:

Curran Associates, Inc.
57 Morehouse Lane
Red Hook, NY 12571

Some format issues inherent in the e-media version may also appear in this print version.

Printed by Curran Associates, Inc. (2017)

For permission requests, please contact the Institute of Physics
at the address below.

Institute of Physics
Dirac House, Temple Back
Bristol BS1 6BE UK

Phone: 44 1 17 929 7481
Fax: 44 1 17 920 0979

techtracking@iop.org

Additional copies of this publication are available from:

Curran Associates, Inc.
57 Morehouse Lane
Red Hook, NY 12571 USA
Phone: 845-758-0400
Fax: 845-758-2633
Email: curran@proceedings.com
Web: www.proceedings.com

Table of contents

Volume 159

IDDRG2016 conference on "Challenges in Forming High-Strength Sheets".
12–15 June 2016, Linz, Austria

Accepted papers received: 15 November 2016
Published online: 13 December 2016

Preface

Papers

Manufacturing Issues

Simulation

Forming

Small Scale Testing

Lightweight Structures

Springback

Material Model

IOP Conf. Series: Materials Science and Engineering **159** (2016) 011001 doi:10.1088/1757-899X/159/1/011001

IDDRG^{2016}: Challenges in Forming High-Strength Sheets

Scope

The annual conference of the International Deep Drawing Research Group (IDDRG) is devoted to experimental and computational contributions with industrial relevance. The intent of the conference is to present new work and ideas from industry and academia and to provide opportunities for discussion among researchers in sheet metal formability. The main topics of the IDDRG^{2016} were:

- Lightweight Structures
- Manufacturing Issues
- Press Hardened Steels
- Full and Small Scale Testing
- Tribology
- Material Models
- Simulation
- Springback Simulation & Compensation

Hard Facts

The IDDRG^{2016} conference was organized by the Austrian Society of Metallurgy and Materials (ASMET) and the voestalpine Stahl Gmbh. The event took place at the voestalpine Stahlwelt in Linz, Austria, from 12-15 June 2016.

Figure 1. IDDRG^{2016} conference site at the voestalpine Stahl Welt, Linz, Austria

IOP Conf. Series: Materials Science and Engineering **159** (2016) 011001 doi:10.1088/1757-899X/159/1/011001

In total 165 participants from 21 countries all over the world joined the conference. The mix of researchers from industry and academia was well equilibrated (see Figure 2).

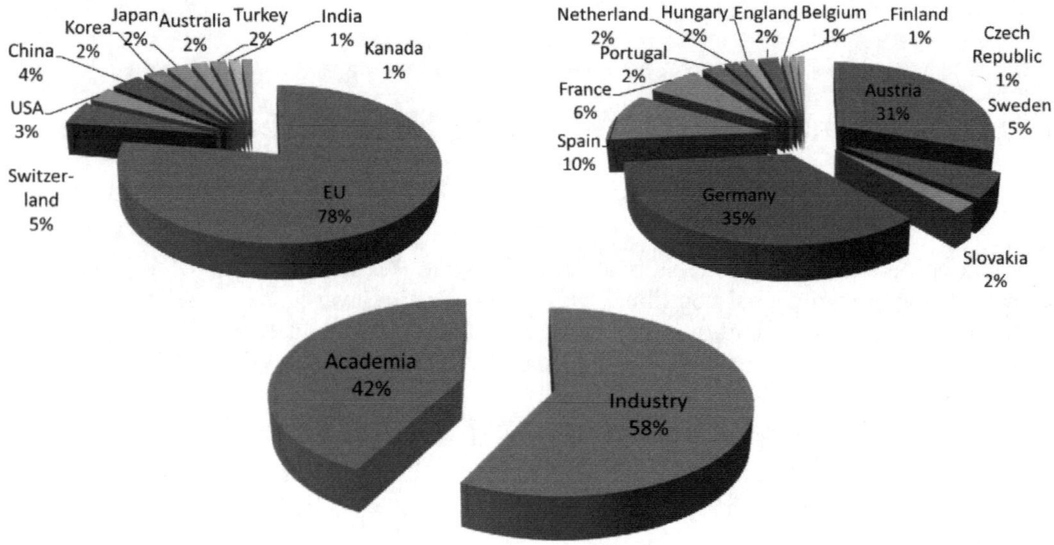

Figure 2. Distribution of IDDRG[2016] participants by country and profession

Summing up the number of the talks, including the key note lectures and the mini symposium on fracture modeling there were 93 presentations at the IDDRG[2016]. The distribution of the talks into the different topics is shown in Figure 3.

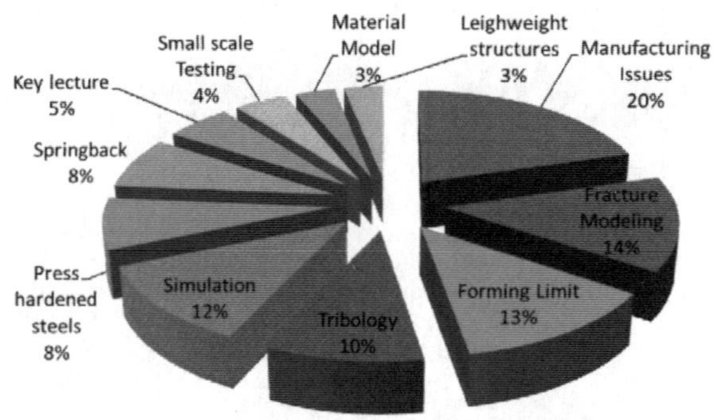

Figure 3. Distribution of the topics of the IDDRG[2016]

For the first time at an IDDRG conference the authors had the choice to publish their results at the IOP Conference Series: Materials Science and Engineering. Almost 50% of the 81 paper contributions at the IDDRG[2016] conference opted for the peer review process.

Soft Facts

In general the atmosphere at the IDDRG[2016] was very good, open and friendly. Especially at the mini symposium on fracture modeling issues where addressed openly and honestly. Discussions proceeded at coffee breaks, lunches and dinners. Besides the generally well prepared and presented talks, the participants enjoyed the side events a lot. There not only technical issues where discussed but personal bonds were founded. Finally the IDDRG[2016] conference felt like a big club of friends.

IOP Conf. Series: Materials Science and Engineering **159** (2016) 011001 doi:10.1088/1757-899X/159/1/011001

Figure 4. Group photo of IDDRG^{2016} participants

Dedication
Edwin Till, the chair of the organizing committee for the IDDRG^{2016} conference and long term member of the executive committee of the IDDRG, died on 22 November 2015. This IDDRG^{2016} conference is dedicated to him.

Figure 5. Edwin Till together with his wife Margit Till

Acknowledgment
The organizers want to thank all key note speakers, speakers, session chairs, the scientific committee, technicians, services and the participants (see Figure 4). Because of you this IDDRG^{2016} conference became a full success.

Scientific Committee

- Dorel Banabic, Technical University of Cluj-Napoca, Romania
- Frédéric Barlat, Pohang University of Science and Technology, Korea
- Jose Cesar de Sa, University of Porto, Portugal
- Patrick Duroux, ArcelorMittal, France

IOP Conf. Series: Materials Science and Engineering **159** (2016) 011001 doi:10.1088/1757-899X/159/1/011001

- Lars Greve, Volkswagen AG, Germany
- Joachim Hecht, Audi AG, Germany
- Pavel Hora, IVP, ETH Zürich, Switzerland
- Lutz Keßler, ThyssenKrupp Steel Europe, Germany
- Nico Langerak, Tata Steel, Netherlands
- Christian Leppin, Suisse Technology Partners, Switzerland
- Kjell Mattiasson, Chalmers University of Technology, Sweden
- Marion Merklein, Friedrich-Alexander University, Germany
- Dirk Mohr, ETH Zürich, Switzerland
- Hans Mulder, Tata Steel, Netherlands
- Karl Roll, Daimler AG, Germany
- Erman Tekkaya, Technical University Dortmund, Germany
- Miklós Tisza, University of Miskolc, Hungary
- Kuwabara Toshihiko, Tokyo University of Agriculture and Technology, Japan
- Ton Van den Boogaard, University of Twente, Netherlands
- Chester Van Tyne, Colorado School of Mines, USA
- Wolfram Volk, Technical University of Munich, Germany
- Klaus Wiegand, Daimler AG, Germany
- Jeong-Whan Yoon, Swinburn University, Australia
- Shi-Hong Zhang, Chinese Academy of Sciences, China

The editors

Edwin Till, voestalpine Stahl GmbH
Melanie Baumgartner, ASMET
Benjamin Hackl, voestalpine Stahl GmbH
Bruno Hribernik, ASMET
Christian Walch, voestalpine Stahl GmbH

IOP Conf. Series: Materials Science and Engineering **159** (2016) 011002 doi:10.1088/1757-899X/159/1/011002

Peer review statement

All papers published in this volume of *IOP Conference Series: Materials Science and Engineering* have been peer reviewed through processes administered by the proceedings Editors. Reviews were conducted by expert referees to the professional and scientific standards expected of a proceedings journal published by IOP Publishing.

The interaction between short-term heat-treatment and the formability of an Al-Mg-Si alloy regarding deep drawing processes

M Machhammer and C Sommitsch

Institute Tools and Forming, Graz University of Technology, Graz, Austria

E-mail: michael.machhammer@tugraz.at

Abstract. Research conducted in recent years has shown that heat-treatable Al-Mg-Si alloys (6xxx) have great potential concerning the design of lightweight car bodies. Compared to conventional deep drawing steels the field of application is limited by a lower formability. In order to minimize the disadvantage of a lower drawability a short-term heat-treatment (SHT) can be applied before the forming process. The SHT, conducted in selected areas on the initial blank, leads to a local reduction of strength aiming at the decrease of critical stress during the deep drawing process. For the successful procedure of the SHT a solid knowledge about the crucial process parameters such as the design of the SHT layout, the SHT process time and the maximum SHT temperature are urgently required. It also should be noted that the storage time between the SHT and the forming processes affects the mechanical properties of the SHT area. In this paper, the effect of diverse SHT process parameters and various storage time-frames on the major and minor strain situation of a deep drawn part is discussed by the evaluation of the forming limit diagram. For the purpose of achieving short heating times and a homogenous temperature distribution a one side contact heating tool has been used for the heat treatment in this study.

Keywords: Short-term heat treatment, Al-Mg-Si alloy, storage time, contact heating

1. Introduction

The use of aluminum components in car bodies, powertrains, and chassis offer a high potential of weight saving in the automotive sector [1]. Thus, to increase the number of aluminum parts in the field of car body construction, the material's disadvantage of a degraded formability compared to deep drawing steel qualities must be improved [2]. A promising method for enhancement of the forming behavior of heat-treatable Al-Mg-Si aluminum alloys (6xxx) is the local reduction of strength, in order to reduce stress in the critical zone, when failures occur during forming processes. This can be achieved by a partial short-term heat-treatment (SHT) on the aluminum sheet in a process state before the actual forming process [3, 4]. The improvement of the formability of 6xxx through SHT depends considerably on the design of the SHT layout, the maximum SHT temperature, and the storage time between the heat-treatment and the forming process. The heat-treatment is mostly applied through conduction by contact plates, or through laser technology. One of the advantages using laser heating technology is a high flexibility regarding modifications of the SHT layout during prototyping. When short SHT process cycle times and homogeneous temperature distributions are required, heating via conduction offers improved performances, as compared to laser heating concepts. The present study shows the modifications of the forming limit diagram (FLD) of a deep drawn 6xxx material through the use of different SHT layouts, SHT temperatures and SHT process times. Furthermore, it illustrates the influence of storage time between SHT processing and forming.

2. Material

The result of natural aging after solution heat treatment and the quenching of Al-Mg-Si alloys is Si- and Mg-atoms grouped in clusters and dispersed within the aluminum lattice. These clusters distort the material lattice and hinder the dislocation motion [5]. SHT leads to a partial or complete dissolution, depending on the maximum SHT temperature of the MgSi-clusters within the material lattice. To be precise, the initiate softening of the material's yield strength through the dissolution of the clusters starts at an SHT temperature of 200°C (T) and increases to 400°C. The profile of the yield strength regarding the SHT temperature is not decreasing linear. A renewed increase of strength appears at a temperature near to 300°C and ends shortly after 300°C. In addition to the softening of 6xxx, at the same time a reduction of ductility in the SHT temperature range between 200 °C and 350 °C occurs [6]. In this research an Al-Mg-Si aluminum alloy 6016 in T4 temper (Si - 1.13% Mn – 0.15 Fe – 0.28 Zn – 0.02 Mg – 0.48 Ti – 0.01 V – 0.01 Cu – 0.06 Cr – 0.06) was used in the SHT and during the deep drawing process.

3. Experimental Procedure

3.1. Heating strategy

In this experiment/research a contact heating tool (Figure 1) was used to provide near-series conditions during the SHT process. Moreover, the heat sources of the tool are evenly distributed heating cartridges, integrated in a steel plate. In order to obtain partial modifications of the material's flange area, the design of the heat-treatment layout was adapted to the shape of the heat plate. In addition, heating nearby the critical zone is achieved through steel cylinders equipped with heating cartridges (Figure 1). The heating cylinders are preloaded with compression springs in order to provide enough contact pressure during the heating process. The upper unit of the tool is lifted with a portal crane, and the closing of the tool can be measured with a position sensor. The pressure that was needed to improve the heat transfer between the heat plate and the blank was applied by an additional weight on the upper tool. Moreover, the 6016 material is placed on an insulating multilayer textile (ThermTex 1100SVG) in the shape of the heat-treatment layout to minimize the heat loss on the bottom side of the blank (Figure 2 and 3).

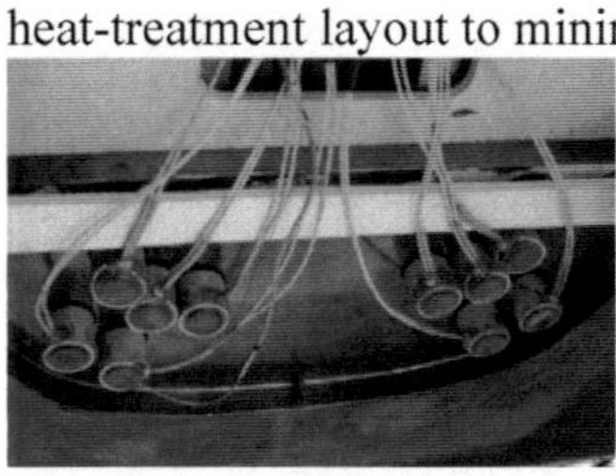

Figure 1. Heating tool

Figure 2. Insulating textile

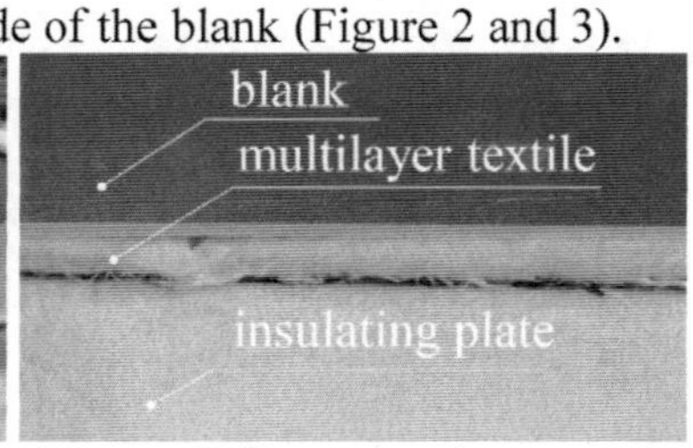

Figure 3. Bottom unit

3.2. SHT process settings

To investigate the influence of different SHT temperatures on the forming limit diagram (FLD) of a deep drawn part, the SHT layout 1 shown in figure 4 was used. In this test series maximum temperatures in a temperature range of 200°C to 360°C were conducted during the SHT process. The modification of formability, caused by the design concept of the heat treated area, was evaluated with the SHT layouts presented in figures 4, 5 and 6. Therefore, a maximum SHT temperature of T + 70°C was used for each layout.

Figure 4. SHT layout 1

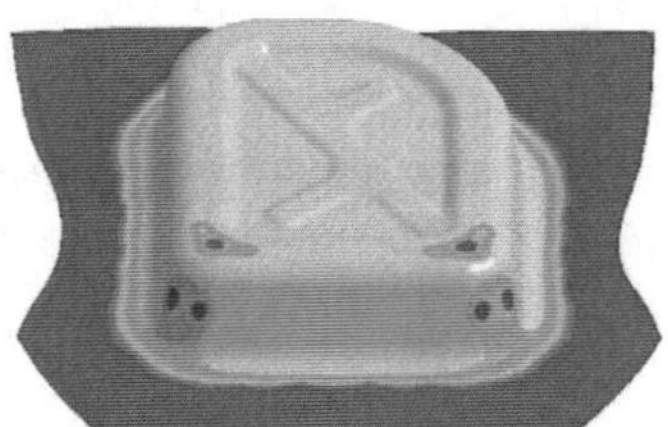

Figure 5. SHT layout 2

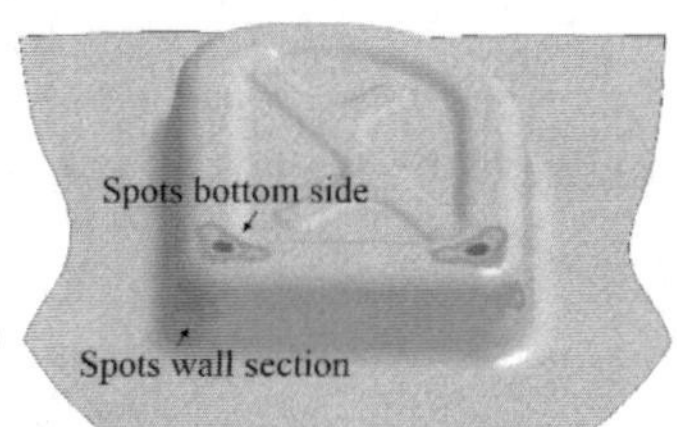

Figure 6. SHT layout 3

In brief, not only the maximum SHT temperature and the design of the SHT layout, but also the time-temperature profile affect the forming behavior. Due to the aim of dissolving the MgSi-clusters, and not enlarging them, the heating and holding time should be as short as possible [7]. In this study an SHT process time of eight seconds was mainly used. The time temperature profiles for each target temperature were evaluated in preliminary tests with blanks which have been equipped with thermocouples (figure 7). To analyze the influence of the heating time on formability, a test series with SHT process times (tx, ty, tz) between 2 and 14 seconds were conducted at the SHT process. In order to achieve fast heating times (tx) of a maximum SHT temperature of T + 70°C, the settings of the SHT process with an SHT temperature of T + 160°C was used (figure 7).

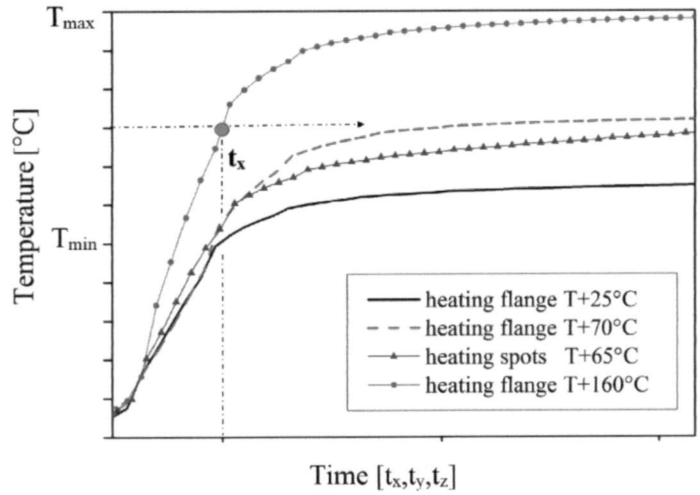

Figure 7. Profiles of the temperature over time measured in the blank for different heating configurations

3.3. *Deep drawing*

The performed deep drawing processes were conducted with a rectangular shaped tool with different radius combinations and an additional stamping on the bottom side (figure 4). The forming process was controlled with distance plates (thickness = 1.15mm) and a blankholder force of 200 kN. With reference to the coefficient of friction between tool and blank, former investigations have identified that dry lubricants are changing tribological properties by the use of SHT [8]. Thus, the applied dry lubricant, on the material's surface during manufacturing, was completely removed before SHT processes. Thus, to minimize adhesive wear, galling, and abrasion of the tool, an oil-based lubricant (WISURA ZO 3107/180) was applied on both surfaces of the blank after the SHT process. Considering natural aging of the SHT area, it has to be pointed out, that the forming processes of SHT layouts 1, 2 and 3 were achieved shortly after the heat treatment. In the analyses of how the storage time in process cycle effects the deep drawing process, the forming of SHT layout 2 (figure 5) was performed after additional time frames of 16, 40, 100 and 164 hours. To evaluate the potential of SHT in a manufacturing process, a die with a maximum drawing depth of 115 mm was used. Moreover, to minimize wrinkling the blank outline of the 6016 was optimized and a blank holder force of 420 kN was applied during the deep drawing process.

3.4. *Deep drawing Measurement and evaluation of formability*

In order to assess the deep drawing results the major and minor strain distribution on each test part was measured with the optical forming analysis ARGUS. Figure 8 shows the measured area in this analysis. Also, to present the improvement and potential of the SHT on deep drawing processes, a

IOP Conf. Series: Materials Science and Engineering **159** (2016) 012001 doi:10.1088/1757-899X/159/1/012001

forming limit curve (FLC) of the 6016 material in T4 and SHT (T + 70°C) state was determined through the Nakajima test.

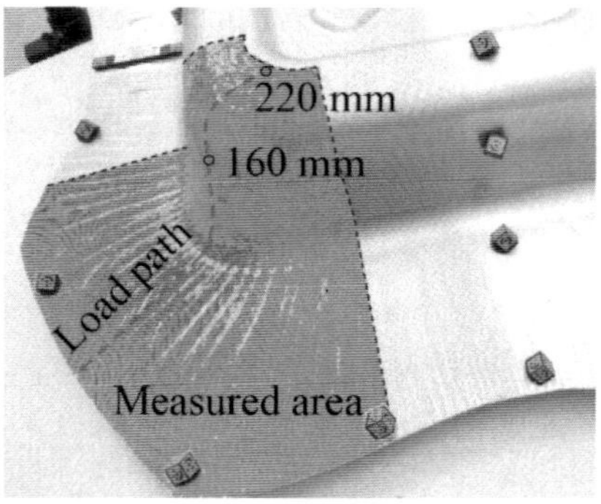

Figure 8. Area of measured strain situation

4. Results

4.1. Deep drawing of SHT layout 1 – SHT temperatures T+25°C, T+70°C, T+160°C – ty

Figure 9 clearly shows the influence of SHT before the forming process in regard to the material failure. In a manufacturing process, without SHT, the material cracks at a drawing depth of 73 mm. A drawing depth of 85 mm could be produced, without crack deformations, by including SHT in the process chain with temperatures of T + 25°C, T + 70°C, T + 160°C and a SHT process time of ty (figure 9).

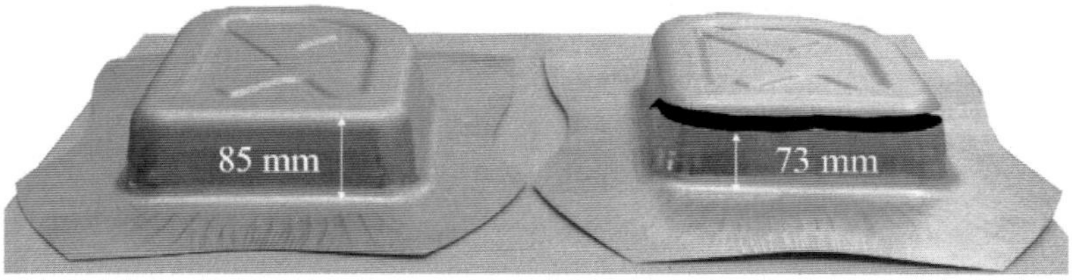

Figure 9. Deep drawn test part with and without SHT (drawing depth 85mm)

As illustrated in figure 10, the critical zone in the radius area reaches the minimum of resistance to necking at the deformation mode near plane strain tension, by the use of the SHT temperature of T + 25°C. Due to the increase of the SHT temperature from T + 25°C to T + 70°C, the strain distribution in the critical zone was lowered to the FLC's safety margin of 20 % (figure 11). A further increase of the SHT temperature from T + 70°C to T + 160°C, leads to the highest decrease of major strain in the critical zone (figure 12).

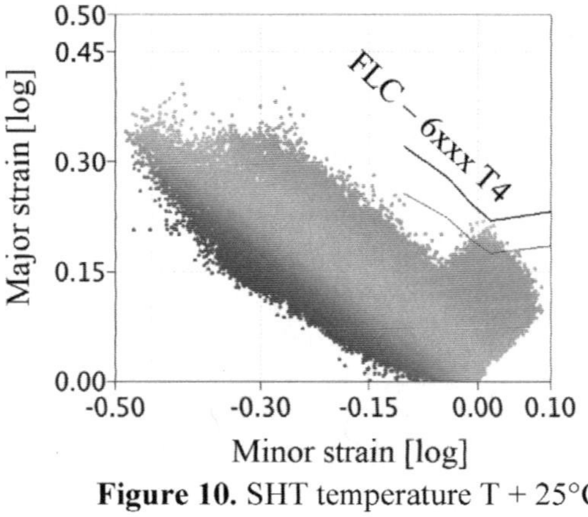

Figure 10. SHT temperature T + 25°C

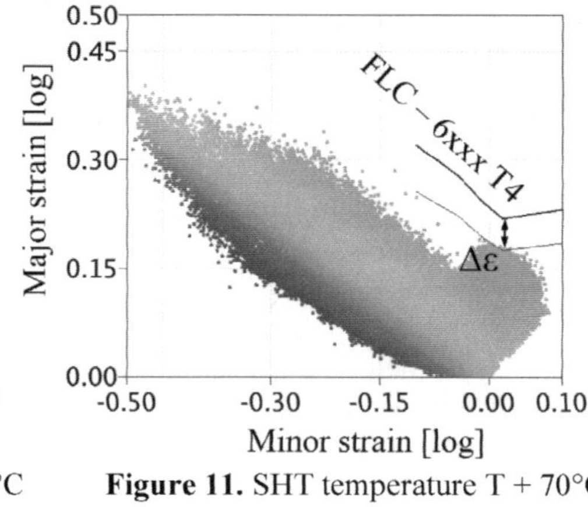

Figure 11. SHT temperature T + 70°C

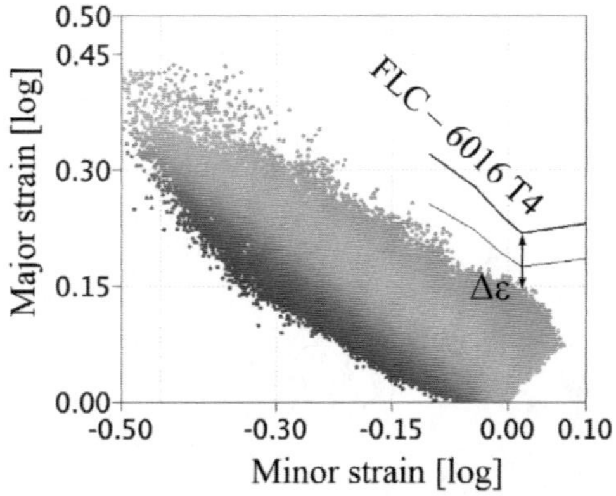

Figure 12. SHT temperature T+ 160°C

4.2. Deep drawing of SHT layout 1, 2, 3 – SHT temperature T+70°C – ty

Figures 11, 13 and 14 show the modification of the forming limit diagrams caused by SHT layout 1, 2 and 3. The SHT process was performed with an SHT temperature of T + 70°C and an SHT process time of ty. Thereby, deep drawing without material failure was achieved by layout 1 and layout 2.

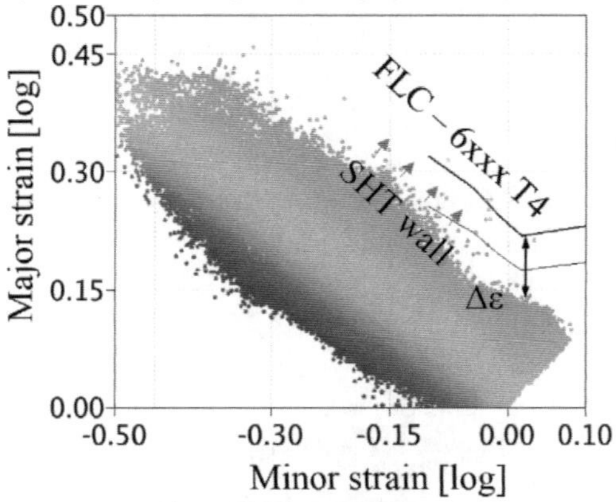

Figure 13. SHT layout 2
SHT temperature T+ 70°C

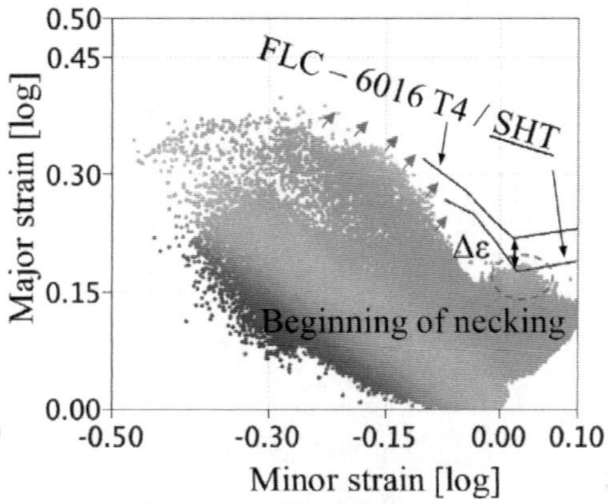

Figure 14. SHT layout 3
SHT temperature T+ 70°C

The forming results when using layout 3 show two parts at a number of three tests with material failure and different oriented crack formations, compared to manufacturing processes without SHT (figure 9 and 15). This outcome can be explained by considering the reaction forces along the load path (figure 8) and consequently the stress situation nearby to the critical zone (figure 9). It is vital to point out that literature argues that a reduced averaged flow stress in flange sections leads to a reduction of the maximum drawing force [9]. Furthermore, the drawing force affects the reaction forces along the load path and therefore the stress within the SHT areas of layout 3. To be more specific, figure 16 shows the measured strain margin Δε to the FLC (T4 - state) along the path marked in figure 8. The use of layout 3 shows that the determined strain margin Δε is distinctly smaller within the part's bottom SHT area than when using layout 1 or 2 (figure 16). The smaller Δε is related to the reduced flow stress through SHT and a simultaneous larger stress situation due to the untreated flange section. Also, important to mention is that the lower forming limit in SHT state (figure 14 and 16) provides an earlier beginning of necking during deep drawing. Accordingly, the crack formation starts on the part's bottom SHT area and can be also detected by the beginning of local necking, at the only test part, conducted with layout 3, without material failure. Even though cracks occurred in the SHT area of layout 3, also a reduction of the plane strain in the critical radius could be measured, as shown

IOP Conf. Series: Materials Science and Engineering **159** (2016) 012001 doi:10.1088/1757-899X/159/1/012001

in figure 14 and figure 16. Hence, it can be concluded that the combination of layout 1 (reduction of drawing force) and layout 3 (local reduction of strain through soft zones) provides the most desired result in formability.

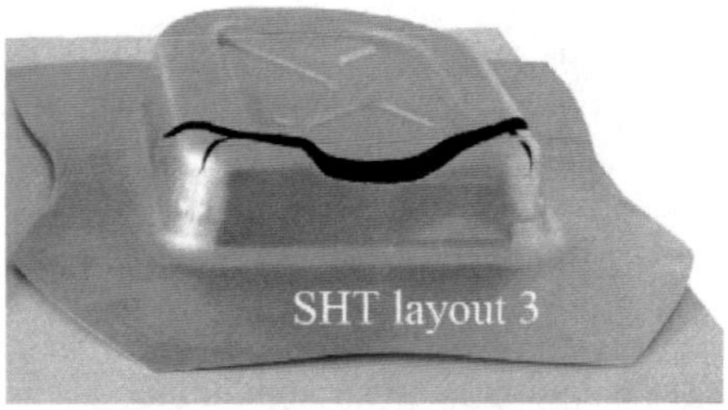

Figure 15. Crack formation layout 3

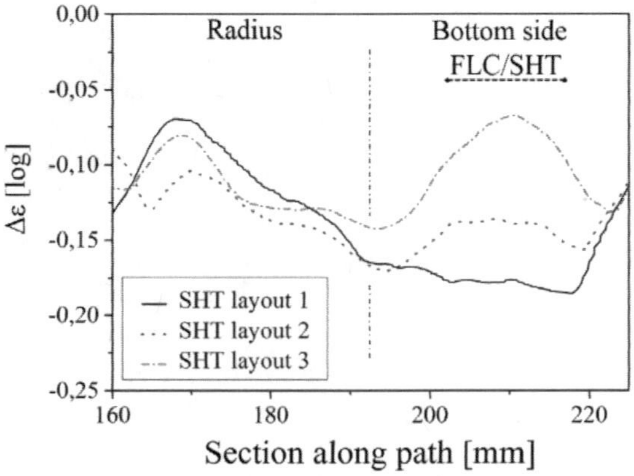

Figure 16. Δε along path of layout 1, 2 and 3

4.3. Deep drawing of SHT layout 1 – SHT temperature T+70°C – tx, ty and tz

In the following section, the influence of SHT process times on deep drawing processes was analyzed. The material, used in this study was heated up with SHT layout 1, at a temperature of T + 70°C and SHT process times of tx, ty and tz. The outcome of the measured forming limit diagrams in figure 11, 17 and 18 show a marginal deviation of the strain distribution with regard to the SHT process times. Moreover, figure 19 illustrates the strain margin Δε to the FLC depending on the particular SHT process time applied.

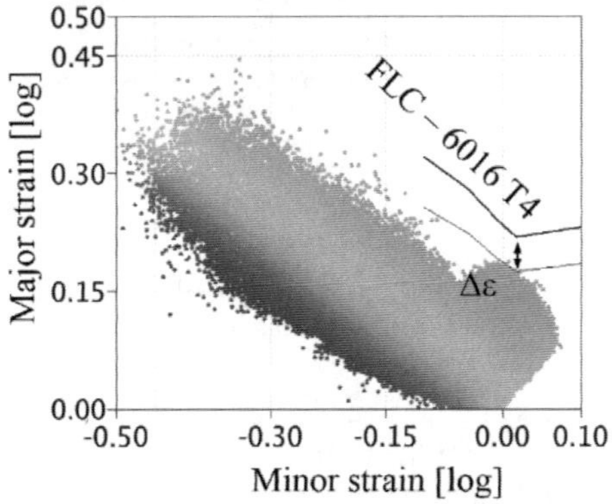

Figure 17. Heating time t_x

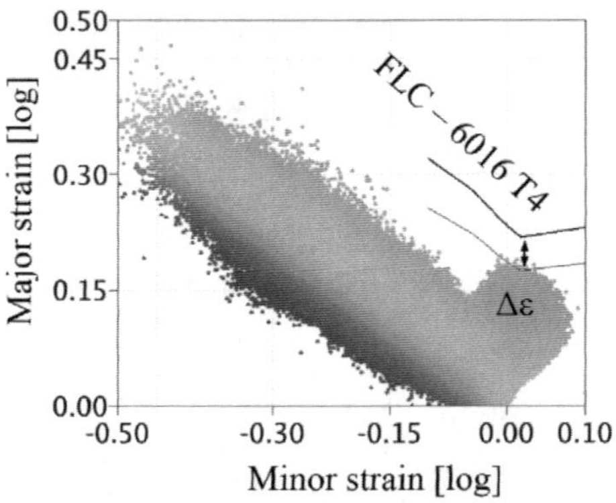

Figure 18. Heating time t_y

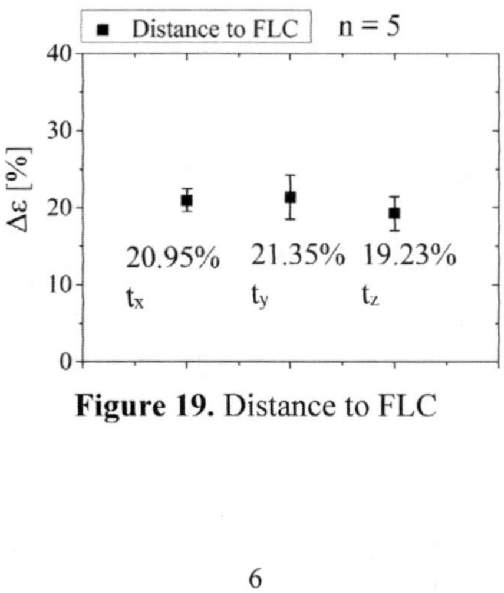

Figure 19. Distance to FLC

4.4. Natural aging – SHT layout 2 – SHT temperature T+70°C – storage time 0, 16, 40, 100, 164 hours – ty

To investigate the effect of natural aging on the formability of 6016, the blank was heat treated with layout 2, T + 70°C SHT temperature and ty SHT process time. This is illustrated in figure 20, which shows an increase of 5.5 % plane strain at a storage time of 16 hours compared to a continuous manufacturing process (0 hours storage time, figure 13). Additionally, figure 21 shows a considerable increase of major strain in the FLD when a storage time of 164 hours is applied. The distance between the maximum plane strain and the FLC of the analyzed storage times is presented in figure 22.

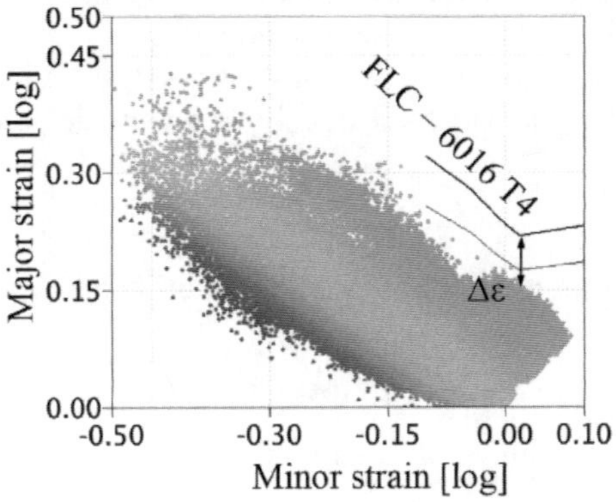

Figure 20. Natural aging 16h

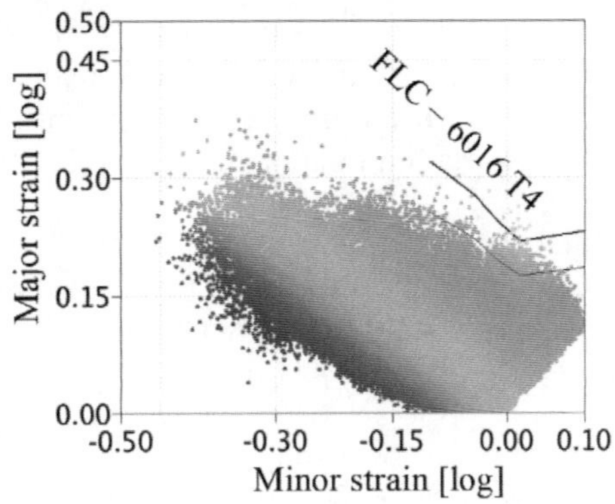

Figure 21. Natural aging 164h

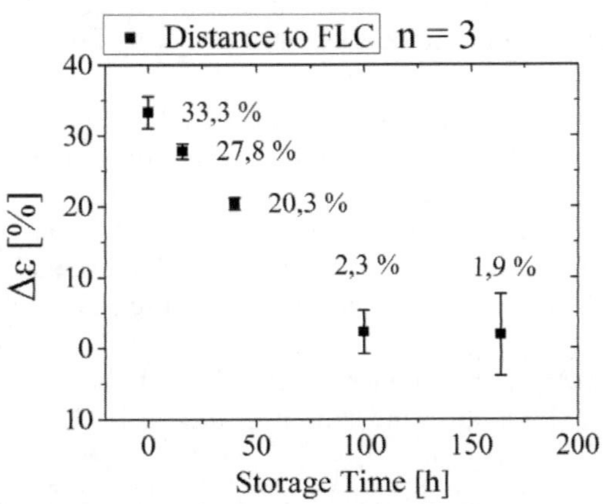

Figure 22. Distance to FLC

4.5. Deep drawing 115mm – SHT layout 2 – SHT temperature T+70°C – ty

To achieve the deep drawing depth of 115 mm, the SHT process was conducted with layout 2, the SHT temperature of T+70 °C and the SHT process time of ty. During the manufacturing process without SHT, cracks occurred at a certain drawing depth of 63mm (figure 23). Due to SHT, the target depth of 115 mm was reached without any material failure (figure 23).

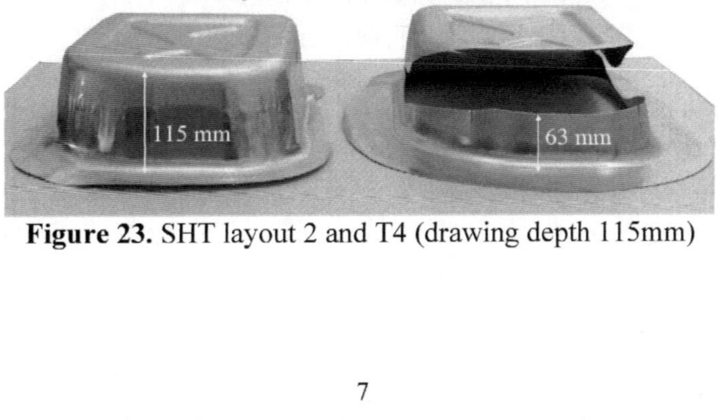

Figure 23. SHT layout 2 and T4 (drawing depth 115mm)

5. Conclusion

Summing up the results, it can be concluded that:

- Short-term heat-treatment (SHT) with the objective of modifying the formability of heat-treatable Al-Mg-Si aluminum alloys (6016) can be successfully performed through conduction heating.
- Varied SHT process times showed marginal deviations in the forming limit diagram (FLD).
- The SHT process temperature of T + 160°C showed a much greater decrease of plane strain in the critical area compared to an SHT process temperature of T + 70°C
- The design of the SHT layout plays an essential role in terms a robust forming process. Therefore, the findings suggest a combination of a reduced drawing force through SHT in flange sections and the reduction of strain in the critical areas due to soft zones.
- The storage time between heat treatment and forming process has a significant impact on the deep drawing result. In fact, storage times higher than 16 hours showed considerable changes in the forming limit diagram.

Acknowledgements

This work was supported by the Österreichische Forschungsförderungsgesellschaft (FFG), Cosma Engineering Europe GmbH and the Austrian Institute of Technology.

References

[1] Ostermann, F. Anwendungstechnologie Aluminium. 2nd edn. Berlin Heidelberg New York: Springer Verlag, 2007.
[2] Haller, D. Anforderungen neuer Materialien an das Presswerk von morgen. In: Vortragstexte zur Veranstaltung Neuere Entwicklung in der Blechumformung, Stuttgart, 2006. pp.79-99.
[3] Vollertsen, F, and Lange, L. Enhancement of Drawability by Local Heat Treatment. CIRP Annals, 1998, 47, pp. 181-184.
[4] Hofmann, A. Erweiterung der Formgebungsgrenzen beim Umformen von Aluminiumwerkstoffen durch den Einsatz prozessangepasster Platinen. University of Erlangen, 2002.
[5] Vissers, R, Huis, M.A, Jansen, J, Zandbergen, H.W, Marioara, C.D, Andersen, S.J. The Crystal Structure of the β'' Phase in Al-Mg-Si Alloys. Acta Materialia, 2007, 55, pp. 3815-3823
[6] Vogt, U. Seriennahe Auslegung von Aluminium Tailored Heat Treated Blanks. University of Erlangen, 2009.
[7] Vogt.U, Merklein, M. Design Principles of Tailored Heat Treated Blanks for the Manufacturing of Complex Car Body Parts. In: Proceedings of the IDDRG, Olofström, 2008, pp. 557-569.
[8] Machhammer, M, Kolleck, R. The effect of short-term heat-treatment on the tribological characteristic of heat-treatable aluminum alloys. In: Proceedings of the 7th ICTMP, Phuket, 2016, pp. 149-158.
[9] Lange, k, Liewald, M. Umformtechnik Handbuch für Industrie und Wissenschaft. Band 3. 2nd edn. Berlin Heidelberg: Springer Verlag, 1990

IOP Conf. Series: Materials Science and Engineering **159** (2016) 012002 doi:10.1088/1757-899X/159/1/012002

Data-driven inline optimization of the manufacturing process of car body parts

S Purr[1], A Wendt[1], J Meinhardt[2], K Moelzl[1], A Werner[1], H Hagenah[3] and M Merklein[3]

[1] BMW Plant Regensburg, 93055 Regensburg, Germany
[2] BMW Group, 80788 Munich, Germany
[3] Chair of Manufacturing Technology, Friedrich-Alexander University, 91058 Erlangen, Germany

E-mail: stephan.purr@bmw.de

Abstract. The manufacturing process of car body parts needs to be adaptable during production because of fluctuating variables; finding the most suitable settings is often expensive. The cause–effect relation between variables and process results is currently unknown; thus, any measure taken to adjust the process is necessarily subjective and dependent on operator experience. To investigate the correlations involved, a data mining system that can detect influences and determine the quality of resulting parts is integrated into the series process. The collected data is used to analyze causes, predict defects, and optimize the overall process. In this paper, a data-driven method is proposed for the inline optimization of the manufacturing process of car body parts. The calculation of suitable settings to produce good parts is based on measurements of influencing variables, such as the characteristics of blanks. First, the available data are presented, and in the event of quality issues, current procedures are investigated. Thereafter, data mining techniques are applied to identify models that link occurring fluctuations and appropriate measures to adapt the process so that it addresses such fluctuations. Consequently, a method is derived for providing objective information on appropriate process parameters.

1. Introduction

The car body part production process is affected by several variables [1] that cause quality fluctuations. It faces major challenges in relation to decreasing process windows and lot sizes. When a quality defect appears, the process needs to be manually adjusted using control variables until parts of acceptable quality are produced. The search for suitable settings often results in significant downtime. Improving the strategy for identifying settings would therefore offer a potential increase in stamping plant productivity.

The current problems are related to a lack of knowledge about influential factors and existing correlations in the overall process. Digitization in the context of the so-called "industry 4.0" offers the potential to solve these problems. One aspect of this digitization involves the automatic exchange of information between machines and products, which creates production processes that continuously produce large amounts of data [2]. The analysis of this data enables the understanding and the modeling of complex relations. Results that until now have appeared to be random can be predicted, and measures that have depended on intuition and operator experience can be determined based on these relations. Consequently, defects can be prevented and time needed for trial and error efforts can be reduced considerably.

A method for predicting defects was previously created by Purr et al. [3]. Now a method is presented that calculates objective recommendations for adapting the production process to avoid defects. The determination of these recommendations is based on influences and relations derived from measured and collected data using data analysis and learning algorithms.

2. Available Data

The proposed method is based on the analysis of data generated during series production in stamping plants. To acquire the data, a system was developed that documents uncontrollable influences, process parameters and process results for each car body part. This system was integrated in a BMW stamping plant. Details on the data acquisition system have been published in [4]. To track and trace parts, each blank is labelled by a laser marking unit. A component thus gets a unique identity to which the acquired data is assigned. As a result, the system generates a feature vector, $\boldsymbol{f}(i)$, for each produced component, i, which includes the material properties, $\boldsymbol{m}(i)$, control variables during processing, $\boldsymbol{c}(i)$, and an associated description of the part quality, $\boldsymbol{q}(i)$. Definition (1) shows the structure of a feature vector $\boldsymbol{f}(i)$. When considering the system for "producing a car body part," $\boldsymbol{m}(i)$ and $\boldsymbol{c}(i)$ contain the input variables, and the elements of $\boldsymbol{q}(i)$ are the output variables. The quantities of $\boldsymbol{c}(i)$ can be controlled and set during production. Material properties $\boldsymbol{m}(i)$ cannot be influenced during series production. It has to be mentioned that unlike all other variables, quality data $\boldsymbol{q}(i)$ is manually recorded. It is therefore not available for each part.

$$\boldsymbol{f}(i)=\begin{bmatrix}\boldsymbol{m}(i)\\ \boldsymbol{c}(i)\\ \boldsymbol{q}(i)\end{bmatrix} \tag{1}$$

The material properties vector $\boldsymbol{m}(i)$ consists of measured values that describe the characteristics of blanks as precisely as possible. These variables are determined during the blank cutting step and are assigned to the appropriate part. The elements of $\boldsymbol{m}(i)$ are partially vectors themselves. The structure of $\boldsymbol{m}(i)$ is defined in (2). The value $s(i)$ is an indicator of the strength of a blank, $\boldsymbol{p}(i)$ and $\boldsymbol{u}(i)$ provide roughness parameters, Ra and Rpc, in five tracks in a transverse manner to the strip running direction. $\boldsymbol{o}_{\text{top}}(i)$ and $\boldsymbol{o}_{\text{bottom}}(i)$ specify the amount of lubricant in 20 tracks across the strip on the top and bottom side respectively. In addition to the measured oil amount of the single tracks, their average and standard deviation are calculated to describe the overall oil layer. The variable $\boldsymbol{d}(i)$ includes three thickness measures transverse to the strip running direction.

$$\boldsymbol{m}(i)=\begin{bmatrix}s(i)\\ \boldsymbol{p}(i)\\ \boldsymbol{u}(i)\\ \boldsymbol{o}_{\textbf{top}}(i)\\ \boldsymbol{o}_{\textbf{bottom}}(i)\\ \boldsymbol{d}(i)\end{bmatrix}=\begin{bmatrix}s(i)\\ Ra_{\text{track }1}(i),\dots,Ra_{\text{track }5}(i)\\ Rpc_{\text{track }1}(i),\dots,Rpc_{\text{track }5}(i)\\ o_{\text{top,track }1}(i),\dots,o_{\text{top,track }20}(i),o_{\text{top,average}}(i),o_{\text{top,deviation}}(i)\\ o_{\text{bottom,track }1}(i),\dots,o_{\text{bottom,track }20}(i),o_{\text{bottom,average}}(i),o_{\text{bottom,deviation}}(i)\\ d_{\text{track }1}(i),d_{\text{track }2}(i),d_{\text{track }3}(i)\end{bmatrix} \tag{2}$$

The vector of control variables $\boldsymbol{c}(i)$ includes settings during part processing. Its structure is defined in (3). The control variables include the adjustment of a drawing cushion, $\boldsymbol{bf}(i)$, setting of an oiling machine, $\boldsymbol{ro}(i)$, and setting of blank holder spacers, $\boldsymbol{bs}(i)$. In addition, other settings such as the stroke rate are summarized in the vector $\boldsymbol{os}(i)$. Each setting contains multiple variables; for example, $\boldsymbol{bf}(i)$ consists of the settings of six cylinders from the multipoint drawing cushion, and $\boldsymbol{ro}(i)$ provides the spray pattern of the oiling machine, which consists of the settings of 40 valves on each side of a blank.

The vector $\boldsymbol{bd}(i)$ includes the state of all existing spacers. The total number of spacers, n, depends on the tool used.

$$\boldsymbol{c}(i)=\begin{bmatrix}\boldsymbol{bf}(i)\\\boldsymbol{ro}(i)\\\boldsymbol{bs}(i)\\\boldsymbol{os}(i)\end{bmatrix}=\begin{bmatrix}bf_{cyl.1}(i),\ldots,bf_{cyl.6}(i)\\\boldsymbol{ro}_{valve1}(i),\ldots,\boldsymbol{ro}_{valve80}(i)\\bs_{spacer1}(i),\ldots,bs_{spacer\,n}(i)\\stroke\ rate(i),\ldots\end{bmatrix} \quad (3)$$

3. Adjustment of the process to solve quality problems

3.1. Current procedure

When quality problems occur during series production, usually a readjustment of the process via control variables is required. Various options are available to adapt the forming process. The most important adjustment options are the blank holder forces, $\boldsymbol{bf}(i)$, the setting of the oiling machine, $\boldsymbol{ro}(i)$, and the setting of the blank holder spacers, $\boldsymbol{bs}(i)$. Figure 1 shows the impact on the process of the three main settings in a simplified representation. The effect on the part is based mainly on the retention force for all settings. Mechanisms used to affect the quality of certain settings are described as follows:

- **Blank holder forces** determine the normal force acting on the flange of a metal sheet. Usually these forces are generated by an adjustable hydraulic multipoint drawing cushion. A value can be specified for each existing displacer cylinder, or at times a curve can be used.
- The **oiling machine** enables application of additional lubricant on the surface of the blanks. The valves of the oiling machine can be controlled separately. In consequence, lubricant can be selectively applied to critical part regions.
- The height of the **spacers** can be adjusted, which has an indirect effect on the normal forces via a force bypass. Therefore, spacers have a local impact on the material flow [5]. An increase in spacer height leads to a reduction in normal forces acting on a metal sheet.

However, no clear procedure is specified on how to use these mechanisms in the everyday activity of a stamping plant. When a quality problem occurs, the employees decide which and how existing parameters should be adjusted. This decision is based on individual experiences. Each plant operator has an approximate idea of the relation between the control variable adjustment and its effect on quality. Because of the subjective experience of each employee, which varies between individuals, in practice the adjustment procedure is not uniform.

The measures currently taken to fix quality problems during series production were recorded and analyzed to obtain an overview of the applied procedures. The production of 15,000 parts from multiple production lots were recorded. During the recording 25 quality problems occurred. Data were collected during different shifts with respect to the various parts. Figure 2 shows the relative frequency of usage as a first measure of the different control variables discussed above and the relative frequency of successful usage, where "successful" means that a quality problem was fixed after the adjustment.

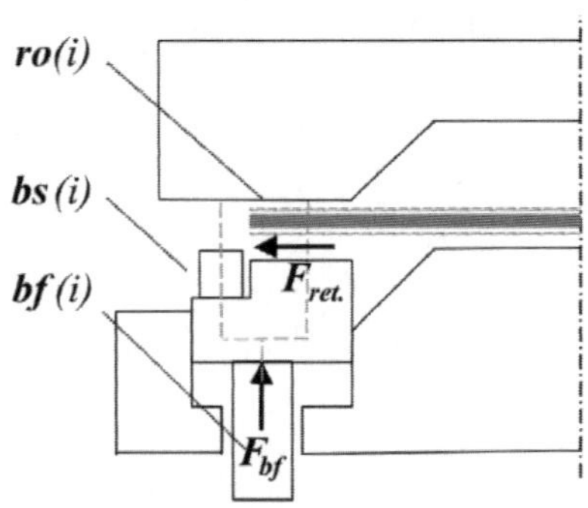

Figure 1. Simplified representation of the effect of the three main control variables.

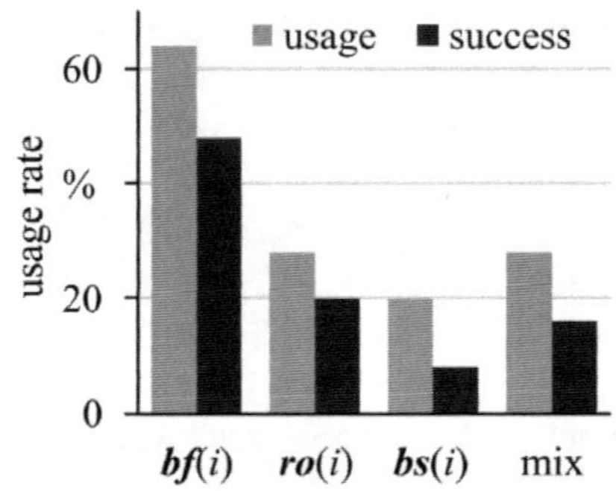

Figure 2. Rates of usage and successful usage of settings determined during series production.

All of the main variables were actively used. In addition, in some cases multiple variables were adjusted simultaneously. The blank holder forces were the most commonly used settings. In 48% of all cases the defects were fixed after an adjustment of the blank holder forces. Consequently, the blank holder forces are considered to be the most important setting used for process adjustment.

In summary, it was documented that many possible settings exist and that the procedures used when defects occur are diverse, although the settings have a similar effect. Reasons for the use of diverse procedures could be the varying experience of operators, a lack procedural standardization, and a lack of knowledge about the exact effect of control variables. In addition, the settings of control variables are subject to restrictions and interactions; for example it is only possible to increase the amount of lubricant using the oiling machine (the amount of oil on a blank cannot be reduced). Currently the effectiveness of the different settings and influences is not documented.

3.2. Analysis of the effect of control variables

To overcome this lack of knowledge, the effect of the existing control variables is analyzed and a simplified conceptual model is developed to describe the relation between the condition of settings, $\boldsymbol{c}(i)$, and the quality of parts, $\boldsymbol{q}(i)$. The model is based on the physical effects of the settings described in section 3.1. Equation (4) shows this simplified model for the effect of the control variables:

$$\boldsymbol{F}_{\text{ret.}} = \boldsymbol{F}_{\text{others}} + \boldsymbol{F}_{\text{friction}} = \boldsymbol{F}_{\text{others}} + \mu \boldsymbol{F}_{\text{N}} = \boldsymbol{F}_{\text{others}} + \mu(\boldsymbol{F}_{\text{bf}} - \boldsymbol{F}_{\text{spacer}}) = \boldsymbol{F}_{\text{others}} + \underline{\mu}(\underline{\boldsymbol{F}_{\text{bf}}} - \underline{\Delta H}D) \quad (4)$$

It is assumed that all settings shown in figure 1 primarily act on the retention force, and thus act on the material flow (see section 3.1). The model therefore describes the effect of each of the control variables on the retention force. It is expected that the material flow is proportional to the retention force, which is the sum of the friction force, $\boldsymbol{F}_{\text{friction}}$, and other forces such as the forming force produced by draw beads. Only the friction force can be controlled using the different settings introduced in section 3.1. The Coulomb friction law is the most suitable approach to describe the friction in the flange [6]. In the model, it is therefore assumed that the Coulomb friction law is valid and the friction force is the product of the normal force, $\boldsymbol{F}_{\text{N}}$, and the friction coefficient, μ. The local normal force is determined by the blank holder force, $\boldsymbol{F}_{\text{bf}}$, and the spacer force, $\boldsymbol{F}_{\text{spacer}}$, while the friction coefficient, μ, depends on the amount of lubricant and the material being processed. The spacer forces reduce the normal force, $\boldsymbol{F}_{\text{N}}$, depending on their set height. They are considered to be preloaded springs, and Hook's law is applied. The spacer force, $\boldsymbol{F}_{\text{spacer}}$, is thus a result of the product of the deflection, ΔH, and a spring constant, D. The deflection of the spacer is also correlated with the thickness of the sheet [7], but this is not considered in this model. Adjustable variables in the model are: the blank holder force, $\boldsymbol{F}_{\text{bf}}$; the deflection of the spacers, ΔH; and the coefficient of friction, μ (based on the amount of oil). These are underlined in (4).

According to the model, these settings should linearly affect the retention force and material flow. The conceptual model was validated by an experiment wherein parts were produced using different settings, and the effect of adjustments on the product was determined. One control variable was changed at a time, and the position of the impact lines resp. their distance to a reference was measured to quantify results. These lines result from the movement of the metal sheet under tensile stress over a tool edge [8]. The location of the impact line offers information about the material flow which is correlated with part quality [9]. To investigate the influence of control variables independently from the material properties, components with similar material properties, $\boldsymbol{m}(i)$, were chosen for the experiment. Interactions between different settings were not considered. In addition, the experiments were limited to typical settings used within a critical component region of a sample part. A possible influence of the component geometry on the local effect of the control variables is therefore not considered. In this case, a side frame of the current BMW X1 was regarded. The most critical region of the part is the C-pillar area. This is the area where most defects occur during series production. Figure 3 shows the critical area of the example part, together with the settings that were changed during the experiment. Firstly, the pressure of cylinder 2 (the influence area of this cylinder on the part can be found in figure 3) of the drawing cushion was altered using steps documented during the investigations in 3.1, then the amount of additional oil in the marked region was altered, and finally the heights of three marked spacers were simultaneously varied.

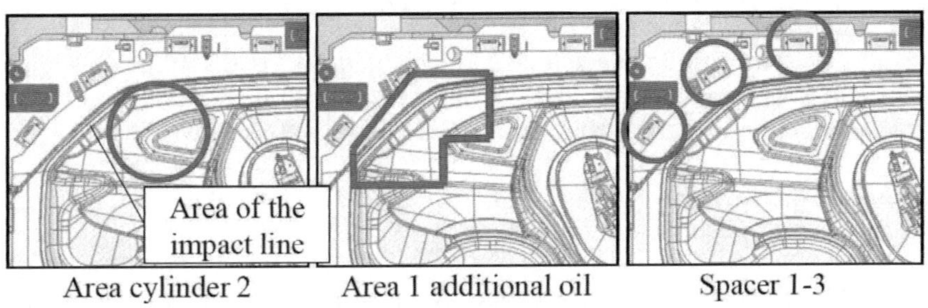

Figure 3. Setting area of the drawing cushion (left), setting region of the oiling machine (middle), and adjusted spacers (right).

Through this the relation between control variables and material flow and the assumed linear correlations of the model were investigated. Figure 4 shows scatterplots of the correlation between settings and the location of the impact line. In all cases a linear correlation between the examined control variables and the material flow could be detected. The left scatterplot shows that a higher blank holder force results in a larger distance between the impact line and the reference; this indicates that the impact line is further outside of a component and that the material flow is lower. The middle plot shows that an increase in the amount of lubricant leads to a smaller distance and thus to an increased material flow, and the right plot indicates that the increase of the spacer height has an analog effect on the material flow than better lubrication. The Pearson correlation coefficient between the adjustment value and the measured material flow is higher than 0.7 for each of the main control variables, and is therefore in the range of a strong linear relation [10]. It is evident that the influence of the control variables can be described using this simplified model with only a small error.

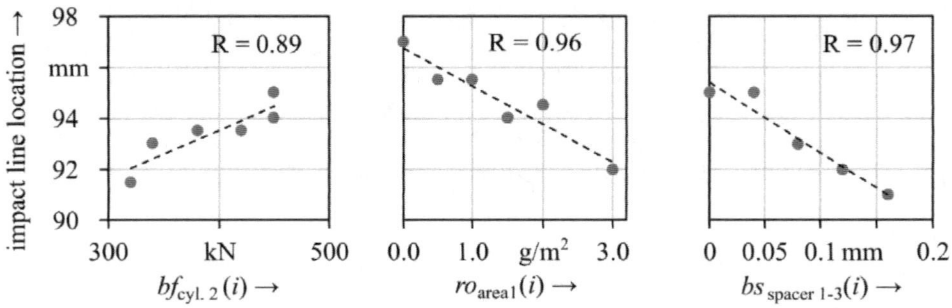

Figure 4. Scatterplots and correlation coefficients for relations between settings of control variables and location of impact line.

This experiment demonstrates that all control variables affect the retention force and therefore the material flow. In addition, the assumed linear relations are proven and the simplified model is validated in this respect. Based on these results, it can be assumed that the response of one control variable to quality problems would actually be sufficient in most cases and that the use of multiple settings is usually not necessary.

It is thus proposed that a main control variable should be defined and that the use of other variables should be prevented. Any effect reached by one control variable can also be reached by adjusting another one accordingly. Therefore the remainder of this paper focusses on developing a method to recommend the settings of one control variable. If a necessary adjustment is not possible due to restrictions, the model can be used to convert the measure to the setting of another control variable.

4. Data-driven inline optimization

4.1. Method

According to Bräunlich [11], it is always possible to adjust the manufacturing process of car body parts by altering the control variables to produce good parts. In practice, the difficulty lies in the identification of suitable measures. Using the results of 3.2 a method is proposed that efficiently determines the necessary adjustments of one control variable. Common optimization techniques for determining appropriate parameter settings would presuppose that the system behavior in the entire parameter space is described by a function [12]. If all variables of a feature vector $\boldsymbol{f}(i)$ are considered, this is not possible due to the high number of dimensions. Therefore a data-based method for calculating the appropriate settings is presented. The basic idea is that historical data can be used to extract direct connections and patterns between uncontrollable material properties, $\boldsymbol{m}(i)$, and the settings used for the production of good parts, $\boldsymbol{c}_{\text{good}}(i)$. Using these patterns and models, suitable parameters are calculated based on the properties of the incoming material. Figure 5 shows the basic principle of this concept. Since all main control variables affect the retention force and the blank holder force is the most commonly used one, the initial investigation is limited to the settings of the 6 cylinders of the drawing cushion. According to figure 6, the cylinder settings to produce good parts, $\boldsymbol{bf}_{\text{good}}(i)$, than are the output variables of the model.

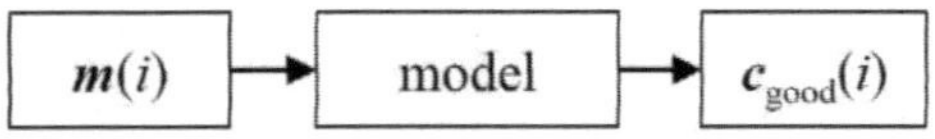

Figure 5. Basic principle of the model to predict appropriate adjustments.

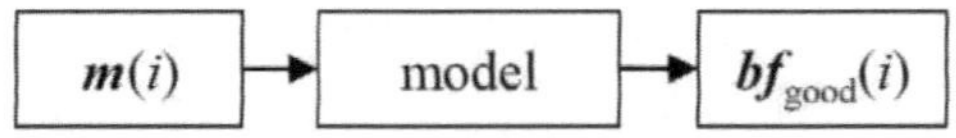

Figure 6. Structure of the model to predict the setting of the drawing cushion.

A description of part quality, $\boldsymbol{q}(i)$, is not required in this method. The models are generated by application of learning algorithms to the data described in section 2. The multistage procedure for creating models to calculate handling instructions is defined in the following. This approach can be applied using different components and varying amounts of data.

4.1.1. Data Selection and Pre-processing

To train models, records are required that include the properties of the materials involved as well as the corresponding settings that produce good parts. As such, a matrix acquired in series production with part-specific information is used. The data sets $\boldsymbol{f}(i)_{i=1,\dots,z}$ include different settings of the blank holder force. Since the process parameters are not adjusted for each of the z produced parts, there are a smaller number of different settings than there are parts. Therefore, in the first step, the settings of the drawing cushion $\boldsymbol{bf}(i)_{i=1,\dots,z}$ are taken, and duplicate rows of the matrix are deleted. The result is the matrix $\boldsymbol{bf}_{\text{unique}}(u)_{u=1,\dots,y}$, which includes only y unique settings of the drawing cushion. Unique control variable settings are then assigned to the parts that were processed using the respective control variable settings. This step results in groups of part records that have been produced with the same drawing cushion setting. Subsequently, the average material properties, $\overline{\boldsymbol{m}}(u)$, are calculated for each of the unique settings $\boldsymbol{bf}_{\text{unique}}(u)_{u=1;\dots;y}$. The resulting dataset contains unique settings, together with the average properties of the blanks that have been processed with these settings. The number of parts, $n(u)$, processed using these settings is added as supplementary information. (5) defines the resultant matrix, $\boldsymbol{ru}$.

$$\boldsymbol{ru} = \begin{Bmatrix} [\boldsymbol{bf}_{cyL1}(u=1),\dots,\boldsymbol{bf}_{cyL6}(u=1)] & \overline{\boldsymbol{m}}(u=1) & n(u=1) \\ \vdots & \vdots & \vdots \\ [\boldsymbol{bf}_{cyL1}(u=y),\dots,\boldsymbol{bf}_{cyL6}(u=y)] & \overline{\boldsymbol{m}}(u=y) & n(u=y) \end{Bmatrix} \tag{5}$$

As the matrix should only include settings that lead to the production of good parts, rows that contain less than a specified minimum of parts (n < minimum) are removed, and the resulting matrix is labelled $\boldsymbol{ru}_{\text{red}}$. It is assumed that settings used to produce only a few components have not led to the production of good parts. The remaining records are supposed to have led to the production of good parts. These assumptions are based on the current procedure in production where parameters are changed until good parts are produced again. This process leads to mistakes if there are small numbers of parts with very different characteristics $\boldsymbol{m}(i)$. Due to the structure of the press line, after an adjustment, at least five press strokes are carried out until the measures are proved to be effective. The minimum m is therefore set to five parts. For double parts the minimum is set to 10 records.

4.1.2. Data Analysis and Modelling

The data obtained in step a) can now be examined to determine correlations between the average material properties and the unique settings of the control variables. Therefore, a correlation analysis and visual examination are firstly performed; the results are the relations between single uncontrollable variables and single cylinder forces.

After identifying correlations, relations between the averaged uncontrollable input variables and each of control variables are modeled. The aim is not to interpret and explain, but to accurately forecast necessary adjustments. Modeling is always carried out with respect to the relation between at least one input variable and exactly one output variable/setting, and various algorithms are applied and evaluated. Examples include support vector machines, linear models, polynomial models, and neural networks.

4.1.3. Validation

The validation step is iteratively with the modeling step. If the validation result is not satisfying then it is necessary to repeat the modeling step to result in a better model. It is essential to prove that the model is able to generalize. To validate results, the available datasets of $\boldsymbol{ru}_{\text{red}}$ are divided into training and test data sets [13]; the records are split in a ratio of 2:1. The model is trained using the larger set, and it is then applied to both datasets. The correlation coefficient between the predicted and actual settings is then considered and scatterplots are used to compare the prediction with the reality. This allows an evaluation of model fitting to training data, as well as the model's ability to generalize using test data. This process is repeated several times, each time with new divisions of the data. To avoid finding spurious correlations, the data is divided so that the training and test sets do not include data from the same production orders.

4.2. Experimental results

For validation, the proposed method is applied to the series process data from a data set involving the production of 52,753 bottom plates of a BMW 3 series sedan. These parts are manufactured as double parts side-by-side (two blanks per stroke). The objective was to derive models for predicting the appropriate adjustments of the drawing cushion. The pre-processing step resulted in a matrix, $\boldsymbol{ru}_{red}(u)$, with 23 unique cylinder settings. Each of them was used for the production of at least 10 components (due to the double part manufacturing, the minimum is set to 10). Some settings were applied for the manufacture of several thousand components.

Through a correlation analysis of the matrix $\boldsymbol{ru}_{red}$, several parameters were identified that show a correlation with the cylinder settings (figure 7 shows an excerpt from the correlation matrix). These correlations allow to derive tendencies. For example, a high correlation arises between the setting of cylinder 1, $bf_{cyl. 1}(u)$, and the amount of lubricant at measuring track 7, $\bar{o}_{top, track 7}$. According to this correlation, the blank holder forces need to be increased with an increasing amount of lubricant in this area. According to Section 3 material flow rises with an increased amount of lubricant and this effect can be compensated by applying a higher blank holder force. The correlation can therefore be physically explained.

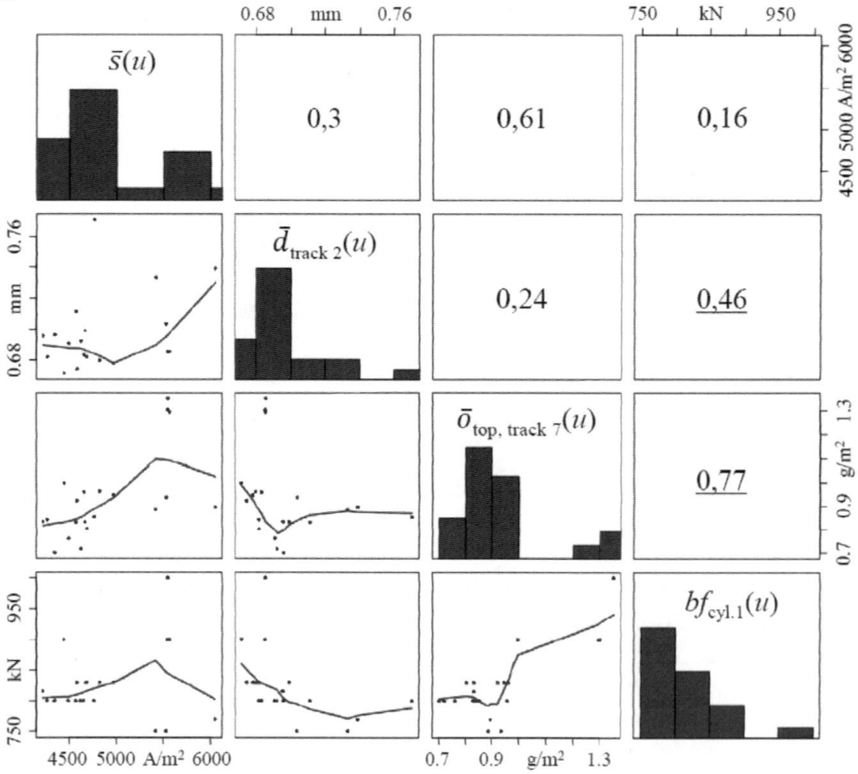

Figure 7. Extract from correlation matrix for matrix $\boldsymbol{ru}_{red}$, with scatterplots, correlation coefficients, and frequency distribution.

A low correlation was found between the average blank thickness on track 2, $\bar{d}_{\text{Track 2}}(u)$, and the setting of cylinder 1, $bf_{\text{cyl. 1}}(u)$. When the strip thickness is higher, the blank holder force settings tend to be lower. This is also explained using the conceptual model in Section 3. When there is an increase in the blank thickness there is a reduction in the deflection of the active blank holder spacer, ΔH. This reduces the fraction of the normal force derived from the spacer, and increases the fraction acting on the metal sheet. Consequently, a lower blank holder force is needed to achieve the same retention force. This physically explains the correlation.

After a manual search for correlations, models were developed that use several material properties to calculate the necessary adjustment of one cylinder of the drawing cushion. One model was derived for each of the six cylinders. In this case, the best-performing models were found to be support vector machines and linear models. From these, linear models were selected due to their simple functionality. Function (6) shows a linear model used to calculate the setting for cylinder 1 based on six ***m*** variables. The calculated coefficients and the selection of variables are not generally valid, and are re-established for other cylinders and components.

$$bf_{\text{cyl.1,good}}(i) = C_1 s(i) + C_2 o_{\text{top,average}}(i) + C_3 o_{\text{top,deviation}}(i) + C_4 o_{\text{top,track 1}}(i) + C_5 o_{\text{top,track 7}}(i) + C_6 d_{\text{track2}}(i) + C_7 \quad (6)$$

It was possible to derive six linear models to determine the required setting. Figure 8 shows the validation results for the example of cylinder 1, and figure 9 shows the results for cylinder 6; the Pearson correlation coefficients are higher than 0.7 for both the training and test data for both cylinders. A strong linear correlation [10] is thus proven between the predicted and actual settings used for both cylinders. For cylinder 1, the slopes of the trend line for the training data and test data are similar, which indicates that a quantitative forecast of the necessary adjustment could be possible. In contrast, the gradients at cylinder 6 are different for the test and training data, which indicates that only a qualitative recommendation can be calculated.

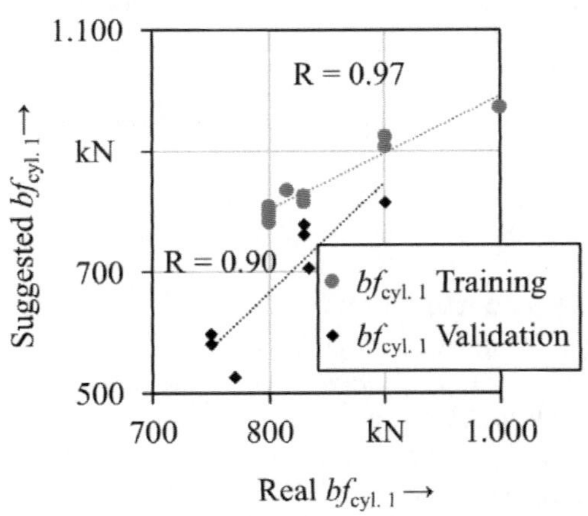

Figure 8. Scatterplot of correlation between suggested and actual settings for cylinder 1 for training and test data.

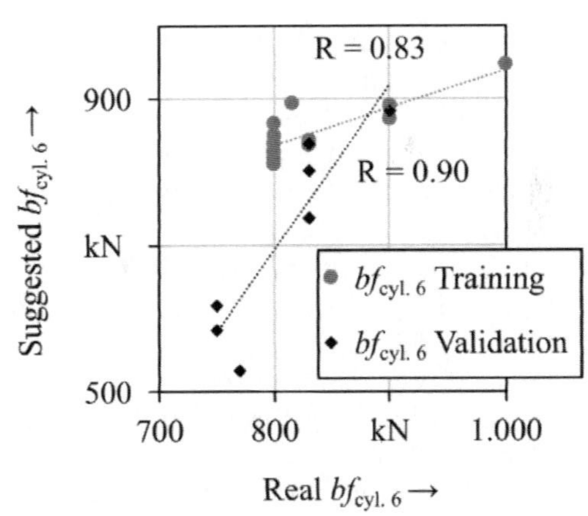

Figure 9. Scatterplot of correlation between suggested and actual settings for cylinder 6 for training and test data.

In conclusion, it is not yet possible to reliably calculate exact values. But the high correlations between suggestions and real settings indicate that tendencies can already be suggested with a high degree of certainty.

5. Conclusion

A method to provide process adjustments was developed and tested. To develop this method, current practices and the effect of control variables were investigated. A simple model that could be validated was introduced to describe the effect of control variables. Subsequently, a data-driven method to predict suitable settings was demonstrated. This method is based on the use of algorithms to generate models and patterns using historical data and their application to future data. To generate the models, a multi-step process has been defined which is applied to available series process data. The functionality of the proposed method was confirmed using a dataset containing more than 50,000 produced parts. The results prove that the suggestion and reality are well correlated and are already sufficient for qualitative recommendations. Considering that even historical data usually does not include ideal settings (settings are only changed after defects occur), these results are already very promising. In order to apply the model to upcoming new parts it will be mandatory to smoothen the input variables of the next stack of blanks. If a calculated measure cannot be carried out due to restrictions, it should be converted to an adjustment of another control variable. As shown in section 3 this is possible.

The success of this method depends on the nature of historical data. The forecast precision will be improved, when the uniformity and effectiveness of the procedure in the training data increases. Therefore, to improve predictions, it is necessary to specify the procedure accurately. Furthermore, to improve results, one control variable should always be preferred over the others during series production. If, other control variables are used instead of the preferred one within the training data, the measures are not considered by the model and accuracy is reduced. If the instructions calculated by this method (which are limited to one control variable) are followed and this data is used in future training, it is expected that predictions will automatically improve.

References

[1] Singh C P, and Agnihotri G 2015 Study of deep drawing process parameters: a review *International Journal of Scientific and Research Publications* IJSRP-P3864(2015)02

[2] Lee J, Bagheri B, and Kao H A 2015 A cyber-physical systems architecture for industry 4.0-based manufacturing systems *Manufacturing Letters* **3** pp 18-23

[3] Purr S, Meinhardt J, Moelzl K, Ostermair M, Hagenah H and Merklein M 2015 *Proceedings of the 5th International Conference on Accuracy in Forming Technology* vol 88, ed D Landgrebe W G Drossel and M Putz (Zwickau: Verlag Wiss. Scripten) pp 623-42

[4] Purr S, Meinhardt J, Lipp A, Werner A, Ostermair M and Glück B 2015 Stamping plant 4.0– basics for the application of data mining methods in manufacturing car body parts *Key Engineering Materials* **639** pp 21-30

[5] Faass I, Palaniswamy H, Hoffmann H, Altan T, Golle M and Griesbach B 2008 Effect of spacers on the material flow for process control in automotive stamping *Advanced Technology of Plasticity* **2008** pp 1711-16

[6] Giuliano G 2015 Evaluation of the Coulomb friction coefficient in DC05 sheet metal forming *Strojniški vestnik-Journal of Mechanical Engineering* **61** pp 709-13

[7] Girschewski B 2004 *Optimierung des Umformprozesses ziehkritischer Blechteile* ed H Hoffmann (München: Hieronymus Buchreproduktions GmbH)

[8] Hahn M 2004 *Beurteilung der Ausprägung von Anhau- und Nachlaufkanten mit Hilfe der FEM* ed H Hoffmann (München: Hieronymus Buchreproduktions GmbH)

[9] Neumann A, Hortig D and Merklein M 2011 Measurement of material flow in series production *Key Engineering Materials* **473** pp 137-44

[10] Asuero A G, Sayago A and Gonzalez A G 2006 The correlation coefficient: an overview *Critical reviews in analytical chemistry* **36** pp 41-59

[11] Bräunlich H 2002 *Blecheinzugsregelung beim Tiefziehen mit Niederhalter-ein Beitrag zur Erhöhung der Prozessstabilität* ed R Neugebauer (Zwickau: Verlag Wiss. Scripten)

[12] Myers R H, Montgomery D C and Anderson-Cook C M 2016 *Response surface methodology: process and product optimization using designed experiments* vol 4 (New Jersey: John Wiley & Sons)

[13] Payam R, Lei T and Huan L 2009 *Encyclopedia of Database Systems* ed L Liu M Tamer Özsu (USA: Springer US) pp 532-38

IOP Conf. Series: Materials Science and Engineering **159** (2016) 012003 doi:10.1088/1757-899X/159/1/012003

Pulse current assisted drawability of AZ31B magnesium alloy sheets

J H Song[1], S Choi[1], M J Kang[1], D Kim[2], M-G Lee[3], and C Y Lim[4]

[1] Research Institute of Advanced Manufacturing Technology, Korea Institute of Industrial Technology, Incheon 21999, Korea
[2] Materials Deformation Department, Korea Institute of Materials Science, Changwon 51508, Korea
[3] Department of Materials Science and Engineering, Korea University, Seoul 02841, Korea
[4] Automotive R&D Team, SHIN YOUNG CO. Ltd., Yeongcheon 38899, Korea

E-mail: jhsong@kitech.re.kr

Abstract. The thermal effect and athermal effect such as electro-plastic effect of metallic materials induced by high density current can dramatically reduce the flow stress, which is beneficial to the forming process of less formable metal. In this paper, pulse current-assisted deep drawing of the magnesium alloy is proposed due to lower energy consumption and higher efficiency. In this process, the metal sheet is designed in series in a pulse current circuit and heated directly by the pulse current. In addition, the insulated mould is employed to avoid the current leaking. Experiments were conducted to demonstrate the feasibility and advantages of the proposed process. An experimental process system was established and the electrical-assisted Erichsen cupping tests and rectangular cup drawing tests were performed. The experiments showed that the forming load was reduced and the cupping height and associated principal strains were increased in the Erichsen cupping and deep drawing process assisted by high-density electric current.

1. Introduction

Recent efforts in the automotive industry to improve fuel efficiency are reflected in the increasing use of lightweight materials. Magnesium alloys have been investigated to verify their high potential for use as light weight structural components in auto-motive applications, which results from their excellent properties, such as a low density and a high specific strength. However, the poor formability of magnesium alloy sheets, especially at room temperature, is one of the main factors hindering their industrial application [1-3]. In the traditional industry, warm or hot forming was commonly employed to manufacture the magnesium parts with a desired shape, because elevated temperature decrease the flow stress and increase the ductility of magnesium alloys. The traditional thermal forming method of heating the sheets via a furnace has many disadvantages, such as the decrease in the temperature of the sheets removed from the furnace prior to forming [4-5]. The sheets usually need to be heated to a higher temperature because of the heat dissipation. In addition, the sheets exposed to the air suffer a significant amount of oxidation due to the longer heating time. Therefore, reducing the process time and improving heat rates remain the major challenges.

IOP Conf. Series: Materials Science and Engineering **159** (2016) 012003 doi:10.1088/1757-899X/159/1/012003

As a developing technique, pulse current auxiliary forming technique has arrested the attention of researches for high heating rate and electro-plastic effect. The thermal effect and athermal effect such as electro-plastic effect of metallic materials induced by high density current can dramatically reduce the flow stress, which is beneficial to the forming process of less formable metal [6-10]. In this paper, pulse current-assisted deep drawing of the magnesium alloy is proposed due to lower energy consumption and higher efficiency. In this process, the metal sheet is designed in series in a pulse current circuit and heated directly by the pulse current. In addition, the insulated mold is employed to avoid the current leaking. Experiments were conducted to demonstrate the feasibility and advantages of the proposed process. An experimental process system was established and the electrical-assisted Erichsen cupping tests and rectangular cup drawing tests were performed.

2. Experimental procedure

2.1 Experimental materials

A commercial AZ31B Mg alloy sheet was provided by POSCO using a strip casting process. The thickness of the sheet was 1.4 mm. The chemical composition of the AZ31B Mg alloy sheet is listed in Table 1. The microstructure of the as-received material is shown in Fig. 1, which shows equi-axed grains with an average size of 10 μm. A strong basal texture, which is a typical rolling texture of magnesium alloys, produced extension twin boundary in the grains. Engineering stress–strain curves at various are also represented in Fig. 2.

Table 1. Chemical composition of the AZ31B Mg alloy sheet (wt%)

Al	Mn	Zn	Ni	Fe	Mg
2.5-3.5	0.2-1.0	0.6-1.4	<0.005	<0.005	Bal.

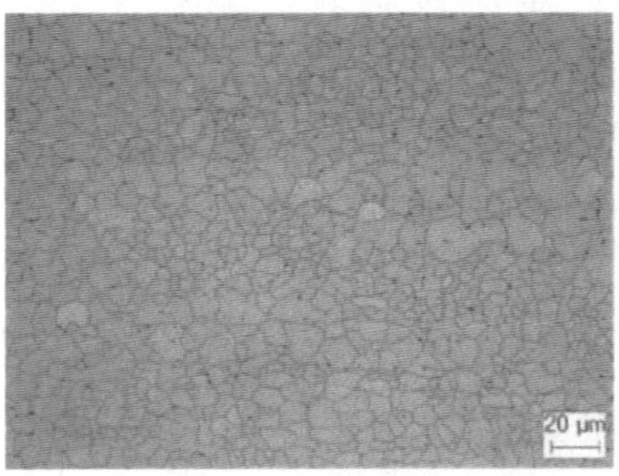

Figure 1. Optical microscopy of the as-received AZ31B sheet

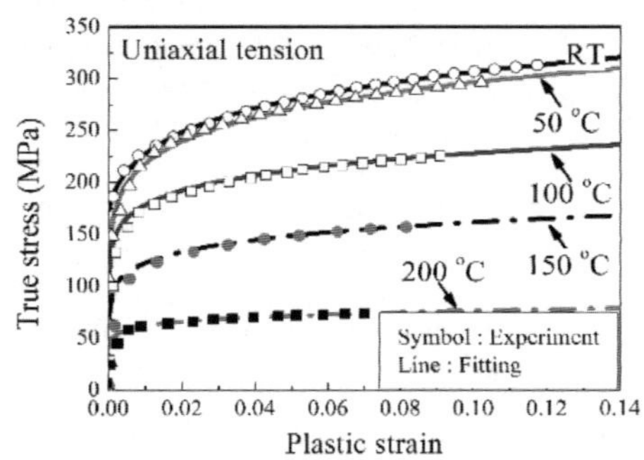

Figure 2. Engineering stress–strain curves of AZ31B at various temperatures

2.2 Experimental apparatus

Pulse current assisted forming apparatus was shown in Fig. 3. The rectangular sheet was directly set into the dies. The two edges of the sheet were held between two electrodes, and enough pressure was applied to guarantee sufficient contact for the electrifying. A laboratory-made electropulsing generator was used to acquire a maximum electro-pulsing of 10,000 A. The pulse current source was applied to discharge positive direction pulse with various parameters, including pulse width, frequency and amplitude. A typical wave form used in experiments is shown in Fig. 3(b), where t_d represents the pulse width, t_p is the pulse period. Hydraulic universal testing machine (UTM) was used for conducting the Erichsen and rectangular cup drawing tests with periodically pulsed currents. In order to prevent current leaking, electrical insulated bakelite were used to insulate every screw in the electrode and in the hydraulic testing machines.

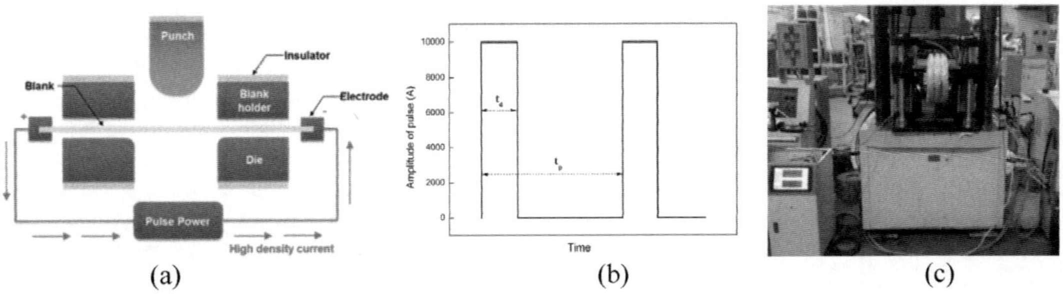

Figure 3. Apparatus of pulse current assisted forming (a) schematic diagram; (b) pulse current waveform; (c) device photo

3. Results

3.1 Pulse current assisted Erichsen tests

Pulse current assisted Erichsen tests for AZ31B were conducted with various pulsing conditions. For the quasi-static Erichsen tests under a pulsed electric current, the specimen was firstly set into the dies at room temperature and periodically pulsed with a constant amplitude of electric current over a selected duration while the specimen was continuously deformed by the movement of punch. In order to investigate the effect of the electrical pulsing on AZ31B, current density was changed from 50.0 A/mm^2 to 100.0 A/mm^2 while the pulsing duration and period were held constant value of 0.1 sec and 1.0 sec, respectively.

Fig. 4(a) represents typical load–displacement curves during conventional and pulse current assisted Erichsen tests. The punch load decreased instantly once a pulsed electric current was applied. After removing the pulsed electric current, the load increased with a transient behavior. The magnitude of load reduction increased because the electric current density increased as the cross-sectional area decreased.

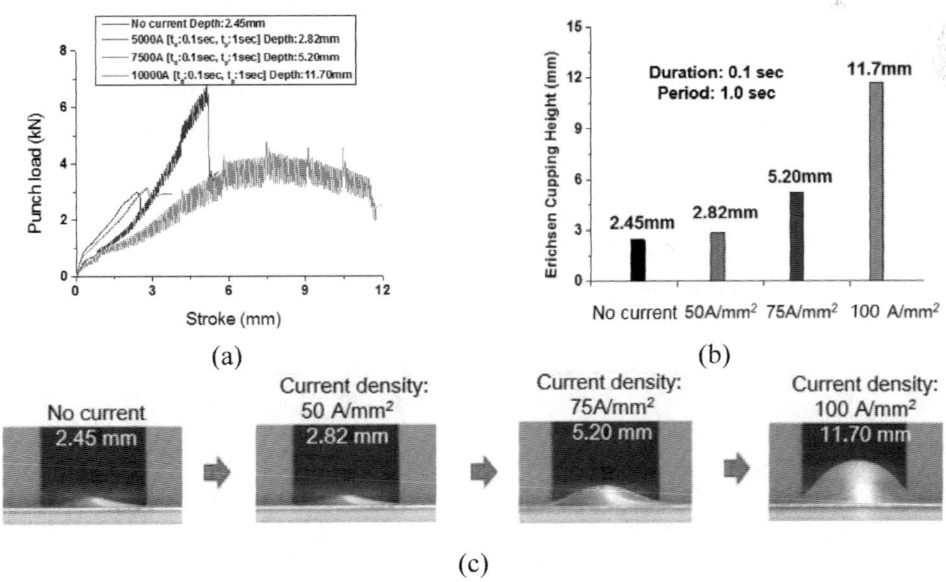

Figure 4. Pulse current assisted Erichsen tests for AZ31B with various electric current densities; (a) punch loads and displacement curves; (b) Erichsen cupping heights; (c) deformed shape of samples.

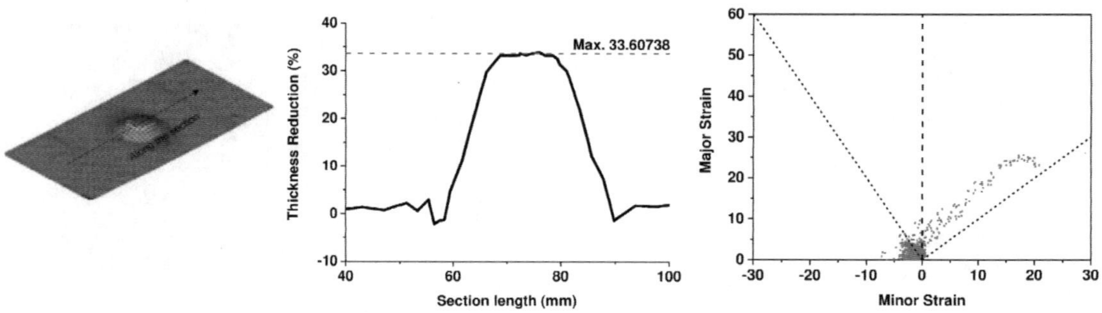

(a) (b)

Figure 5. Erichsen tests with current densities of 100 A/mm^2: (a) thickness reduction; (b) major and minor strain distributions.

Fig. 4(c) illustrates the samples of AZ31 Mg alloy produced by tradition Erichsen tests and pulse current assisted tests at applying pulse current density of 50, 75 and 100 A/mm^2, respectively.

It is found that the Erichsen index (EI) increases when the electric current density increases while the other parameters keep the same. The maximum EI of 11.7 is obtained when the current density increases up to 100 A/mm^2. Thickness reduction and principal strains are also analysed as depicted in Fig. 5. The results indicate that AZ31B samples are bi-axially stretched with the maximum principal stain of 30 % without failure under electric pulsing conditions.

3.2 Pulse current assisted deep drawing tests

Pulse current assisted rectangular cup drawing tests for AZ31B were also conducted with various pulsing conditions. Punch and blank sizes are assigned as 60 × 60 mm^2 and 90 × 170 mm^2, respectively. In order to investigate the effect of the electrical pulsing on AZ31B, current density was changed from 50.0 A/mm^2 to 75.0 A/mm^2 while the pulsing duration and period were held constant value of 0.1 sec and 1.0 sec, respectively.

Fig. 6(a) represents typical load–displacement curves during conventional and pulse current assisted cup drawing tests. Similarly with Erichsen tests, The punch load decreased instantly once a pulsed electric current was applied. It is found that the cup drawing height increases when the electric current density increases while the other parameters keep the same. The maximum drawing depth of 10.01 mm is obtained when the current density increases up to 75 A/mm^2 while that is 2.85 mm under conventional drawing tests at room temperature.

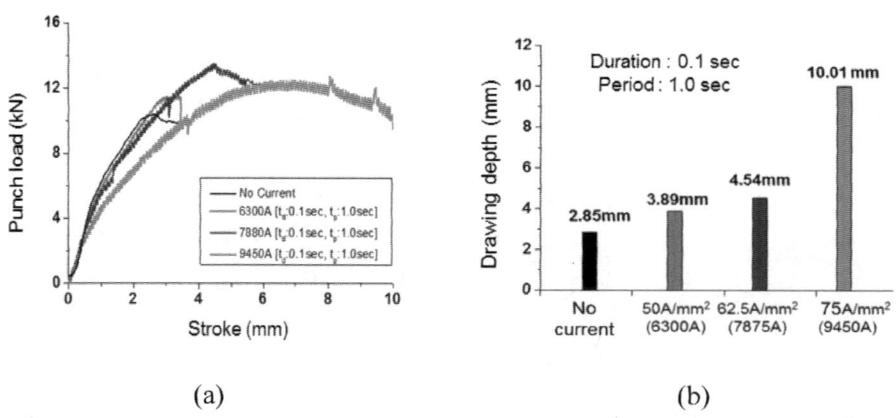

(a) (b)

Figure 6. Pulse current assisted cup drawing tests for AZ31B with various electric current densities; (a) punch loads and displacement curves; (b) change of drawing depth

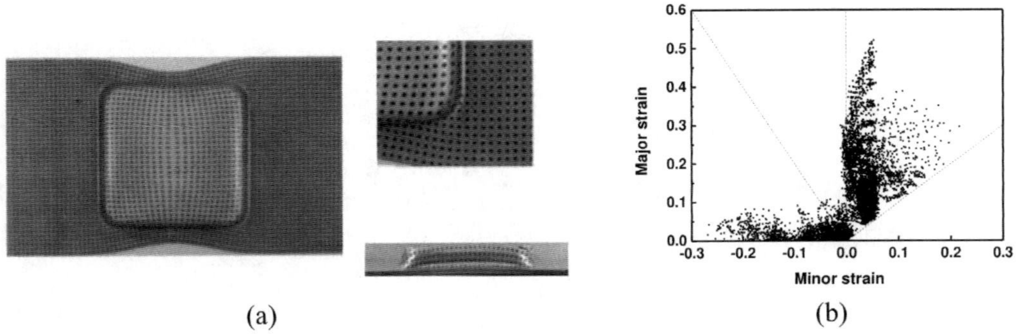

Figure 7. Cup drawing tests with current densities of 75 A/mm^2: (a) photo of deformed sample; (b) major and minor strain distributions.

Fig. 7 illustrates the samples of AZ31 Mg alloy produced by pulse current assisted cup drawing tests at applying pulse current density of 75 A/mm^2. Major and minor strains are measured using DIC apparatus (ARAGUS). As shown in Fig. 7(b), the sample deformed under plane strain and stretch modes with maximum major strain of 40%.

4. Conclusion

In this paper, the effect of a pulsed electric current on the mechanical behaviors of a typical magnesium alloy, AZ31B, are investigated. Pulsed current assisted Erichsen cupping and rectangular cup deep drawing test of AZ31B are conducted with various pulsing conditions. The effect of pulsing conditions on the Erichsen index and cupping heights are discussed. The maximum EI of 11.7 is obtained when the current density increases up to 100 A/mm^2. It is found that the cup drawing height increases when the electric current density increases while the other parameters keep the same. The maximum drawing depth of 10.01 mm is obtained when the current density increases up to 75 A/mm^2 while that is 2.85 mm under conventional drawing tests at room temperature. It can be said that electrically-assisted forming technology for Mg alloy sheet is candidate to replace the warm forming technology. It is possible to preoccupy practical application technology to automotive component and commercialize its technology.

Acknowledgments

This research is financially supported by Ministry of Trade, Industry and Energy (MOTIE) under R&D programs (Grant No. 20152020106500). This work was supported by a grant (No.B551179-13-02-03) from the Co-operative Research Project funded by the Korean government (MSIP) and KITECH projects.

References

[1] Lou XY, Li M, Boger RK, Agnew SR and Wagoner RH 2007 *Int. J. Plast.* **23** 44
[2] Yukutake E, Kaneko J and Sugamata M 2003 *J. Mater. Trans.* **44** 452
[3] Deng J, Lin YC, Li SS, Chen J and Ding Y 2013 *Mater. Des.* **49** 209
[4] Mori K, Makia S and Tanaka Y 2005, *CIRP Ann. Manufact. Technol.* **54** 209
[5] Toros S, Ozturk F and Kacar I 2008 *J. Mater. Process. Technol.* **207** 1
[6] Roh J-H, Seo J-J, Hong S-T, Kim M-J, Han HN and Roth JT 2014 *Int. J. Plast.* **58** 84
[7] Conrad H 2002 *Mater. Sci. Eng. A* **322** 100
[8] Jones JJ, Mears L and Roth JT 2012 *J. Manuf. Sci. Eng.* **134** 34504
[9] Liu K, Dong X, Xie H and Peng F 2015 *Mater. Sci. Eng. A* **623** 97
[10] Zhu R, Tang T, Shi S and Fu M 2013 *J. Mater. Process. Technol.* **213** 30

IOP Conf. Series: Materials Science and Engineering **159** (2016) 012004 doi:10.1088/1757-899X/159/1/012004

Virtual method for the determination of an optimum thermal design of hot stamping tools

W Weiß[1], M Koplenig[1], M Alb[1,3] and J Graf[2]

[1] VIRTUAL VEHICLE Research Center, Inffeldgasse 21/A, 8010 Graz, Austria
[2] weba Werkzeugbau Betriebs GmbH, Weba-Straße 2, 4407 Dietach, Austria
[3] Dipartimento di Ingegneria Industriale, Università di Padova, Padova, Italy

E-mail: wolfgang.weiss@v2c2.at

Abstract. This work presents a new virtual method for the optimised thermal design of hot stamping tools. It provides optimal positions of the tool's tempering ducts with respect to the average working temperature and its homogeneous distribution on the surface of a tool. It consists of a specific procedure for hot stamping tool design and a software framework in order to interconnect three domains: (I) a parametrised CAD tool model, (II) a linear thermal solver using a fast boundary element method and (III) an optimisation algorithm. This enables the automated set-up, simulation and optimisation of a duct topology. The boundary conditions for the simulations are derived from a reduced model of the thermal loading of the tool. The virtual method proposed is demonstrated on simplified tool segment geometries. The results are transferred to complex tool designs used in industry. For a selected use case, the number of ducts could be reduced by 50% through the application of the proposed method. These results are validated virtually based on an existing design. Hence, the new virtual method contributes to a CAE-driven tool design and a more efficient tool manufacturing.

1. Introduction

The process of hot stamping is a wide-spread process for the manufacturing of ultra-high strength steel (UHSS) parts in the automotive industry. Since its first applications in the mid-1980s the number of press hardened components in car bodies continuously increased to a level beyond 500 million parts per year. This evolution was accompanied by the development of enhanced simulation models in order to capture the complex thermo-mechanical-metallurgical process for the purpose of virtual process and component design. The phase transformation of the steel grade 22MnB5, commonly used for hot stamping applications, was in the special focus of the scientific community during the last 10 years (refer to [1]). Also aspects of tribology (e.g. [2], [3]) and tool wear (e.g. [4], [5]) have been addressed extensively in order to accurately describe the forming and quenching process of the blank. Many of these scientific developments are available in commercial simulation codes. The consideration of the stamping tools as elastic solid bodies with thermal conduction rather than shell surfaces with constant temperature further contributes to accurate descriptions of the stamping and quenching process. Therefore the quality of the stamped components in terms of mechanical and geometrical properties can be predicted with good approximation to reality. These detailed simulations require complex model set-up and expert knowledge on both the stamping process and the FEM-software (refer to [6] for further details on forming process simulation). Further, these detailed stamping simulations rely on tool designs and duct layout concepts that are normally not available in early design phases of the stamping process. So an iterative procedure becomes necessary for the design of an optimal duct layout. This is true for both cooled and heated stamping tool segments, often referred to as "soft zone" tools (see figure 1).

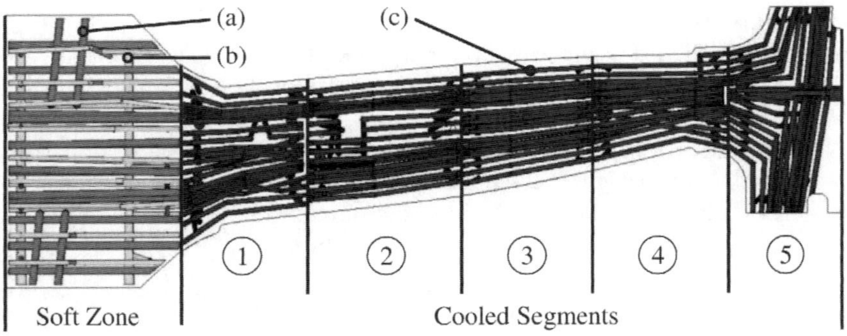

Figure 1. Tempering duct design of a segmented lower hot stamping tool with one soft-zone segment and 5 cooled segments; (a) heating ducts, (b) air cooling ducts; (c) water cooling ducts. *source: weba GmbH*

2. Tool Design

The design of a tool and its specific duct layout is performed on individual standards and knowledge of tool making companies. There exists no single, optimal tool design but there are a multitude of feasible tool designs possible. Normally, the stamping process claims homogeneous quenching of the full component in order to reach the desired high strength properties. This standard process requires a tool surface temperature lower than 200°C [7], evenly distributed over the surface of the tool. High contact pressure enables good contact for adequate heat transfer between blank and tool, leading to a homogeneous quenching process. Whereas the contact pressure is mainly influenced by the spot-grinding process during run-in and the capacity of the press, the working temperature of the tool mainly is determined by process parameters (e.g. cycle time, blank's initial temperature, tool cooling performance, heat losses to environment) and by the tool design regarding the duct layout. Since the process parameters are fixed for a given series process, the duct layout remains as the most critical parameter for the tool's average working temperature.

There are several designs of a cooling duct layout in stamping tools described in [8], e.g. straight, transversal, parallel, serpentine. Regarding the manufacturing of these layouts there are three common possibilities: (i) drilling of cooling bores, (ii) cast-in pipes and (ii) near-surface milling of shells (refer to [9]). Additive manufactured tools with arbitrary near-surface duct layouts are possible too and also discussed in the stamping community [10], although this type of production process is mainly applied with injection moulding and related processes [11]. The most common solution for hot stamping tools is the drilling of straight cooling ducts into the tool body (see figure 1), which is widely applied among the tooling industry.

For the standard process of cooled segments, the duct layout is mainly determined by a maximum density of bores to reach maximum cooling at minimum cycle times in series processes. This is achieved through the specific expertise of the CAD engineer who designs the duct layout, the inlets and outlets and the flow path based on experience and company standards. At early design phases, the design engineer has no feedback on the effects of the duct layout regarding the thermal status of the tool during the series process. This leads to trial and error loops between the design and the simulation process. For more complex stamping tools with e.g. soft-zones with different average temperature levels at different segments, these iterative loops become even more important since the temperature level of the tool is not determined by maximum cooling any more. A combination of integrated heating elements, mainly to heat up the segment at start-up, and cooling ducts to maintain the segment's nominal temperature during the series process becomes necessary. Through this, a different

IOP Conf. Series: Materials Science and Engineering **159** (2016) 012004 doi:10.1088/1757-899X/159/1/012004

duct layout design is required due to the different tempering strategy of soft-zone segments. Within this work, a virtual tool design method and an optimisation framework are presented for the efficient design of duct layouts for hot stamping tools.

3. Virtual Tool Design Method

The virtual tool design method developed within this work utilises a generalised software framework developed at Virtual Vehicle Research Center (VIF Optimisation Framework) that interconnects (i) a CAD-software for duct layout design (e.g. CATIA®), (ii) a solver based on boundary element methods BEM (e.g. HyENA[1]) and (iii) an optimisation algorithm (Particle Swarm Optimisation, PSO). The structure of this Virtual Tool Design Software Framework and its information flow is presented in figure 2, followed by a description of the single information paths.

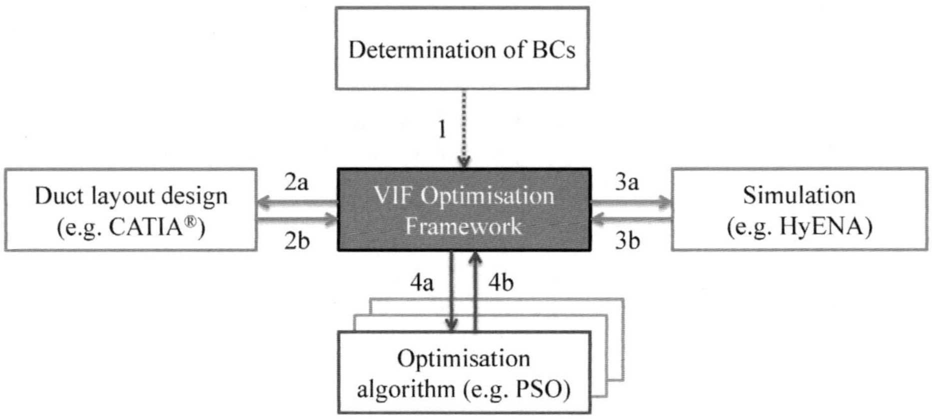

Figure 2. Structure of the Virtual Tool Design Software Framework.

(1) The thermal boundary conditions (BCs) for the later simulation are calculated based on analytical models for surfaces to the environment and the tempering ducts. The heat input from the blank is reduced to a time-averaged heat flux into the active surface.

(2a) Design parameters (duct diameter, distance to surface and spacing) are sent from the VIF Optimisation Framework to CATIA® which uses this input to set up a single duct design based on a parametric design template and a software script (CATScript®).

(2b) The surface of this duct design is automatically meshed with triangular elements based on an user-defined element size using Gmsh software [12] and merged with a predefined mesh of the outer surfaces of the design template.

(3a) The VIF Optimisation Framework generates a boundary condition file based on process parameter inputs from (1) and feedback from the optimisation algorithm (4b), sends this data to a computer cluster and starts the simulation job.

(3b) The solver returns the arithmetic average of the surface temperature of the active surface T_{av} together with its spatial range dT or rather its homogeneity over the active surface.

(4a) The optimisation algorithm uses these values to evaluate an objective function (OF) and constraints, which can be defined freely within the framework. The VIF optimisation framework enables the variation of each boundary condition (or rather process parameter) and geometry parameters.

(4b) The Particle Swarm Optimisation (PSO) algorithm returns all generated parameter sets (boundary condition and geometry parameters) while converging towards an optimal solution.

[1] Schanz M 2016 Graz University of Technology HyENA - Hyperbolic and Elliptic Numerical Analysis Online: http://www.tugraz.at/en/institute/am-bm/forschung/software/

3.1. *Determination of boundary conditions (BCs)*

The boundary conditions for the simulation model are derived from analytical models of free convection (for the outside surfaces), forced convection (for the cooling duct surfaces) and radiation (for outside surfaces). The heat input from the blank is calculated from the transient results of a one-dimensional thermal Finite Difference model of the system blank-tool-duct. The blank thereby is reduced to a mixed boundary condition, representing the time-averaged heat flux into the active surface during a quasi-stationary series process. It is assumed, that the contact heat transfer between the blank and tool is homogeneous over the whole active surface, thus a one-dimensional model is applicable.

3.2. *VIF Optimisation Framework*

The solver, the optimisation algorithm and the design template are controlled by the VIF Optimisation Framework, a modular software prototype realised in Matlab®. It enables the set-up and control of the optimisation jobs within this work. Its graphical user interface enables the definition of an objective function OF, constraints and parameters based on the geometry and process parameter input of the underlying simulation model. The VIF Optimisation Framework uses Dynamic Link Libraries to enable a direct communication with the CAD software (CATIA®). Simulation jobs are started on a remote server using secure shell communication. Graphical job monitoring and basic error handling capabilities enable the design engineer to set up and supervise the optimisation run without specific expert knowledge.

3.3. *Simulation (HyENA)*

To provide information on the expected time-averaged working temperature of hot stamping tools in the series process, a simplified simulation method has been developed and validated in [9] and presented in [13]. This method uses the energy equilibrium at the stamping tool between the heat input from the blank and the heat withdrawal to the cooling fluid and the environment in order to simulate the time-averaged working temperature of a hot stamping tool. This simplified simulation method is useful to predict the temperature status of the stamping tool during a series production process already at a very early design phase. Besides the boundary conditions, the minimum input requirements are the active surface design model and the duct layout of the hot stamping tool (refer to [9], p.97]). Through this, the effect of duct layout designs on the stamping tool temperature can be predicted. The main characteristic of the used simulation method is the utilization of a solver based on the Boundary Element Method (BEM) rather than on the Finite Element Method (FEM). This enables a simplified, automated model set-up due to the dimensional reduction from a volume model (FEM) to a surface model (BEM).

3.4. *Duct layout design - CAD*

The VIF Optimisation Framework offers two possibilities to consider geometric variations of the duct layout design in the optimisation procedure: (i) design templates and (ii) design variants, illustrated in figure 5.

3.4.1. *Design templates*

A general duct layout design as an input for the VIF Optimisation Framework is generated through a design template in CATIA® software. The design template represents a generic design of a stamping tool segment with bounding surfaces and basic geometric items for a parametric duct layout design. The parametric design rules are programmed in the CATScript®. Figure 3 shows the principle procedure of the parametric duct design and model development for a simple block-segment. First, a duct design is built up based on parameters controlling the distance to the active surface and the duct spacing. Second, the surface model of the duct design is exported and automatically meshed using the open source meshing surface Gmsh with a batch procedure. Third, the resulting mesh of the duct design is merged with the boundary mesh of the design template. Figure 3 (right) shows one result of this procedure.

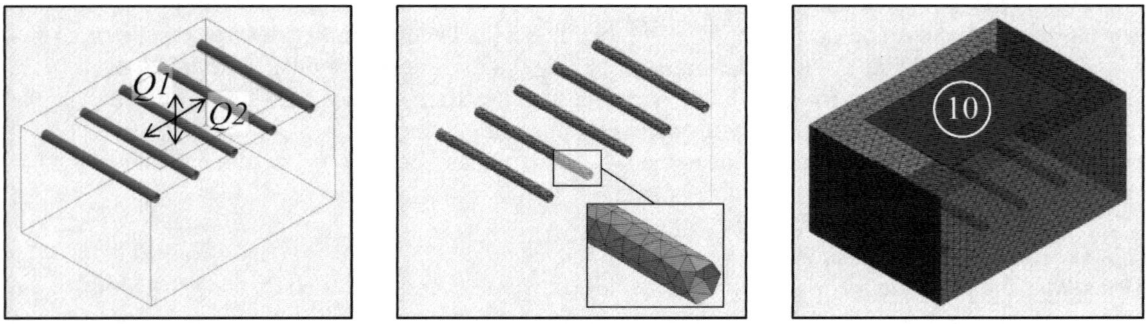

(a) (b) (c)

Figure 3. Principal procedure of the parametric duct design and model development: (a) parametric design of duct layout with geometric parameters *Q1, Q2*; (b) batch-meshing of duct layout; (c) merging of duct mesh with boundary mesh, (10)...active surface.

This model is then used for the optimisation of the tempering duct positions with respect to an average surface temperature of the active surface. The simple block-segment design is very efficient and enables the variation of the principal geometric model parameters. The design template can be modified to meet any demands of the user, depending on the complexity of the design rules.

3.4.2. Design variants

A second method of tool design is applicable within the workflow of the Virtual Tool Design Software Framework: the usage of arbitrary and highly complex duct layouts enables highest flexibility for the duct design. The design engineer can model various configurations of duct layouts, representing specific geometric inputs for the optimisation procedure. Examples of such design variants *V* are shown in figure 4 for the use case stamping tool segment of this work. Inlet and outlet ducts are neglected because they do not contribute to the cooling of the segment.

Using arbitrary design variants affects the choice of the optimisation algorithm. As described in section 3.5, gradient based algorithms cannot be used for such problems.

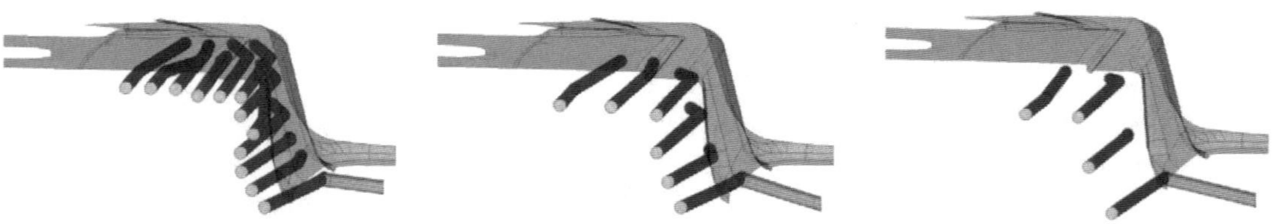

(a) (b) (c)

Figure 4. Three different duct layout variants of the use case stamping tool segment: (a) *V=9*, (b) *V=5*, (c) *V=1* according to table 2.

3.5. Optimisation algorithm (PSO)

The heart of an efficient optimisation set is an optimisation algorithm which is tailored to the problem and can be adapted in many ways. Many different algorithms are available and there is no overall general purpose algorithm which can be used in any case. The stated problem of stamping tool cooling uses duct layout design variants which need to be treated as discrete variables. Because of lacking a functional relationship between the duct designs, the designer has the freedom to rank them differently. Therefor the structure and the appearance of the design space can change according to the ranking. That makes it impossible to calculate meaningful gradients for an algorithm which uses gradient information or deduced gradients. Particle Swarm Optimisation (PSO) introduced by Kennedy and Eberhard [14] is an optimisation algorithm which does not need any gradient information. It is derived from the behaviour of bird swarms searching for food by flying through the space. The communication between the particles (i.e. process and design variants) is described mathematically and used for searching good solutions in a given design space without the necessity of gradient information. Because of possible high numbers of degrees of freedoms (geometry and process parameters), the design space can be high-dimensional. Therefore it is necessary to adopt the number of initial particles to the size of the design space. However, increasing the population size slows down the convergence rate. The solution for this contradiction is an adaptive algorithm which reduces the population size based on the concept of an artificial potential of each evaluated particle. It uses a special routine of moving barriers, calculated with Bernoulli's equation of colour temperature [15]. The combination of PSO and adaptive reduction of population results in a suitable algorithm for optimising hot stamping tool segments, as presented in this work.

4. Application of the Virtual Tool Design Software Framework

The optimisation procedure of the duct layout design of a cooled stamping tool segment is presented. Figure 5 shows the simulation model surfaces of the design template (left) and a design variant of the stamping tool segment (right), considered in this work. The design template represents a simplified model of the stamping tool segment and hence shows slightly different boundary conditions, given in table 1.

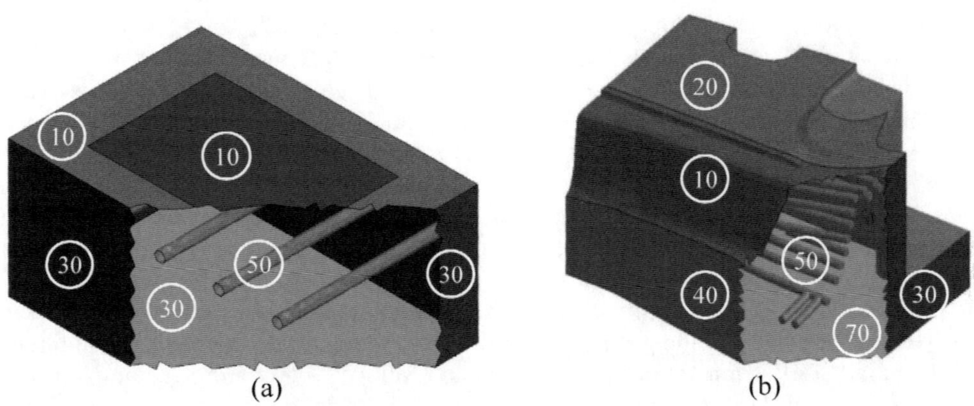

Figure 5. Simulation model surfaces of the design template (a) and a design variant of the stamping tool segment (b), considered in this work: (10)...active surface; (20)...outer surface; (30)...segmentation surface/base surface; (40)...surface to environment; (50)...duct surface; (70)...base surface isothermal.

The parameters for the boundary conditions (*P1, P2*) in table 1 are calculated as described in chapter 3.1. Looking at the parameters for surface 10 (active surface) it is obvious that for the simulation of the time-averaged working temperature, the transient heat input from the blank is replaced by a mixed BC. The determination of the respective parameter values based on an energy balance at the stamping tool is described in [9]. The segmentation surfaces and also the base surface of the design template (surface 30) are assumed to be isolated. Heat losses to the environment are considered at the stamping tool segment (surface 20 and 40) only. The convection heat transfer coefficient at the cooling ducts (surface 50) is calculated with the analytical expression for forced, turbulent convection in pipes of Gnielinsky [16]. The base surface (surface 70) of the stamping tool

IOP Conf. Series: Materials Science and Engineering **159** (2016) 012004 doi:10.1088/1757-899X/159/1/012004

segment is assumed to be at environmental temperature; this assumption is true for cooled segments and needs to be proven from case to case.

Table 1. Boundary condition types and values for the surfaces of the design template and the stamping tool segment considered in this work: boundary conditions: R...mixed BC (Robin, $q=P1*(T-P2)$, N...constant flux (Neumann, $q=P1$), D...constant temperature (Dirichlet, $T=P2$) with q as specific heat flux and T as Temperature, SI units are used for $P1$ and $P2$.

surface	BC type	P1	P2
10	R	350	850
20	R	16	390
30	N	0	-
40	R	16	390
50	R	14307	25
70	D	-	25

4.1. *Design template*
For the optimisation with the design template, the objective function $OF=minimise[abs(T_{av}-T_{nom})]$ is used, whereas the objective temperature for the active surface is set to $T_{nom}=115°C$ for this particular stamping tool segment. The constraints were set to $10K<dT<50K$, meaning that a specific spatial temperature range dT is tolerable. Otherwise the solution would be trivial: a homogeneous temperature distribution will be achieved through a very dense duct layout. Two variable geometric parameters $Q1$ and $Q2$ are considered with the optimisation: (i) the normal distance between the active surface and the duct axes ($Q1$, continuous) and (ii) the distance in between the ducts axes or rather the duct spacing ($Q2$, vector). The values for the optimisation are set to $0.008 < Q1 < 0.07$ and $Q2 = [0.0105\ 0.013\ 0.0168\ 0.024\ 0.042\ 0.056\ 0.084]$, representing feasible duct layouts. SI units are used for the parameters ($[Q1, Q2] = m$).

For the presented use case, the optimisation of the geometric parameters $Q1$ and $Q2$ is of interest, but in principal any parameter of the geometry and the process (e.g. duct diameter, cycle time, closed time of tool, initial blank temperature, temperature and velocity of cooling medium, heat loss to environment, conductivity of tool material, environmental temperature) can be considered and optimised.

4.2. *Design variants*
Besides the described optimisation procedure using a design template, the approach using design variants of a stamping tool segment is applied too. Therefore, nine cooling variants, defined through the vector $V = [1\ 2\ 3\ 4\ 5\ 6\ 7\ 8\ 9]$, were designed. Some of these design variants are illustrated in figure 4, figure 9 and figure 10. Each of them represents a single, feasible duct layout. The single variants of this use case mainly differ with their average normal distance between the active surface and the duct axes $Q1$ and the average distance in between the duct axes $Q2$, similar to the parameters of the design template before.

For the optimisation of the presented use case, the parameter $P1$ of surface (50) in table 2 is set continuous in the range of $10000<P1<20000$, defining feasible values of the convective heat transfer coefficient at the duct surface. Equal to the optimisation with the design template, any process parameter could be set variable and optimised.

An optimisation run is performed with the two parameters *V* (discrete) and *P1* (continuous). The objective function *OF=minimise[dT]* is used in order to optimise towards a homogeneous temperature distribution on the active surface of the stamping tool segment. The constraints are set to $75°C<T_{av}<155°C$, meaning that an average temperature of the active surface T_{av} in the given range is tolerable.

Table 2. Design variants *V* of the use case with their average distances *Q1* and *Q2*; initially realised variant: *V=9*.

V		duct axes - duct axes, *Q2*		
		0.014	0.028	0.043
surface - duct axes, *Q1*	0.012	<u>*9*</u>	*6*	*3*
	0.024	*8*	*5*	*2*
	0.036	*7*	*4*	*1*

5. Results

In the following, the results of the two optimisation approaches based on (i) a design template and (ii) design variants are presented and discussed.

5.1. *Results using the design template*

The evolution of the PSO algorithm towards an optimal solution using a design template is shown in figure 6. The movement of the particles in the design space *Q1* and *Q2* towards an optimal solution is illustrated. The optimal parameters are *Q1=0.008* and *Q2=0.024*, leading to an objective function value of $abs(T_{av}-115)=6K$ and a constraint value of $dT=21.3K$.

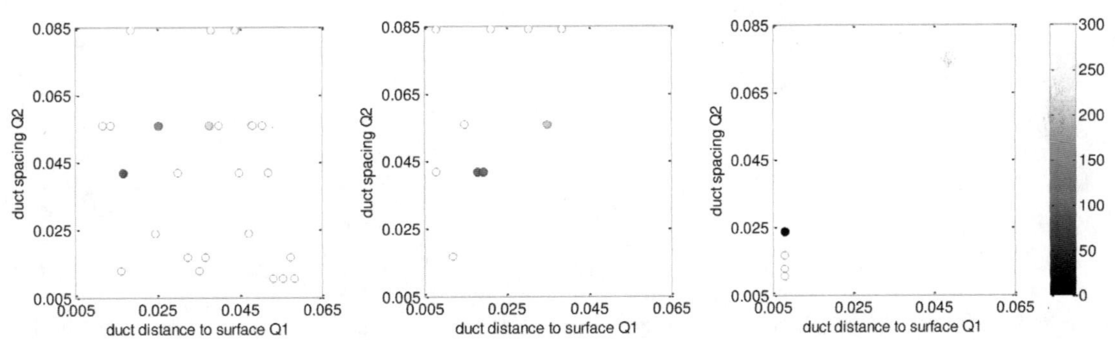

(a) (b) (c)

Figure 6. Evolution of Particle Swarm towards optimal solution: iteration 1 (a), iteration 4 (b) and iteration 10 (c). The colours represent the *OF* values in Kelvin.

The optimal parameter *Q1=0.008* is rather small for cooling ducts of industrial stamping tools. Raising *Q1* to a feasible level, the geometric solution from the design template is very close to the design variant *V=6* (see table 2; *Q1=0.012, Q2=0.028*) of the stamping tool segment. Compared to the initially realised variant *V=9* (see table 2; *Q1*=0.012, *Q2=0.014*) this optimal solution leads to a significant reduction of the number of cooling ducts due to the double average distance in between the duct axes *Q2*.

The respective simulation result of the optimal solution is shown in figure 7, representing a non-trivial solution of the constraint optimisation problem with an average active surface temperature of $T_{av}=121°C$ and a spatial temperature range of $dT=21.3K$.

Figure 7. Temperature simulation result of the optimal solution of the PSO algorithm using the design template; temperatures in °C

5.2. Results using design variants

The evolution of the PSO algorithm towards an optimal solution using design variants is shown in figure 8. The movement of the particles in the design space *V* and *P1* towards an optimal solution is illustrated. The optimal parameter set is $V=6$ and $P1=20000$, leading to an *OF* value of $dT=190K$ and a constraint value of $T_{av}=145°C$. The respective simulation result of this optimal solution is shown in figure 9.

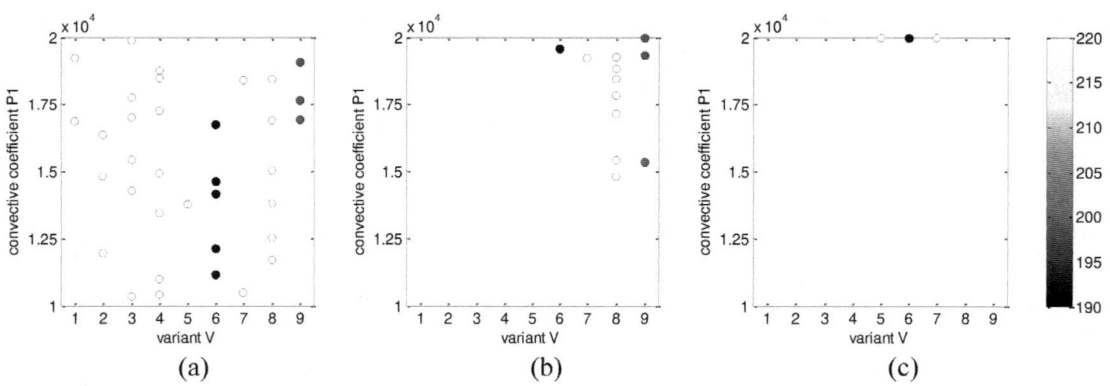

Figure 8. Evolution of Particle Swarm towards optimal solution: iteration 1 (a), iteration 2 (b) and iteration 9 (c). The colours represent the *OF* values in Kelvin.

The common result of both approaches is, that the design variant $V=6$ offers an optimal solution for the stated use case in terms of surface temperature and temperature homogeneity. The initial duct layout contains 12 ducts (see figure 10) while the optimised solution only contains 6 ducts (see figure 9). Hence, the optimisation procedure leads to a design with a reduced number of cooling ducts by 50%.

IOP Conf. Series: Materials Science and Engineering **159** (2016) 012004 doi:10.1088/1757-899X/159/1/012004

(a) (b)

Figure 9. Cooling duct design (a) and simulation result of the optimal duct layout (b) of the cooled stamping tool segment as a result of the optimisation procedure; $V=6$ according to table 2.

6. Validation

The reduced simulation method applied within this work has been validated on cooled stamping tool segments in [9]. A further validation with the particular stamping tool segment used within this work has not been performed.

The validation of the optimisation results is performed virtually by comparing the initial, non-optimised cooling duct design, shown in figure 10, with the optimised cooling duct design, shown in figure 9. The difference in the average surface temperature of the active surface is $\Delta T_{av}=11K$ ($T_{av\_init}=134°C$, $T_{av\_opt}=145°C$), the difference in the spatial temperature range is $\Delta dT=-11K$ ($dT_{init}=201\ K$, $dT_{opt}=190K$). Considering this rather small differences which, of course, are related to the specific use case of this work, the optimised design reduces the number of cooling ducts significantly by 50% compared to the initial design while maintaining the cooling performance of the stamping tool segment.

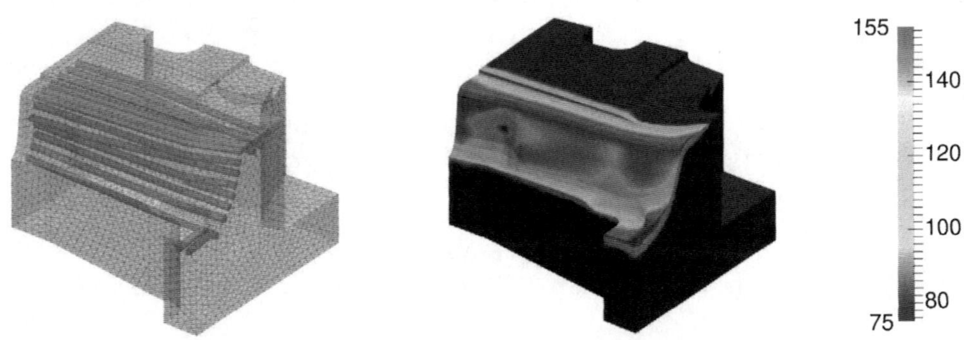

(a) (b)

Figure 10. Cooling duct design (a) and simulation results of the initial duct layout (b) of the cooled stamping tool segment; $V=9$ according to table 2.

7. Conclusion

The presented method for virtual tool design using a software framework that interconnects a CAD-software, a numerical solver and an optimisation algorithm was successfully applied to a cooled segment of a hot stamping tool. Two approaches were presented: (i) a design template to optimise the principal geometric configuration of a cooling duct layout and (ii) design variants to optimise process parameters based on arbitrary cooling ducts designs. Both approaches can be utilised individually. Within this work, both approaches led to similar results on an industrial stamping tool segment.

The result of the optimisation with the design template was a geometric configuration, similar to the design variant that is finally determined as optimal (*V=6*). For this first optimisation approach, the objective function was related to a nominal average surface temperature of the active surface. The final optimal design variant *V=6* was also obtained with the second optimisation approach based on nine individual design variants. For this optimisation, the objective function was related to homogeneity of the surface temperature. It was shown, that both approaches lead to the same optimal and feasible design, reducing the number of cooling ducts by 50%.

Acknowledgments
This work was accomplished at the VIRTUAL VEHICLE Research Center in Graz, Austria. The authors would like to acknowledge the financial support of the COMET K2 - Competence Centers for Excellent Technologies Programme of the Austrian Federal Ministry for Transport, Innovation and Technology (bmvit), the Austrian Federal Ministry of Science, Research and Economy (bmwfw), the Austrian Research Promotion Agency (FFG), the Province of Styria and the Styrian Business Promotion Agency (SFG).

They would furthermore like to express their thanks to their supporting industrial and scientific project partners, namely weba Werkzeugbau Betriebs GmbH and Graz University of Technology.

References
[1] Merklein M, Wieland M, Lechner M, Bruschi S and Ghiotti A 2016 *Journal of Materials Processing Technology* **228** pp 11–24
[2] Pelcastre L, Hardell J and Prakash B 2013 *Tribology International* **67** pp 263–271
[3] Azushima A, Uda K and Yanagida A 2012 *Journal of Materials Processing Technology* **212** pp 1014–1021
[4] Pujante J, Pelcastre L, Vilaseca M, Casellas D and Prakash B 2013 *Wear* **308** pp 193–198
[5] Ghiotti A, Bruschi S and Medea F 2015 *Wear* **332–333** p 810–821
[6] Neugebauer R, Schieck F, Meinel S, Polster S, Scheffler C and Lausch M 2011 Aspects of simulation based process and tool design for heat supported sheet metal forming processes *Proceedings of the 10th International Conference on Technology of Plasticity ICTP2011 (Aachen)* pp 426–431
[7] Hoffmann H, So H and Steinbeiss H 2007 *CIRP Annals - Manufacturing Technology* **56** pp 269–272
[8] Hu P, He B and Ying L 2016 *Applied Thermal Engineering* **96** pp 338–351
[9] Weiß W 2013 *Thermische Auslegung von Werkzeugen für Presshärtprozesse* Ph.D. thesis Graz University of Technology
[10] Edelmann K 2015 Generative Fertigung im Werkzeugbau - Laserstrahlschmelzen ermöglicht neue Funktionen und Mehrwerte *Workshop Blechwarmumformung*
[11] Faßnacht P and Overmeyer L 2011 *ZWF–Zeitschrift für wirtschaftlichen Fabrikbetrieb* **106** pp 868–872
[12] Geuzaine C and Remacle J F 2009 *International Journal for Numerical Methods in Engineering* **79** pp 1309–1331
[13] Weiß W, Kolleck R, Schanz M and Messner M 2012 Application of the Boundary Element Method to the Thermal Analysis of Hot Forming Tools *Lightweighting: Possibilities & Challenges - Proceedings of the IDDRG Conference (Mumbai)* pp 105–110
[14] Kennedy J and Eberhart R 1995 Particle swarm optimization *Proceedings of IEEE International Conference on Neural Networks* vol 4 pp 1942–1948
[15] Feoktistov V 2006 *Differential evolution: In Search of Solutions* Optimization and Its Applications (Springer US)
[16] Incropera F P and DeWitt D P 2007 *Fundamentals of heat and mass transfer* 6th ed (Hoboken, NJ: Wiley)

IOP Conf. Series: Materials Science and Engineering **159** (2016) 012005 doi:10.1088/1757-899X/159/1/012005

Numerical analysis of the flexible roll forming of an automotive component from high strength steel

B Abeyrathna[1], A Abvabi[2], B Rolfe[3], R Taube[2], and M Weiss[1]

[1] Deakin University, Geelong, Australia, Institute for Frontier Materials, Waurn Ponds, Pigdons Rd., VIC. 3216
[2] Ford Motor Company of Australia Ltd, 1735 Sydney Rd, Campbellfield VIC. 3061
[3] Deakin University, Geelong, Australia, School of Engineering, Waurn Ponds, Pigdons Rd., VIC. 3216

E-mail: buddhika.abeyrathna@deakin.edu.au

Abstract. Conventional roll forming is limited to components with uniform cross-section; the recently developed flexible roll forming (FRF) process can be used to form components which vary in both width and depth. It has been suggested that this process can be used to manufacture automotive components from Ultra High Strength Steel (UHSS) which has limited tensile elongation. In the flexible roll forming process, the pre-cut blank is fed through a set of rolls; some rolls are computer-numerically controlled (CNC) to follow the 3D contours of the part and hence parts with a variable cross-section can be produced. This paper introduces a new flexible roll forming technique which can be used to form a complex shape with the minimum tooling requirements. In this method, the pre-cut blank is held between two dies and the whole system moves back and forth past CNC forming rolls. The forming roll changes its angle and position in each pass to incrementally form the part. In this work, the process is simulated using the commercial software package Copra FEA. The distribution of total strain and final part quality are investigated as well as related shape defects observed in the process. Different tooling concepts are used to improve the strain distribution and hence the part quality.

1. Introduction

Roll forming is an incremental sheet forming process which was introduced more than one hundred years ago. The process has generally been used for manufacturing longitudinal profiles for different applications such as structural, household and automotive components with uniform cross-sections until early 2000. Since then the conventional roll forming technology has been advanced to produce parts with variability in width and depth. This process is called “flexible or 3D roll forming” and as this technology is fairly new, there have been some recent research papers published [1, 2]. The main advantage of the process is the ability to form high strength and low ductility materials and it is attractive for manufacturing automotive parts from Ultra High Strength Steel (UHSS) and Advanced High Strength Steel (AHSS). The use of 3D roll forming presents an opportunity to reduce tooling cost and global tooling footprint by simplifying tool designs and potentially re-using tooling for multiple components and derivatives. Whilst this has been well documented for traditional roll forming (linear constant cross section components), the opportunity to roll from components with variable geometry significantly broadens the application and hence potential benefits.

In the flexible roll forming operation some rolls are kept stationary with an axial rotation while the others have both axial rotation and translational movement to generate the variable width or depth of the parts. Several researchers have investigated flexible roll forming [1, 3, 4] and the first validated flexible roll forming line for automotive parts manufacturing was developed by the European R&D project PROFORM [2, 5]. Accuracy of the final shape is critical in flexible roll forming of automotive components. The main defects in flexible roll formed products are wrinkling and web warping [6]. Excessive compressive stresses in the flange at the transition region cause wrinkling [7] while web warping is caused by the lack of longitudinal plastic strains in the flanges of the transition region [6]. Larranaga [6] introduced heat-assisted flexible roll forming to minimise the web warping. Later Sedlmaier et al. [8] introduced a blankholder system which helped to eliminate the occurrence of web warping by applying some pressure on the web region while flange forming. Groche et al. [9] introduced an analytical model to predict the critical stress level at the flange which causes wrinkles in variable width flexible roll forming. Park et al. [10] introduced an analytical model to predict the longitudinal strain in flexible roll forming and discovered that it can be controlled by changing the forming parameters. Jiao et al. [11] developed an analytical model to predict the amount of web warping in flexible roll forming and extended this study to obtain an analysis to predict the longitudinal strain in this process. Yu et al. [12] showed that the finite element model of the flexible roll forming process can be improved by supplemented material input based on the Swift's law.

Even though a number of investigations have been carried out on flexible roll forming, careful process design is needed to achieve precise profiles. The main challenge is to obtain the best profile with the minimum tooling cost. In this paper, a new variable width flexible roll forming technique is introduced similar to variable depth forming of the type presented in [13]. This process allows significant cost reduction through tooling optimisation. Several forming concepts are introduced and material behaviour under each forming condition is discussed.

2. Part dimensions and materials

The CAD part of an automotive component was provided by the Ford Motor Company, Australia and its simplified isometric view is shown in Figure 1(a); the cross-sectional dimensions are shown in Figure 1(b). It can be seen that this profile has a uniform cross-section, but is curved.

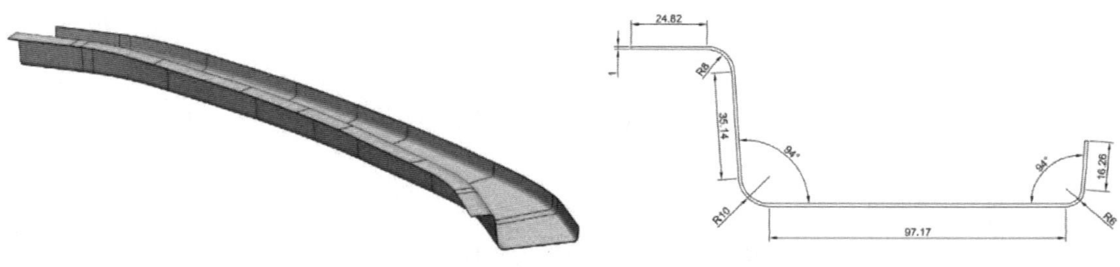

(a) (b)

Figure 1. (a) Isometric view of the profile (b) Cross sectional dimension of the profile

The strip thickness is 1mm and a Dual Phase steel (DP1000) was used as the material in the finite element models. The tensile test was carried out from the bone shaped samples oriented along the rolling direction of the strip and the standard given by ASTM E8/E8M [14] was followed. An Instron 5967 with a 30kN load cell was used for the test and the test speed was $0.025mms^{-1}$ giving a strain rate of $0.001s^{-1}$. The true stress–true strain curve of the material is given in Figure 2.

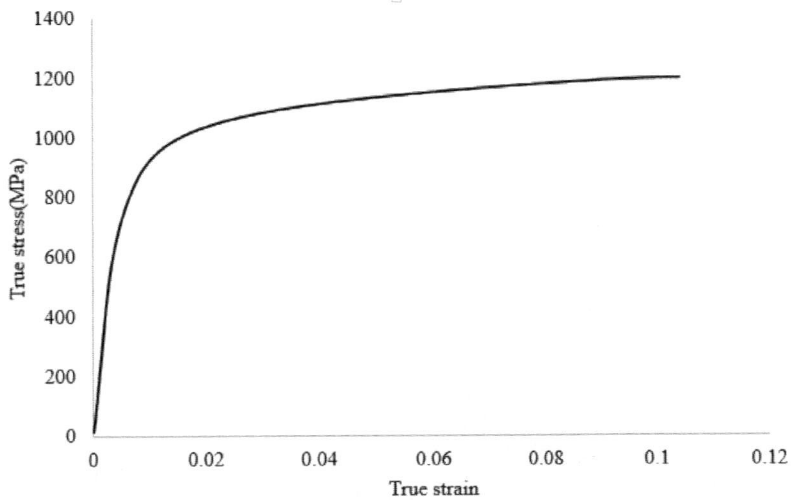

Figure 2. Average true stress strain curve for the samples tested along the rolling direction

Other material parameters were obtained from the Hollomon's power law fitted to the true stress-plastic strain curve of the material and are given in Table 1.

Table 1. DP1000 material data

Material	Yield Strength (MPa)	Ultimate tensile strength(MPa)	Elastic limit	n	K (MPa)
DP1000	764	1194	0.004	0.12	1633

3. Forming strategy

The part shown in Figure 1 is of uniform section, but because of the curved shape in the plan view, the deformation is similar to that of a part of varying width. The proposed forming method employs a die to hold the blank and the whole die set is driven back and forth by a lead screw. A tool holder carries the tooling which is CNC controlled; so that rotational and translational movement can be introduced to the rolls to form the sides as shown in Figure 3. The blank is pre-cut to a shape obtained by unfolding the profile in a CAD system. The forming rolls follow a selected path to form both sides and flanges. The forming sequence is designed to have the minimum number of tooling changes to keep tooling cost and tool changing time to a minimum. In this paper two different tooling concepts will be discussed and the strain distribution under each condition indicated.

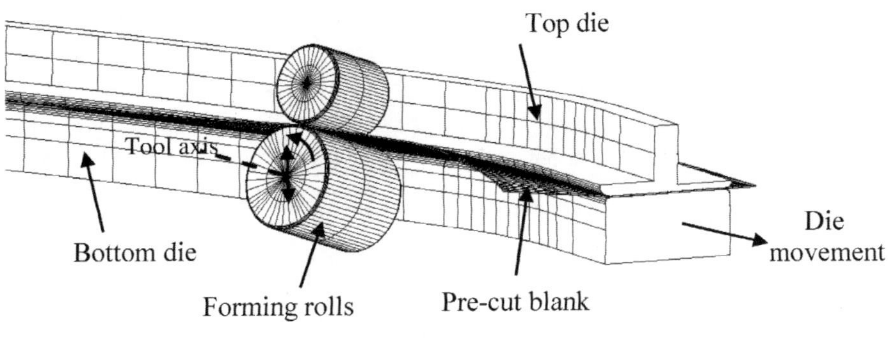

Figure 3. Flexible roll forming methodology

IOP Conf. Series: Materials Science and Engineering **159** (2016) 012005 doi:10.1088/1757-899X/159/1/012005

4. Flower pattern and tooling

The forming sequence used in this investigation is given by the flower pattern shown in Figure 4. Two tooling sets were used for the forming sequence.

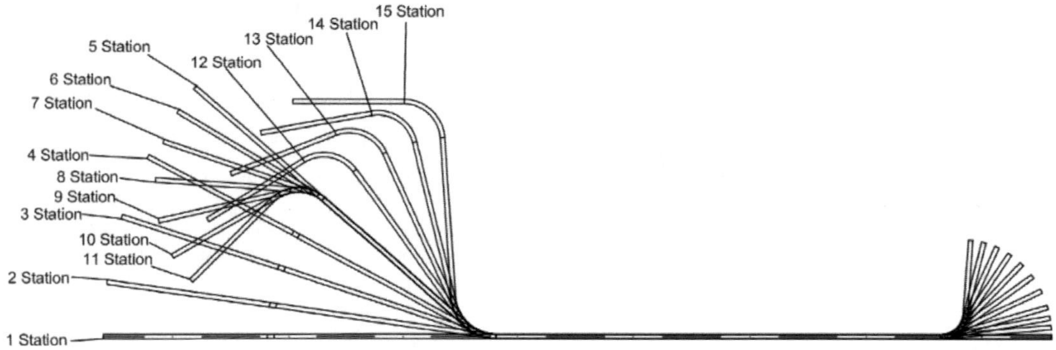

Figure 4. Flower pattern

The dimensions of the pre-cut blank were obtained by unfolding the simplified CAD profile given in Figure 1(a) and are shown in Figure 5.

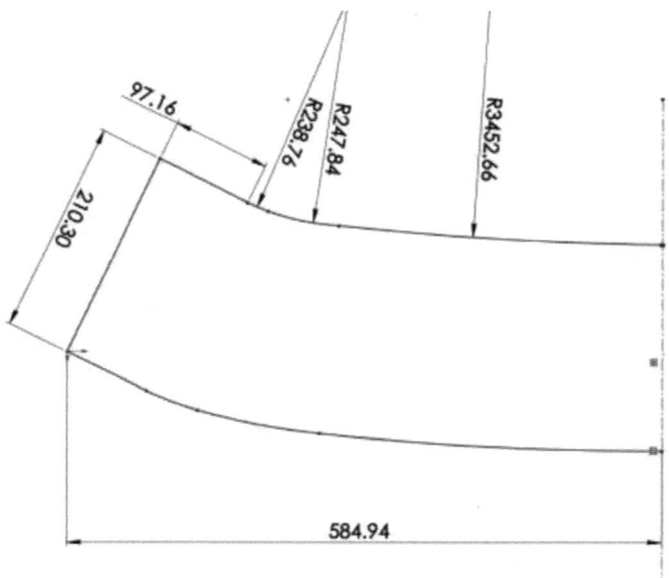

Figure 5. Symmetric half of the pre-cut blank

In this investigation, two different tooling concepts were used to form the left-hand side of the part shown in Figure 4 and a single tooling concept for the right side of the profile (here we mainly focused on the left-hand side of the part since it introduces the complexity to the profile). However the same tooling configuration was used in both concepts from 1st - 11th stations and is shown in Figure 6.

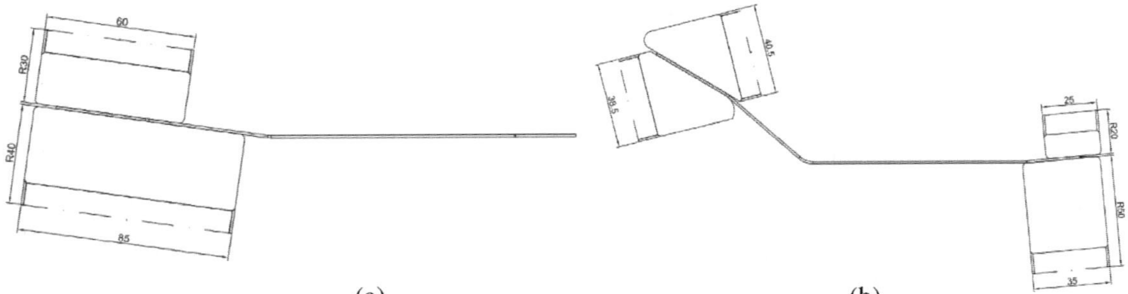

(a) (b)

Figure 6. Tooling configuration used in both concepts for (a) 1st - 5th stations (b) 6th - 11th stations

Two different tooling configurations were used for forming from the 12th to the 15th station as shown in figure 6. In the first tooling concept only one cylindrical bottom roll was used to form the flange to the final angle as shown in Figure 7 (a). In concept 2, top and bottom rolls were used to form the flange as shown in Figure 7 (b).

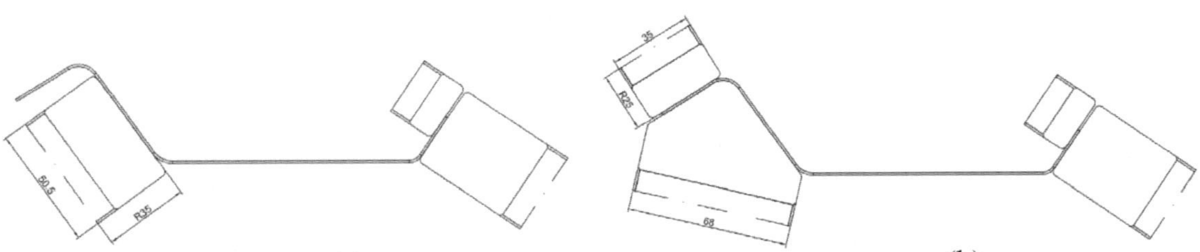

(a) (b)

Figure 7. Tooling configuration used in 12th - 15th station for (a) tooling concept 1 and (b) tooling concept 2

5. Finite element model

Unlike in the traditional roll forming process, in the variable width flexible roll forming process, tooling is moving in a 3D space to generate the required part geometry. In this work, a new concept of variable width flexible roll forming is introduced and is similar to the variable depth roll forming process introduced in [13]. In the finite element model, first the pre-cut blank is held between the clamping dies and the forming tools pass a certain contour to form the blank incrementally to the desired shape. The die set then opens and the finished product can be removed.

The tool design and numerical simulation were carried out with the commercial software package COPRA® FEA RF [15] which uses built-in MSC.Marc and an implicit solver. Half of the length of the part was modelled due to the symmetry and full integration, hexahedral, type 7 arbitrarily distorted brick elements available in the software package were used to model the pre-cut blank. Only one element through the thickness was used and a refined mesh was introduced at the bending regions. The effect of friction was neglected in the FEA model. Material input for the FEA model was taken from the tensile test data given in Figure 2 while the Poisson's ratio was taken as 0.3. Given that several previous studies have shown that in conventional and flexible roll forming a first adequate representation of material behaviour can be made using a simple isotropic hardening model [16, 17] the von Mises Criterion was used to define plastic material behaviour.

Three boundary conditions are applied on the pre-cut blank as shown in Figure 8. An X lock is introduced on the nodes along the transverse centre of the blank which restricts the transverse movement of the part during the forming operation. A Y lock is applied on three nodes of the bottom of the pre-cut blank (see Figure 8) which limits the vertical movement of the strip when it is not supported by the die set. The longitudinal movement of the strip is avoided by a Z lock boundary condition which is applied on some nodes at the front edge and mid-section of the pre-cut blank (see Figure 8).

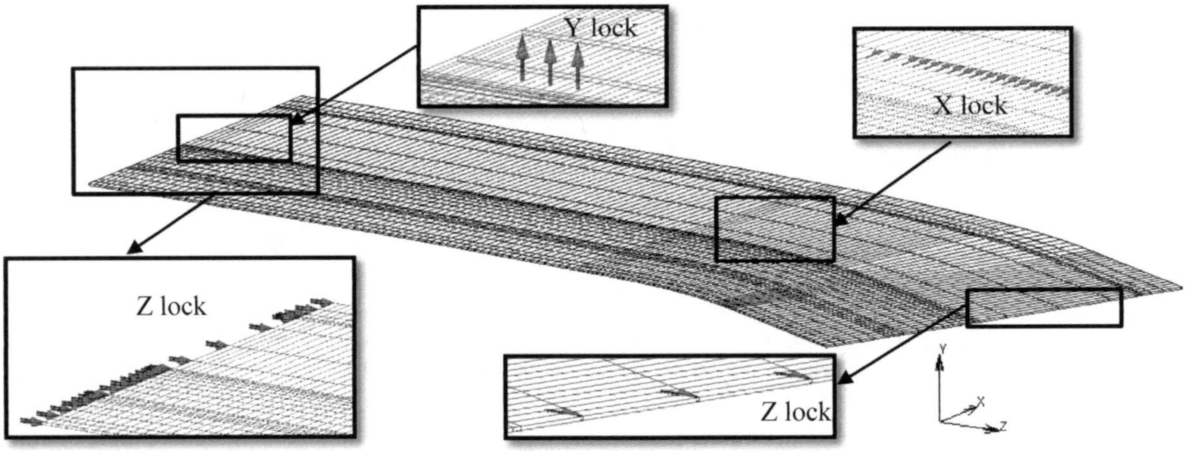

Figure 8. Boundary conditions on the pre-cut blank

6. Results

Different views of the final part simulation with the two tooling concepts are shown in Figure 9. There is some distortion in the horizontal flange at the ends of the both parts. Significant longitudinal stretching takes place in that region when the pre-cut is formed to the final geometry; since this stretching is irreversible, the additional material generates some waviness in the flange; however the waviness generated by the tooling concept 2 is significantly smaller than that of tooling concept 1 (see side view in Figure 9.

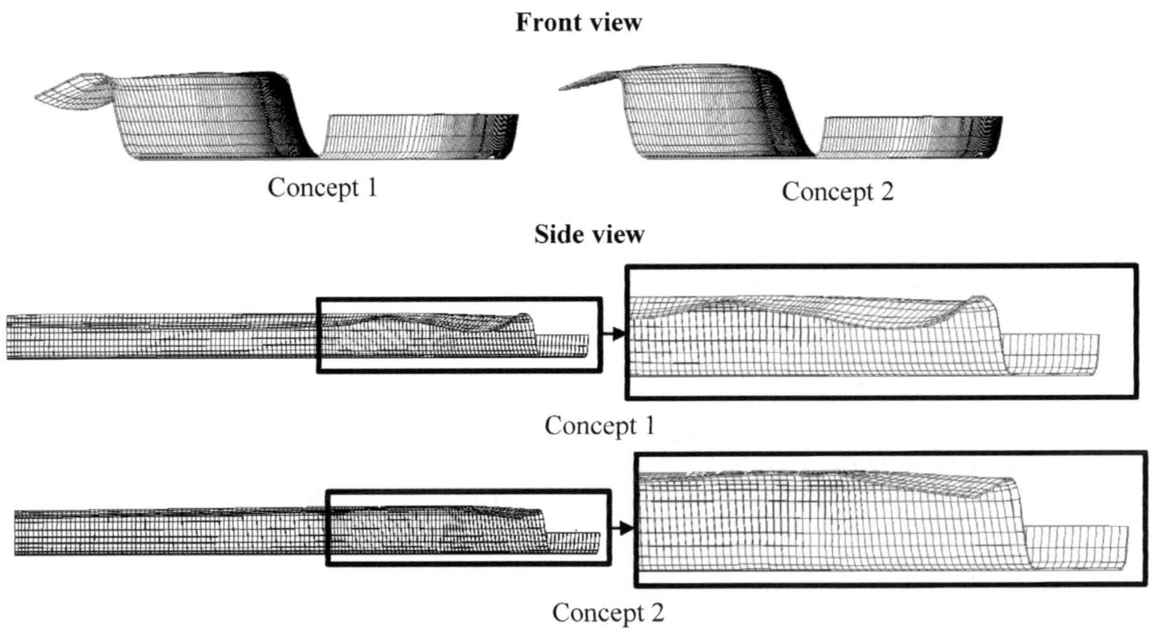

Figure 9. Final part quality

IOP Conf. Series: Materials Science and Engineering **159** (2016) 012005 doi:10.1088/1757-899X/159/1/012005

To understand the material deformation in the process further, the strain history was analysed and for this a node in the critical region of deformation was taken as shown in Figure 10(a). The longitudinal and transverse strain during the process for both tooling concepts are shown in Figure 10(b). Between station 1 and 11, the processes are identical and as this is an edge node the deformation is uniaxial. The peak membrane strains lie between approximately +5% and -2.5% and this is much greater than strains encountered in conventional roll forming. During forming of the side, stations 1 – 5, and forming the flange, 6- 11, the edge is stretched plastically to residual strain of about 3%. During the reforming of the side, stations 12 – 15, the residual longitudinal strain is reduced, i.e. it is compressed to 2 and 1% residual longitudinal strain in tooling concept 1 and 2 respectively. In practical metal forming parlance it would be said that the metal is “crowded” and hence buckling occurs. To illustrate this more clearly, the longitudinal strains in stations 12 – 15 are shown to a larger scale in Figure 10(c). It is seen that the compression is higher in tooling concept 1 (2%) than in tooling concept 2 (1%). The lower longitudinal compressive strain in tooling concept 2 can be attributed to the additional support provided by the top roll as this suppresses material movement.

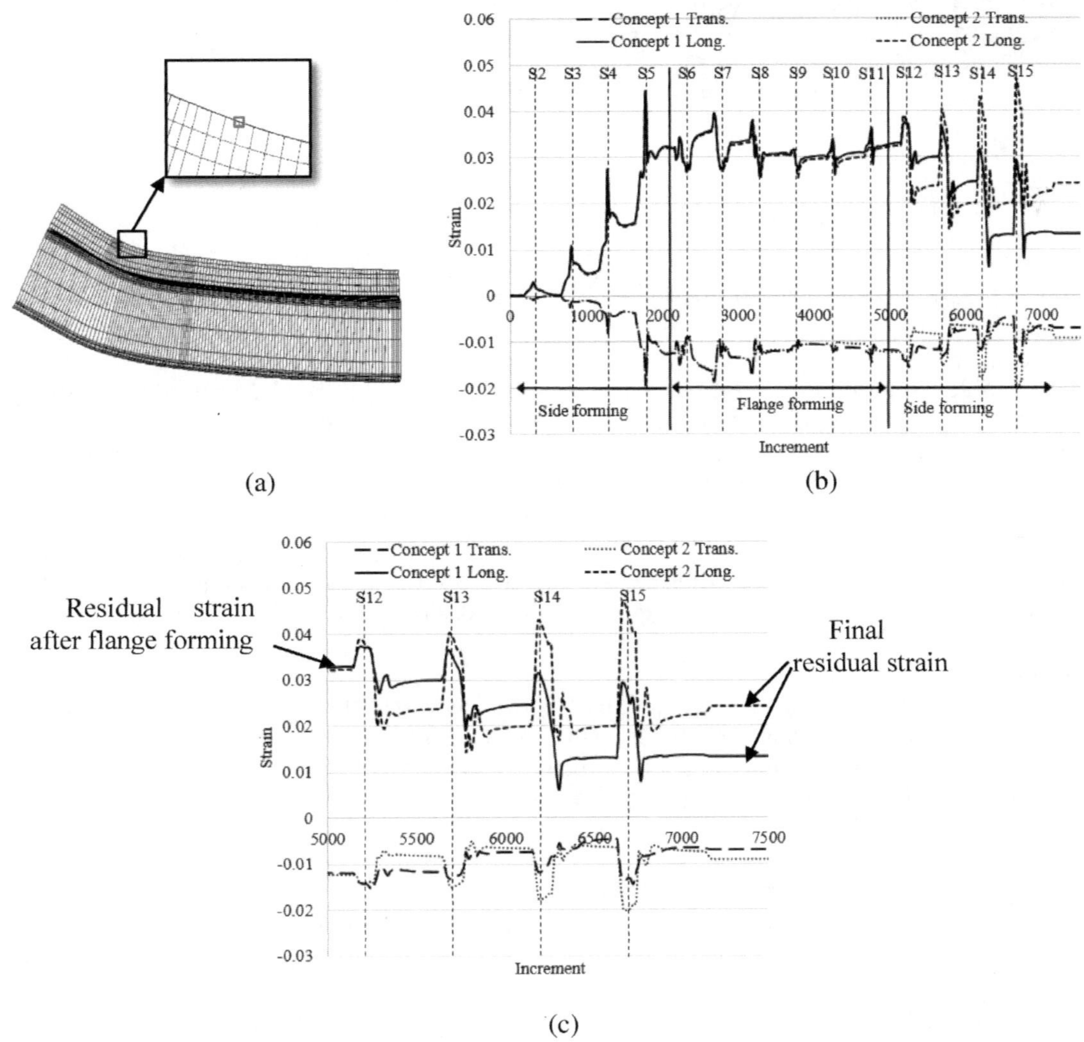

Figure 10. (a) Location of the node which considers for the strain plot (b) Different strain components (c) Different strain components from S12 to S15

The strain distribution of the whole part along the inside edge is shown in Figure 11; the strain in Figure 11(b) is not the peak strain, but that after forming the complete section. It may be seen that most of the areas do not undergo severe deformation, but between an arc length of 100mm and 200mm there is significant residual strain in the final part. The tooling concept 2 shows higher longitudinal residual strain than that of the tooling concept 1. This comes about because in concept 1 the reduction in edge length is achieved by buckling, which is undesirable, while in concept 2 the reduction is achieved by plastic compression of the metal. Further, it may be seen from Figure 10(b) that concept 2 shows less reduction of longitudinal strain compared to tooling concept 1, because, as mentioned, the top supporting roll suppresses the material movement. As a result tooling concept 2 shows higher longitudinal residual strain in the flange edge (see Figure 11(b)) and improved part quality compared to tooling concept 1.

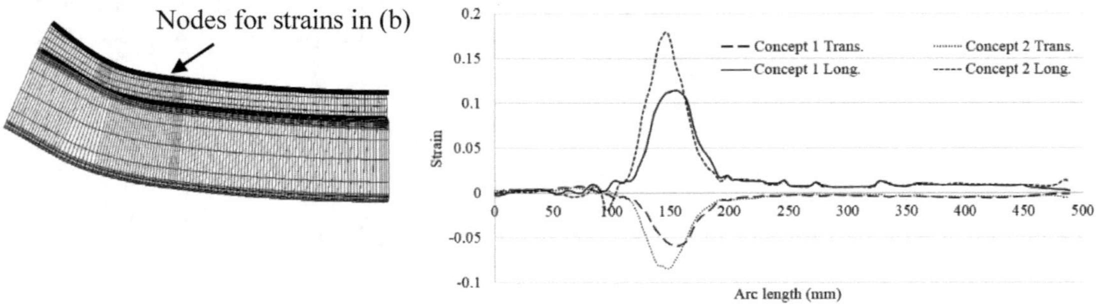

(a) (b)

Figure 11. (a) Node path considered for the strain plot (b) Different strain components

7. Conclusion

A new flexible roll forming technique for a curved channel is introduced and was successfully simulated with the COPRA® FEA RF software package. Two tooling concepts are studied and the final part quality investigated. Improved quality was indicated with a tooling concept which provided additional support for the flange during final side forming. In both tool designs, buckling of the flange occurred where the curvature of the part in the plan view is greatest; buckling was less in the preferred design even though higher final strains due to the suppression of material where found. The work indicated a direct relationship between flange distortion and curvature of the part in the plan view and suggests that this distortion can be minimised by wise part and tooling design.

Acknowledgment

The authors would like to thank Emeritus Professor J.L. Duncan for his assistance in writing this paper and acknowledge the financial support made by the Ford Global - University Research Program – “2012-5089R Flexible roll forming of automotive structures” as well as the technical and software support provided by data M Sheet Metal Solutions GmbH.

References

[1] Abee A, Sedlmaier A, and Stephenson C, 2009, "Development of new 3D roll forming applications by means of numerical analysis as a part of a quality control methodology," presented at the 1st International Congress on Roll Forming, Bilbao, Spain.

[2] Guitierrez M A, Eguia I, Berner S, Hennigs C, Sedlmaier A, Agnello A, and Bahillo J, 2010, "Proform: Profile Forming Innovation " presented at the IDDRG 2010, Graz, Austria.

[3] Istrate A, 2002, "Verfahrensentwicklung zum Walzprofilieren von Strukturbauteilen mit über der Längsachse veränderlichen Querschnitten," TU Darmstadt,

[4] Abee A, Sedlmaier A, and Stephenson C, 2009, "Development of Roll Forming Applications by means of Numerical Analysis as a part of Quality Control Methodology," presented at the 1st International Congress on Roll Forming, Bilbao, Spain.

[5] PROFORM, 2009, "An innovative manufacture process concept for a flexible and cost effective production of the vehicle body in white: Profile Forming," presented at the Sixth framework programme of the European Commission.

[6] Larranaga J, 2011, "Geometrical accuracy improvement in flexible roll forming process by means of local heating," PhD, Department of Mechanical and Industrial Production, Mondragon University,

[7] Berner S, Storbeck M, and Groche P, 2011, "A Study on flexible roll formed products accuracy by means of FEA and experimental tests," in AIP Conference Proceedings, p. 345.

[8] Sedlmaier A, Hennig R, and Abee A, 2011, "Finite Element Analysis of 3D Profiles with Changing Cross Sections," presented at the 10th International Conference on Technology of Plasticity, Aachen, Germany.

[9] Groche P, Zettler A, Berner S, and Schneider G, 2011, "Development and verification of a one-step-model for the design of flexible roll formed parts," International journal of material forming, vol. 4, pp. 371-377,

[10] Park J C, Yang D Y, Cha M H, Kim D G, and Nam J B, 2014, "Investigation of a new incremental counter forming in flexible roll forming to manufacture accurate profiles with variable cross-sections," International Journal of Machine Tools and Manufacture, vol. 86, pp. 68-80,

[11] Jiao J, Rolfe B, Mendiguren J, and Weiss M, 2015, "An analytical approach to predict web-warping and longitudinal strain in flexible roll formed sections of variable width," International Journal of Mechanical Sciences, vol. 90, pp. 228-238,

[12] Yan Y, Wang H, Li Q, and Guan Y, 2015, "Finite element simulation of flexible roll forming with supplemented material data and the experimental verification," Chinese Journal of Mechanical Engineering, pp. 1-9,

[13] Abeyrathna B, Rolfe B, Hodgson P, and Weiss M, 2015, "Flexible roll forming of an automotive component with variable depth," in Advances in Materials and Processing Technologies Conference, Madrid, Spain.

[14] ASTM, "Standard test methods for tension testing of metallic materials," ed.

[15] (2014, June 2014). COPRA® RF - Software for Roll Forming. Available: http://www.datam.de/en/products-solutions/roll-forming/

[16] Abeyrathna B, Rolfe B, Hodgson P, and Weiss M, 2016, "A first step towards a simple in-line shape compensation routine for the roll forming of high strength steel," International Journal of Material Forming, vol. 9, pp. 423-434,

[17] Jiao J, Rolfe B, Mendiguren J, Galdos L, and Weiss M, 2013, "The effect of tooling design parameters on web-warping in the flexible roll forming of UHSS," in NUMISHEET 2014: The 9th International Conference and Workshop on Numerical Simulation of 3D Sheet Metal Forming Processes: Part A Benchmark Problems and Results and Part B General Papers, pp. 892-895.

IOP Conf. Series: Materials Science and Engineering **159** (2016) 012006 doi:10.1088/1757-899X/159/1/012006

Inline feedback control for deep drawing applications

P Fischer[1], D Harsch[2], J Heingärtner[2], Y Renkci[3] and P Hora[1]

[1] ETH Zurich, Institute of Virtual Manufacturing, Technoparkstrasse 1, 8005 Zurich, Switzerland
[2] Inspire-ivp, Technoparkstrasse 1, 8005 Zurich, Switzerland
[3] Franke Technology and Trademark Ltd, Franke-Strasse 2, 4663 Aarburg, Switzerland

E-mail: fischer@ivp.mavt.ethz.ch

Abstract. In series production of deep drawing products the quality of the parts is significantly influenced by material scatter. To guarantee a robust manufacturing the processes are designed to have a large process window. As the different material properties can lead to a drift in the process, the press settings have to be adjusted to keep the quality. In the scope of the work a feedback control system is proposed to keep the operation point inside the process window. The blank draw-in measured in predefined points is used as the primary indicator of the expected part quality. A simulation based meta model is then used to design the control algorithm with the blank holder forces as control variable. As the draw-in measurements are carried out punctually, their positioning within the tool becomes of critical importance. A simulation based study is therefore presented for the identification of sensor positions with the highest significance in relation to the process outcome. The baseline calibration of the controller is also based on the meta model. The validation of the proposed control system is illustrated based on experiments in a production line.

1. Introduction

The robustness in series production of deep drawing parts is greatly influenced by tool design, as well as scattering material properties and process noise. In literature different approaches can be found to visualize the working windows [1] as well as to widen the windows [2], [3]. Due to the increasing complexity of deep drawing parts, the process windows cannot be kept as wide as wanted, therefore a different approach has to be added to keep the processes inside the narrow windows. To account directly for the changing material properties like yield strength or tensile strength, as well as to reduce the influence of the process noise, like changes in friction due to the heating of the tools, a closed loop feedback control has to be introduced. In order to achieve a robust control, the measured feedback value should be controllable by the possible actuators as well as it should be representative for the process noise which should be canceled out. To keep the investment costs as low as possible, no additional actuators should be introduced to the tool. Therefore the blank holders are chosen as actuators, as well as the blank position which can be easily adjusted in the production of the chosen demo part. As demo part, a kitchen sink made from 1.4301 stainless steels is chosen. In the following the different stages for the design of the control algorithm are show, beginning with the classical finite element analysis of the demo part, followed by building a meta model from stochastically distributed simulations. In the end the model is validated as well as the control approach is tested in the production line.

IOP Conf. Series: Materials Science and Engineering **159** (2016) 012006 doi:10.1088/1757-899X/159/1/012006

2. Modelling the process for virtual analysis

For the development of control approaches it is indispensable to generate a model for the purpose of testing and evaluation of different algorithms. As the demonstrator is already in series production, the first step in building up the model is the generation of a finite element model at the nominal working point. In the next step stochastically distributed simulations inside a defined multidimensional space (friction, forces, etc.) are evaluated and based on the evaluations, the control variables as well the as the sensor positions are chosen.

2.1. Nominal FE-model

The nominal simulation of the forming process is generated in agreement with the process settings in the series production. For a better approximation of the process, one charge of the material is thoroughly tested. As baseline material model for the stainless steel 1.4301, the Hänsel model [4] is chosen and therefore tensile tests at six different temperatures are carried out, as well as the bulge test. For purpose of fitting of the model, the martensite content and forming rate during the tensile test are measured as well. In order to reduce the calculation time as well as to suit the needs of the industry partner, the model was collapsed to a single yield curve at 30 degree Celsius (figure 1a) for the calculation of the forming process in AutoForm. As model for the yield surface (figure 1b), the BBC 2005 implementation in AutoForm[5] was chosen. Compared with the other available yield surfaces, the BBC model has the highest flexibility and can therefore be fitted best to the extensive test data. The BBC yield surface is fitted to the three Lankford coefficients (0.861, 1.333, 0.807), the biaxial R-value of one, as well as the as the yield stresses σ_0=264 MPa, σ_{45}=247.7 MPa, σ_{90}=250.6 MPa and the biaxial yield stress σ_b=262.9 MPa

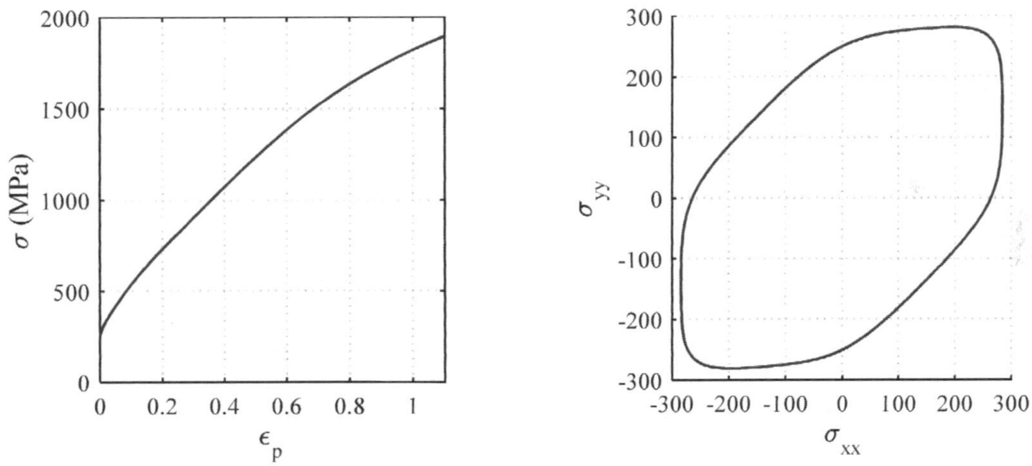

(a) (b)

Figure 1. Yield curve(a) and Yield locus (b)

As opposed to the try-out phase in automotive industries, the tools in the kitchen sink production are usually not modified and therefore the nominal milling geometry can be used for the simulation, instead of digitalized tool. With the experience of the series production a force curve for the blank holder force is chosen, which shows a decreasing ramp of the force between a drawing depth of 80 mm and 100 mm, while the overall drawing depth for the given part is 180 mm. Therefore a force of 2000 kN until 80mm and a force of 750kN from 100 mm on is chosen. The simulation shows acceptable agreement in thinning and is therefore chosen as the basic model for the upcoming variant simulation.

2.2. Meta model for draw-in

For the generation of an accurate model, the process has to be varied in a wide area to cover all possible process windows. A special focus in variation has to be laid on the possible actuators for the process control, therefore the two different force settings are varied independently to have the option of shifting them correlated or of changing them independently in the control algorithm. The forces are varied asymmetric to their nominal value, as the experience shows that shifting the forces to a higher level as the 2500 kN can lead to base fractures in the process as well as in the simulation, which have to be rejected as outliers in the meta model. The shift in the column position might be another possibility of controlling the symmetry of the draw-in and is therefore varied as well. As the positioning of the blank is one of the control parameters in manual press adjustment, the variation for the model is specified a bit wider than it usually occurs in the manual search for a robust process. The last three variation parameters cannot be directly influenced by the operator and therefore considered as process noise. For the calculation in AutoForm, the yield stress is directly coupled with yield strength and the variation band is based on pilot tests of the eddy-current measurement for stainless steel [6], while the variation of the Lankford coefficients is chosen to be between plus and minus ten percent of the values determined in the tensile tests. As the friction in the process is greatly influenced by the tool temperature and the amount of lubrication which is left in the tool, the friction coefficient is varied over an extremely wide band to map all possible states of the tool. The finally chosen variation values can be seen in table 1.

Table 1. Input variations for meta model

Parameter	Lower limit	Nominal value	Upper limit
Force until 80 mm	1000 kN	2000 kN	2500 kN
Force after 100 mm	400 kN	750 kN	1000 kN
Column position x-direction	-50 mm	0	50 mm
Blank position x-direction	-20 mm	5 mm	30 mm
Coulomb friction μ	0.05	0.07	0.15
Yield stress / yield strength delta	-25 MPa	0	25 MPa
R_0	0.7749	0.861	0.9471

The next step after choosing the variation variables and the variation range, is setting up an appropriate design of experiments. As the influences of the different parameters might be highly nonlinear, a latin hypercube design is chosen instead of a factorial design. Compared with the Monte Carlo design, the latin hypercube design has the advantage of iterating the design under certain boundary conditions to optimize the result. For the variant simulation with 90 simulations, the design was optimized under the condition of maximizing the minimal distance between the experiments and as second condition a limitation of the correlation between the different parameter is introduced.

The evaluation of the final correlation between the parameters shows that the Pearson correlation coefficient calculated for the design of experiments results in a maximal value of 0.14. As the correlation between the parameters is low, the results should not show any correlation which is introduced through the design of experiments. The draw-in at certain sensor positions, as displayed in figure 2, is evaluated and the correlation coefficient between draw-in and parameter is calculated.

The evaluation of the coefficients in table 2 shows that the possible actuators have acceptable correlation with certain sensor positions. The change in the blank holder forces can be recognized in the draw-in at sensor S03 and sensor S04, while the change of the blank position can be detected in all other sensors. As sensor S03 and sensor S04 show a good agreement with the friction coefficient, the measurement at these positions can be used together with the blank holder force to account for a change in friction. The change in the material is difficult to detect through the sensors, but the evaluation of different criteria like the maximum failure criteria shows, that a change in material does not significantly influence the process in the simulation. After choosing the sensor positions, it is necessary to build an appropriate model of the sensor input relation to generate a virtual testing environment for the control algorithms.

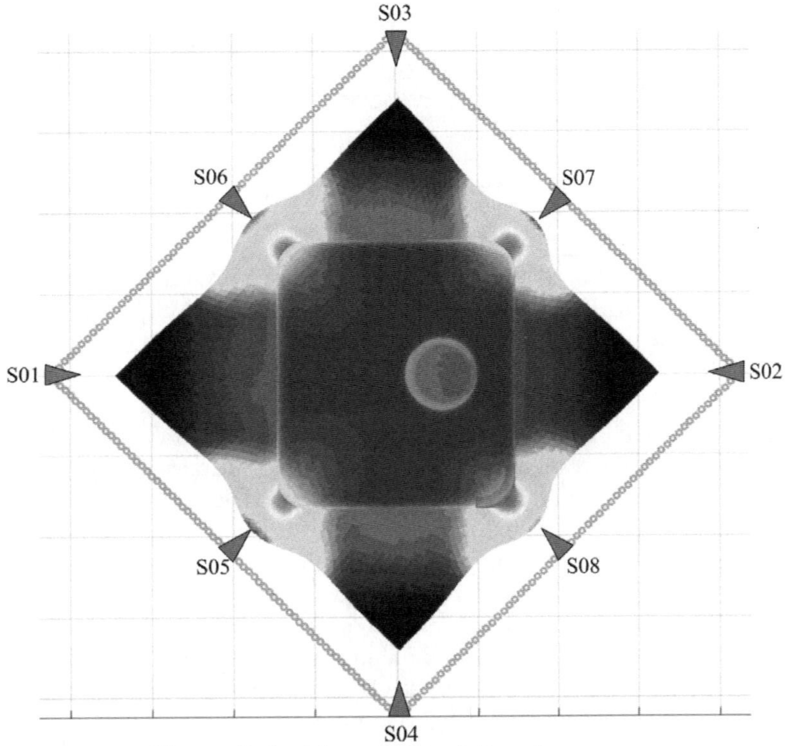

Figure 2. Position of virtual draw-in sensors

In general two different modelling approaches for the meta models can be used, the first approach uses approximating models, like response surfaces, while the second approach is based on interpolating models, like Kriging [2]. The usability of the different models is checked by cross validation. As the model is directly evaluated in the testing environment and not used in the controller, the complexity of the model can be neglected and model can be chosen based on the best fit. Therefore Kriging is used for sensor S01, while radial basis functions are used for sensor S02 and S03. Sensor S04 on the other hand showed the best result with the much simpler approximation by a quadratic function with interactions. In the evaluation of the correlations and the fitting of the functions, the inputs are normalized to reduce the influence of the different scales of the parameters. The knowledge about the current state of the part cannot be extended by the usage of sensor S05 to S08 and therefore the sensors are not used for the control.

Table 2. Correlation between results and variation parameters

	S01	S02	S03	S04	S05	S06	S07	S08
Force until 80 mm	0.031	-0.211	-0.226	-0.223	0.071	0.075	-0.216	-0.213
Force after 100 mm	-0.065	-0.187	-0.410	-0.405	0.068	0.064	-0.121	-0.123
Column position x-direction	0.421	-0.324	0.185	0.203	0.274	0.275	-0.170	-0.170
Blank position x-direction	0.813	-0.827	-0.074	-0.054	0.904	0.905	-0.883	-0.884
Coulomb friction μ	-0.295	-0.300	-0.844	-0.845	-0.216	-0.214	-0.266	-0.264
Yield stress / yield strength	0.038	0.075	-0.114	-0.113	-0.120	-0.124	-0.101	-0.103
Lankford coefficients	0.072	-0.065	0.047	0.056	0.049	0.049	-0.074	-0.071

Based on these models, the virtual testing environment can be designed.

3. Design of control model and definition of control parameters

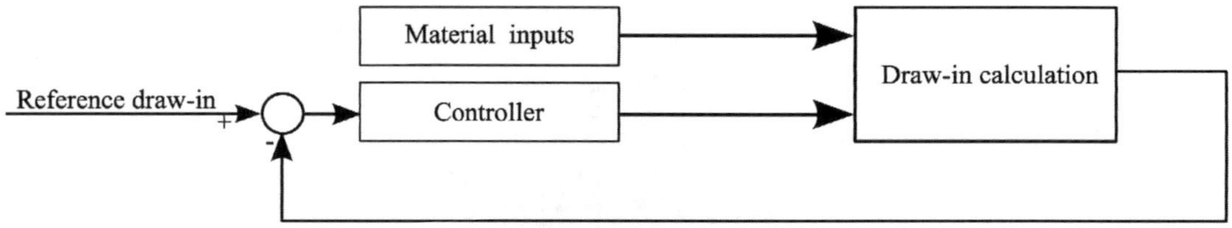

Figure 3. Scheme of the virtual test environment

The virtual try-out system for different control approaches is built up in MATLAB Simulink and consists of the three sub-systems shown in figure 3. In the material inputs section the process noise like material and friction can varied in the range of the meta model, while the actuating parameters are set and calculated in the controller section. Finally the values of both sections are taken for the calculation of the current draw-in which is fed back to the controller as deviation from the reference part.

The independency between the response of the sensors S01, S02 and S03, S04 results in a decoupled MIMO system. As the sensors S01 and S02, respectively S03 and S04, contribute to the same actuator, their deviations can be combined by simple arithmetic to a single control input. These simplifications lead to two independent SISO systems. The most common control algorithm for SISO systems, is the PID-controller. As the part to part variation in friction and material properties is low, the PID-controller has no need to react extremely fast and accurate, therefore it can be simplified to a proportional controller. The second reason for the application of a proportional controller is the low curvature of the models which reduces the risk of getting wind up in a local minima of the process, as the P-controller is similar to the Newton optimisation method which would not be able to find the global minimum in a function with local minima.

For finding the controller gains, different parameter settings have been tested. A high gain leads to a faster response of the controller, but can lead to overshoots, while a small gain reduces the risk of overshoots, but on the other hand it might be too slow to influence the process significantly. The chosen parameter settings aim to reach the reference draw-in within five to ten parts, while the overshoot should be marginal. Finally a shift of 0.4mm in blank position per mm of the combined deviation in the draw-in of the sensors S01 and S02 is chosen. The correction factor for the blank holder force is chosen to be 60kN per mm of the combined deviation of sensor S03 and S04. With the definition of the controller gains, the requirements for testing the process control are fulfilled.

4. Experimental set-up

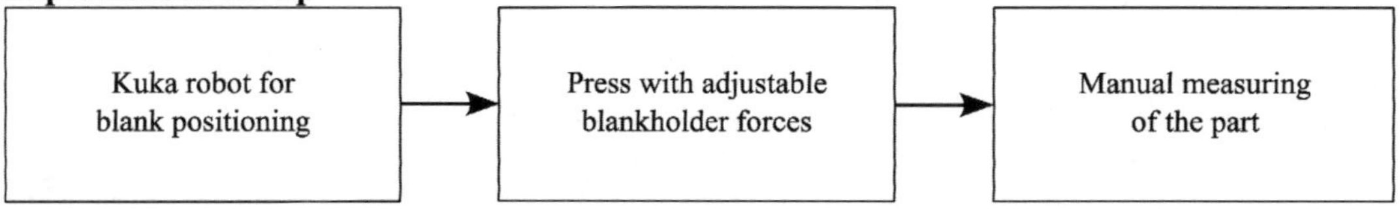

Figure 4. Scheme of experimental set-up

In the series production line, the implementation of the control would require direct control of the first two blocks in the scheme shown in figure 4 and is therefore highly complex. For that reason the settings calculated by the control algorithm are transferred by an operator for the purpose of testing the system. As the draw-in is nearly impossible to measure in the deep drawn part, the remaining flange is measured instead and then converted for the controller. As the measurements are done manually the repeatability and also the accuracy lies within the error of 1mm. The set-up is chosen to demonstrate the functionality of the control algorithm without the need of the complex integration.

5. Results

5.1. Correlation in experiments

Before the control can be tested it is necessary to determine if the calculated meta model corresponds to the real process behaviour. Therefore a design of experiment is carried out, where the blank holder force and the blank position are set to the values shown in figure 5. The shown force is the force until 80mm, while the force at 100 mm is the force of 80 mm divided by 2.66. The factor 2.66 is based on the experience of the press operators. The experimental design of experiments leaves the virtual design space, due to the fact that a scaling factor is introduced after the first comparison of the simulation with an experiment. The correlation between the chosen sensor and the corresponding actuators lies around 0.98 for the blank position and around 0.92 for the blank holder force. This leads to the conclusion that the draw-in at the sensor position is controllable with the chosen actuators. The results of these nine parts compared with the expected draw-in, shows that the influence of the blank position is approximated quite well, while the overall model seems to have a small offset even with the previous introduced scaling. The influence of the blank holder forces on the over hand looks underestimated. All in all the trends are predicted well enough and therefore the tests with the control algorithm can be carried out.

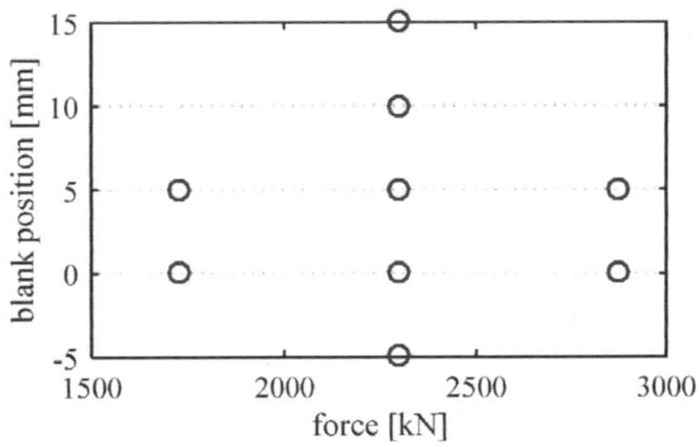

Figure 5. Input scatter of experimental try-out

5.2. Feedback control

With the boundary condition of using the series production line the decision is taken to use four parts per run. As no change in material is possible in such a small number of runs, the decision is taken to prove the concept by starting from a random point in the design space with the aim of finding back to the reference settings.

The reference settings are 2300kN and a shift in the blank position of 0mm from the center. These results in the measured remaining flange which can be seen as S01-S04 nom in figure 6 and figure 7. For the first controlled run a starting position with a too low force of 1725kN and a positive shift in the blank position of 5 mm is chosen. The graph in figure 6a shows the influence in the shift of the blank, which is taken back by the control algorithm until the nominal value is reached at part four. In figure 6b the influence of the adaption of the blank hold forces can be seen, also it is clearly visible that the part is already missing the symmetry in the reference value and therefore reacts differently than it would be expected. All in all at part four the nominal geometry is reached inside the measurement error.

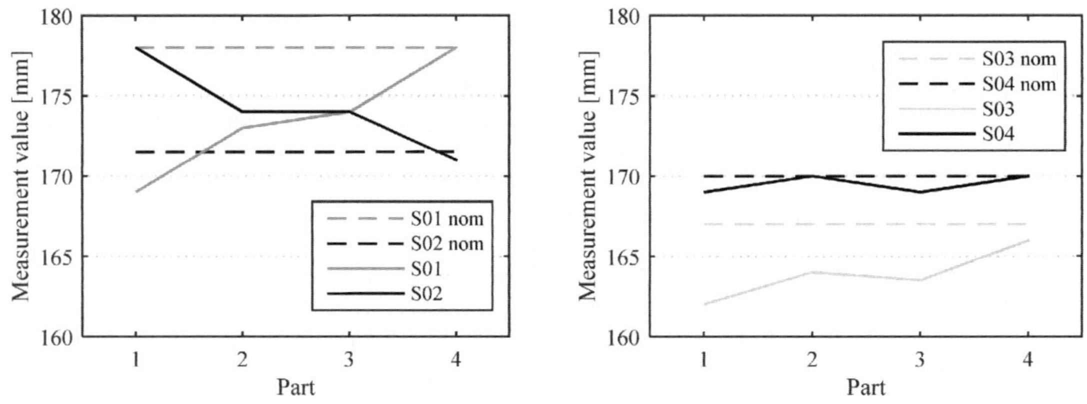

(a) (b)

Figure 6. Results of the first run with feedback control

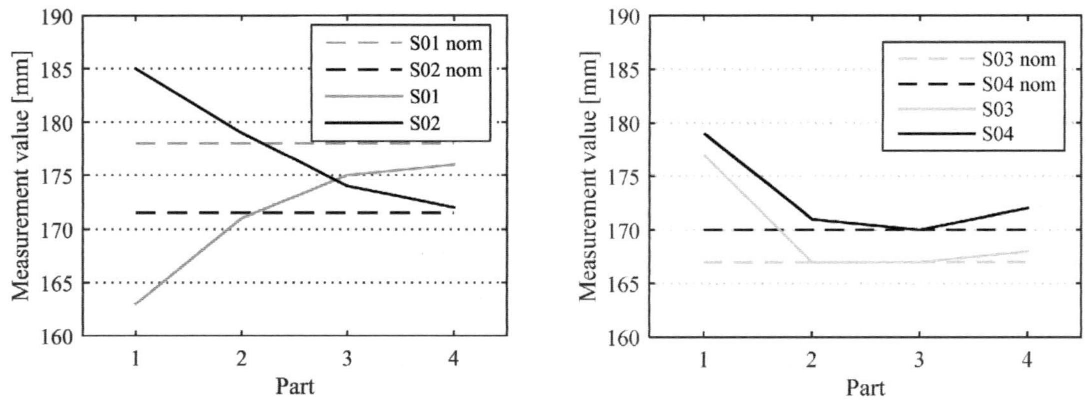

(a) (b)

Figure 7. Results of the second run with feedback control

For further validation of the control algorithm, a second run from a different starting point is carried out. The second starting point with a force of 2875 kN and a shift in the blank position of 10 mm lies further away than the one in first run. In figure 7b it can be determined that the blank holder force correlated sensor values reach their nominal value even quicker that in the first run, but diverge insignificantly at part 4, as the divergence is still close to the measurement error. The blank position related values in figure 7a on the other need longer to reach their nominal values as the distance to the reference value is larger. In sensor S01, the changing slopes can be explained directly by the proportional controller which reduces the change proportional to the distances between nominal value and current value. The same behaviour might not be visible in the other figure due to some side effects of friction or the combined change of both parameters. All in all the controller reaches again the reference point.

6. Conclusion

The results clearly show that the controller is able to improve the performance in the production of kitchen sinks. It also shows that developing a control algorithm based on a solid meta model reduces the calibration time for the controller as the shown results are the first experiments carried out with the controller. The defined meta model can not only be used for the design of the controller but also used generally in the search for a working point and therefore the extended computation time is used wisely.

The next step in the development of the feedback control is the integration into the production line as industry 4.0 application.

References

[1] C. Annen, "Entwicklung einer neuen Methode zur Ermittlung und Visualisierung von robusten Prozessfenstern in der Blechumformung," Zürich : ETH, Zürich, 2012.

[2] M. H. A. Bonte, "Optimisation strategies for metal forming processes," Enschede, 2007.

[3] J. H. Wiebenga, "Robust design and optimization of forming processes," Enschede, 2014.

[4] M. Hans Christian Andreas Hänsel Deutschland, 1969 -, "Nichtisothermes Werkstoffmodell für die FE-Simulation von Blechumformprozessen mit metastabilen austenitischen CrNi-Stählen," Düsseldorf : VDI Verlag, Düsseldorf, 1998.

[5] D. Banabic, D. S. Comsa, M. Sester, M. Selig, W. Kubli, K. Mattiasson, and M. Sigvant, "Influence of constitutive equations on the accuracy of prediction in sheet metal forming simulation," in *7th International Conference NUMISHEET*, 2008.

[6] J. Heingärtner, Y. Renkci, and P. Hora, "Non-destructive testing of material properties," in *6th Forming Technology Forum*, 2013.

IOP Conf. Series: Materials Science and Engineering **159** (2016) 012007 doi:10.1088/1757-899X/159/1/012007

Virtual tryout planning in automotive industry based on simulation metamodels

D Harsch[1], J Heingärtner[1], D Hortig[2] and P Hora[3]

[1] inspire AG - ivp, Technoparkstrasse 1, 8005 Zürich, Switzerland
[2] Daimler AG, Käsbrünnlestrasse, 71059 Sindelfingen, Germany
[3] ETH Zürich, Institute of Virtual Manufacturing,
Tannenstr. 3, 8092 Zürich, Switzerland

E-mail: harsch@inspire.ethz.ch

Abstract. Deep drawn sheet metal parts are increasingly designed to the feasibility limit, thus achieving a robust manufacturing is often challenging. The fluctuation of process and material properties often lead to robustness problems. Therefore, numerical simulations are used to detect the critical regions. To enhance the agreement with the real process conditions, the material data are acquired through a variety of experiments. Furthermore, the force distribution is taken into account. The simulation metamodel contains the virtual knowledge of a particular forming process, which is determined based on a series of finite element simulations with variable input parameters. Based on the metamodels, virtual process windows can be displayed for different configurations. This helps to improve the operating point as well as to adjust process settings in case the process becomes unstable. Furthermore, the time of tool tryout can be shortened due to transfer of the virtual knowledge contained in the metamodels on the optimisation of the drawbeads. This allows the tool manufacturer to focus on the essential, to save time and to recognize complex relationships.

1. Introduction

Given the increasingly tighter tolerance requirements, the fluctuation of process and material properties, as well as changing environmental conditions, often lead to robustness problems during series production [1][2]. In the current state of the art the factors influencing robustness are not systematically measured and online action is limited to a manual intervention on a trial-and-error basis. The quality of the outcome is thus strongly correlated to the experience of the staff and eventual corrections are costly. Additional circumstances, such as the reduction of the material thickness, complex geometries with sharp radii and the flexible choice of the press, limit the process stability even further.

In order to increase the significance and illustration accuracy of the simulations, a variety of different arrangements are taken into account, such as the use of accurate material models based on experiments, the consideration of the press construction and the implementation of digitised tool geometries. The goal of the project is to inspect the critical regions at an early stage regarding robustness problems and to propose improvement measures. Furthermore, the virtual knowledge of the simulation should be transferred to the tool tryout in order to support the staff with purposeful actions. In this work, a methodological approach based on simulation metamodels is developed to find a suitable operating point and to generate virtual tryout maps around the operating point. Thus, cost-intensive rework could be reduced or even avoided.

2. Process description

The trunk lid of the actual B-class by Daimler shows diverse imperfections during serial pressing, such as splits, wrinkling, hardening through minimal thinning, sink marks and skid impact lines. The simulations in the planning stage did not class all areas as critical, due to insufficient reproduction of the real process conditions. Therefore, the illustration accuracy is enhanced through a variety of different arrangements.

2.1. Material experiments and material model

The part is made out of DC06 steel. In AutoForm predefined material cards can be used for simulations, which are an average of different batches and suppliers. To achieve the best accuracy possible of the material model, several experiments with the specific material used for the trunk lid were done. From this data an optimized material was generated and used in the simulations. The characteristic flow behaviour of the material is measured by doing tensile experiments in rolling direction of the material. The anisotropy coefficient, yield stress, tensile strength and equivalent strain are measured in 0°, 45° and 90° to rolling direction with additional tensile experiments.

The tool temperature is increasing monotonously during the serial pressing. Depending on the part, part number and measuring point, large temperature differences can be observed. Hence, the material behaviour is measured between 20°C and a maximal temperature of 80°C in steps of 20°C.

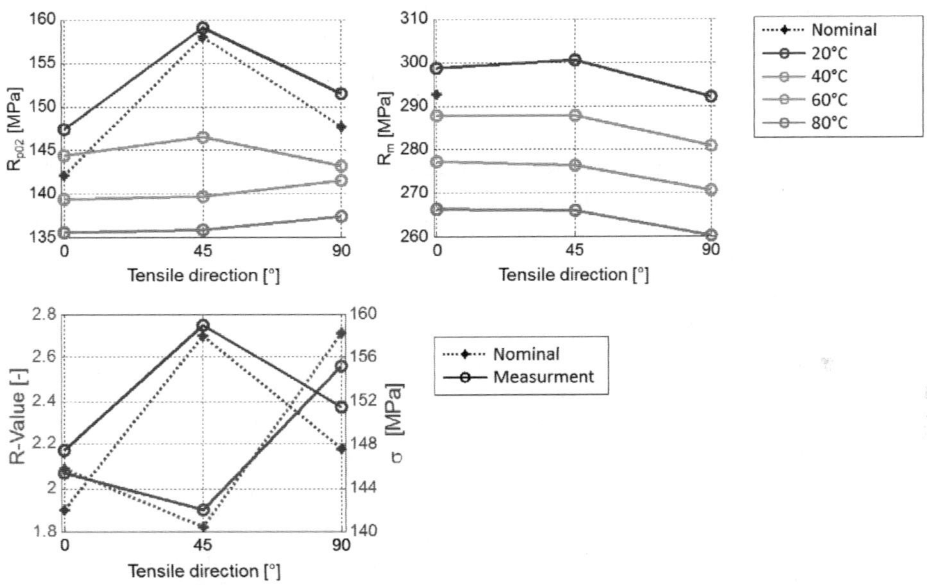

***Figure 1**: Measured material coefficients at different temperatures*

The yield stress shown in figure 1 above decreases in average of about 5 MPa with a temperature increase of around 20°C. Except the yield stress in 45° tensile direction at room temperature decreases much faster for higher temperatures, which implies a decreasing anisotropy. However, the measurement of the yield stress is quite sensitive in evaluation, although the values are averaged over three experiments. Thus, the interpretations of the absolute values have to be taken with caution. On the other hand the tensile strength is quite robust in evaluation. In each tensile direction the tensile strength drops down about 10 to 11 MPa per temperature increase. The reduction of the tensile strength is about twice of the yield stress. Consequently the growth of tension through hardening declines with increasing temperature. In other words the material holds out less and therefore has a worse formability. The comparison between the anisotropy coefficients (R-values) from the nominal values of the supplied material data sheet and the measured R-values at 20°C shows quite similar values. The differences are all in the range of variation of changing coils. Also the differences in yield stress between the nominal data and the measured data result from the fluctuation between the coils. The deviations of the individual experimental values to the averaged values (averaged values are shown in figure 1 and figure 3a) are all in the range of 1.5%.

By using a Bulge-Test the biaxial stress is evaluated. Furthermore, the Bulge-Test is used to enhance the prediction of the flow behaviour at strains greater than the equivalent strain. By means of the principle of equivalent work the biaxial stress state is transformed into a uniaxial stress state [6] (see figure 2). The transformation of the curve is based on the evaluated ratios of σ_b/σ_0 and $\varepsilon_b/\varepsilon_0$ [3]. For the extrapolation range the last values are used.

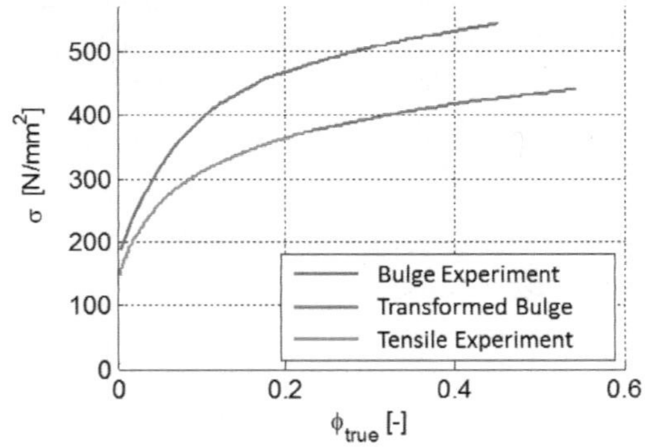

***Figure 2**: Transformed Bulge curve to uniaxial stress state*

The Forming Limit Curve (FLC) is measured by doing Nakajima experiments. The experiments are evaluated with the Method of Volk and Hora (Volk 2011b), which is based on the strain rate evolution [4]. The determined values of the experiments with identical specimen width are averaged and connected with each other. If the line is exceeded, localized necking may occur in the sheet metal. A safety margin of about 20 to 25% is deducted from the measured FLC to take uncertainties in measurement and material deviations into account.

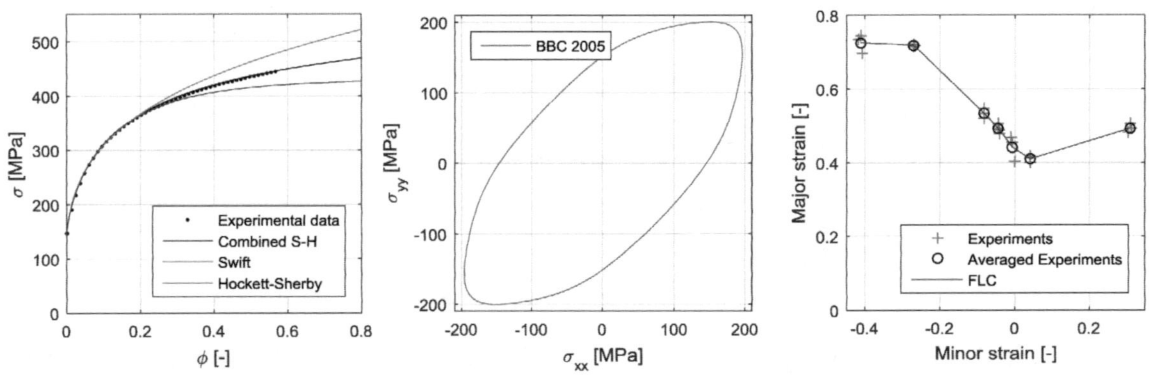

***Figure 3**: Experimental and fitted flow curve (a), yield locus (b) and FLC (c)*

The material model is then created with the AutoForm Material Generator R3.1. The hardening behaviour is fitted at 20°C with the combined S-H approach [8]. The yield locus is fitted with the BBC model for the measured data at 20°C. The biaxial r-value ($r_b = 0.809$) and the biaxial stress ($\sigma_b = 187.9\ MPa$) are measured in the Bulge experiment.

2.2. *Press construction*

Most of the major quality problems such as splits, wrinkles, sink marks and skid impact lines occur in the first press operation. Therefore, the press description is restricted to the deep drawing operation.

The mechanical, single-acting press can be controlled over four die cushion forces. To transfer the forces from the die cushions to the binder, 15 cushion quills are arranged symmetrically on each of the four die cushion. For the simulative illustration the forces are reduced from the 15 cushion quills to the corresponding die cushion. This can be calculated with the mechanical equilibrium condition (1).

$$(x_S|y_S) = \frac{\sum_{p=1}^{n}(x_p|y_p) * F_p}{\sum_{p=1}^{n} F_p} \tag{1}$$

Hence, four coordinates of force application result for the hydraulic four-point drawing cushion (see figure 4).

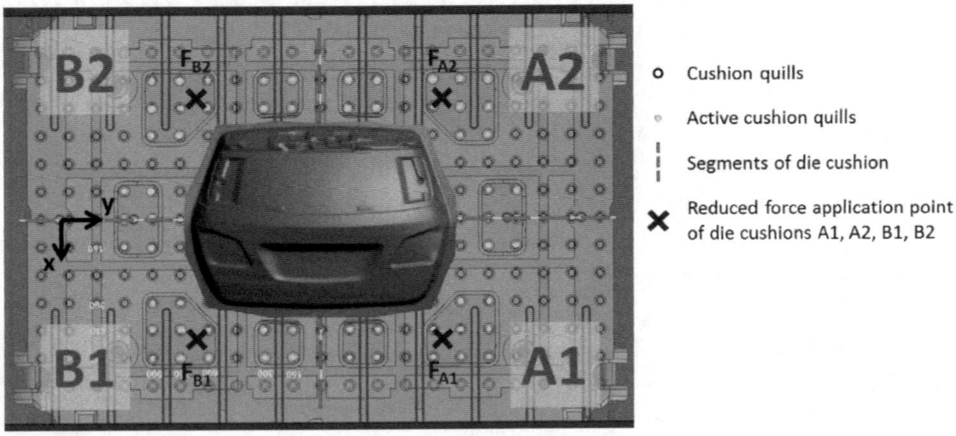

***Figure 4**: Binder suspension on cushion quills and segments of die cushion*

As in the state of the art, which is also implemented in AutoForm, the tools are assumed to be rigid bodies. Therefore, the deflection of any tool is locked. If a certain flexibility is required, a tool stiffness factor can be defined which allows a local homogenisation of the pressure distribution [8]. Because of the equilibrium condition all acting forces can be reduced to one total force with one specific force application point respectively coordinates. This simplification does not influence the simulation result. However, in reality major differences would appear because of the deflection, in particular for large tools. The total force and the coordinates of the binder can be implemented in AutoForm with the function *Columns (User def.)*. In case of the trunk lid, y-coordinate can be neglected because of symmetry. So the x-coordinate and the total force remain.

2.3. *Tool geometries*

The designed tool geometries of the method plan are used as a reference for the milling geometry. Additionally to the designed geometry, drawbead contours are added to the reference. After milling, the geometries have to be processed by hand in tool tryout to rework the tool regions where splits, wrinkles, sink marks or skid impact lines occur. Thereby, originally designed radii in the punch and in the die are harmonised or increased. Furthermore, most of the efforts are put into the tryout of drawbeads to adjust the draw-in of the blank. Typically the tryout is done in many loops. After each loop the influences of the made adjustments on the part quality are checked in the tryout press.

The simulation of a part with digitised tool geometries strongly differs from the simulation coming from the method plan with exact the same simulation settings. Hence, the tools are digitised after the first tryout loop. Thus, the major problems are remedied, but the complex fine-tuning is still pending.

The digitalization is made with a GOM ATOS measurement system. The finalized and re-meshed geometries have a tolerance of maximum 10 microns.

3. Finding the operating point

The metamodels, which describe the simulation behaviour in terms of different process fluctuations and process settings, should be evaluated around a suitable operating point. First of all, this point is required. It can be found by optimising all parameters of process design, such as binder forces or blank position. The remaining parameters of process fluctuation such as friction and material properties as well as the design parameters will be varied around the selected point.

The suitable configuration of the process design parameters is optimised with a design of experiments. The parameters are the total binder force, the x-coordinate of the force application point (to allow different force distributions), the force on the die inlet and the x-coordinate of the blank position. The die contains a die inlet to avoid skid impact lines. The y-coordinates of the force application point and the blank position stay constant because of the symmetry. Thereby, the four design variables are varied independently over 30 simulations.

The simulation results are then exported from AutoForm and imported into Matlab. Then the critical quality criteria are selected by specifying a result variable and the corresponding region of the criterion. For each defined quality criterion a metamodel is evaluated. Each quality criterion is limited by a specified value, e.g. *Max. Failure* in AutoForm (ratio between the simulated major strain and the major strain from FLC for a given minor strain [8]) at 0.8 or *Thinning* at 2%. By keeping the blank position and the die inlet force constant at a certain value, the contour lines of the metamodels at the specified values can then be visualized in the plane "total force – force distribution".

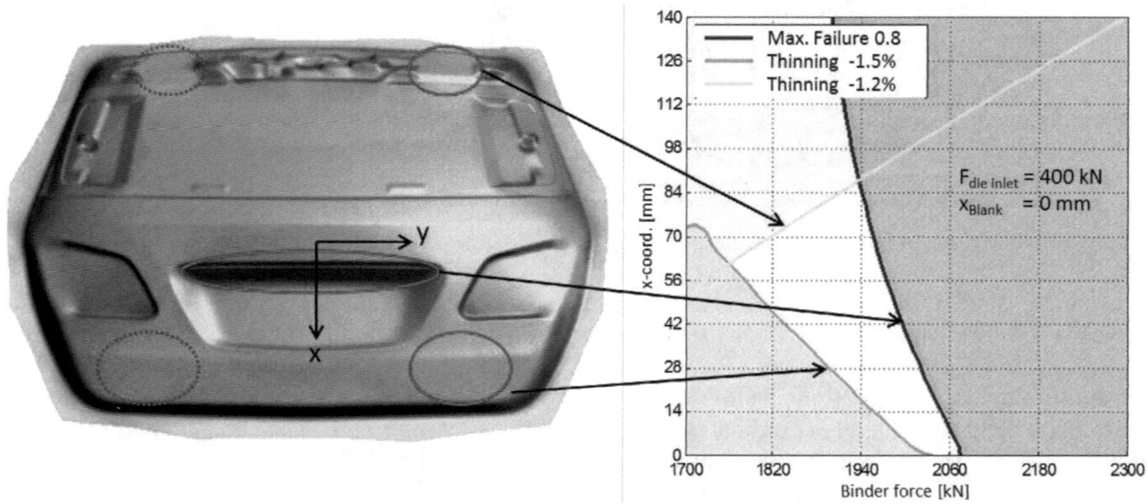

***Figure 5**: Definition of quality criteria (a) and their process limit in the process window (b)*

The coloured regions in figure 5 (b) show the area where the limit of the corresponding quality criteria is exceeded. The remaining white triangle represents the valid process window. A reduction of the force on the die inlet would extend the process window considering the three quality criteria. At the same time the skid impact line is getting worse. Therefore, the force on the die inlet is reduced as much as possible without getting skid impact lines. The determined force is 400 kN. The blank position is then also optimised to enlarge the process window. The best parameter configuration is reached for the initial blank position at 0 mm. The resulting process window for the constant blank position at 0 mm and the constant die inlet force at 400 kN is shown in figure 5 (b).

4. Virtual tryout map

Once the operating point of the simulation with digitised tool geometries is found, a new design of experiment considering the major process fluctuations and the process settings is created. It contains the variation of six independent and three dependent variables, which are shown in Table 1. The total binder force is deliberately varied for higher forces because the drawbeads are tending to be reduced which results in less critical risk of splits and at the same time in lower thinning in the critical regions shown before. Additionally, lower forces on the die inlet, which is also varied in the new design of experiment, lead to the same effects. The x-coordinate of the force application point (similar to a force ratio) is again varied in a range of 140 mm. However, to optimise the prediction behaviour the centre of the range is moved to the level of the chosen operating point, which leads in a minimum value of minus 60 mm and a maximum value of 80 mm. The friction coefficient reflects the influence of the tool temperature. At the beginning of the serial pressing the tools are cold, which results in lower friction coefficients (nominal value). During serial pressing, the tools are heating up and will reach a steady state value after a certain number of parts. Due to the increasing temperature the lubricant changes its physical behaviour (particularly the viscosity), resulting in an increased friction coefficient [10].

The variations of the material parameters have been measured from ten different batches of material. The measured minimum and maximum of the material coefficients have been taken as upper and lower variation values respectively. It is assumed that the ratios between the r-values (r_0, r_{45}, r_{90}) as well as the ratios between the yield stress and the tensile strength stay constant (see figure 6).

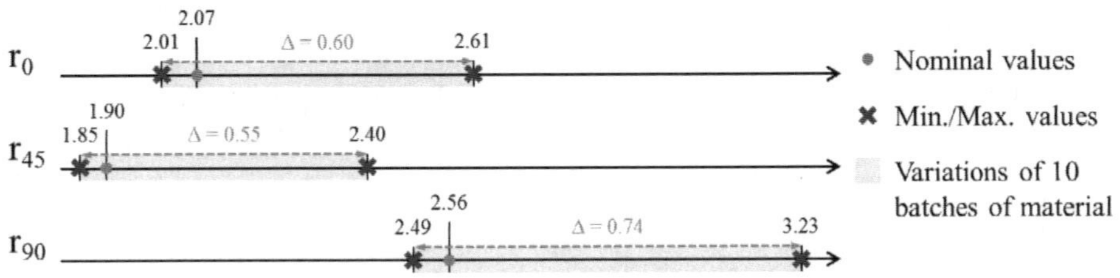

***Figure 6**: Variations of measured r-values*

***Table 1**: Variation parameters and variation boundaries*

Variation parameter	Dependence	Minimum	Maximum
Binder force F_{tot}	Independent	1800 kN	2400 kN
Force distribution x_{Pos}	Independent	-60 mm	80 mm
Force on die inlet $F_{die\ inlet}$	Independent	0 kN	400 kN
Friction μ	Independent	0.10	0.16
Yield stress R_{p02}	Independent	137.7 MPa	151.5 MPa
└ Tensile strength R_m	Dependent from R_{p02}	270.6 MPa	310.4 MPa
R-value r_0	Independent	2.01	2.61
├ R-value r_{45}	Dependent from r_0	1.85	2.40
└ R-value r_{90}	Dependent from r_0	2.49	3.23

A quadratic metamodel based on the six influence variables is fitted with 13 parameters. Therefore, 32 simulations are carried out with a Latin-Hypercube-Design to avoid over-fitting and to allow a cross-validation of the metamodels. For each quality criterion, which already have been defined in figure 5 (a), a metamodel is calculated and validated with the leave-one-out cross validation method. Furthermore, the influences of the varied parameters on the defined quality criteria are quantified by using a sensitivity analysis. For this the FAST-method (*Fourier Amplitude Sensitivity Test*) is applied. Figure 7 shows the percentage of the sensitivities on the AutoForm criteria *Max. Failure* (license plate) and *Thinning* (spoiler region). The remaining quality criteria based on Thinning demonstrate a very similar behaviour as the criterion in figure 7 (b).

The calculated sensitivities are directly related to the variation range of the parameters. Therefore, the upper and lower variation values of the forces are based on the typical press settings during a serial pressing and the lubrication values are based on experience values [7]. The sensitivities for *Max. Failure* (around the license plate) in figure 7 (a) suggest that an increasing friction, due to rising tool temperature during serial pressing, cannot only be corrected by a reduction or a redistribution of the forces to stabilise the process. For this reason the influences of the forces seem to be too small in comparison to the friction. This leads to an increasing risk of splits. One of the possible relieving actions is using a higher amount of lubrication as it is confirmed in the real process behaviour, whereby higher costs are the consequence [9]. For *Thinning* criteria, friction is also the most influential variable. The influences of the forces are in total 36%, so that changing friction behaviour can be partially corrected only by adjusting the press forces. The identified actions are restricted to local behaviours. Therefore, they could influence the part globally, wherefore it is necessary to analyse the effects separately.

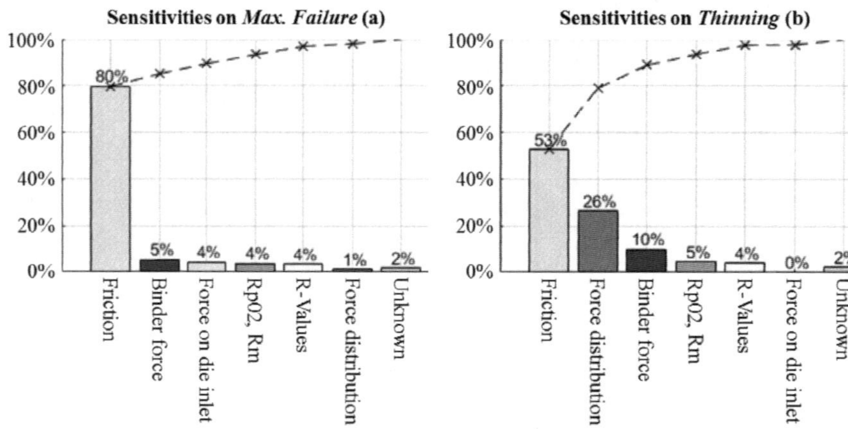

***Figure** 7: Sensitivities on Max. Failure (a) and Thinning (b)*

Furthermore, correlations are calculated between the draw-in and the different quality criteria defined in chapter 3. . For this the draw-in values of each simulation of the design of experiments are correlated with the simulation results of the predefined quality criteria. Each draw-in point on the contour around the part in figure 8 represents a vector of draw-in results of the 32 simulations. Also each quality criterion has 32 results from the different simulations. The draw-in values for each point around the part are correlated with the respective quality criteria and coloured depending on the resulting correlation value (see Figure 8).

Draw-in lengths for point *i*: (*n* simulations) $$l_i = \begin{bmatrix} l_{i,1} \\ l_{i,2} \\ \vdots \\ l_{i,n} \end{bmatrix}$$

Values of quality criterion *q*: (*n* simulations) $$v_q = \begin{bmatrix} v_{q,1} \\ v_{q,2} \\ \vdots \\ v_{q,n} \end{bmatrix}$$

Correlation: $c_{q,i} = corr(l_i, v_q)$

- For each quality criterion q correlation values c_q are calculated.
- Each entry of the correlation array describes the correlation of the draw-in at one point i of the part boundary.

This procedure is done exemplarily for the two quality criteria, *Max. Failure* and *Thinning* in the lower part region (see figure 5). The coloured contour is located on the part boundary in the undeformed stage right after closing the binder with the die. The gap between the correlation points and the deformed part boundary (grey) depicts to the draw-in of the part.

The draw-in correlations detected to be relevant for the Max. Failure criterion in the middle of the part are surprisingly not as expected in the lower part region (see figure 8 a). The draw-in correlating the best with the risk of splits in this region is located sideways of the part (circled in blue). For the

Thinning criterion the best correlations occur in the lower part boundary. The correlation values are nearly symmetric because of the intrinsic symmetry of the part and the fact that the force distribution is only varied in direction of x-axis.

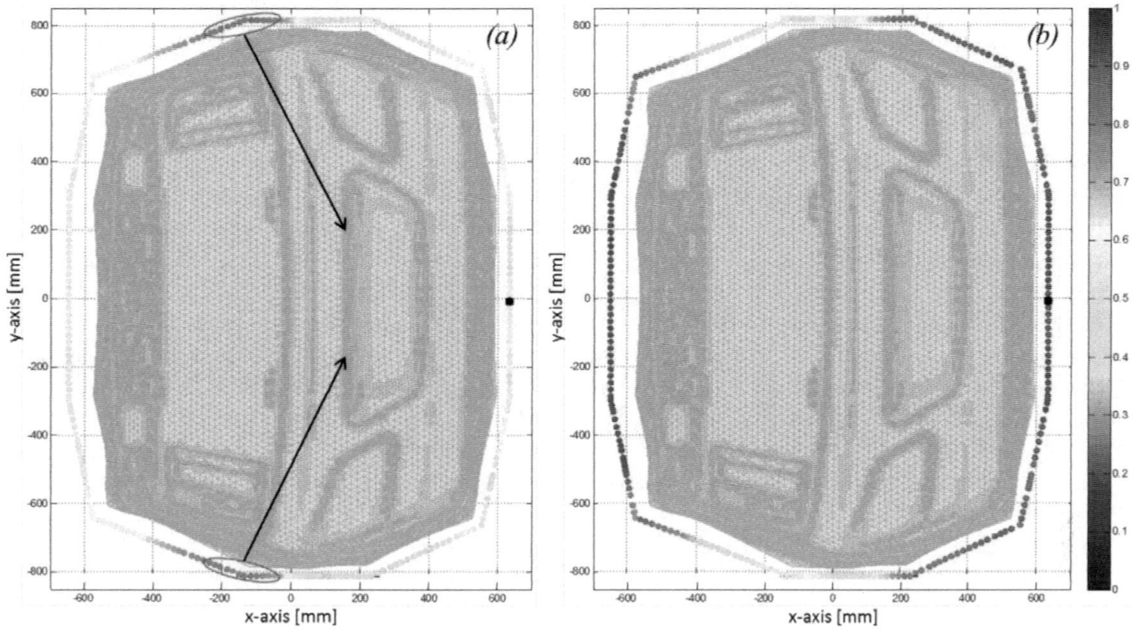

***Figure 8**: Draw-in correlations with Max. Failure (a) and Thinning (b) criteria*

The identified draw-in regions are supposed to be the relevant drawbeads, which have to be reduced (or intensified) to improve the quality of the part in the corresponding quality criterion. In some cases a reduction of a drawbead is not only improving a quality criterion, but it can also deteriorate another. To avoid this effect all relevant correlations (values over 0.75) are determined and transferred into one global tryout map (see figure 9). In case the superimposition shows overlapping segments the workforce of the tool tryout is warned to change these drawbeads with caution. For the presented part example no overlapping influences could be detected. Therefore, the three quality criteria can be worked in separately. Instead of intensifying the drawbeads for the two *Thinning / Insufficient Stretch* criteria by welding, the binder force could be increased and in return the sensitive drawbeads to the Max. Failure criterion (bordered in green) could be reduced.

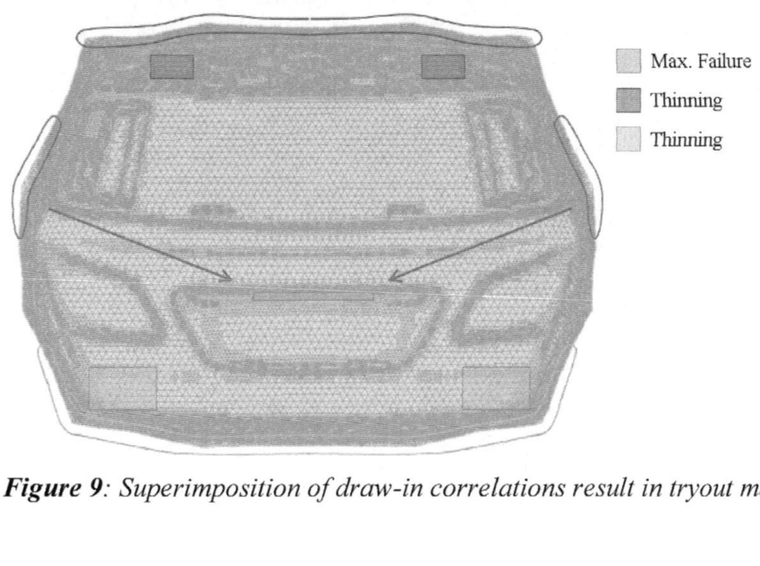

***Figure 9**: Superimposition of draw-in correlations result in tryout map*

The developed methodological approach of virtual tryout planning has been compared with the real tryout steps that have been done on the presented tool. A very good accordance between the virtual tryout and the real behaviour of the part could be determined [9].

Moreover, the virtual draw-in correlations are very useful to identify the position of a sensor, which is relevant enough to observe the process or in particular the accordant quality criterion. Such a sensor is able to quantify a potential split, wrinkles or insufficient stretch, depending on where the sensor is positioned. In serial production the quality of the part is identified after the last step of the transfer press.

5. Conclusion and Outlook

The simulative agreement with the real process condition is essential to judge quality criteria. Already small simulative changes, such as appropriate material modelling and the force transmission of the binder, contribute significantly to a proper reproduction of the behaviour. The virtual drawbead models used in the die face engineering are sufficient to evaluate the feasibility of the process and to detect the potential critical zones in the part. To obtain more accurate information about the part, to show realistic process windows and to generate virtual tryout maps, the tool geometries used in simulation have to be digitised. Because these tool geometries display the modifications and tryout effects done so far on the surfaces. Especially the draw-in is mapped much better, if real drawbead geometries are used in comparison to the virtual drawbead models implemented in AutoForm.

The virtual draw-in correlations allow the illustration of tryout maps and the transfer of the virtual process understanding to the tool tryout workforce. Thus, the tool tryout will not be based on personal experience of the workforce but on a methodological approach. The comparison between the virtual determined measures and the real performed tryout steps has shown a very good accordance with the results. By means of the performed analyses and developed illustrations, tool optimisations in the methodology planning and in tool tryout can be shortened. Moreover, the amount of scrap parts can be reduced due to more robust process design.

The determined correlations between draw-in and quality criteria are particularly useful for the assessment of the process observation of a subsequent process control. The presentation of the correlations around the part in colour show immediately where a draw-in sensor should be positioned, which is significant enough, to evaluate a quality criterion qualitatively. Therewith, based on a few draw-in and the deposited metamodels, a process control can be realised. In case the draw-in differs from original condition the controller reacts and regulates the process back into the valid operating point.

References

[1] D. Hortig. *Experiences with the Robustness of sheet metal forming processes*. Forming Technology Forum 2011, Proceedings, Zürich, 2011.

[2] P. Hora, J. Heingärtner, N. Manopulo, L. Tong. *On the way from an Ideal Virtual Process to the Modelling of the Real Stochastic*. Forming Technology Forum 2011, Proceedings, Zürich, 2011.

[3] P. Hora, B. Hochholdinger, A. Mutrux, L. Tong; *Modeling of anisotropic hardening behavior based on Barlat 2000 yield locus description*; FTF 2009

[4] W. Volk, P. Hora. *New algorithm for a robust user-independent evaluation of beginning instability for the experimental FLC determination*. International Journal of Material Forming, Volume 4, Issue 3, pp 339-346, 2011.

[5] C. Annen. *Entwicklung einer neuen Methode zur Ermittlung und Visualisierung von robusten Prozessfenstern in der Blechumformung.* Diss. ETH Nr. 20573, ISBN 978-3-906031-35-4, 2012.

[6] P. Peters. *Yield functions taking into account anisotropic hardening effects for an improved virtual representation of deep drawing processes*. Diss. ETH Nr. 22707, 2015.

[7] J. Krauer. *Erweiterte Werkstoffmodelle zur Beschreibung des thermischen Umformverhaltens metastabiler Stähle.* Diss. ETH Nr. 19070, ISBN 978-3-18-367602-6, 2010.

[8] AutoForm. *Helpviewer 11.5.4*. AutoForm Engineering GmbH, Version AutoForm R3.1, 2011.

[9] N. Schelhammer, Mercedes-Benz Cars Operation – Press line 3 & 4 (TF/PW4), Sindelfingen; *Private communications*, 2014.
[10] H. Watter. *Hydraulik und Pneumatik.* Springer Link, ISBN 978-3-658-01310-3, 2013.

IOP Conf. Series: Materials Science and Engineering **159** (2016) 012008 doi:10.1088/1757-899X/159/1/012008

Numerical Simulation of Rectangular Deep Draw Containers Made from Thin Molybdenum Sheet

M Žmindák[1], T Donič[2] and P Jurík[3]

[1] Department of Applied Mechanics, Faculty of Mechanical Engineering , Univerzitná 8215/1, University of Žilina, 01026 Žilina, Slovakia
2 Department of Design and Mechanical Elements, Faculty of Mechanical Engineering, Univerzitná 8215/1, University of Žilina, 01026 Žilina, Slovakia
3 Department of Applied Mechanics, Faculty of Mechanical Engineering , Univerzitná 8215/1, University of Žilina, 01026 Žilina, Slovakia

E-mail: milan.zmindak@fstroj.uniza.sk

Abstract. The paper presents numerical simulation-FEM variations in the technological and design parameters of deformation at different levels during the plastic deformation of complexly shaped hollow parts of a pure molybdenum sheet in terms of the thermo forming. Variations of the process parameters such as deformation temperature, blank holder force, shape and geometry of the drawbeads, the shape and geometry of the blank thin sheet of the molybdenum metal allow making optimisation procedures for hot forming of the thin molybdenum sheet when the products of these processes have no rotational symmetry. The results of the numerical simulations in programming environments PAM-STAMP 2015 enable implementing the validation of each technological and design elements of the deformation system for an optimal process of plastic deformation of the rotationally asymmetrical hollow containers of the thin molybdenum metal as well as determining the critical deformation and stress zones in the discrete volumes of the molybdenum sheet metal. In the area of allowance, attention is to be paid to the shape and geometry of the drawbeads, friction ratios in the critical zones of deformation and to the aspect of their positive effect on the flow of molybdenum metal to unbalanced hole-forming tool to apply the criterion of the maximum allowable wall thinning yield.

1. Introduction

One of the features of the modern technology is the dramatic increase in demands for service characteristics of the material used. These materials include molybdenum sheet, used as a construction material working in conditions of heating at high temperatures. With reliable protection from oxidation the products made of molybdenum sheet have a sufficient stability of shape and size and a relatively high specific strength at high temperatures. Due to the mentioned quality parameters and due to the use of molybdenum sheet in the growth of optically transparent sapphire single crystals of refractory they are melted by the horizontal directional solidification - Horizontal directed crystallization (HDC) method [1].

The process of growing single crystals is characterised by the refractory critical values of the temperature and temperature gradient, reaching more than 2000°C. In these conditions the formation of the real structure of a sapphire single crystal is not completed and continues with their cooling. The HDC method is technically simple to create controlled temperature field necessary for growing a large

IOP Conf. Series: Materials Science and Engineering **159** (2016) 012008 doi:10.1088/1757-899X/159/1/012008

single crystal weighing up to 80 kilos with desired optical properties. In this regard the form and the production technology of the crucibles are of a particular importance. The current design and manufacturing technology of containers is based on the bent molybdenum sheet using flame heating. This procedure has a number of disadvantages as to the formation of the desired crystallisation front and production of the high quality sapphire single crystal. The main disadvantages of the construction and existing technology manufacturing crucibles are as follows: when you move the crucible with the melt through a temperature gradient there is a distortion of the form of crystallisation front connected with a variable heat transfer between the melt and the crucible walls. The flame heating is an intense oxidation of the molybdenum sheet, which leads to the formation of cracks and leakages of the melt.

However, it is difficult to work with the hot forming processes due to the extensive surface and edge cracking. Therefore, it is necessary to study the fracture behaviour of the pure molybdenum sheet and to establish the fracture condition to predict the crack initiation [2].

It is generally accepted that Finite Element Method (FEM) has been widely used as a tool to design a product and develop technological processes such us forming, induction hardening, machining, ECAP processes, etc. [3,4]. The main benefit of FEM is to shorten a time for a product to the market, to minimise a risk of die failure during a production, to reduce a number of trial-and-error in production and to reduce unnecessary down-time (maintenance) due to the uncertainty happening during productions [5]. FEM is being gradually adopted by industry to predict the formability of sheet metals. Sheet metal forming operation involves complex physical mechanisms that give rise to a high order non-linear problem [6]. A general nonlinear structural analysis can include the effects of large displacements, large strains and nonlinear material conditions [7,8]. FEM can provide not only the final results, but also the information of intermediate steps, like the distributions of displacement, stress, strain and other internal variables.

The objective of the presented work is to investigate the formability of the molybdenum sheet in deep drawing by FEM simulation, friction analyses, the effect of the process variables such as BHF, forming rate and blank diameter on the drawing depth, thickness distribution and positive or negative influence of drawbeads configuration on the molybdenum sheet plastic deformation.

The simulation is verified on a simple rotationally symmetric blank. Subsequently more complex containers using the same parameters are investigated. Here we monitor the influence of the container on the change of the technological and design properties with the goal to find the best suitable parameters. Wang et al. deals in work [9] with the influence of the thermal deep drawing on the analysis of the pure molybdenum sheet for the axially symmetric yields and this work will be compared just with these results. The deep drawing process is affected by a large number of factors and friction belongs to the most important ones. Meng et al. [10] deals with the influence of the pure molybdenum sheet friction during the deep drawing process at an elevated temperature. The results show that an increasing forming temperature of the Mo sheets leads to a reduced drawing force and an increased friction coefficient. Therefore it is effective to use a lubricant for improving the Mo sheets´ formability.

2. Molybdenum

Molybdenum is the lightest available heat-resistant material. It is typical by its excellent strength properties at increased temperatures, high resistance against corrosion except for the resistance against oxidation at a temperature higher than 973K, with high thermal conductivity, high flexibility module and a low coefficient of the thermal expansion [2]. The low formability of molybdenum significantly affects its utilisation. However, the formability of the Mo sheets can be improved by increasing the forming temperature before or during the forming process. Heat forming is an effective technology for manufacturing the Mo thin-wall components which also brings better friction properties between the sheets and tools [10].

We can observe that the tensile strength and the yield point of molybdenum at the room temperature is 980MPa and 690MPa with an extension of 8%, which shows its bad plastic properties at the room temperature, see the figure 1.

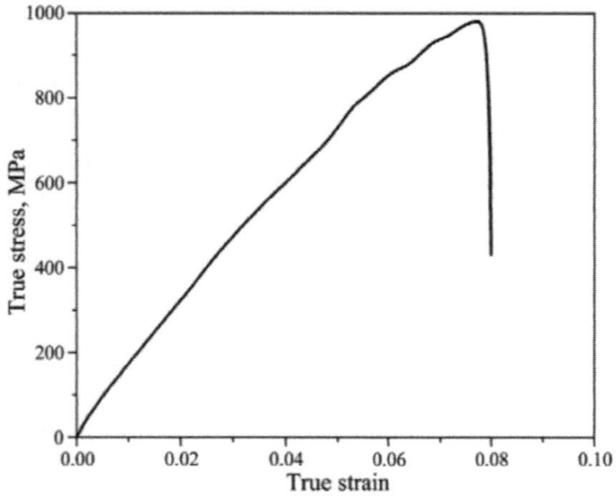

Figure 1. The true stress-strain curve of pure molybdenum at the room temperature [11].

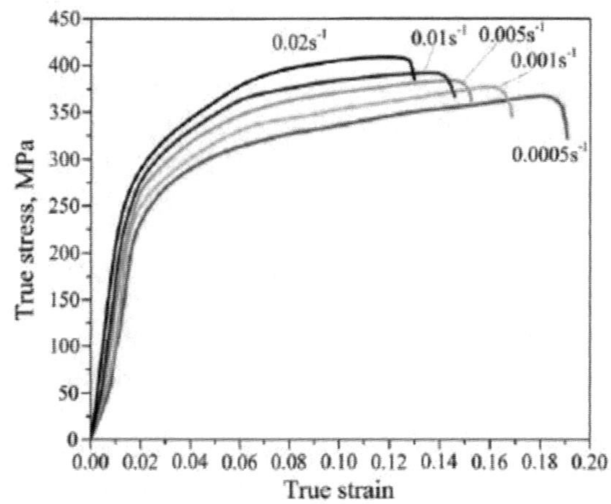

Figure 2. The true stress-strain curves for pure molybdenum at the temperature of 1043K for various strain rate [11].

Based on the experimental measurements of the single-axis tensile tests we can observe the influence of deformation and temperature on the deformation strengthening. The deformation strengthening declines with the increasing temperature and decreasing strain rate (figure 2). The dynamic softening of the behaviour can be seen when the temperature is 1093K and higher, when the strain hardening predominates at a temperature lower than 1093K [11].

3. Used material model

Due to a better operation of the programme PAM-STAMP it is more suitable to use the material model or to approximate the curve acquired during the tensile strength test up to the value of true strain 1 than to use the discrete values in dependence on the true stress-true strain. [12] The tensile tests were realised at various temperatures (993K, 1043K, 1093K and 1143K) Because of the high molybdenum sheet oxidation at increased temperatures the test samples were sprayed by the anti-oxidant coating. This coating did not affect the strength properties. The tests were realised on a universal test machine Yonekura. Several material models were created from the experimental results of the tensile tests (figure 3). The Hollomon´s equation describes strain hardening as a power law function of stress and strain after yielding

$$\sigma = \mathrm{k}.\,\varepsilon^{\mathrm{n}} \tag{1}$$

where k and n are material constants known as the strength coefficient and the coefficient of deformation strengthening. The constants k and n are acquired from the curves of dependence of the true stress on true strain by calculating the logarithm of both equation sides

$$ln\,\sigma = n\;ln\,\varepsilon + ln\,k \tag{2}$$

$$Y = A(x) + B \tag{3}$$

This equation is an equation of the straight line whose inclination is n and whose intersection point with the axis y is (ln k). [13]

Figure 3. Tensile test at elevated temperature.

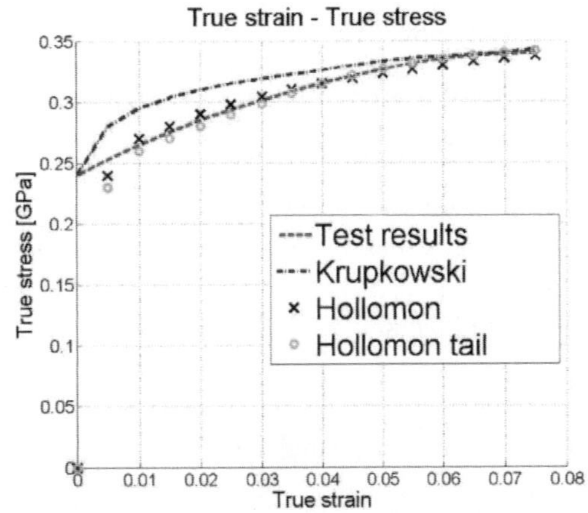

Figure 4. Comparison of various material models at the temperature of 1093K.

4. Friction Coefficient

The friction coefficient it is a very important factor affecting the deep drawing process. At present there is not at disposal any generally accepted method of determining the friction coefficient under plastic deformation of the thin sheets. The particularity of the presented determination method of the friction coefficient lies in the fact that the indicator is the sample of the tested sheet itself [14]. The investigation of Thiébaut et al. [15] also deals with the friction coefficient of molybdenum and his work shows that the coefficient value grows with the increasing temperature. For the Tresca law it says the friction coefficient is $f = 55.4/(1353\text{-}T)$ and for the Coulomb law $g = 23.8/(1332\text{-}T)$. For the temperature $T =$ 1023K we obtain $f = 0.17$ and $g = 0.08$. The friction coefficient between the steel tool and molybdenum sheet used for this work was achieved by an experimental tool (figure 5) and its value was stated as $f =$ 0.21.

Figure 5. The experimental tool for analysing the friction coefficient.

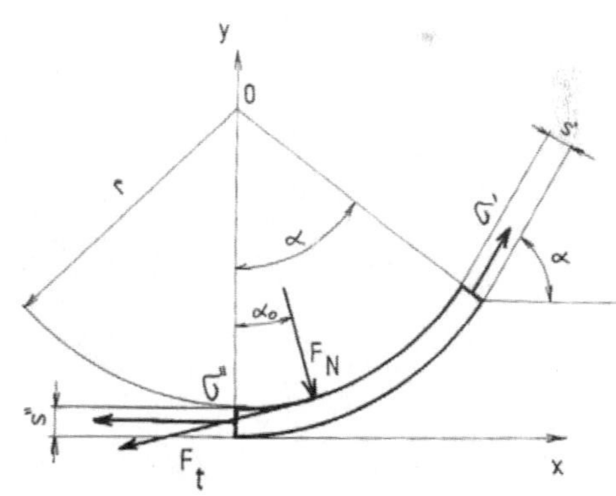

Figure 6. A scheme of the method for analysing the friction coefficient.

The calculation of the friction coefficient results from the equilibration conditions of the forces and moments acting on the tested element of a unit width.

$$\sum F_x = \sigma'' s'' - \sigma' s' \cos\alpha + F_t \cos\alpha_0 - F_N \sin\alpha_0 = 0 \tag{4}$$

$$\sum F_y = \sigma' s' \sin\alpha - F_t \sin\alpha_0 - F_N \cos\alpha_0 = 0 \tag{5}$$

$$\sum M_0 = \sigma'' s'' - \sigma' s' + F_t = 0 \tag{6}$$

IOP Conf. Series: Materials Science and Engineering **159** (2016) 012008 doi:10.1088/1757-899X/159/1/012008

where s', s" are elements of thickness in the end parts in a deformed condition,

F_N - a normal force component by which the tool acts on the sample element,
F_t - the friction force,
A - the half angle,
α_0 - the angle between y axis and force of normal components F_N,
σ' - normal stresses, determined from the tensile test diagram.

The actual deformations were determined from the equations:

$$\varepsilon' = \ln\frac{s_0}{s'} \ ; \ \ \varepsilon'' = \ln\frac{s_0}{s''} \tag{7}$$

where s_0 – the initial sample thickness. After adjusting we determine the force F_t from the following equation

$$F_t = \left(\sigma' s' \frac{\sin\alpha}{\cos\alpha_0} - F_N\right)\frac{\cos\alpha_0}{\sin\alpha_0} \tag{8}$$

Then we determine the friction coefficient under the sheet plastic deformation from the equation

$$f = \frac{F_t}{F_N} \tag{9}$$

5. Deep drawing of cylindrical cup

The simulation was dealing with the correctness of the material model, the friction coefficient and marginal conditions in the programme PAM-STAMP 2015 and with comparing the results with the experimental results of the Beihang University [10]. The scheme of the tools and their geometry is shown in the figure 7. One plane of symmetry was used (figure 8).

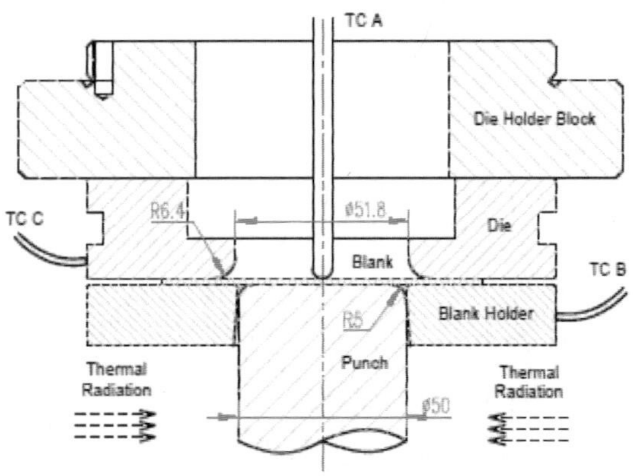

Figure 7. The forming tools and geometry [10].

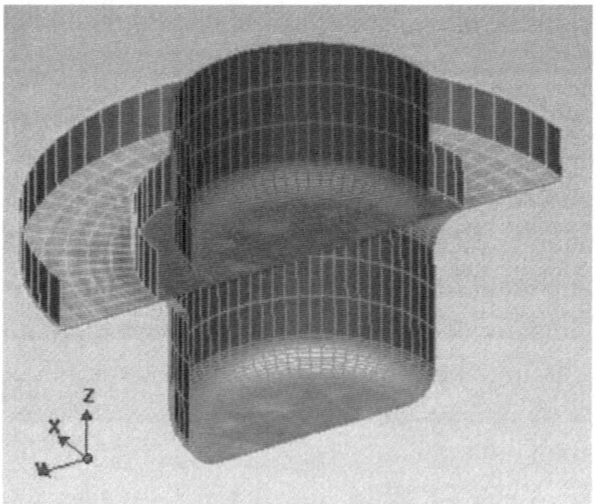

Figure 8. A model of forming system.

Due to the low formability of the Mo sheet at the room temperature the experiments were carried out at an elevated temperature and the sheets were prevented from oxidation with an anti-oxidation coating. The thickness of the pure Mo sheet is 0.5 mm. The simulations were carried out at several temperatures and the optimal temperature is 1123K. A blank with diameter of 95 mm and temperature of 1023K can be taken as an example. The blank holder force (BHF) varies from 12kN to 2kN and the results show that wrinkling occurred on the blank at the BHF of 2kN and fractures occurred on the blank at the BHF of 2 and 12 kN, while the blank under the BHF from 4 to 10 kN can be successfully deep drawn. Thus, the reasonable range for BHF in the thermal deep drawing simulation can be assumed from 4 to 10kN [10] and the simulation results are presented in the figures 9 and 10.

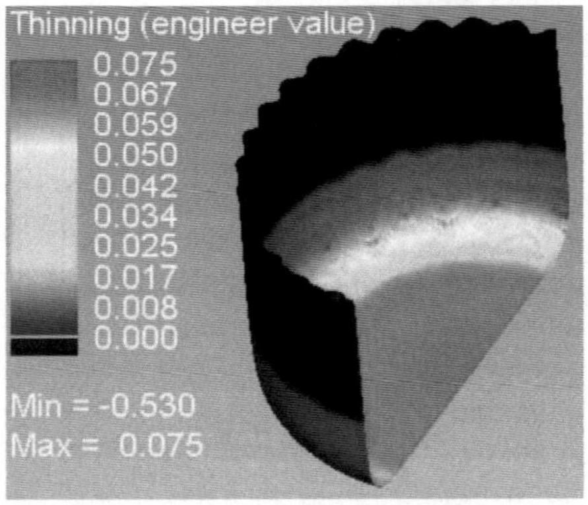

Figure 9. The course of thinning at BHF 2kN.

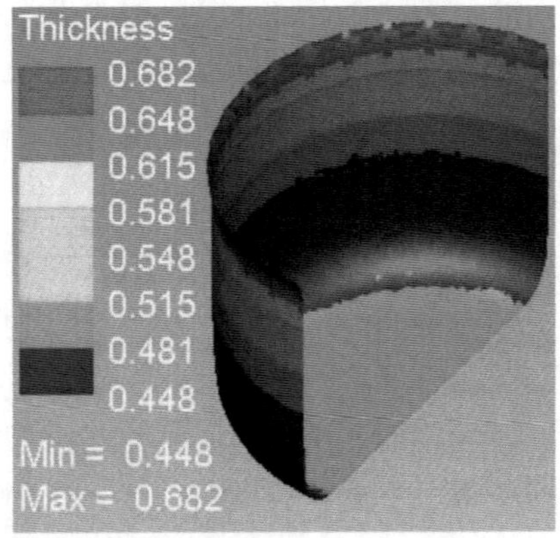

Figure 10. A successfully deep drawn cup.

All these experiments were realised as a FE simulation and in this way the correct marginal conditions in the PAM-STAMP were verified. The Hill 48 was used as the plasticity model. The thinning method was used as a criterion for assessing the yield. The maximal thinning of 0.14 was considered for the given material.

6. Deep drawing of the rectangular container

The simulation of an angular yield with the same marginal conditions used for the simulation of the axisymmetric containers with a temperature of 1023K was carried out. The various shapes and geometry of the drawbeads was used and we also monitored their influence on the thinning of the rectangular parts. The shapes of the drawbeads are shown in the figure 11 and we varied the dimensions for each rib. The drawbeads were placed to those areas where the material has to be decelerated as much as possible. The figures 12 and 13 show the geometry and due to saving the computer capacity only two symmetry levels were used. The blank in the simulation had dimensions of 160 x 160 mm.

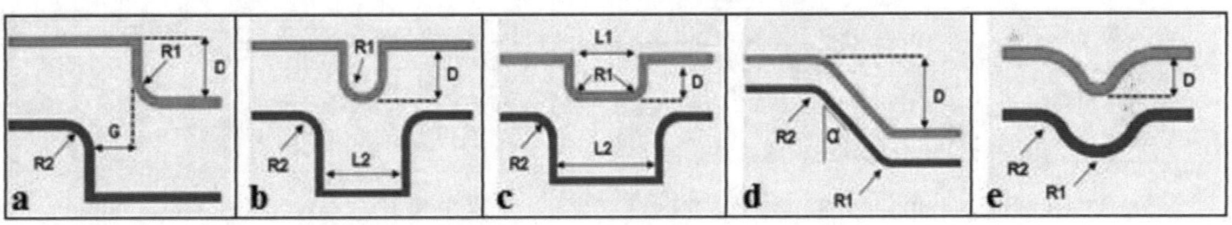

Figure 11. The various shapes of the drawbeads

The figure 14 shows the result without using a drawbead – the maximal achieved depth of the yield was 36 mm. The subsequent implementation of the drawbead for all types resulted in increasing the thinning. Only in the case of the type **a** there developed a reduction of the thinning and in this way the depth of the yield was increased by a value of 3 mm – see the figure 15 for R1=2.0, R2=2.0, D=2.0 and G=1.3. We note that the results presented in this paper are supported by experimental tests of the cylindrical container with a diameter of 80 mm at a temperature of 20°C (figure 16). This creates real assumptions for mathematical simulations of the optimal deformation conditions for forming crystallisation containers from molybdenum sheet for the sapphire monocrystal which is realised by the horizontal method of crystallisation according to Bagdasarov (figure 17).

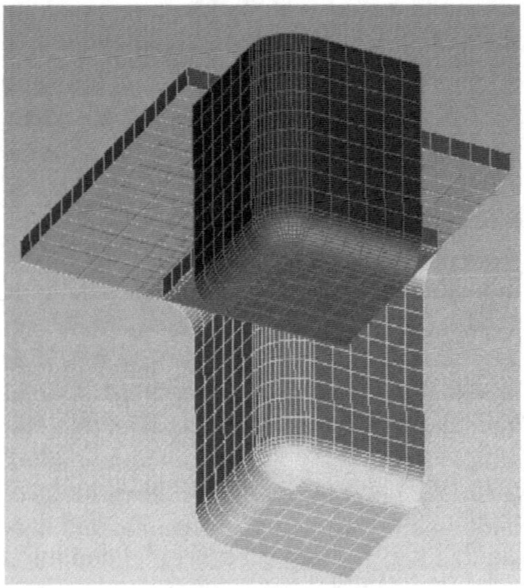

Figure 12. A model of the forming system.

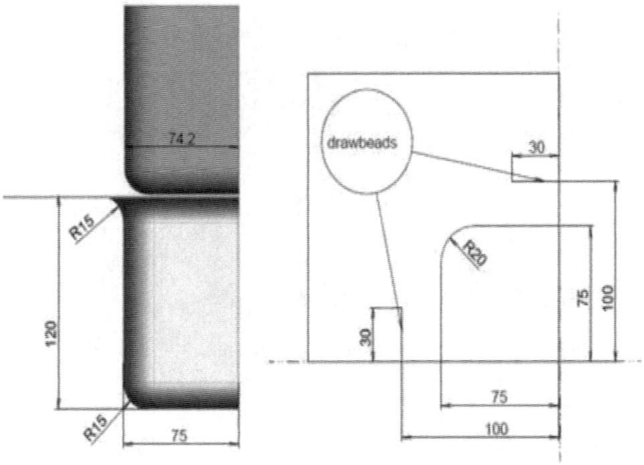

Figure 13. The geometry of the forming system and placement of the drawbeads.

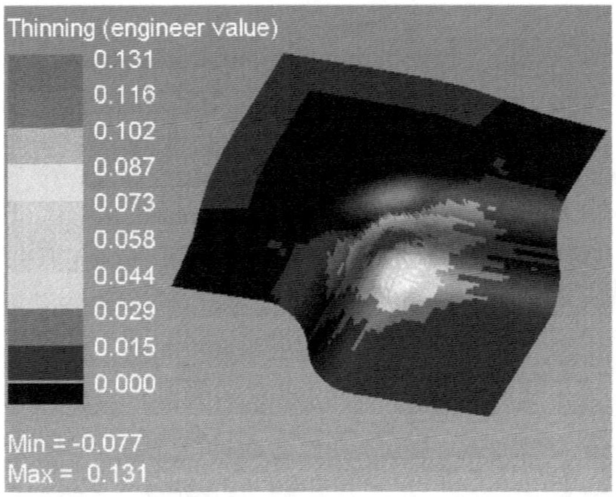

Figure 14. The rectangular container without using the drawbeads.

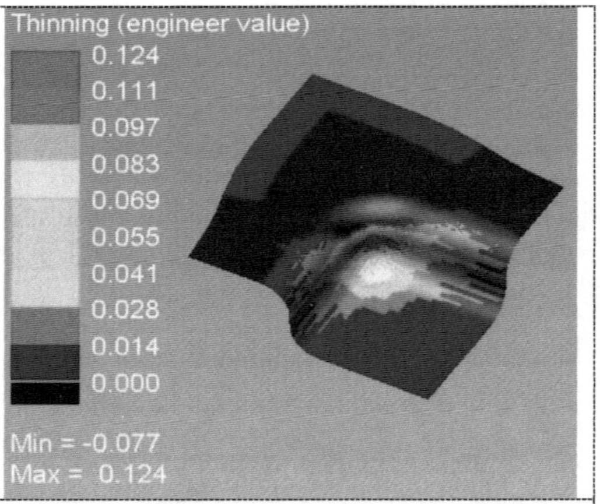

Figure 15. The rectangular container with using the drawbeads of the **a** type.

Figure 16. Molybdenum cylindrical containers.

Figure 17. FEM implementation of the crystallisation container - HDC method.

7. Conclusion

The results of the simulation of the rotational symmetric container show that the selected boundary conditions appropriately describe the really measured values. A model of the rectangular container was created by using the same boundary conditions. The maximal depth of the yield acquired from the given blank was 36 mm. The objective of the simulation was to increase the maximal depth of the rectangular container without any damage by using the drawbeads. The simulation results show decrease in ductility due to increasing the value of the rectangular container thinning. Only when we used the **a** rib shape a reduction of the thinning developed and in this way the depth of the rectangular container was extended by a value of 3 mm. This simulation can serve for the future design of the forming system which will already consider only the selected shapes of the drawbeads and this will result in achieving significant economic savings of a new design. Another important factor detected during simulations is that the friction coefficient fundamentally affects the deformation process of the Mo sheet and due to this fact it is necessary to influence its maximal reduction during the real experimental analyses.

Acknowledgments

This work was supported by the Slovak Research and Development Agency under the contract No. APVV-14-0284 and this paper presents results of work supported by the Slovak Scientific Grant Agency of the Slovak republic under the project No. VEGA 1/0077/15.

References

[1] Bagdasarov H S High-temperature crystallisation from melt 2004 (Moscow)
[2] Wang C, Wan M, Zhou Y, Wu X 2015 A ductile fracture criterion with Zener-Hollomon parameter of pure molybdenum, sheet in thermal forming *MATEC web of conferences* 21, 07006
[3] Tropp M, Handrik M, Kopas P, Sága M 2012 *Pipeline and Gass Journal* 239/11
[4] Donić T, Martikán M Hadzima B 2014 New unique ECAP system with ultrasound and back-pressure *IOP Conf. Series. Materials Science and Engineering* 63 012047 doi:10.1088/1757-899X/63/1/012047
[5] Anaraki A P, Shahabizadeh M, Babaee B 2012 Finite element simulation of multi-stage deep drawing process & comparison with experimental results. World Academy of Sciences *Engineering and Technology* 61
[6] Kim H K and Hong S K 2007 *Journal of Material Processing Technology* 184 pp 354–362
[7] Kojić M, Bathe K J 2005 *Inelastic Analysis of Solids and Structures* (Springer-Verlag Berlin Heidelberg)
[8] Kompiš, V., Toma, M., Žmindák, M., Handrik, M. 2004 *Computers and Structures* 82 (27) pp 2351-2360
[9] Wang C, Wan M, Donič T, Wu X, Zhou Y, Meng B 2015 *The 34th IDDRG Conference* (Shanghai, China)
[10] Meng B, Fu M W, Wan M 2015 *Int. Adv Manuf Technol* vol 78 p 1005
[11] Meng B, Wan M, Wu X, Zhou Y, Chang C 2014 *Int. Journal of Refractory Metals and Hard Materials* 45 p 41
[12] Lin Y, Hsu k, Lee P 2010 *China Steel Technical Report* 23 (China) p 32
[13] Wagoner R H, Chenot J 1997 *Fundamentals of metal forming* (USA: Wiley)
[14] Brončck, J., Fabian, P., Radek, N. 2015 *Materials Science Forum* 818 www.scientific.net/MSF.818.209, p.209.
[15] Thiébaut C, Bonnet C, Morey J M 1998 *Journal of Material Processing Technology* 77 p 240

IOP Conf. Series: Materials Science and Engineering **159** (2016) 012009 doi:10.1088/1757-899X/159/1/012009

Determination of Forming Limit Diagrams of AA6013-T6 Aluminum Alloy Sheet using a Time and Position Dependent Localized Necking Criterion

S Dicecco[1], C Butcher[1], M Worswick[1], E Boettcher[2], E Chu[3], C Shi[4]

[1] Department of Mechanical and Mechatronics Engineering, University of Waterloo, Canada
[2] Honda R&D Americas, Inc.
[3] Alcoa Technical Centre
[4] Promatek Research Centre

E-mail: sdicecco@uwaterloo.ca

Abstract. The forming limit behaviour of AA6013-T6 aluminium alloy sheet was characterized under isothermal conditions at room temperature (RT) and 250°C using limiting dome height (LDH) tests. Full field strain measurements were acquired throughout testing using *in situ* stereoscopic digital image correlation (DIC) techniques. Limit strain data was generated from the resulting full field strain measurements using two localized necking criteria: ISO12004-2:2008 and a time and position dependent criterion, termed the "Necking Zone" (NZ) approach in this paper, introduced by Martínez-Donaire *et al.* (2014). The limit strains resulting from the two localization detection schemes were compared. It was found that the ISO and NZ limit strains at RT are similar on the draw-side of the FLD, while the NZ approach yields a biaxial major limit strain 14.8% greater than the ISO generated major limit strain. At 250°C, the NZ generated major limit strains are 31-34% greater than the ISO generated major limit strains for near uniaxial, plane strain and biaxial loading conditions, respectively. The significant variance in limit strains between the two methodologies at 250°C highlights the need for a validation study regarding warm FLC determination.

1. Introduction

Current legislative requirements introduced by the Obama administration require Corporate Average Fuel Economy (CAFÉ) of 54.5 mpg (4.3 litre/100 km) by the year 2025 [1]. To achieve this goal, high strength aluminum alloys, such as the 6xxx and 7xxx series alloys, are being aggressively pursued for vehicle light-weighting applications. These alloys are heat-treatable and achieve peak strength through a multi-step aging process. In the peak aged condition, known as the T6 temper, these alloys have specific strengths comparable to ultra-high strength steels, however, their room temperature formability is limited, making the production of complex components difficult. To mitigate these formability issues, the warm forming processing route has been suggested [2].

Warm forming is a processing route in which the sheet material is deformed at temperatures above 150°C, but below the recrystallization temperature of the alloy. In this temperature range, the ductility of aluminum alloys is greatly improved, while flow resistance generally decreases, which has prompted studies [2, 3] on the benefits of the warm forming processing route on aluminium sheet formability.

While warm forming offers formability improvements in aluminum alloys, the exact quantification of the improvements in formability is less clear, since universally accepted standards for the warm formability characterization of aluminium sheet alloys have yet to be established. As a result, forming limit curves (FLC) of the same alloy and processing route may differ based on the adopted FLC methodology.

The concept of an experimentally determined FLC was first investigated by Keeler & Backhofen [4]. In their work, parts were incrementally formed under different loading conditions. Prior to localization, principal strain pairs at eventual localization points were recorded as 'safe' strain points. After many tests, various principal strain pairs were plotted as a function of the strain state, leading to the first experimentally derived FLC. If strain pairs on a formed part are found to lie above the FLC, the part is said to have exceeded the 'safe' strain limits of the sheet in that region. The FLD is useful for describing the formability of a sheet as a function of a processing route parameter, such as temperature. Decades after the work of Keeler & Backhofen [4], the FLC has seen wide acceptance within industry, although the methods for generating FLCs are not as widely accepted.

While the concept of an experimental FLC is useful, the testing procedures outlined by Keeler and Backhofen [4] were too resource intensive. Thus, practical standardized approaches have been developed over time for experimental FLC determination, such as ISO 12004-2:2008 [5]. The ISO approach was primarily developed for circle grid analysis of different limiting dome height (LDH) tests, where only the strain distribution of the dome post-failure is available. In the ISO approach, the post failure strain distribution of a dome specimen and an idealized assumption of the parabolic strain distribution about the dome pole prior to localization are utilized to calculate 'safe' principal strains for the dome. This assumption is valid for symmetric strain distributions with a single neck, however, in applications where multiple necks, or peaks in the major strain distribution, are present, the ISO approach is thought to be overly conservative [6, 7]. Additionally, the ISO standard was developed for room temperature (RT) applications and may not be applicable to warm forming. Consequently, alternative approaches for developing experimental FLCs are still actively pursued in the literature. Nevertheless, the ISO standard has been significant in standardizing FLC generation in an approach that is simple to implement and automate.

With the widespread adoption of commercial digital image correlation (DIC) techniques [6, 8] a surge in the number of alternative FLC methodologies has been observed. Digital image correlation allows for the tracking of the time-spatial evolution of dome strain distribution and dome topology over the duration of an LDH test. Thus, new FLC techniques have no longer been restricted to post-failure dome analysis. Some prominent DIC-based FLC techniques involve the work of Eberle *et al.* [9] and Li *et al.* [10], both of which are time-dependent methods that examine changes in the evolution of the major strain at the peak strain location over the duration of the dome test. Other FLC techniques include the work of Wang *et al.* [8] and Martínez-Donaire *et al.* [6], which are based on dome topology evolution and strain rate evolution, respectively, and are classified as both time and position dependent methodologies. The time and position dependent methodologies of Wang [8] and Martínez-Donaire [6] are developed based on the physical behaviour of sheet metals during the localization process. Therefore, these methodologies should be independent of temperature and strain rate effects and may potentially be applicable to warm forming FLC development.

In the current work, two different FLC methodologies are considered to generate an elevated temperature forming limit diagram for the sheet aluminium alloy AA6013-T6. The purpose of this work is twofold. First, the applicability of the ISO12004-2:2008 standard and the time-position dependent methodology of Martínez-Donaire *et al.* [6] are investigated with respect to warm formability characterization. Second, both methodologies are applied to the AA6013-T6 alloy to quantify the formability of the alloy in the warm forming processing route. Experimental procedures, DIC techniques, and FLC approaches are discussed in detail. Analysis between the two FLC methodologies is primarily comparative in nature, with validation work proposed for future study.

2. Necking Criteria

Necking criteria, or 'safe' limit strain criteria, encompass the various methodologies used for generating forming limit curves. Here, a brief discussion on the background, implementation, and automation of the two necking criteria that will be used to generate FLCs is presented: (a) ISO 12004-2:2008 and (b) the time-position method proposed by Martínez-Donaire *et al.* [6]. The ISO method was selected for its prominence in industry, while the time-position dependent approach was selected for its strong physical basis. As previously mentioned, both approaches have been automated for this study in an effort to reduce user subjectivity. To the knowledge of the authors, (b) has not been applied to warm forming applications. Other methods of FLC determination, such as those proposed by Eberle *et al.* [9] and Li *et al.* [10], are not considered in the current work but may also be suitable for warm FLC determination and are planned for implementation in a comprehensive future warm FLC determination study.

2.1. *FLC Method 1 - ISO12004-2:2008 [5]*

The FLC identification algorithm in ISO12004-2:2008 (hereafter, referred to as the ISO method) is well documented in the literature and within the standard so only a brief discussion of the method is presented here. Using the ISO method, the strain distribution perpendicular to the crack location of a failed dome specimen is extracted. This will be referred to as 'line slice data' for the remainder of this study. Line slice data can be extracted using DIC post-processing techniques or from circle grid analysis although only DIC measurements were used in this study. A DIC extracted line slice consists of a set number of points along the line where the major and minor strain data and deformation data are known at each location. For the ISO method, line slice data from multiple line slices are extracted from the DIC image before the onset of visible cracking, shown in **Figure 1**.

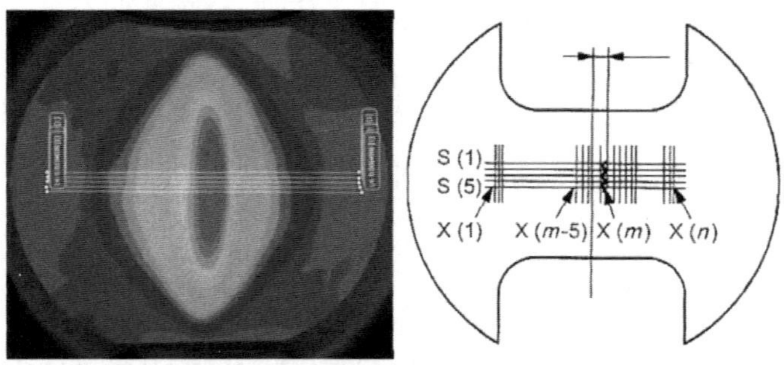

Figure 1: Major strain distribution of an AA6013-T6 plane strain dome geometry one image before the onset of cracking (left) and the procedure used for extracting strain information (right) [5]

With the extracted strain distribution, the peak major strain (ϵ_I) location should be extracted. A fitting window is then established based on the peak strain location. This window typically omits $\pm 4\ mm$ from the failure location, and is meant to encompass the inflection points in the ϵ_I strain distribution. Strain data within the fitting window is then used to fit an inverse parabola, with the peak of the parabola generally coinciding with the location of the ϵ_I failure strain, as shown in **Figure 2**. The 'safe' strain limits for FLC generation are subsequently obtained from the inverse parabola fit.

The ISO method is frequently used in industry due to its ease of automation, and its ability to be used with both DIC and circle grid strain acquisition techniques. Furthermore, the method is relatively straight forward. Potential weaknesses of the method centre on the assumption of the inverse parabolic fit. If there is significant asymmetry in the ϵ_I distribution, for example, such as from multiple peaks, a poor or ill conditioned parabolic fit will result, which, in turn, can affect the calculated 'safe' strain values [6, 7].

IOP Conf. Series: Materials Science and Engineering **159** (2016) 012009 doi:10.1088/1757-899X/159/1/012009

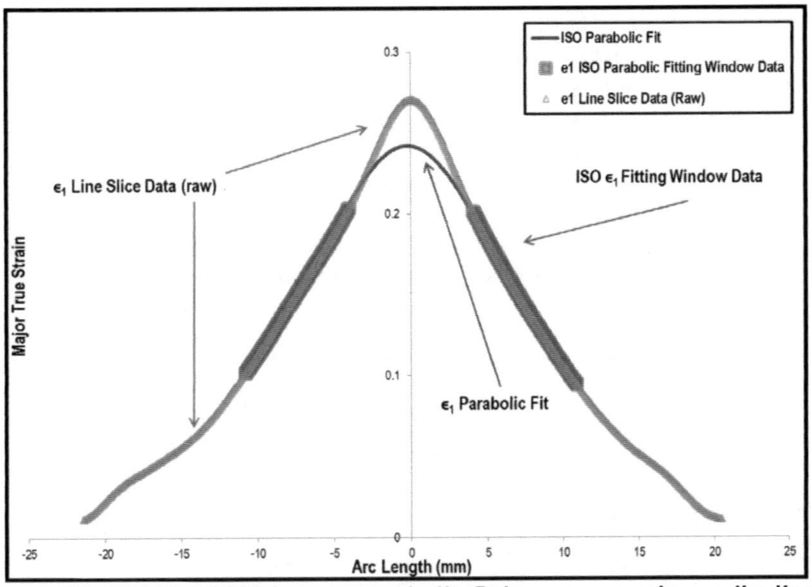

Figure 2: Sample ISO12004-2:2008 inverse parabolic fitting process using ϵ_1 distribution acquired from DIC dataset of a RT plane strain specimen

2.2. *Method 2 - Necking Zone Approach [6]*

The so-called Necking Zone (NZ) approach was proposed by Martínez-Donaire *et al.* [6] where they discuss the ϵ_1 (or ϵ_3) strain rate evolution at different locations on the surface of the dome. The motivation for the NZ method is that the local strain rate at the centre of the localization zone should continually increase until fracture. Outside of the localization region, the strain rates should begin to decrease as the material begins to unload as deformation concentrates within the neck. An example of this behaviour is shown in **Figure 3**. In the NZ method, the detection of a decreasing strain rate at the boundary of the neck indicates that the forming limit has been reached and the strains at the centre of the neck are assumed to be the limit strains for the time at which a decreasing strain rate is detected.

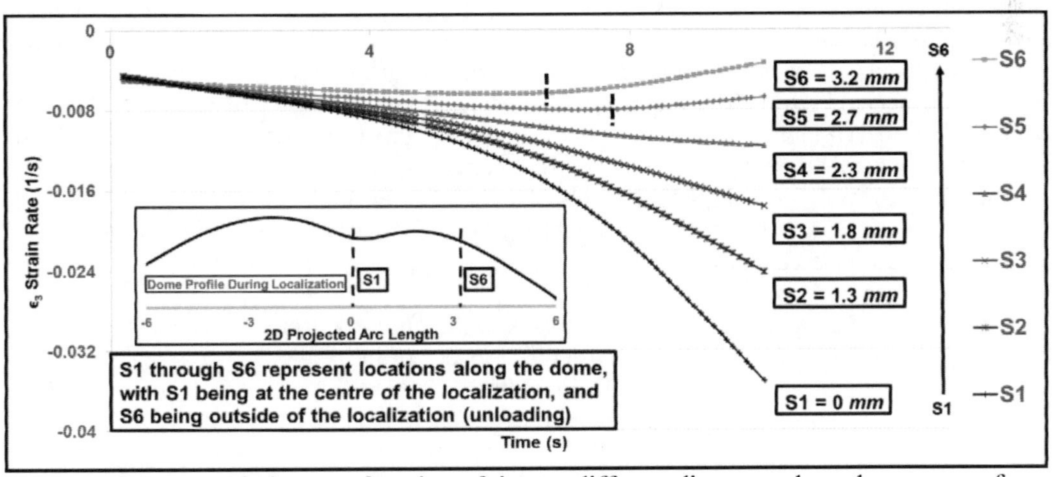

Figure 3: Strain rate evolution as a function of time at different distances along the upper surface of a uniaxial tension dome specimen subjected to RT LDH testing.

The Necking Zone approach is attractive in that it is was developed from physical observations of localization and failure. Furthermore, the NZ approach requires no extra processing within the digital image correlation software relative to the ISO approach, since the same line slice data can be used for calculating strain rates (ϵ_1 or ϵ_3). However, there are some challenges with the NZ method that introduce subjectivity and hinder its automation compared to the ISO method which has a well-documented implementation. First, smoothing of the DIC strain rate data is required. This smoothing is sensitive to the operator and smoothing algorithm. The second issue lies in the determination of the location for observing strain rates outside of the localization region, since there are many points where an 'unloading' in strain rate can occur (**Figure 4**), most of which can result in slightly different unloading times. This variation can also lead to differences in FLC curves based on user application. Finally, the NZ approach can be applied to each side of the crack location, which, in turn, results in the determination of two unloading times.

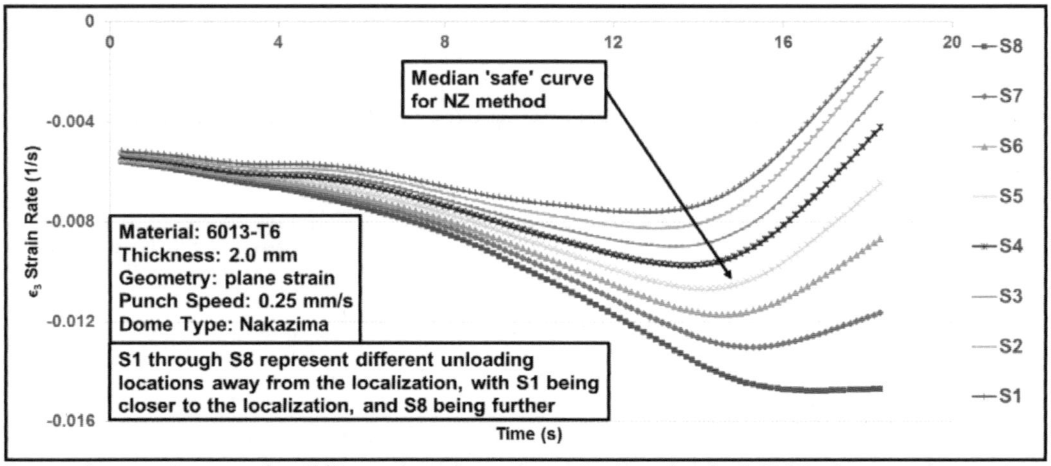

Figure 4: ϵ_3 strain rates for different locations along the length of a 250°C plane strain specimen

For the NZ approach used in this work, ϵ_3 strain rates were calculated for all spatial points ± 10 *mm* away from the location of the minimum ϵ_3 strain in the image before visible cracking. In this work, the term 'spatial point' refers to a point taken along the previously described 'line slice'. The ± 10 *mm* range corresponds to five times the thickness of the sheet material. The average spacing between points ranged from 0.05 *mm* to 0.10 *mm*. For each spatial point, the strain rate time history is searched for a point where the strain rate increases to a maximum value before decreasing some time before the final processed DIC image, herein referred to as an 'unloading' point. The first spatial point that contains unloading is marked as the spatial start point (S1 in **Figure 4**). From the spatial start point, spatial points up to 2 *mm* (one sheet thickness) away from the spatial start point are searched for unloading times in their strain rate histories (S8 in **Figure 4**). All unloading times are recorded and the median of all recorded times is considered to be the unloading time for 'safe' strain calculation (**Figure 4**). This process is completed for each side of the neck. The average of the two principal strain pairs is taken as the 'safe' strain set for the test.

The influence of the searching width on either side of the neck from first unloading, one sheet thickness (2 *mm*) in this instance, can have a large impact on the resulting safe strains and requires validation through future study. Although a fixed width is used in the current work, a more appropriate search width may also consider the stress state and hardening behaviour of the alloy.

3. Experimental Procedures

In this section, a brief description of the limiting dome height testing parameters is provided. This includes, but is not limited to, sheet material properties and geometries, lubrication conditions, and forming parameters. The image acquisition equipment and procedures are also described, as well as the corresponding DIC analysis parameters.

3.1. *Limiting Dome Height Testing*

The material used for testing in this work was 2.0 mm gauge sheet aluminium alloy AA6013 in the T6 temper provided by Alcoa. The T6 temper is considered to be fully age hardened by Alcoa [11]. The sheet is specified to have a nominal yield strength of 317 *MPa*, a nominal ultimate tensile strength of 379 *MPa*, and a total elongation of approximately 8% [11]. All specimens were machined lengthwise parallel to the rolling direction of the sheet.

Three geometries were machined from the sheet with approximate stress states of uniaxial tension, plane strain, and balanced biaxial loading. These geometries are discussed in the ISO12004-2:2008 standard for Nakazima LDH testing. The first two geometries are similar in shape to the specimen depicted in **Figure 1**, with reduced width sections of 25.4 *mm* and 76.2 *mm*, respectively. The balanced biaxial specimen is a 200 *mm* by 200 *mm* square blank.

Prior to testing, a random black and white speckle pattern was applied using aerosol based paint to the center regions of all blanks, on the side of the blanks not in contact with the punch. The speckle pattern was used to enable DIC analysis. **Figure 5** shows a typical speckle pattern with an artificially increased contrast to highlight the speckle pattern. It is important to note that the samples were cleaned with acetone and dried with compressed air before speckling, to promote good adhesion between the speckle paint and the samples.

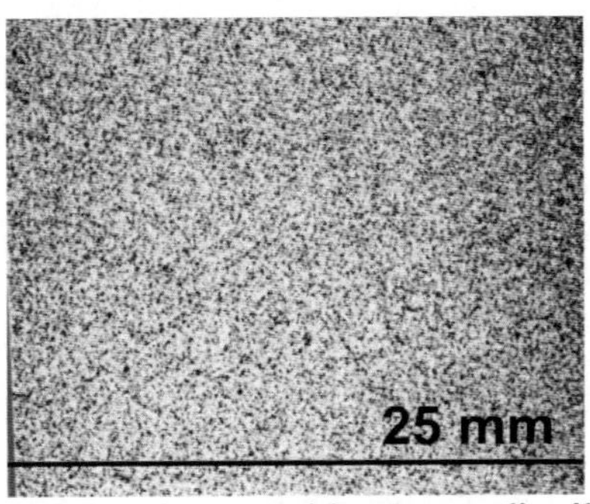

Figure 5: Random speckle pattern distribution

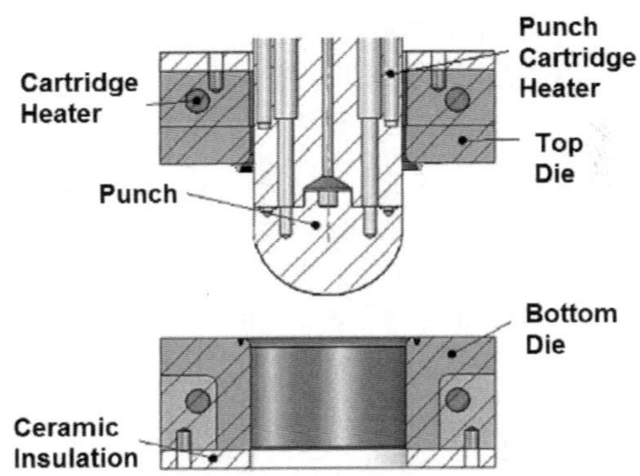

Figure 6: Cross-sectional illustration of heated LDH tooling

Isothermal testing was completed at RT and 250°C. During testing, a flat die set was used with a clamping load ranging from 300 *kN* to 600 *kN*. The variation in loading served only to mitigate material draw-in. In warm forming, the blank was uniformly heated to the desired test temperature within tooling that had embedded cartridge heaters, ceramic insulation, and PID control, as shown in **Figure 6**. Given PID control associated with the heated tooling and insulation, the process is considered isothermal. To achieve a temperature within 2% of the target of 250°C, roughly 200 seconds of blank heating was required from the time the cool blank was clamped by the warm tooling. With the current setup, it was possible to heat the blank for a significantly longer period of time to achieve exactly the desired temperature, however, the T6 temper condition of the alloy is sensitive to time at temperature and the authors wished to minimize the over-aging of the alloy.

The punch geometry used for LDH testing was a 100 *mm* diameter hemispherical (Nakazima) punch. Lubrication between the tooling and the blank consisted of different layers of Polytetrafluoroethylene (PTFE) for 250°C testing, and a combination of PTFE and Vaseline for room temperature testing. Using this lubrication scheme, failure was almost always found to coincide with the maximum dome height location. The punch was operated at a speed of 0.25 $mm{\cdot}s^{-1}$.

3.2. *Image Acquisition and Post-Processing*

Two cameras were operated at 5 to 15 frames per second for stereoscopic image acquisition, which corresponds to 20.0 $images{\cdot}mm^{-1}$ to 6.67 $images{\cdot}mm^{-1}$ of punch displacement for a punch speed of 0.25 $mm{\cdot}s^{-1}$. A single LED light source was used to illuminate the blanks during testing.

Image processing was completed using the Correlated Solutions software program Vic-3D. Relevant analysis parameters in Vic-3D 7 consist of: (a) step size, (b) subset size, and (c) strain filter. The subset size represents how the area of interest on the dome is discretized for analysis. Each subset must contain a sufficiently random paint speckle distribution. The step size corresponds to the number of pixels within a subset that are to be analysed for correlation purposes. If a step size of three is selected, every third pixel within the subset will be analysed. Finally, the strain filter is a type of smoothing operation applied to the raw strain distributions. In this study, the DIC processing parameters consisted of a step size of 5, a subset size ranging from 25 to 35, and a strain filter of 11. The resolution of each DIC image was approximately 0.09 mm/pixel.

Following processing, line slice data was extracted for the last 100 images before observations of cracking. Five line slices were extracted perpendicular to the crack location for each dome test. Each line slice was processed using the FLC methodologies outlined in Section 2.

Smoothing of the DIC data is generally necessary, as is often the case when computing time derivatives from experimental datasets. In this study, a simple and general two-stage smoothing process was employed within the Matlab software environment for calculating ϵ_3 strain rates. First, a LOESS filter with a span of five data points was applied to the datasets to be smoothed. Following this, a quintic b-form smoothing spline was generated from the LOESS filtered datasets [12]. While it is beyond the scope of this work to discuss the b-form smoothing spline, it should be noted that this type of spline can essentially be thought of as a weighting (p) between a least squares polynomial approximation ($p \sim 0$) and a piece-wise cubic spline ($p \sim 1$). In this work the weighting (or smoothing parameter) of the spline was 0.5. A smoothing parameter of 0.5 was utilized for time sensitive data because it is inherently noisier, notably at warm forming temperatures where heat waves can cause minor fluctuations in the measured strain. Nevertheless, smoothing of the time series data was still relatively minor, as illustrated in **Figure 7**, where an example of the level of smoothing of two ϵ_3-*time* datasets for different spatial locations on a dome is presented.

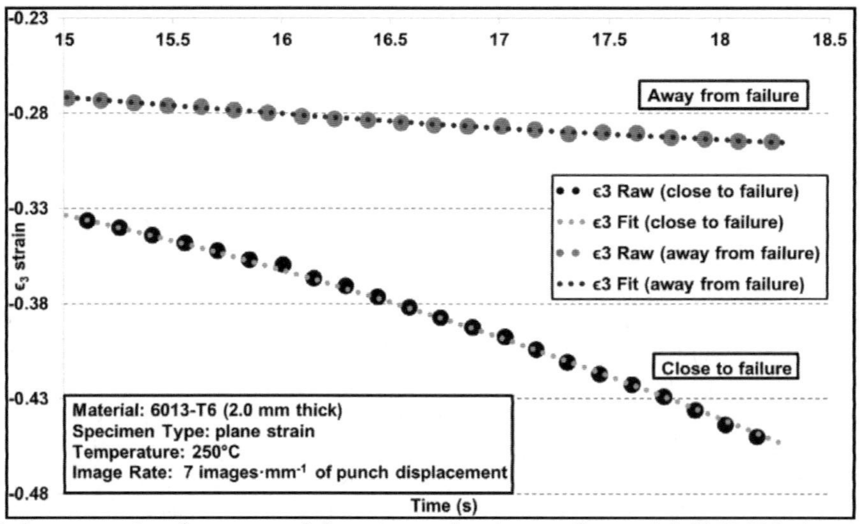

Figure 7: Fitting example of a raw partial ϵ_3-*time* series dataset for an AA6013-T6 plane strain LDH specimen at 250°C

4. Experimental Results and Discussion

In this section, forming limit curves generated using the ISO approach and the NZ approach for AA6013-T6 at RT and 250°C are presented. Both localized necking detection schemes are compared based on the observed differences in FLCs.

Average limit strains and standard deviations from the LDH testing of the AA6013-T6 alloy for the two localization detection methodologies discussed in this study are included in **Table 1**. Standard deviations greater than or equal to 0.02 are bolded for easier identification. These results were generated from five LDH tests per specimen geometry and temperature combination for all conditions, except for the room temperature biaxial and 250°C uniaxial tension datasets which were generated from only three tests, respectively. Five line slices were extracted from the DIC data of each test. The average strains to failure and standard deviations of each geometry and temperature combination were also extracted from the line slice data, and have been included in **Table 1**. These failure strains do not necessarily represent the fracture behaviour of the alloy and have only been included to provide guidance on the potential variation of each dataset. Limit strains from **Table 1** are plotted to generate forming limit curves based on the two localization methodologies in **Figure 8** for both room temperature testing and 250°C testing. All strain data reported in this section and the remainder of the study will be strictly true strains.

Table 1: AA6013-T6 true limit strains generated using different necking detection criteria

Temp.	Specimen	ε1 Failure Strain		ε2 Failure Strain		ε1 ISO Strain		ε2 ISO Strain		ε1 NZ Strain		ε2 NZ Strain	
		Avg	Std	Avg	Std	Avg	Std	Avg	Std	Avg	Std	Avg	Std
RT	25.4 mm	0.34	**0.04**	-0.08	0.01	0.22	**0.02**	-0.06	0.01	0.21	**0.02**	-0.05	0.00
	76.2 mm	0.27	0.01	0.02	0.00	0.16	0.01	0.02	0.00	0.16	0.00	0.02	0.00
	200 mm	0.37	0.00	0.26	0.01	0.27	0.01	0.26	0.01	0.31	0.01	0.24	0.01
250°C	25.4 mm	0.55	0.01	-0.12	0.00	0.28	0.01	-0.06	0.01	0.36	**0.03**	-0.09	0.01
	76.2 mm	0.46	**0.02**	0.01	0.00	0.20	0.01	0.02	0.00	0.26	0.01	0.02	0.00
	200 mm	0.49	**0.03**	0.28	**0.03**	0.28	**0.02**	0.25	**0.02**	0.37	**0.03**	0.25	**0.03**

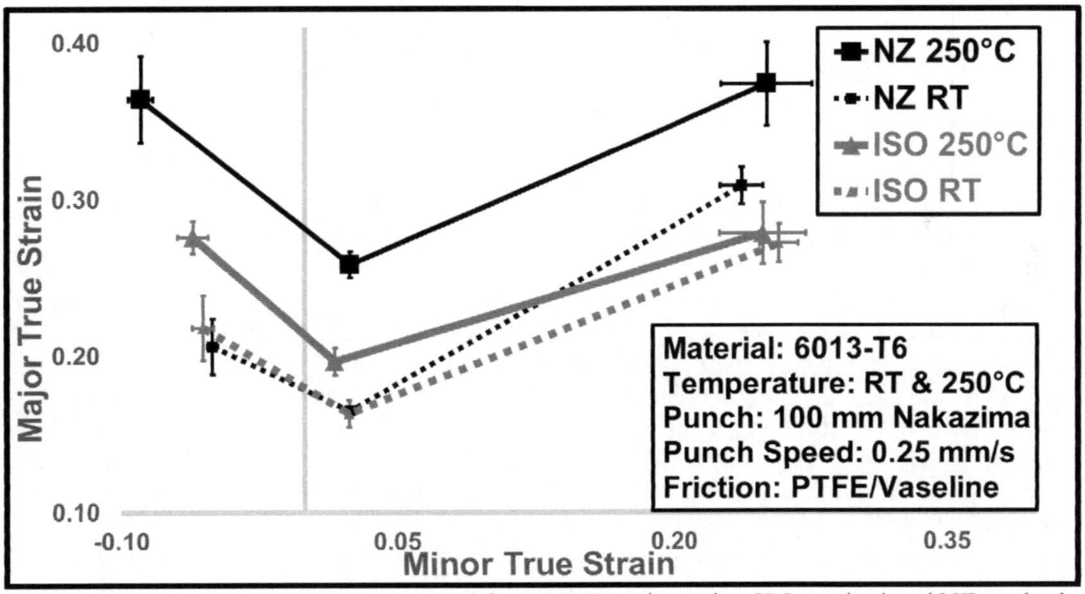

Figure 8: FLD of AA6013-T6 generated from LDH testing using ISO method and NZ method

In **Figure 8**, it can be seen how the two localization methodologies compare at RT and 250°C. At room temperature, it can be seen that the ISO and NZ methods produce similar limit strains for the uniaxial geometry, with major strain limits of 0.22 and 0.21, respectively. A similar trend is also observed for the room temperature plane strain testing. Conversely, in the biaxial regime, the ISO method predicts a significantly lower major limit strain (0.27) relative to the NZ method (0.31), while both methods produce standard deviations about the major biaxial limit strains of only 0.01.

At 250°C, the differences between the ISO and NZ method FLCs are profound relative to that observed for the RT FLCs. The major limit strains at 250°C for uniaxial tension, plane strain, and biaxial loading are 0.28, 0.2, and 0.28 for the ISO approach and 0.37, 0.26, and 0.37 for the NZ approach. For the three geometries in the same previously stated order, the ISO 250°C limit strains correspond to increases of 26.7%, 20.6%, and 2.35% relative to the ISO RT limits, while the NZ 250°C limit strains correspond to increases of 76.8%, 56.7%, and 21.2% relative to the NZ RT limits.

On the draw side (negative minor strain) of **Figure 8**, it can be seen that both the ISO and NZ methods capture increases in formability at 250°C in the 6013-T6 sheet, however, the NZ method shows significantly better formability than the ISO method. While the accuracy of the NZ method in detecting the onset of localization at temperature cannot be confirmed until a validation study is complete, some guidance on the lower ISO limit strains is offered in **Figure 9**.

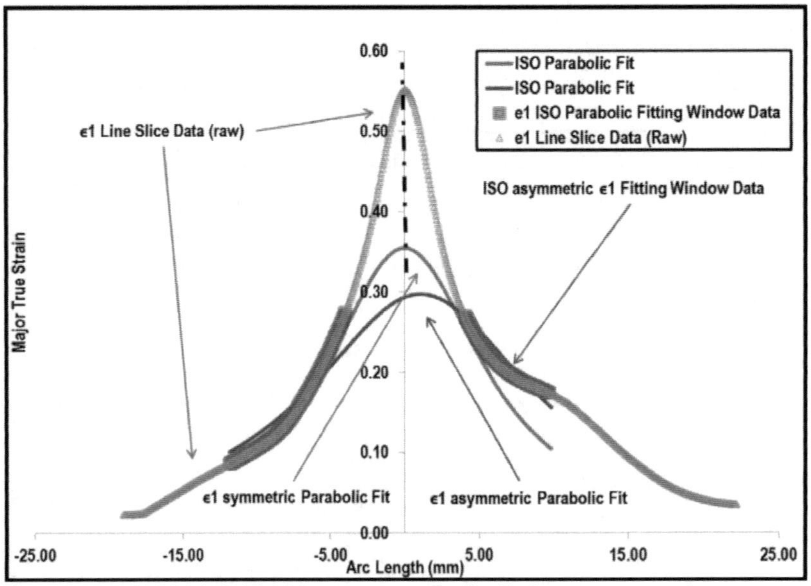

Figure 9: Sample ISO12004-2:2008 inverse parabolic fitting process using an asymmetric ϵ_1 distribution acquired from DIC dataset of a 250°C uniaxial tension dome specimen

In **Figure 9**, an example of a measured ϵ_1 strain distribution for a 25.4 *mm* Nakazima dome specimen tested at 250°C is presented. Note the asymmetry in the distribution and its effect on the parabolic fit. This asymmetry was frequently observed in uniaxial and plane strain specimens, despite attempts to improve lubrication conditions and achieve a central failure location. As seen in **Figure 9**, and previously discussed in Section 2, asymmetry in the strain distribution can cause an ill conditioned parabolic fit, which can subsequently lead to overly-conservative limit strain predictions. To illustrate the effect of strain distribution shape on the parabolic fit, a second parabolic fit, generated by mirroring the left of the distribution, is also shown in **Figure 9**. The second parabolic fit results in an increase in the major limit strain of approximately 0.06 strain, and has a peak which corresponds to the peak of the experimental strain distribution. Note that no recommendation is being made to mirror strain distributions in this manner. Instead, this mirrored fit is merely shown here to illustrate the dependency of the fit to the shape of the strain distribution. Furthermore, it is recognized that the ISO standard anticipates this issue by specifying that "good tests" must fail at the centre of the specimen, thereby avoiding this skewing effect, however, a central failure requires low friction conditions between the sheet and punch which is not always possible, particularly for warm forming conditions.

On the stretch side of **Figure 8**, a negligible increase in biaxial formability at 250°C relative to RT conditions is observed, as determined using the ISO method. This result is uncharacteristic of warm forming and does not coincide with the observed increases on the draw-side of the FLD. The reason for this behaviour may be the result of a poor inverse parabolic fit caused by asymmetry in the ϵ_1 strain distributions in the elevated temperature biaxial specimens, although further investigation is required.

Figure 10 shows sample strain paths extracted from the failure region of test specimens at RT and 250°C, as well as the ISO and NZ limit strains. It can be seen that the NZ limit strains generally closely follow their corresponding strain paths, as is expected for the NZ method. For the ISO method, the same is also true, except for the biaxial limit strains, which deviate significantly from their respective sample strain paths, notably at 250°C.

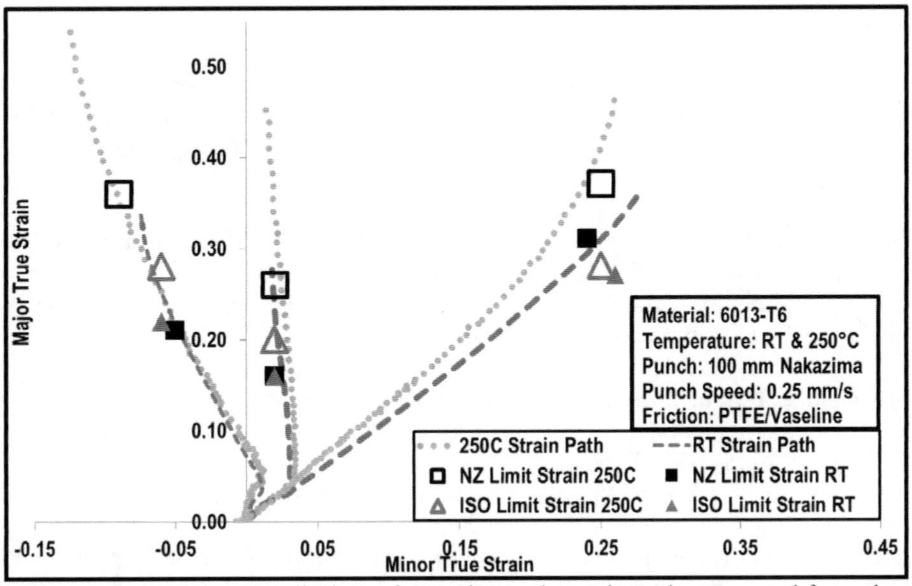

Figure 10: ISO and NZ forming limit strains and sample strain paths extracted from the peak ϵ_1 locations of AA6013-T6 LDH specimens tested at RT and 250°C

The observations made herein demonstrate that the NZ method, as implemented in this study, generally results in higher FLC values than those obtained when applying the ISO method to the asymmetric strain distributions obtained for the plane strain and draw specimens tested herein. On the stretch side of the FLD, it was found that the limit strains calculated using the ISO method deviate from the experimental strain path whereas the NZ method does not. The deviation in the strain path for the limit strains in the ISO method is related to its reconstruction of the strain distribution and its determination of the minor strain from the ϵ_3 (thickness) and ϵ_1 strain distributions. The limit strains in the NZ method lie on the experimental strain path because the limit strains are extracted directly from the DIC measurements. Nonetheless, further experiments and validation are required to identify the appropriate FLD detection algorithm for warm forming conditions.

5. Conclusions

The ISO method and NZ method show almost identical limit strains at RT for AA6013-T6 on the draw-side of the forming limit curve. On the stretch side of the RT FLC, the major limit strain using the current NZ method is 14.8% above the value calculated by applying the ISO method. At 250°C, the NZ method FLC lies anywhere from 31% to 34% above the FLC calculated by applying the ISO method to the measured (asymmetric) strain distributions. While neither approach has been validated at this point in time for warm forming applications, the summarized results highlight that different localization detection methodologies can yield very different warm forming limit strains. Therefore, it is paramount that a validation study be completed prior to the application of either approach to generating warm FLCs for industrial applications. A future validation study using the NZ approach, the ISO approach, the approach of Eberle *et al.* [9], and a topology based approach is planned.

The NZ limit strains fall closely onto corresponding sample strain paths at 250°C. Similarly, the ISO draw and plane strain limit strains fall onto corresponding strain paths, however, in the biaxial loading

IOP Conf. Series: Materials Science and Engineering **159** (2016) 012009 doi:10.1088/1757-899X/159/1/012009

condition, the ISO limit strains deviate from the biaxial strain path. The deviation of the ISO limit strains from the experimental strain path is an artefact of the algorithm in that it reconstructs the strain distributions. The drift in the ISO limit strains from the experimental strain paths under warm forming conditions may be related to the formation of asymmetric strain distributions as multiple localization zones can occur at elevated temperatures, however, further study is needed.

Acknowledgments

The authors greatly appreciate financial support from the Honda R&D Americas, Promatek Research Centre (Cosma International), Alcoa Technical Center, the Natural Sciences and Engineering Research Council of Canada, the Canada Research Chairs Secretariat, and the Ontario Research Fund.

References

[1] National Highway Traffic Safety Administration, "2017 - 2025 Model Year Light Duty Vehicle GHG Emissions and CAFE standards: Supplemental," 2011.

[2] N. Sotirov, P. Simon, C. Chimani, D. Uffelmann and C. Melzer, "Warm Deep Drawability of Peak-Aged 7075 Aluminium Sheet Alloy," *Key Engineering Materials,* **504-506**, 955-60, 2012.

[3] N. Abedrabbo, F. Pourboghrat and J. Carsley, "Forming of AA5182-O and AA5754-O at elevated temperatures using coupled thermo-mechanical finite element models," *International Journal of Plasticity,* **23**, 841-75, 2007.

[4] S. Keeler and W. Backofen, "Plastic instability and fracture in sheets stretched over rigid punches," *ASM Transactions Quarterly,* **56**, 25-48, 1963.

[5] International Organization for Standardization, "ISO/DIS 12004-2:2008. Metallic Materials - Sheet and strip - Determination of Forming Limit Curves - Part 2: Determination of Forming Limit Curves in the Laboratory," 2008.

[6] A. Martínez-Donaire, F. Garcia-Lomas and C. Vallellano, "New approaches to detect the onset of localised necking in sheets under through-thickness strain gradients," *Materials and Design,* **57**, 135-45, 2014.

[7] D. Vysochinskiy, T. Coudert, O. S. Hopperstad, O. -G. Lademo and A. Reyes, "Experimental detection of forming limit strains on samples with multiple local necks," *Journal of Materials Processing Technology,* **227**, 216-26, 2016.

[8] K. Wang, J. Carsley, B. He, J. Li and L. Zhang, "Measuring forming limit strains with digital image correlation analysis," *Journal of Materials Processing Technology,* **214**, 1120-30, 2014.

[9] B. Eberle, W. Volk and O. Hora, "Automatic approach in the evaluation of the experimental FLD with a full 2d approach based on a time depending method," in *Numisheet 2007 - Proceedings of the 7th International Conference and Workshop on Numerical Simulation of 3D Sheet Metal Forming Processes, Part A.*, 2008.

[10] J. Li, J. Carsley, T. Stoughton, L. Hector Jr and S. Hu, "Forming limit analysis for two-stage forming of 5182-O aluminum sheet with intermediate annealing," *International Journal of Plasticity,* **45**, 21-43, 2013.

[11] Alcoa, "Alloy 6013 Sheet - Higher Strength With Improved Formability," Alcoa Mill Products. Inc., Bettendorf, Iowa, 2001.

[12] C. Reinsch, "Smoothing by spline functions," *Numer. Math,* **10**, 177-83, 1967.

IOP Conf. Series: Materials Science and Engineering **159** (2016) 012010 doi:10.1088/1757-899X/159/1/012010

Influence of hydrogen on formability and bendability of DP1180 steel for car body application

Q Gao [1,3], F Han [2], D Wortberg [1], W Bleck [3] and M Liewald [2]

[1] Daimler AG, Process Development and Materials, Benzstr., 71063 Sindelfingen, Germany
[2] Institute for Metal Forming Technology, Stuttgart University, Holzgartenstr.17, 70174 Stuttgart, Germany
[3] Department of Ferrous Metallurgy, RWTH Aachen University, Intzestr. 1, 52072 Aachen, Germany

E-mail: qi.q.gao@daimler.com

Abstract. In order to reach future light weight targets, it is increasing necessary to use advanced high strength steels with tensile strength 980 MPa or higher in automotive body-in-white structures. Due to the sensitivity to hydrogen embrittlement and the limited understanding of various aspects of hydrogen embrittlement on processing and function, the wide application of these steels is still limited.

In the current work, the influence of hydrogen on the multiaxial forming behavior was investigated by determining the forming limit curve and bending limit curve of DP1180 steel. Hydrogen concentration in the material was modified by cathodic charging. Then Nakajima tests on hydrogen uncharged and pre-charged samples were carried out in order to adjust and study different strain states resulting in the forming limit curve. In the study of bending limit curve, the steel sheets were pre-strained by Marciniak test. Bending load on the uncharged and pre-charged samples was introduced by VDA238-100 bending tests. The experimental results indicated that the presence of hydrogen affected the formability and bendability of DP1180 steel. A clear difference in the influence of hydrogen at different strain states was observed. When formed in a biaxial strain state via the Nakajima test, the material showed the highest degradation in formability. Moreover, the samples with biaxial pre-loading showed more degradation in bendability comparing to those pre-strained in plane strain and uni-axial paths. Fractography by scanning electron microscope gave evidence of hydrogen-induced cleavage fracture on pre-charged Nakajima samples. Thus this investigation improves the understanding of influences of hydrogen on forming processes and provides important evidence for further studies on HE susceptibility of AHSS for the application on car body constructions.

1. Introduction

Today the automotive industry is utilizing more high strength steels to enable lightweight construction. However, the wide application of Advanced High Strength Steels (AHSS) with tensile strength over 980 MPa is limited by their sensitivity to Hydrogen Embrittlement (HE). HE in steels is defined as a loss in ductility and resulting brittle fracture due to the introduction and diffusion of hydrogen in the steel [1]. It can take place during the automotive production process or in-service process, once the applied stress and hydrogen content reach the threshold value of the material. In general, the process of HE is due to the interaction between matrix and atomic hydrogen atoms on a very small scale, the combination of several individual modes probably contributes. So far, two of the mechanisms have been widely accepted in the research field: the Hydrogen Enhanced De-cohesion (HEDE) mechanism [2, 3] and Hydrogen Enhanced Localized Plasticity (HELP) [4, 5] mechanism. The susceptibility to HE in AHSS was already reported in many publications. Nevertheless, the previous studies focused on the loss of mechanical properties in tensile tests and hydrogen induced cracking on deformed samples [6-9]. To understand the role of hydrogen in automotive production process, the impact of hydrogen on formability need to be further investigated.

Forming Limit Curve (FLC) is a most commonly used processing criterion, representing the maximal in-plane formability [10]. FLC describes the first occurrence of membrane instability, but neglects the effect of sheet bending [11]. To fully evaluate the forming behavior the Bending Limit Curve (BLC) is also needed. BLC enables the analysis of drawing operations with superimposed bending and describes the failure due to local shear zones [10-12].

In the present work, the effect of hydrogen on the FLC and BLC was studied. Nakajima tests with uncharged and pre-charged samples were carried out to determine the FLCs of DP1180 steel with different hydrogen concentrations. Impacts of hydrogen on forming limits in different strain states were compared. Fractography by Scanning Electron Microscope (SEM) model was used to determine the hydrogen-induced fractures in pre-charged samples via Nakajima tests. In the study of BLC, Marciniak tests were carried out to provide pre-strains in different strain paths. The pre-loaded samples were pre-charged with hydrogen and afterwards critical strains to bending fracture were studied by bending tests according to VDA 238-100. The results illustrated that the pre-strains in different strain paths lead to a different susceptibility to HE and the bendability degraded distinctly. Thus this work clarifies the influence of hydrogen on the formability and bendability in different strain states.

2. Material and experimental methods

2.1 Material

A commercial advanced high strength steel DP1180 was investigated. The steel sheets were cold rolled to reach a thickness of 1.5 mm. The micrographs of DP1180 by SEM, as shown in Figure 1, illustrate that the microstructure of DP1180 consists of ferrite and tempered martensite phases. In rolling direction the material has a yield strength of 1024 MPa and tensile strength of 1224 MPa. The elongation to fracture is 7.6%.

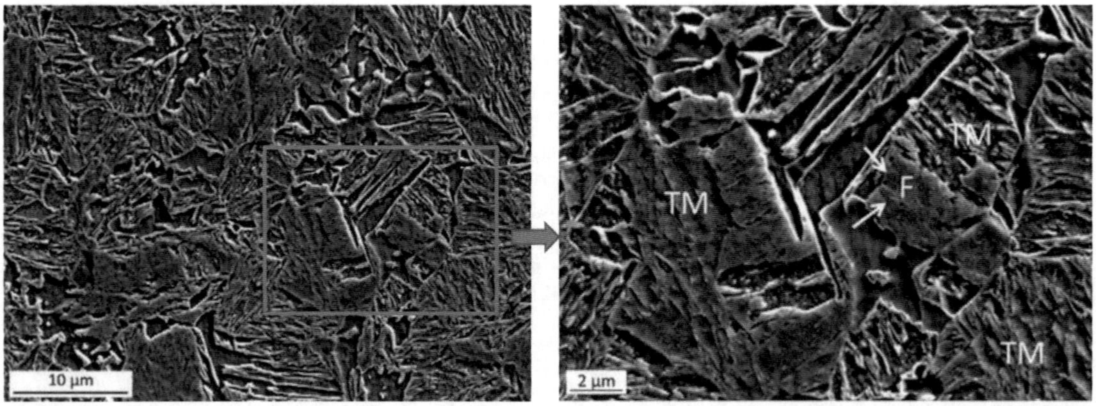

Figure 1. Micrographs of DP1180A in RD-TD plane, characterized by SEM, etched with 3% Nital.

2.2 Hydrogen charging and electro-galvanizing

Pre-charging was carried out to modify the hydrogen concentration in the steel. Figure 2 explains the preparation procedures. The samples for forming and bending tests were cleaned with an acid solution of 18% HCl with 3.5 g/l Hexamethylenetetramine for 10 s and subsequently with distilled water and acetone. The samples were pre-charged with hydrogen and electro-galvanized to prevent the effusion of hydrogen during forming/bending tests. The to-be-charged samples acted as a cathode.

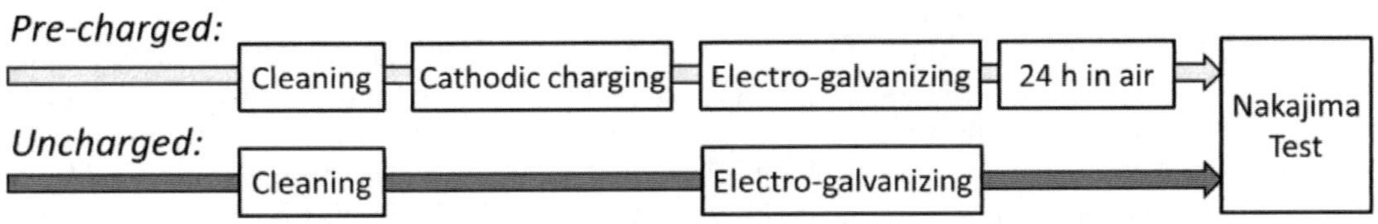

Figure 2. Preparation of pre-charged and uncharged samples.

The relevant parameters in pre-charging and electro-galvanizing processes are listed in Table 1. As reference, the uncharged samples were directly galvanized as same as the pre-charged samples.

Table 1. Parameters in pre-charging and electro-galvanizing processes.

Process	Samples for	Anode	Solution	Current density	Time
Pre-charging	Forming test Bending test	Pt-Ti net	3%NaCl+0.3% NH_4SCN	1 mA/cm^2	30 min
Electro-galvanizing	Forming test Bending test	Zinc bar	$ZnCl_2$	10 mA/cm^2 40 mA/cm^2	30 min 10 min

2.3 Determination of Forming Limit Curve (FLC)

Nakajima tests were carried out with three different sample geometries, as shown in Figure 3. On the surfaces of sample a characteristic stochastic black and white pattern was obtained by spraying. The Nakajima samples were drawn until failure, with a punch velocity of 2 mm/s. The patterns on the surfaces were recorded by a GOM ARAMIS system in a frequency of 10 pictures per second. The test was repeated five times.

The occurrence of necking was detected by analyzing the recorded pictures, according to the "linear best fit" method [13]. Five elements in the necking areas were selected, to calculate the mean values of the representative thinning rate in the last 20 pictures before fracture. In Figure 4 it is revealed that the representative thinning rate increases linearly with a low slope initially. After a transition period, it rises with a high slope. The intersection of the linear fitting curves in these two areas indicates the beginning of the instable necking during forming. The last picture number during stable deformation was found so that the strain distributions on the surfaces of sample, named forming limits, were determined.

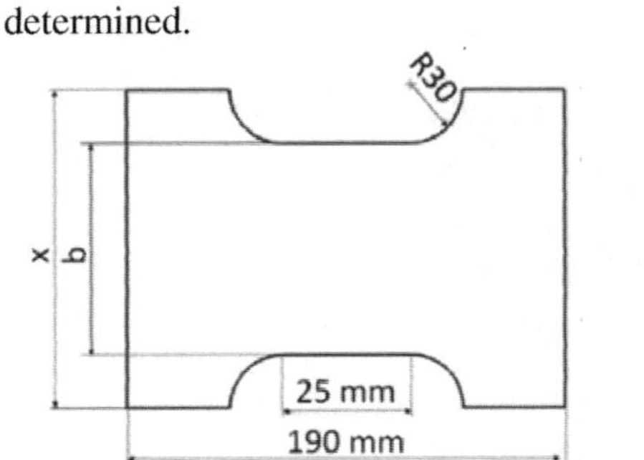

Sample form	Strain state	x [mm]	b [mm]
Form 1	Uni-axial	55	20
Form 2	Plane strain	135	100
Form 3	Biaxial	190	190

Figure 3. Sample geometries for Nakajima tests.

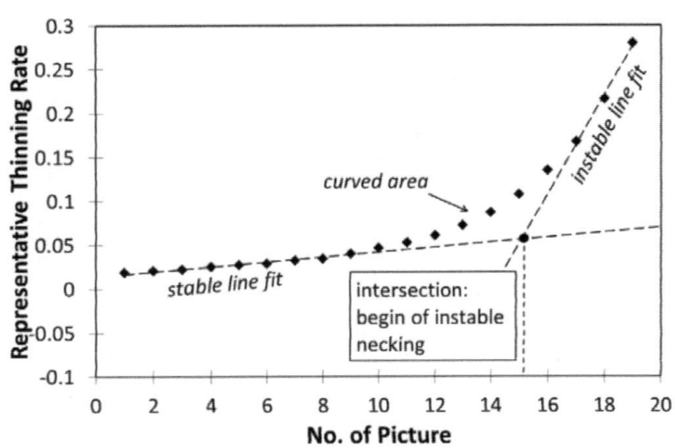

Figure 4. Detection of localized necking in a representative thinning rate diagram.

2.4 Determination of Bending Limit Curve (BLC)

BLC is a strain based criterion to predict failure under bending conditions. To determine the BLC of a material, a drawing operation for pre-straining and a superposed bending process are required [10, 11]. Figure 5 is schematically describes the development of the BLC. In this study a Marciniak test with a punch diameter of 340 mm was applied to introduce the pre-straining in the steel sheets. The blank holder force was 400 kN and the punch velocity was 20 mm/s. The Marciniak samples were drawn in uniaxial, plane strain and biaxial strain states, until the maximal drawing depth was reached. Afterwards, bending samples with dimension of 60 mm×60 mm were cut from the Marciniak samples, and prepared by pre-charging and zinc coating.

Superimposed bending loads were applied by the three-point-bending test according to VDA 238-100. Figure 6 illustrates the schematic bending device. The radius of bending-sword punch was 0.4 mm. The samples were bent with a punch velocity of 20 mm/min. A force drop of 30 N was defined as the failure criterion in bending procedure (Figure 7). In each variant, five tests were repeated.

The strain distributions on the surfaces of Marciniak sample and bending sample were measured by optical metrology: the stochastic black and white patterns before and after deformation were recorded and analyzed by GOM ARAMIS system, in order to calculate the values of major and minor strains. Figure 8 shows the strain distribution on the bending edges [10].

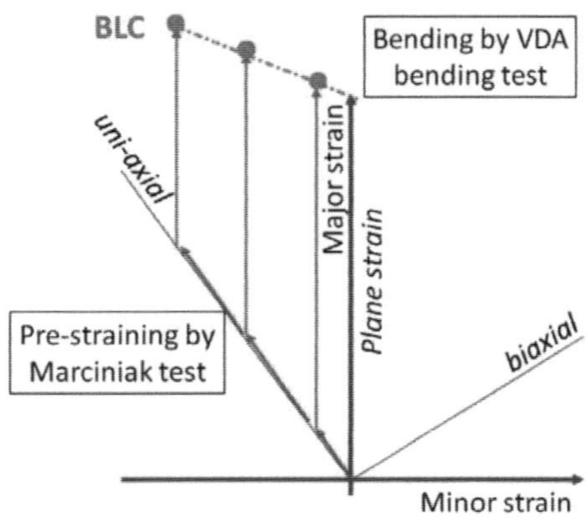

Figure 5. Schematic development of BLC [10].

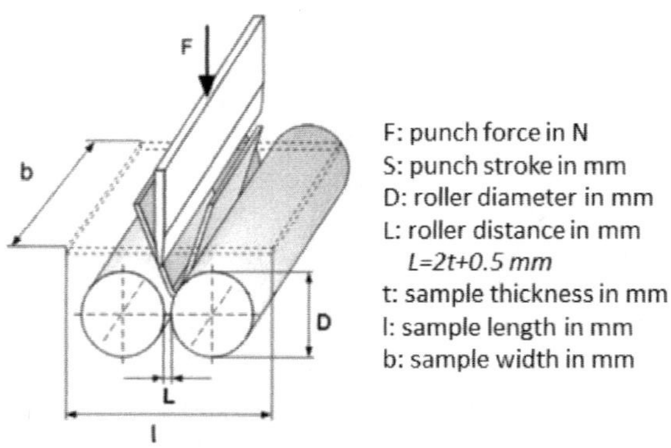

Figure 6. Schematic illustration of the bending device (VDA 238-100).

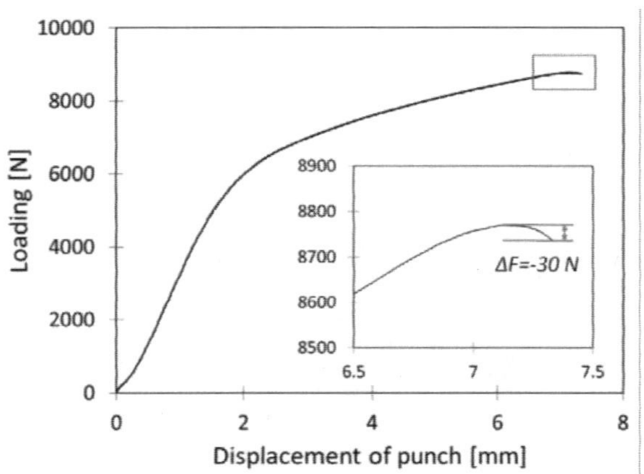

Figure 7. Loading curve in VDA bending test and the defined failure criterion.

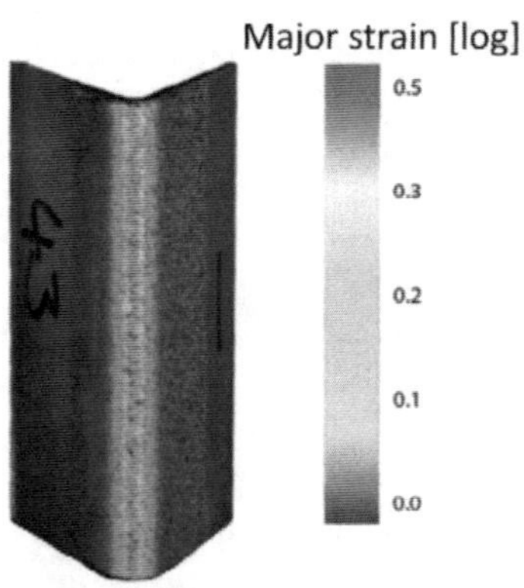

Figure 8. Strain distribution on the surface of the bending samples, measured by optical metrology (GOM ARAMIS) [10].

IOP Conf. Series: Materials Science and Engineering **159** (2016) 012010 doi:10.1088/1757-899X/159/1/012010

3. Results and discussion

3.1 Forming Limit Curves (FLC)

As characterized by SEM, the microstructure of DP1180 is not changed by hydrogen charging process. Figure 9 shows the FLCs of the uncharged and pre-charged DP1180 samples measured via the Nakajima test. It is revealed that, the forming limits in different strain states dropped by different amounts. In the uni-axial strain state, no obvious distinction was found; the major strain dropped from 0.11 to 0.10 while the minor strains stayed the same. In the plane strain condition, the forming limit was lower in the pre-charged samples. The major strain decreased from 0.09 to 0.06. However, in the biaxial strain state, a localized necking period was not found in the pre-charged samples. The representative thinning rate diagrams of biaxial samples are shown in Figure 10: in Nakajima tests the uncharged samples first deformed stably and then deformed unstably (Figure 10 (a)).Conversely, in the pre-charged samples instable deformation was not detected, which means the samples underwent stable deformation up until failure (Figure 10 (b)).

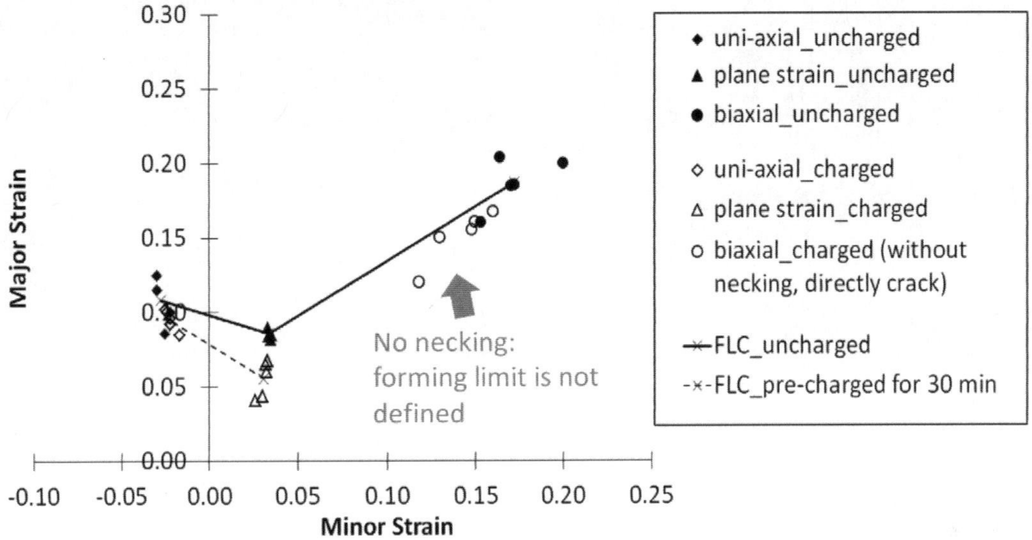

Figure 9. Forming limits of uncharged and charged DP1180 samples in Nakajima tests.

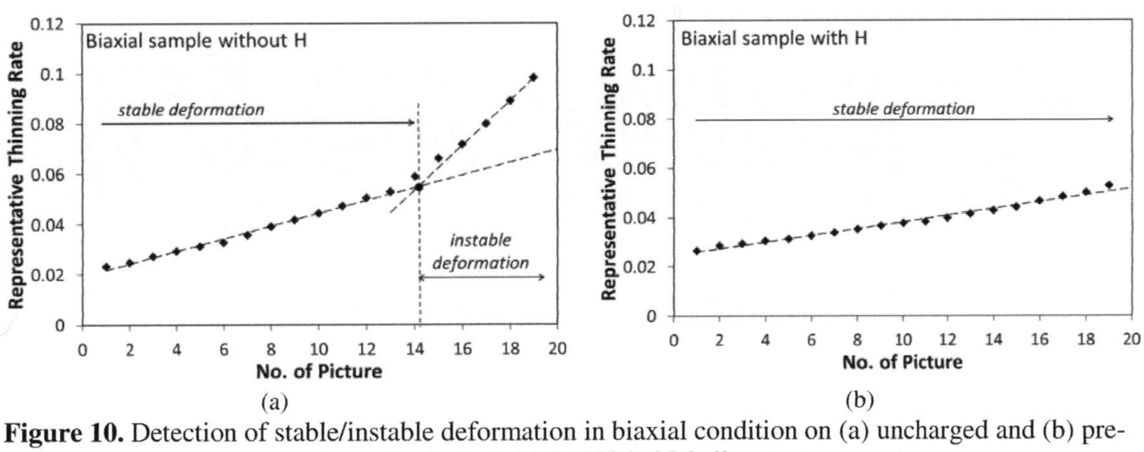

Figure 10. Detection of stable/instable deformation in biaxial condition on (a) uncharged and (b) pre-charged samples of DP1180 in Nakajima tests.

The fracture surfaces of Nakajima samples were analyzed under SEM model, as showed in Figure 11. On the uncharged samples, the fracture surfaces displayed overall ductile fracture with dimples, indicating plastic deformation. Under stress the dislocations pile up at the potential positions where the cavities are firstly created. The multi-axial stress leads to the growth and merging of the cavities and ductile fractures with dimples to result [14]. The fracture surfaces of the pre-charged sample showed quasi-cleavage fracture in transgranular mode inside the large dimples. After pre-charging a high concentration of hydrogen atoms exist in the trapping sites, and the localized stress gradients act as the driving force for its diffusion. The multi-axial stress activates the hydrogen re-contribution process, therefore the hydrogen atoms concentrate at the top of crack tips, which causes local cleavage fracture. In the biaxial strain state, under both major strain and minor strain, the hydrogen re-concentration is more intense because the expanded lattices provide more space for hydrogen atoms than the compressed lattices. The HELP mechanism explains the formability degradation due to HE in aspect of interaction between hydrogen atoms and dislocations [4, 5]. Local hydrogen enrichment weakens the cohesion force of matrix and causes brittle fractures, as explained in the HEDE mechanism [2, 3].

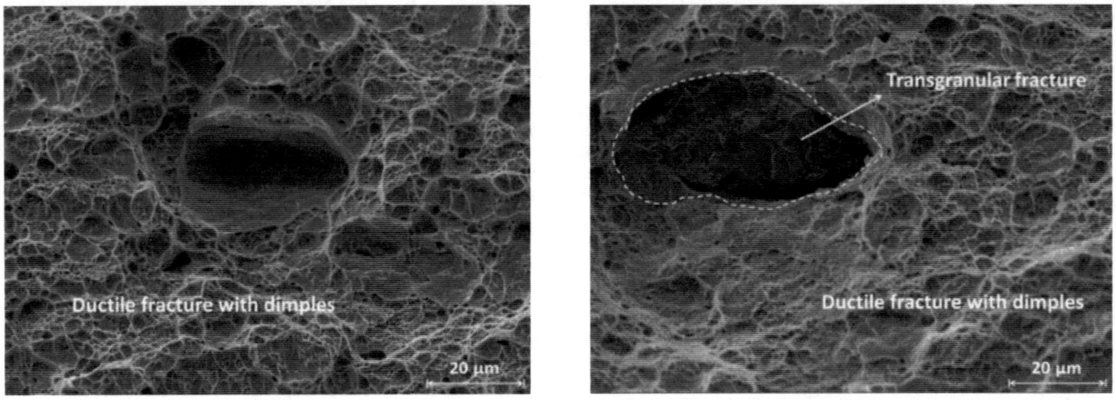

(a) (b)

Figure 11. SEM fracture surfaces of (a) uncharged sample and (b) pre-charged sample of DP1180 after Nakajima test, in biaxial strain state.

3.2 Bending Limit Curves (BLC)

Figure 12 shows the failed samples with ca. 2% pre-straining after bending tests. On the uncharged and pre-charged samples, both materials have comparable quality of bending edges. It was demonstrated that the failure mechanism and the bending fracture were identical, independent of hydrogen pre-charging.

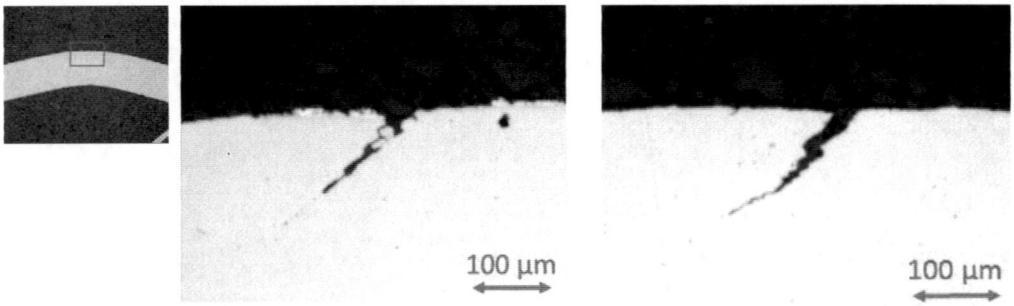

(a) (b)

Figure 12. Microscopic images of bending fracture from (a) uncharged and (b) pre-charged DP1180 samples.

The bending limit curve shown in Figure 13 includes uni-axial, plane strain and biaxial pre-loads, is named enhanced Bending Limit Curve (eBLC) [12]. The pre-straining processes under different strain paths provide the distribution of major and minor strains. By the bending process the major strain in the samples was increased. In Figure 13 it can be seen that after 30 min pre-charging, the critical strains to bending fracture were decreased. This is most likely due to the enrichment of diffusible hydrogen. The comparison of effective strains to bending fracture was illustrated in Figure 14. The degradation of effective strain was defined as an index to evaluate susceptibility to hydrogen, which was calculated according to equation (1):

$$Index\ (HE) = \frac{\varphi_v\ (uncharged) - \varphi_v\ (pre-charged)}{\varphi_v\ (uncharged)} \times 100\% \quad (1)$$

where φ_v is the maximum effective strain without bending fracture. The higher *Index (HE)* stands for higher susceptibility to HE during bending process. For the material without pre-straining, the *Index (HE)* was 31.5%. Therefore, the *index (HE)* of the pre-strained material was higher than that of the material without pre-straining. It is illustrated that after pre-straining, the hydrogen-induced degradation of bendability was higher. Though the hydrogen pre-charging processes of all the samples were same, the samples with pre-straining adsorbed more hydrogen than the samples without pre-straining. Tensile straining played an essential role in providing the trapping sites for hydrogen. Higher hydrogen content leaded to higher *Index (HE)*.

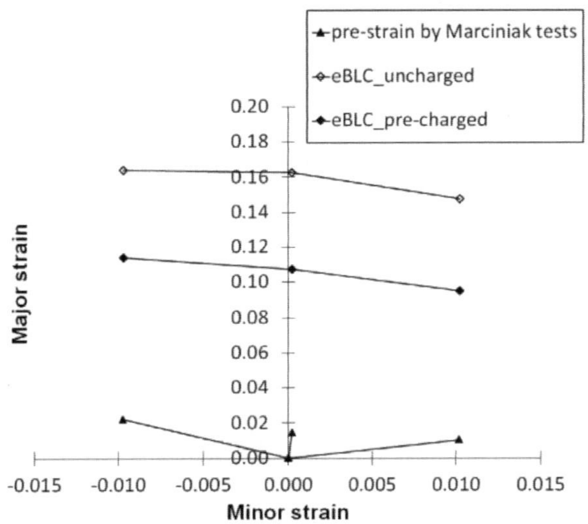

Figure 13. eBLC of uncharged and charged DP1180 samples via VDA-bending tests.

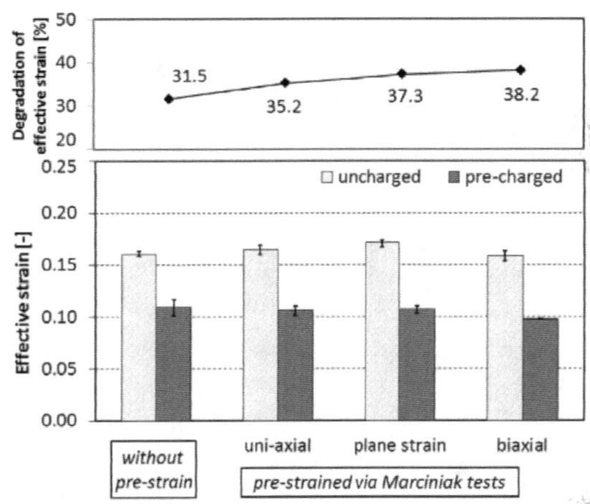

Figure 14. Comparison of effective strains from uncharged and pre-charged bending samples.

4. Conclusion

In the current work, the impact of hydrogen on FLC and BLC in DP1180 steel was investigated via Nakajima tests and 3-point-bending tests with pre-strained samples.

After pre-charging, the forming limits and bending limits were degraded due to the presence of hydrogen. Hydrogen induced degradation of formability is depending on the strain states. In uniaxial strain state, the forming limit is not significantly influenced by hydrogen, while in the plane strain state the forming limit drops to a lower level after pre-charging. The material in biaxial strain state shows the highest susceptibility to HE where the samples fail even before the initiation of necking. The shift of FLC can be explained by HELP and HEDE mechanisms.

Hydrogen plays a critical role when the material is under bending loads. In the study of hydrogen influence on BLC, the bending limits of the as-delivered and pre-strained samples are reduced. The material with pre-straining is more susceptive to HE compared to the material without pre-loads. The pre-strained samples in various strain paths display different in the susceptibility to HE. The susceptibility to HE in different pre-strained material is: biaxial > plane strain > uniaxial > without pre-straining.

In general, FLC and BLC, which are taken as processing criteria for steel sheet materials, are influenced by hydrogen. It means that in actual production processes of car body constructions the impact of hydrogen in AHSS should be considered. The forming and pre-straining processes of hydrogen-existing components should be paid attention to, especially when the strain state is in biaxial strain path. The results of the present work could also be significant for the evaluation of the crash behavior of car body components made of AHSS, in the case of hydrogen presents. In the current work, the punch velocity was set according to the standard. The further work can focus on the hydrogen-induced degradation of formability at different forming speed, to understand the hydrogen diffusion process during the forming process better.

References

[1] H. Kaesche, "Corrosion of Metals", Springer-Verlag Berlin Heidelberg, 2003.
[2] A. Troiano, "The role of hydrogen and other interstitials in the mechanical behavior of metals", Transactions of the American Society of Metals 52 (1960) 54-80.
[3] R. Oriani, "A mechanistic theory of hydrogen embrittlement of steels", Berichte der Bunsengesellschaft für physikalische Chemie, 76(8) (1972) 848-857.
[4] I. M. Robertson and H.K. Birnbaum, "An HVEM study of hydrogen effects on the deformation and fracture of nickel", Acta Metallurgica 34(3) (1985) 353-366.
[5] H. K. Birnbaum and P. Sofronis, "Hydrogen-enhanced localized plasticity – a mechanism for hydrogen-related fracture", Materials Science and Engineering: A 176 (1994) 191-202.
[6] T. Depover and D. Perez Escobar et al., "Effect of hydrogen charging on the mechanical properties of advanced high strength steels", International Journal of Hydrogen Energy 39 (2014) 4647-4656.
[7] J. Rehrl and K. Mraczek et al., "The impact of hydrogen on the mechanical properties of AHSS/UHSS grades at low- and high strain rates" in 2nd Steely Hydrogen Conference 2014, Conference Proceedings 21-32.
[8] X. Zhu and W. Li, et al., "Hydrogen trapping sites and hydrogen-induced cracking in high strength quenching & partitioning (Q&P) treated steel", International Journal of Hydrogen Energy 39 (2014) 13031-13040.
[9] A. Laureys and T. Depover, et al., "Characterization of hydrogen induced cracking in TRIP-assisted steels", International Journal of Hydrogen Energy, 47 (2015) 16901-16912.
[10] R. Denninger and M. Liewald, et al., "Failure prediction in sheet metal forming depending in pre-straining and bending superposition", in ESAFORM 2012, Conference Proceedings 101-106.
[11] R. Shleich and C. Held, et al., "Investigation on the effect of curvature and sheet thickness on forming limit prediction for aluminum sheet metal alloys" International Journal of Material Forming 2 (2009) 69-74.
[12] R. Denninger and M. Liewald, et al., "Investigation on bendability of lightweight materials for various load paths", in ESAFORM 2011, Conference Proceedings 1601-1604.
[13] W. Volk and P. Hora, "New algorithm for a robust user-independent evaluation of beginning instability for the experimental FLC determination", International Journal of Material Forming 4 (2011) 339-346.
[14] W. Bleck, "Material Characterization", Textbook for students, Mainz Verlag, 2009.

IOP Conf. Series: Materials Science and Engineering **159** (2016) 012011 doi:10.1088/1757-899X/159/1/012011

Stretch bending – the plane within the sheet where strains reach the forming limit curve

F M Neuhauser[1,2], O R Terrazas[1], N Manopulo[2], P Hora[2], and C J Van Tyne[1]

[1] Advanced Steel Processing and Products Research Center, Colorado School of Mines, 920 15th St, Golden, CO 80401 USA
[2] Institut für virtuelle Produktion, Department of Mechanical and Process Engineering, Swiss Federal Institute of Technology, ETH Zurich, Tannenstrasse 3, 8092 Zürich, Switzerland

E-mail: cvantyne@mines.edu

Abstract. Finite element analysis (FEA) was used to model the angular stretch bend test, where a strip of sheet metal is locked at both ends and a tool with a radius stretches and bends the center of the strip until failure. The FEA program used in the study was Abaqus. The FEA model was verified by experimental work using a dual phase steel (DP600) and with a simplified analytical analysis. The FEA model was used to simulate the experimental test for various frictional conditions and various radii of an upward moving tool. The primary objective of the study was to evaluate the concave-side rule, which states that during stretch bending the forming limit occurs when the strains on the concave surface plane of the bent sheet (i.e. bottom plane) reach the forming limit curve (FLC). The verification with experimental data indicates that the FEA model represents the process very well. Only conditions where failure occurred on or near the tooling are included in the results. The FEA simulations showed that the actual forming limit of the sheet occurs when the strains on the bottom plane of the sheet (i.e. concave side of the bend) reach the forming limit curve for high friction and small tool radii. For lower friction and for larger tool radii the actual forming limit occurs when strains on other planes in the sheet (i.e. mid planes or top surface plane) reach the forming limit curve. The implications of these results suggest that care must be taken in assessing forming operations when both stretch and bending occur. Although it is known that the FLC cannot predict the forming limit for small bend radii, the common assumption that the forming limit occurs when the strains for the middle thickness plane of the sheet reach the forming limit curve or that the concave side rule is often made. Understanding the limits of this assumption needs to be carefully and critically evaluated.

1. Introduction

The forming limit diagram (FLD) is commonly used in the forming industry to predict localized necking in sheet metals. However experimental and mathematical determination of forming limit curves is based on in plane deformation without taking a bending component into account. In this paper the question is addressed whether a forming limit diagram predicts localized necking too conservative or too late for different amounts of superimposed bending for different strain conditions ranging from plane strain to uniaxial strain. This knowledge is of importance in finding tools, which do not cause failure during production in deep drawing processes via finite element simulations, especially for die and punch radii where bending is superimposed on a stretching component.

Since the forming limit diagram is only valid for membrane deformation *i.e.* from Keeler and Brazier's forming limit curve [1], Tharrett and Stoughton [2] introduced the so called concave-side rule which states, that in bending with tension deformation the neck occurs on the convex-side of the sheet, when the strains on the concave-side reach the plane-strain forming limit. Therefore, Tharrett and Stoughton could detect a clear enhanced formability of stretch bend operations in comparison to in plane only stretch deformation. However, when compared to previous studies, the localization effect seems to dominate the enhancement of formability since shear fracture is still present.

Enhanced formability due to bending could also be found by Atzema et al. [3] who performed Nakazima tests with different punch radii. The smaller the radii, the higher was the observed formability. The effect was highest in plane strain and negligible in the uniaxial region.

Kitting et al. [4] applied the concave-side rule on H340LAD materials and found, that it predicts the formability well for punch radii R $\geqslant$ 10 mm and underestimates material formability for R <10 mm. Further, Kitting et al. [5] expanded the experimental characterization of bending under tension for not only plane strain, but also uniaxial and biaxial strain states.

These previous studies show that bending can enhance the formability of sheet metal, where this enhancement is defined as strains that are above the forming limit curve. From a practical perspective, this enhancement of formability might be exploited during some sheet forming operations. In order to use this extra level of formability requires a better understand of the bending effect.

During stretching and bending the neck formation will be observed on the top (i.e. tensile side due to bending) side of the sheet. The enhancement will allow this top surface to have higher strains than predicted by the forming limit curve. It would be expected that some plane within the thickness of the sheet will be at the forming limit when the top surface experiences a local neck. The objective of the present study was to examine via finite element analysis the angular stretch bend test to determine which plane in the sheet metal reaches the forming limit strain when localization of strain on the top surface occurs during the test.

Although there have been extensive studies reported in the literature on sheet bending under tension, the present investigation as well as the previous studies on the bending enhancement of formability use an angular stretch bend test which is different from the traditional bending under tension test methods. In the traditional test, the sheet metal with some back tension bends onto a roller, slides over the roller and unbends as it comes off the roller. In the angular stretch bend test the tooling simultaneously stretches and bends the sheet without the bending onto and off of a roller. The mechanics and stress states in the sheet are different for these two types of tests.

2. Experimental Procedure

Experimental angular stretch bend tests were performed on DP600. Figure 1 shows a photo and a schematic of the angular stretch bend test. The sheet specimens were 180 mm in length, 25 mm wide and 1.45 mm thick. The lubrication used was Teflon (PTFE) with LPS2 lubricant. The tool geometry was punch radii of 1, 2.5 and 5 mm, die entry radius of 4.7 mm, drawbead radius of 3.2 mm, distance between drawbeads 95 mm, and die opening 76.2 mm. The punch velocity was 5.08 mm/s

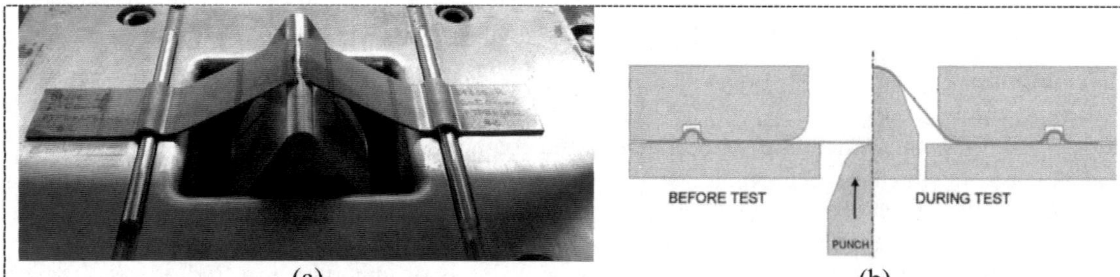

(a) (b)

Figure 1. Angular stretch bend test (a) photograph, (b) schematic.

The tests were run to fracture. Force was measured during the test. The localization of strain was assumed to occur at the maximum force in the force displacement curve.

For the finite element analysis a mixed Hockett-Sherby and Gosh yield curve was used as the constitutive model for the DP600 based on tensile test data. Four tensile tests were performed for DP600 sheet steel in rolling direction at an engineering strain rate of 0.002 1/s in order to describe the material's static stress response in the elastic region and also the hardening in the plastic region. An isotropic hardening model was used in the present study.

Young's modulus was calculated to be E = 195,100 MPa as the mean value of the four tensile tests. A Poisson's ratio of 0.3 is assumed in the elastic region of the flow curve.

In order to determine the dependency of flow stress on the equivalent plastic strains in the plastic part of the flow curve, the following steps were conducted:

1) Engineering stress – strain curve truncated at the uniform elongation strain
2) Removal of the elastic part of engineering stress – strain curve
3) Conversion to true stress – logarithmic strain curve
4) Subtract elastic component of logarithmic strains
5) Fit the experimental true stress – plastic logarithmic strain curve to the Hockett-Sherby hardening law (using a least square fit)
6) Fit the experimental true stress – plastic logarithmic strain curve to the Gosh hardening law (using a least square fit)
7) Fit the last 40 data points of the experimental true stress – plastic logarithmic strain curve to a mixed Hockett-Sherby Gosh hardening law (using a least square fit) which corresponds to a strain range of about 0.08 – 0.12
8) Average all three obtained flow curves

The reason for the rather complicated Hockett-Sherby Gosh fit is that strains occurring in the stretch bend simulation are well beyond the uniform elongation strains. Therefore, it is of crucial importance that the flow curve does not only match the experimental data before the uniform elongation, but also accurately predicts material flow behavior for strains exceeding uniform elongation. The mixed Hockett-Sherby Gosh approach enables the possibility to best capture the slope of the last 40 experimental data points, which correspond to the last 0.04 strain and therefore delivers a reasonable extrapolated region. The constitutive equations are as follows:

$$\sigma_{HS} = A_1 - (A_1 - B_1) * e^{(-m_1 * \varepsilon^{n_1})} \tag{1}$$

$$\sigma_G = A_2 * (B_2 + \varepsilon)^{n_2} - C_2 \tag{2}$$

$$\sigma_{HS-G} = \kappa * \sigma_{HS} + (1 - \kappa) * \sigma_G \tag{3}$$

Equation (1) is the Hockett-Sherby model, Equation (2) is the Gosh model, and Equation (3) is the mixed model. Table 1 gives the model parameters used.

Table 1 – Parameters for the Mixed Hockett-Sherby Gosh Yield Curve

κ []	A_1 [MPa]	B_1 [MPa]	m_1 []	n_1 []	A_2 [MPa]	B_2 [MPa]	n_2 []	C_2 [MPa]
0.54	855.92	406.77	7.26	0.69	1102.1	0.0025	0.17	0

The general purpose software ABAQUS 6.14-1 was used to model the angular stretch bend test. A three dimensional model was implemented, where the tools, including die, holder and punch, are represented by analytical rigid body surfaces. To expand the experimental work, punch radii of 1, 2.5, 5, 10, 15 and 20 mm were used to set the sheet under tension with different superimposed bending components from mild to severe.

The blank was rendered by 3D linear 8 node brick elements with reduced integration (1 integration point per element instead of 2 in each direction) to avoid shear locking. Eleven elements were used through the thickness. Due to symmetry a one-quarter blank was modelled and the blank was locked in position at the drawbeads. The friction coefficient is held constant between the holder and sheet as

IOP Conf. Series: Materials Science and Engineering **159** (2016) 012011 doi:10.1088/1757-899X/159/1/012011

well as between the die and sheet at a value of 0.18. In order to vary the strain paths at the punch nose, the friction coefficient between punch and sheet is varied between 0.0, 0.03, 0.07, 0.15 and 0.2.

An analytical model, based on the assumption of homogenous strains along the ligament and homogenous strains through the thickness was used to verify the finite element results.

In order the compare the strains of the angular stretch bend test sample in different layers of the sheet with forming limits, the FLD introduced by Keeler and Brazier [1] was calculated. The empirical equations consist of a forming limit in plane strain, FLD_0 as:

$$FLD_0^{true} = \ln\left[1 + (23.3 + 14.14 * t) * \left(\frac{n_t}{0.21 * 100}\right)\right], n_t \leqslant 0.21 \tag{4}$$

which is dependent on the n value in the Hollomon hardening law and the sheet thickness t. The left side of the forming limit diagram decreases linearly with ε_{minor} while the right side is a function of a logarithm depending on the values of FLD_0 and ε_{minor} as:

$$\varepsilon_{maj} = \ln\left[0.6 * (e^{\varepsilon_{min}} - 1) + e^{FLD_0^{true}}\right], \varepsilon_{min} > 0 \tag{5}$$

3. Results and Discussion

Figure 2 shows the FEA simulation results compared to the analytical verification and the experimental results. Figure 2a shows a good correspondence between the analytical analysis and the homogeneous portions of the force-displacement curves. The analytical model does not account for localization so there is a difference when the finite element simulations begin to exhibit localized deformation. The analytical model is for the ligament region only, so the model does not have a dependence on the radius of the tooling. Figure 2b shows the comparison of the finite element simulations with no friction and the experimental results with the use of Teflon lubrication. The comparison is quite good. These two verification steps provide good confidence in the finite element simulations, which will be used to determine the plane in the sheet that reaches the forming limit when localization occurs.

Since the scope of the current study was to access the strains in the sample when localized necking occurs and then to compare these strains with a conventional FLC, the punch displacement is needed, where the top surface of the sheets starts to localize. The load maximum generally signals the onset of diffuse necking, not localized necking. However, for a plane-strain condition, localized and diffuse necking fall together to be seen i.e. in the modified maximum force criteria by Hora et al. [6] that extends the maximum force criteria to access localized necking strains.

Three kinds of results where extracted from each FEA simulation: strain paths at the punch nose in different layers of the sheet, major and minor strains along the ligament, also in different layers of the sample and forming limits of the center line when the local neck arises. Figure 3 shows the location of the three surfaces in the sheet from which the strain paths were extracted. Figure 4 shows the strain paths for different surfaces in the steel using two different punch radii. The surfaces are at the punch nose where strain localization would occur. The friction coefficient was 0.03. The touching surface is the surface that reaches the traditional FLC when the localization starts. In Figure 4(a) the touching surface is between the middle surface and the bottom surface and the top surface can accommodate strains well above the FLC before localization occurs. In contrast Figure 4(b) for a larger punch radius the touching surface is between the middle surface and the top surface and the top surface has less strain at localization as compared to the deformation with a 1 mm punch radius. The other results were used to obtain a better understanding of the plane in the sheet that reaches the forming limit;

IOP Conf. Series: Materials Science and Engineering **159** (2016) 012011 doi:10.1088/1757-899X/159/1/012011

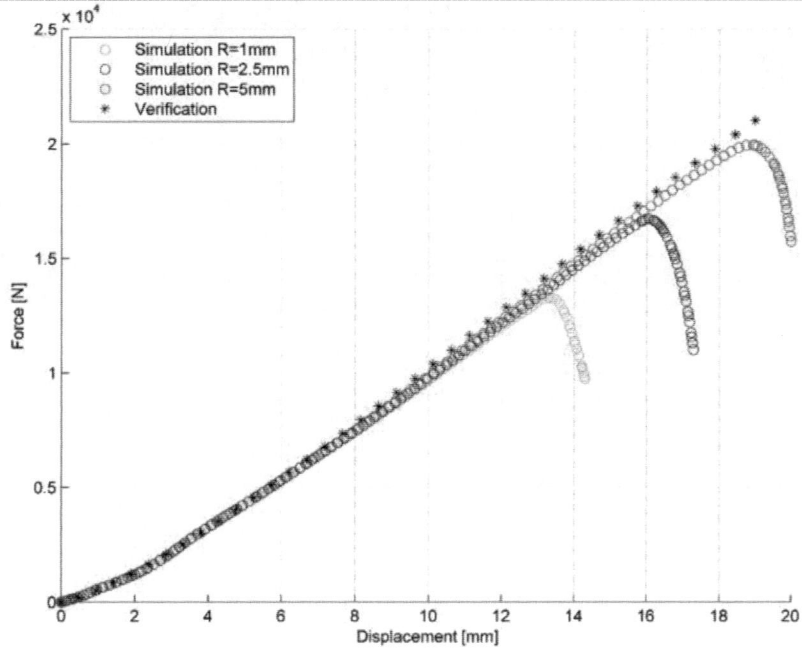

(a)

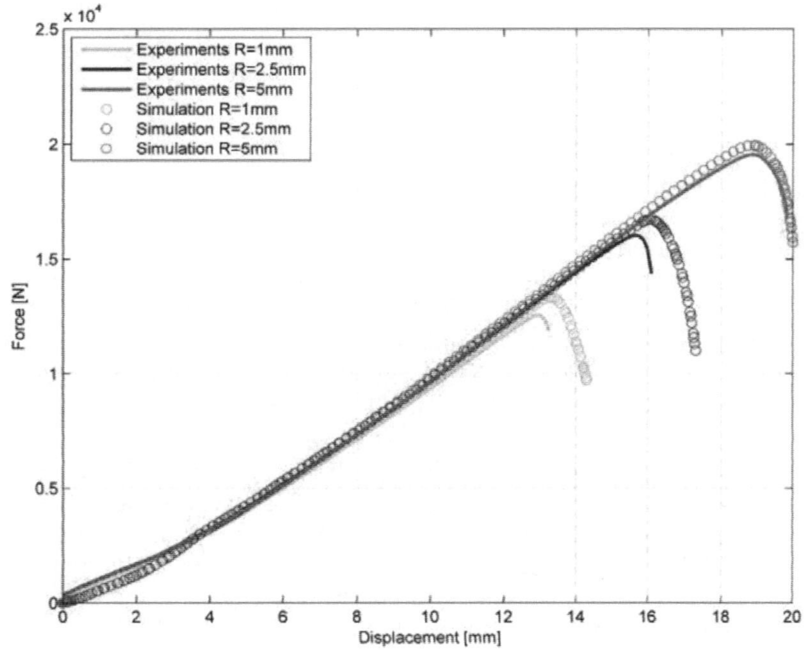

(b)

Figure 2. Verification of finite element model (a) with analytical solution, where the finite element results are shown by open circles and the analytical solution is shown by the black circles (b) with experiments, where the finite element results are shown by the open circles and the experimental results are the solid lines.

IOP Conf. Series: Materials Science and Engineering **159** (2016) 012011 doi:10.1088/1757-899X/159/1/012011

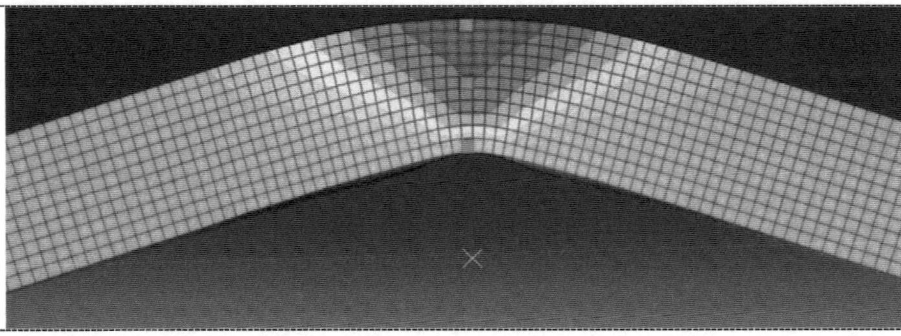

Figure 3. Top, middle and bottom surfaces where strains paths were extracted.

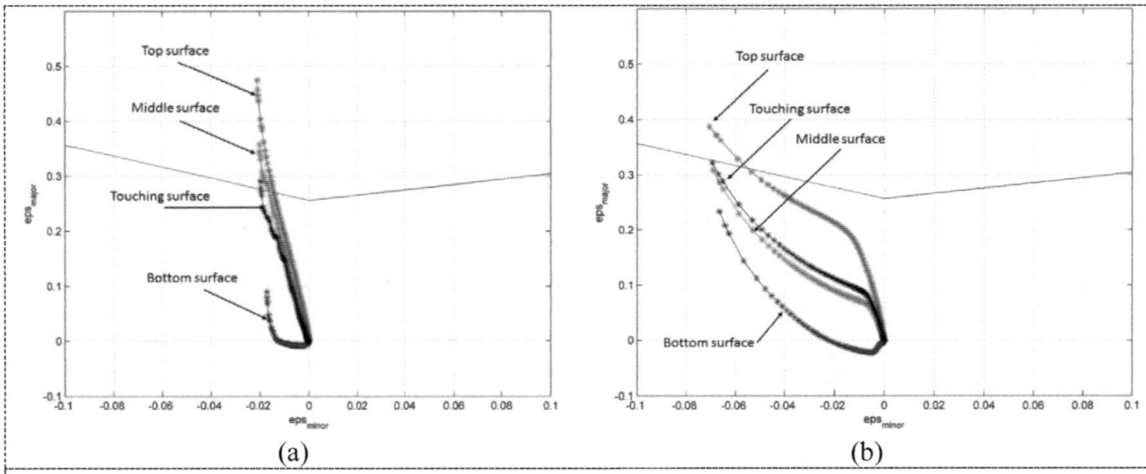

Figure 4. Strain paths for different sheet surfaces at the punch nose (a) with a tooling punch radius of 1 mm, (b) with a tooling punch radius of 5 mm. The top, middle and bottom surfaces of the sheet are tracked in these figures. The touching surface is the one that comes in contact with the FLC at the limit point during the test.

Figure 5 shows the sheet on the 2.5 mm punch at the last stable time increment in the simulation. As can be observed the strains are highly concentrated in the top region of the sheet. Localization of the deformation is observed by the slight valley that is seen in this top region.

Table 2 gives a summary of the position of the sheet plane that touches the FLC when localization occurs. The (–) / (+) means that the touching plane is slightly below/above the respective plane. At large radii and high friction there is sufficient restraint so that necking and a tensile failure occurs in the ligament region rather than on the punch nose. The bending enhanced formability occurs with small punch radii. The results are consistent with the concave-side rule of Tharrett and Stoughton [2] since they used small punch radius and no lubrication. The results in this study show that the concave side rule is valid for limited conditions.

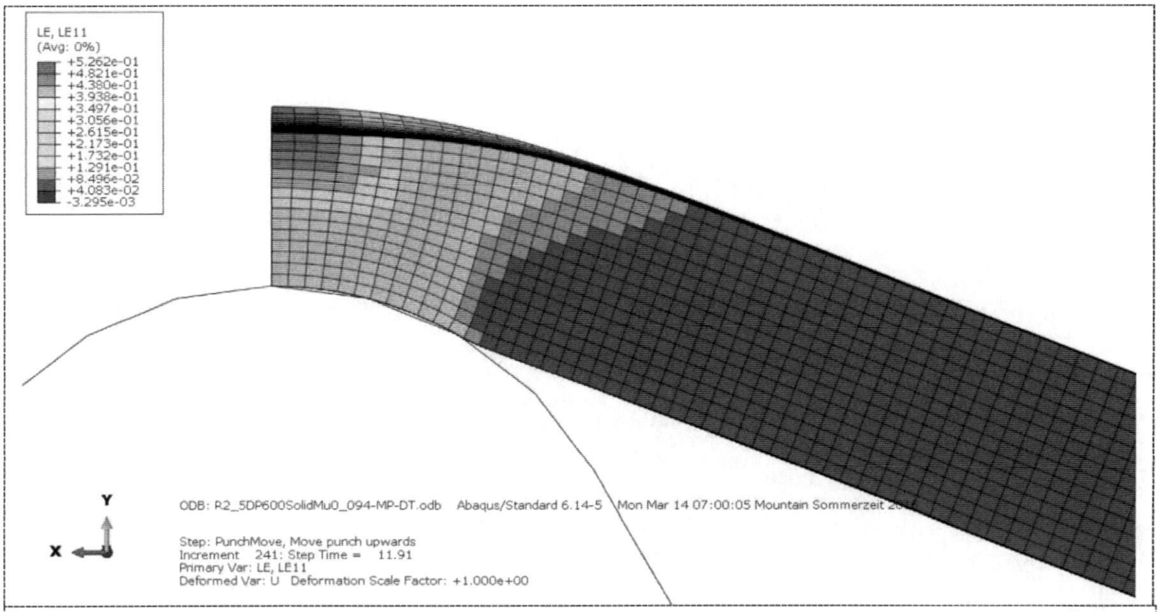

Figure 5. Sheet on the 2.5 mm punch radius at the last stable time increment. Localized necking has started

Table 2 – Summary of Touching Planes in Every Simulation

Punch Radius	μ = 0.00	μ = 0.03	μ = 0.07	μ = 0.15	μ = 0.20
R = 1 mm	Mid (-)	Mid (-)	Bottom (+)	Bottom (+)	Bottom (+)
R = 2.5 mm	Mid	Mid	Mid	Mid	Mid
R = 5 mm	Mid (-)	Mid (+)	Mid (+)	Mid (+)	NL
R = 10 mm	Top	Top	Top	NL	NL
R = 15 mm	Top	Top (+)	Top (+)	NL	NL
R = 20 mm	Top (+)	Top (+)	NL	NL	NL

NL = Necking in ligament

Although both the friction and bend radius influence the location of the failure, examination of Table 2 shows that the value of the coefficient of friction plays a small role in controlling the plane that reaches the formability limit. The punch radius has a more dominant role. It is also observed that the touching plane is lower in the sheet with the smaller punch radius, which indicates that the enhancement of formability due to bending is greater when there is a small radius involved with the bending. Because of the small radius and the very low friction, it is likely that this enhancement of formability could have some use in actual sheet forming operations. Nevertheless, a study, which provides a better quantification of the bending enhancement of formability, can be performed.

4. Summary

The results of the current study show that the plane within the sheet that reaches the traditional FLC can vary depending on the bending radius and the frictional conditions. The bending radius has a much greater effect as compared to friction. At small bending radii the plane that touches the FLC is pushed further from the top surface of the sheet toward the bottom. The concave-side rule of Tharrett and Stoughton occurs with small punch radius and high friction. The common assumptions that the forming limit occurs when the strains for the middle thickness plane of the sheet reach the forming limit curve or that the concave side rule is universally applicable need to be carefully and critically evaluated for the specific forming conditions.

Acknowledgements
We thank the Advanced Steel Processing and Products Research Center (ASPPRC) at Colorado School of Mines for partial support for this project

References

[1] Keeler S P and Brazier W G 1997 Relationship between Laboratory Material Characterization and Press-Shop Formability *Microalloying 75 Proc.* 517
[2] Tharrett M R and Stoughton T B 2003 Stretch-Bend Forming Limits of 1008 AK Steel *Soc. Automot. Eng.* 123
[3] Atzema E H, Fictorie E, van den Boogaard A H and J. M. M. Droog J M M 2010 The influence of curvature on FLC's of mild steel, (A)HSS and aluminium *Int. Deep Draw. Res. Group, IDDRG*
[4] Kitting D, Koplenig M, Ofenheimer A, Pauli H and Till E T 2009 Application of a 'concave-side rule' approach for assessing formability of stretch-bent steel sheets *Int. J. Mater. Form.* **2** 427
[5] Kitting D, Ofenheimer A, Pauli H and Till E T 2011 Experimental Characterization of Stretch-Bending Formability of AHSS Sheets," *AIP Conf. Proc.* **1353** 1589
[6] Hora P, Tong L and Berisha B 2013 Modified maximum force criterion, a model for the theoretical prediction of forming limit curves *Int. J. Mater. Form.* **6** 267

IOP Conf. Series: Materials Science and Engineering **159** (2016) 012012 doi:10.1088/1757-899X/159/1/012012

Investigation of the formability of aluminium alloys at elevated temperatures

M Tisza, D Budai, P Z Kovács and Zs Lukács

Institute of Materials Sciences and Technology,
University of Miskolc, H-3515 Miskolc, Hungary

E-mail: tisza.miklos@uni-miskolc.hu

Abstract. Aluminium alloys are more and more widely applied in car body manufacturing. Increasing the formability of aluminium alloys are one of the most relevant tasks in todays' research topics. In this paper, the focus will be on the investigation of the formability of aluminium alloys concerning those material grades that are more widely applied in the automotive industry including the 5xxx and 6xxx aluminium alloy series. Recently, besides the cold forming of aluminium sheets the forming of aluminium alloys at elevated temperatures became a hot research topic, too. In our experimental investigations, we mostly examined the EN AW 5754 and EN AW 6082 aluminium alloys at elevated temperatures. We analysed the effect of various material and process parameters (e.g. temperature, sheet thickness) on the formability of aluminium alloys with particular emphasis on the Forming Limit Diagrams at elevated temperatures in order to find the optimum forming conditions for these alloys.

1. Introduction

In the past few decades, many investigations were performed on hot forming of aluminium alloys. S. Mahabunphachai and M. Koc investigated the 5082 and 6061 alloys in a hydraulic bulge test at elevated temperatures. They found that the formability continuously increased from room temperature to 300°C [1]. In 2003, D. Li and K. Ghosh published their investigations on 5754, 5182 and 6111 alloys. The investigations were completed at temperatures from 200 to 350°C. The formability of the 5182 alloy has increased significantly from 200°C to 250°C Above this temperature range, the increase of formability was lower than before. At the 6111 alloy a small further increase was experienced in the 200°C to 350°C temperature range [2]. The formability of 5754 alloy was continuously increased up to 300°C, and even above, but the increase was much smaller than in the lower temperature ranges. P.J. Bolt, N.A.P.M. Lamboo and P.J.C.M. Rozier investigated the 5754 and 6016 aluminium alloys at elevated temperatures. The tests were performed from 100 to 250°C. They deep-drawn box-shaped parts, and measured the maximum product height. The 5754 alloy produced larger formability above 200°C, while the 6016 produced a moderate increase until 200°C. Above 200°C, the formability went down. It reveals the presence of an optimum forming temperature [3]. In 2014, M. Kumar, N Sotirov and C.M. Chimani published their investigations on warm forming of 7020 alloy sheet. They performed Cross-die tests at temperatures from 150°C to 250°C. They measured the depth of the specimen, and found that the depth increased highly from 150°C to 200°C, and above 200°C the gaining was much smaller than before [4].

As it can be seen, researchers have conducted their studies at laboratory environment for years, but no well-known technology has been developed for the industry. However, it is very important to transfer warm forming from laboratory to manufacturing plants.

2. Experimental investigations

2.1. Materials

The applied materials were the EN AW 5754 H22 and the EN AW 6082 T6. The EN AW 5754 (AlMg3) alloy is widely used in sports cars like Jaguar XK, Lotus Evora and Chevrolet Corvette. This alloy has medium strength among aluminium alloys. Its composition: 95,8% Al, 2,78% Mg, 0,29% Si, 0,36% Fe, 0,37% Mn. Temper: H22. Mechanical properties: $R_{p0,2}$=180 MPa; R_m=243 MPa; $A_=$17% (Alcoa).

The EN AW 6082 is commonly used for structural elements in most of luxury cars like BMW 6, BMW Z8, Ferrari 548, Jaguar XJ, Jaguar XK, Range Rover and Rolls-Royce Phantom. Its composition: 0,7-1,30% Si, 0,50% Fe, 0,10% Cu, 0,40-1,0% Mn, 0,6-1,2% Mg, 0,25% Cr, 0,20% Zn, 0,10% Ti. Mechanical properties: $R_{p0.2}$=260 MPa, R_m=310 MPa, A_{50mm}=8%.

2.2. Test specimens

During the experiments at different temperatures, the specimens had three different geometries with 1.0 mm thickness. The different geometries are determined in such a way to provide the desired FLC's special strain paths both in the tension and the compression zone. These specimen geometries can be seen in Figure 1. For the investigation of the effect of sheet thickness, we used sheets with 1.0, 1.2 and 2.0 mm thickness.

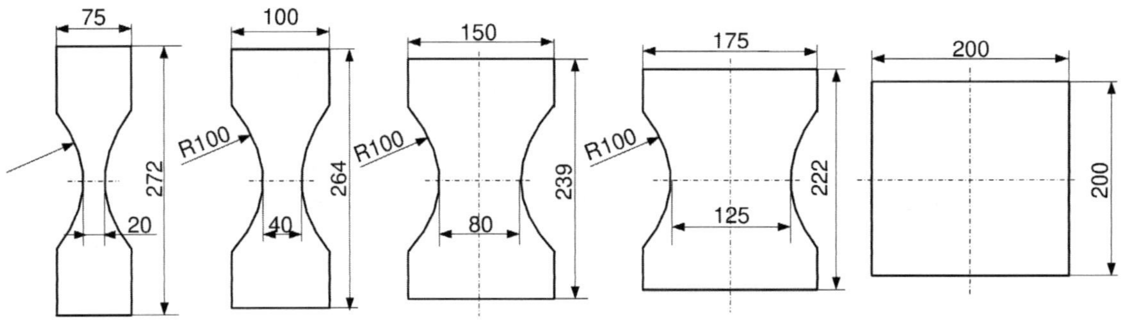

Figure 1. Test specimens used at FLD determination

2.3. Test equipment

In our tests, a computer controlled universal sheet formability-testing machine was used. This is able to perform various formability tests, e.g. Erichsen, Nakazima, Bulge, FLD, FLC tests. The equipment has electrohydraulic powertrain and it is suitable to test up to 3 mm maximum sheet thickness of steel, or 6 mm maximum thickness of aluminium. The forming tool geometry is a hemisphere with a diameter of 100 mm. The maximum load is F_{max}=600 kN, the speed interval of the tool is between v=0 mm/s and v=5 mm/s.

The universal formability-testing machine is equipped with an optical strain measurement system, which records the distortion of the grid, which is painted on the sheet before the tests. The system has 4 CCD cameras to obtain the 3D point cloud from the mesh. The universal formability-testing machine with the optical strain measurement system can be seen in Figure 2. From the measurements, the Vialux-AutoGrid software determines the deformation and strain distribution along the curved surface. With these data, the FLC and FLD can be determined. In our investigations we followed mainly the ISO 12004-2 instructions, to obtain the Forming Limit Diagrams of tested aluminium sheets.

As a difference from the ISO 12004-2 instructions it should be noted that during the investigations no lubrication was used, thus during the experiments at elevated temperatures we could provide similar friction condition between the tool and the sheet. Without lubrication, we could eliminate the

IOP Conf. Series: Materials Science and Engineering **159** (2016) 012012 doi:10.1088/1757-899X/159/1/012012

influence of the different amount and quality of the lubricants, because always the same materials were in contact. But obviously, we have to keep in mind that the lack of lubrication could cause failures by the high friction, and forming results in some cases may completely be dominated by friction and less by the material's behaviour itself. We planned to investigate the effect of dry friction also, with the aim to create a method for steel-based manufacturing to use their tools for aluminium stamping with additional thermo-cells and without lubrication. The lubrication always a hazardous material and using high temperatures like in our investigations, the lubrication oil or paste evaporates and smokes and we would like to avoid these hazardous effects. Next to the temperature the limit of the friction would also defines our process boundary.

Figure 2. Universal formability testing machine with the optical strain measurement system

3. Results

3.1. Experimental results with the EN AW 5754 aluminium alloy at different temperatures

During the experiments, the general conditions of industrial environment were considered. The tests were carried out besides room temperature at the following temperatures: 130°C, 200°C and 260°C. A laser thermometer measured the temperature of the sheets. The focus point was on the top of the dome. Due to the small thickness, the sheet reached the forming temperature within few seconds after closing the die. The speed of the forming tool was 0.5 mm/s in each measurement. No any lubricant was used, thus the lubrication condition was the same in each measurement.

From the experiments, the following main observations were made. The formability increased monotonically with the temperature until 200°C. Above this temperature, the decrease of the formability was observed, and at 260°C, it was even lower than at room temperature. The results show that the optimum forming temperature was closer to 200°C. It is worth mentioning that the formability at 200°C was twice as high as at room temperature. This is mainly due to the changes in the material properties and the microstructure evolution of the tested aluminium alloys.

This significant increase of dome height is well illustrated by the macroscopic figures of tested specimen (Figure 3.). The difference in the dome height of the specimens tested at room temperature and 200°C is very significant. In the tests where the specimens with the width of 80 mm were formed, the dome height was 21.75 mm at room temperature, and 55.82 mm at 200°C. The formability was the highest in this case. In the other cases, the dome height also increases until 200°C. Above this temperature, at 230°C small decrease observed, and above this temperature, the dome height was smaller than at room temperature in most cases. The specimen with the width of 200 mm was only the one which was a little bit higher than at room temperature. The dome height changes also indicates the formability, and in our cases when we did not use lubrication these are the most important characteristics, since the dome height means the maximal elongation of the sheet without lubrication.

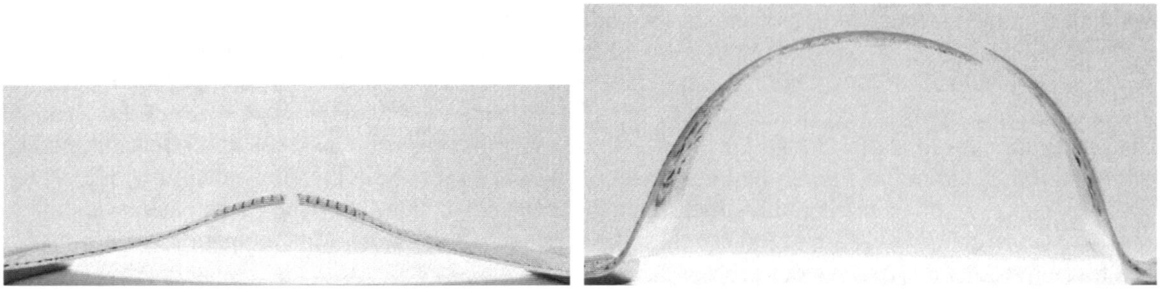

Specimen tested at room temperature Specimen tested at 200°C

Figure 3. Picture of specimens tested at various temperatures (Material: EN AW 5754)

Some results of dome test investigations are summarised in Figure 4. for the tested aluminium alloys at different temperatures. In the figure, the dome height changes are shown for three different specimen width, i.e. for b = 20, 80 and 200 mm. This diagram clearly demonstrates the changes of specimen dome height as an important formability characteristic at elevated temperatures.

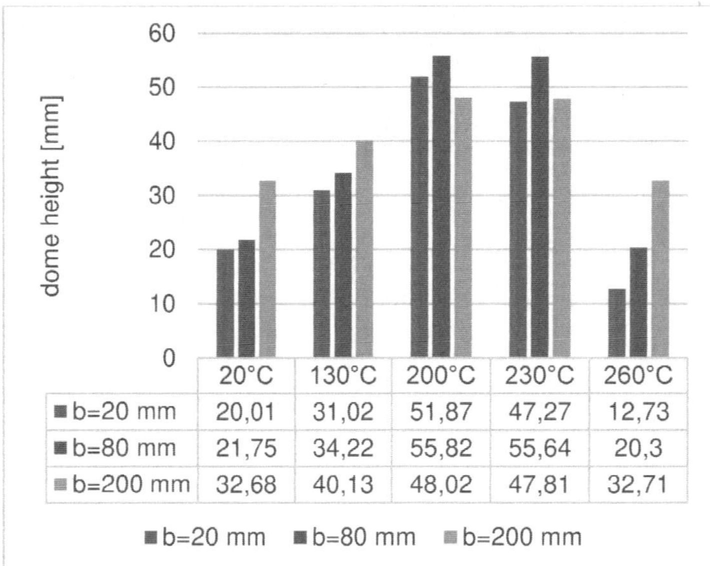

	20°C	130°C	200°C	230°C	260°C
b=20 mm	20,01	31,02	51,87	47,27	12,73
b=80 mm	21,75	34,22	55,82	55,64	20,3
b=200 mm	32,68	40,13	48,02	47,81	32,71

Figure 4. Dome height changes of the specimens after forming at elevated temperatures

After testing all the specimens at the selected various temperatures (T = 20 °C, 130 °C, 200 °C and 260 °C) the Forming Limit Curves were determined at each temperature using the AutoGrid software. These results are shown in Figure 5. for the material grade EN AW 5754. In this figure, the significant increase of the formability up to 200 °C can be well seen. We can also observe in this figure that there is a sharp decrease in the formability at 260 °C. The Forming Limit Curve at 260 °C falls even below of the FLC measured at room temperature.

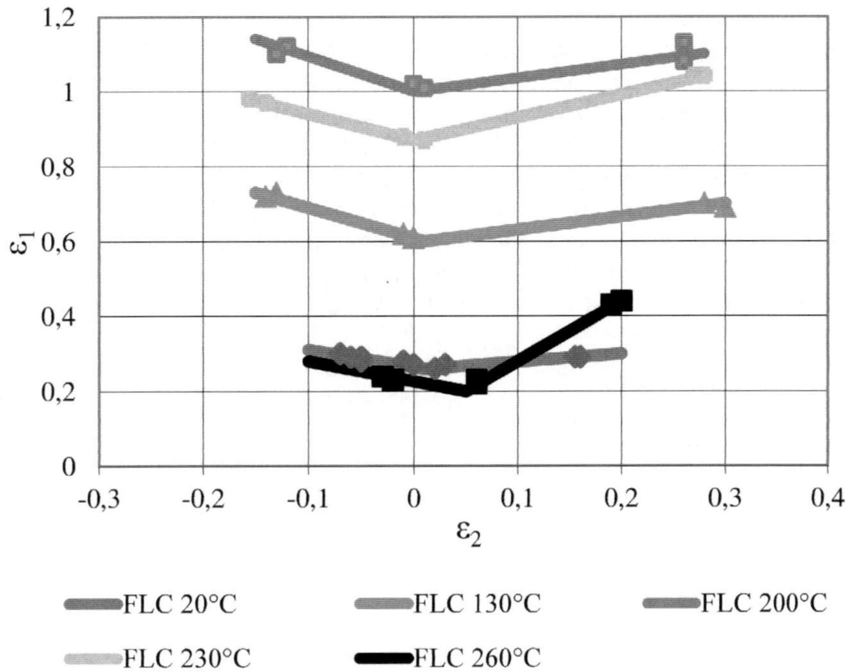

Figure 5. FLC curves for the EN AW 5754 aluminium alloy at different temperatures (sheet thickness t = 1.00 mm)

In the microscopic pictures, (see Figure 6.) it can be also seen that the grain structure at 260°C at this temperature is significantly different from the grain structure obtained at 200°C, which contains more uniform, nearly equiaxed grains.

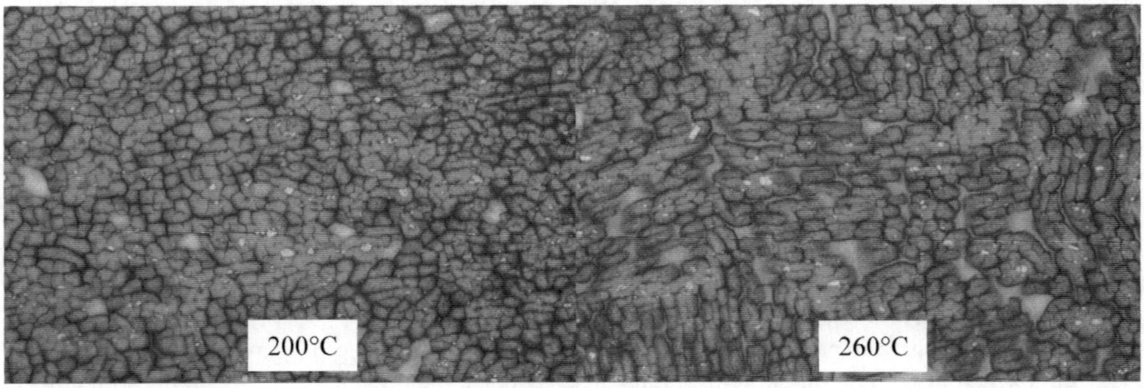

Figure 6. Microstructure at 200°C and above the optimal forming temperature (at 260°C)

IOP Conf. Series: Materials Science and Engineering **159** (2016) 012012 doi:10.1088/1757-899X/159/1/012012

3.2. Experimental results with the EN AW 6082 aluminium alloy at different temperatures

The EN AW 6082 aluminium alloy has higher strength than the EN AW 5754 one and as a consequence it has lower formability. That is the main reason why car manufacturers use this alloy for structural elements. We performed the tests on this alloy at the same temperatures as for the EN AW 5754 alloy, i.e. at room temperature, 130°C, 200°C and 260°C. The increase of the formability was smaller than for the EN AW 5754 alloy – as it was expected – however, the trend of changes was very similar. When we increased the temperature until 200°C, the forming limit curves are shifted up. At 260°C it has gone down to the level near to the values measured at room temperature. This difference is clearly revealed in the FLC curves determined for the various temperature values. The FLC curves for the EN AW 6082 aluminium alloy determined at room temperatures and at various elevated temperatures are shown in Figure 7.

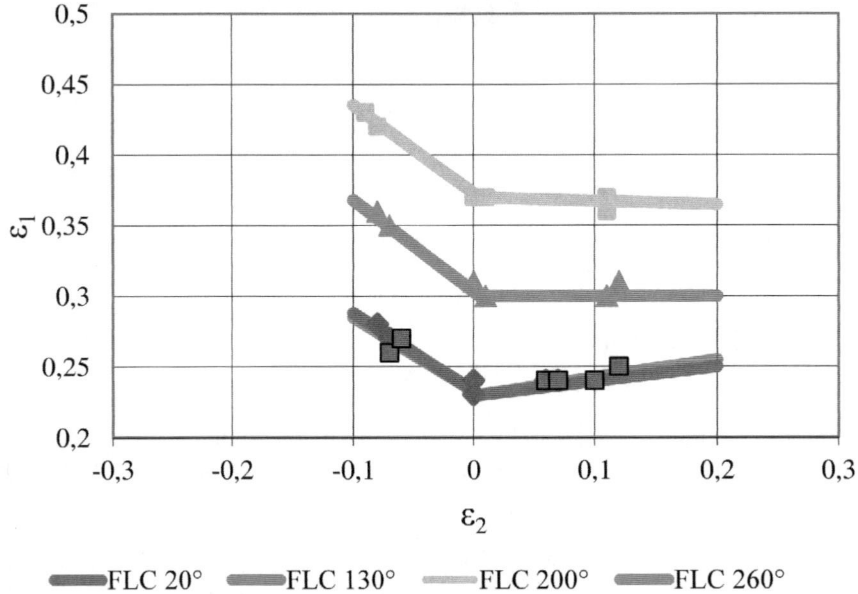

Figure 7. FLC curves for the EN AW 6082 aluminium alloy at different temperatures (sheet thickness t = 1.00 mm)

3.3. Experimental results of EN AW 5754 aluminium alloy with different sheet thicknesses

We also investigated the FLC curves for different sheet thicknesses. From these experiments, we can state that the formability of the EN AW 5754 alloy is continuously increased with the increasing sheet thickness. Concerning the formability we found that it was the best with 2.0 mm thickness and it has significantly lower values with 1.0 mm sheet thickness.

On the basis of these experimental results, we can state that the sheet thickness may significantly affect the formability of aluminium alloys, thus if we increase the thickness of a sheet, the formability will increase with it. These results are shown for the EN AW 5754 aluminium alloy grade in **Figure 8**. for three different sheet thickness. The results are similar in case of EN AW 5754 aluminium alloy with these thickness values.

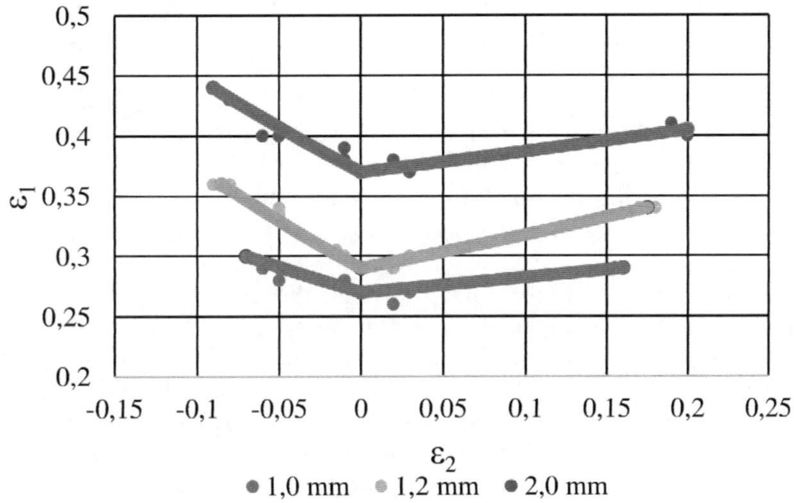

Figure 8. FLC curves belonging to different sheet thicknesses for EN AW 5754 material grade at room temperature

4. Summary and conclusions

In this paper, the formability of two different aluminium alloys (EN AW 5754 and EN AW 6082) was investigated at elevated temperatures. Obviously, the formability of any materials is a quite complex phenomenon influenced by several parameters, besides the material grade like strain path, deformation history, strain-rate effects and failure mode, etc. It cannot be described just selecting one or two of these manifold formability indicators. However, within a limited volume of a paper we have to focus on some of these parameters.

In this paper, we mainly analysed the temperature dependence of aluminium alloys focusing first of all on the Formability Limit Diagram (FLD) and the changes of the Limit Dome Height (LDH) values. The selection of these two parameters may be reasoned by several facts. Using the FLD as a formability issue may be explained by the fact that all the numerical modelling programs that are more and more widely used not only in the academic research but in the everyday industrial practice as well, are based among others on the Formability Limit Diagrams. The selection of Limit Dome Height (LDH) as another formability indicator is also widely used in sheet metal forming mainly due to its simplicity of measurement techniques. These are the main reasons that in our investigations we applied these two formability indicators.

The formability investigations were performed by determining the Forming Limit Diagrams and the Limit Dome Heights for these alloys at various elevated temperatures. For the determination of Forming Limit Diagrams, we applied a universal formability-testing machine equipped with an optical strain measurement system. During the experiments we used the modified Nakajima specimens to be able to determine the whole range of the FLDs (both in the tension and compression side) with relatively small number of experiments.

The FLD tests were performed without lubrication to ensure the same lubrication and friction condition between the surfaces and furthermore to create a dry and more environment-friendly stamping process for the industry. It results in a slight difference concerning the relating ISO 12004-2 standard in the conditions of FLC investigations, but the results can be useful to define the boundaries in dry-hot stamping technology, too.

IOP Conf. Series: Materials Science and Engineering **159** (2016) 012012 doi:10.1088/1757-899X/159/1/012012

It is well known that by increasing the forming temperature the formability is usually increasing. Studying the effect of the temperature on the formability of aluminium alloys, we found that for the investigated aluminium alloys (EN AW 5754 and EN AW 6082) an optimum forming temperature exists where the formability has a maximum value. To find these optimum forming temperatures for the investigated aluminium alloys, we performed large scale formability investigations from room temperature up to the temperature range 130 to 260°C.

We could identify that at the optimum forming temperature, the FLC curves are shifted up significantly along the ε_1 major strain axis (the maximum values of major strain components were about three times higher compared to the room temperature values).

Similar experimental evidences were found analysing the changes of the Limit Dome Height values. The difference in the dome height of the specimens tested at room temperature and 200°C is very significant. In the tests where the specimens with the width of 80 mm were formed, the dome height was more than twice high at 200°C for example for the alloy EN AW 5754 (21.75 mm at room temperature, and 55.82 mm at 200°C). The formability was the highest in this case. In the other cases, the dome height also increases until 200°C. Above this temperature, at 230°C small decrease observed, and above this temperature – e.g. at around 260°C, the dome height was even smaller than at room temperature in most cases.

It was experienced both in the FLD and LDH investigations that over the optimum forming temperature a significant decrease of the formability occurs. Therefore, to find and to hold the optimum forming temperature during the forming processes is of utmost importance. In our investigations, we found that the optimum forming temperature of both investigated alloys (both EN AW 5754 and EN AW 6082) is slightly above 200°C and a sharp deterioration occurs above this optimum value.

In these experiments, we also studied the effect of the material thickness. It was experienced that the formability was increased by the increase of sheet thickness.

References

[1] Mahabunphachai S, Koc M 2010, Investigations on forming of aluminium 5052 and 6061 sheet alloys at warm temperatures, *Materials and Design* **31**, 2422

[2] Li D, Ghosh K, 2004, Biaxial warm forming behaviour of aluminium sheet alloys, *Journal of Materials Processing Technology* **145**, 281

[3] Bolt P J, Lamboo N 2001, Feasibility of warm drawing of aluminium products, *Journal of Materials Processing Technology,* **115**, 118

[4] Kumar M, Sotirov N, Chimani C 2014, Investigations on warm forming of AW-7020-T6 alloy sheet, *Journal of Materials Processing Technology,* **214**, 1769

IOP Conf. Series: Materials Science and Engineering **159** (2016) 012013 doi:10.1088/1757-899X/159/1/012013

Damage characterization of high-strength multiphase steels

S Heibel[1], W Nester[1], T Clausmeyer[2] and A E Tekkaya[2]

[1] Daimler AG, Benzstraße, 71059 Sindelfingen, Germany
[2] Institute of Forming Technology and Lightweight Design, TU Dortmund University, Baroper Str. 303, 44227 Dortmund, Germany

E-mail: sebastian.heibel@daimler.com

Abstract. High-strength steels show an entirely different material behavior than conventional deep-drawing steels. This fact is caused among others by the multiphase nature of their structure. The Forming Limit Diagram as the classic failure criterion in forming simulation is only partially suitable for this class of steels. An improvement of the failure prediction can be obtained by using damage mechanics. Therefore, an exact knowledge of the material-specific damage is essential for the application of various damage models. In this paper the results of microstructure analysis of a dual-phase steel and a complex-phase steel with a tensile strength of 1000 MPa are shown comparatively at various stress conditions. The objective is to characterize the basic damage mechanisms and based on this to assess the crack sensitivity of both steels. First a structural analysis with regard to non-metallic inclusions, the microstructural morphology, phase identification and the difference in microhardness between the structural phases is carried out. Subsequently, the development of the microstructure at different stress states between uniaxial and biaxial tension is examined. The damage behavior is characterized and quantified by the increase in void density, void size and the quantity of voids. The dominant damage mechanism of the dual-phase steel is the void initiation at phase boundaries, within harder structural phases and at inclusions. In contrast the complex-phase steel shows a significant growth of a smaller amount of voids which initiate only at inclusions. To quantify the damage tolerance and the susceptibility of cracking the criterion of the fracture forming limit line (FFL) is used. The respective statements are supported by results of investigations regarding the edge-crack sensitivity.

1. Introduction

The classic tool for failure prediction in metal forming is the forming limit diagram. With this tool the beginning of localized necking can be determined. In nowadays car body design less ductile, high strength materials are used for safety-relevant structural components. For these materials the use of the forming limit diagram is considered to be problematic: For example the failure in the plane-strain region is predicted too early with the forming limit curve, thus the material capacity is underestimated [1]. Another example is the inability of displaying edge cracks correctly. This can be remedied by damage mechanics. Micromechanical motivated damage models for example try to model the damage development that occurs with ductile fracture from void nucleation, growth and coalescence to the initiations of microcracks [2]. These effects are particularly favored by the multiphase nature of high strength steels as has been shown in various studies for dual phase steels [3], [4]. The different formability of the respective phases leads to an inhomogeneous distribution of deformation. Voids develop on the boundary of different phases or within the more brittle phase. If the martensite is finely dispersed in the microstructure, more ferrite is deformed and there are fewer nucleation sites for damage initiation. In [3] it was found that inclusions with an elongated shape rather break, whereas globulistic forms dissolve from the matrix. Furthermore, it was shown that the porosity increases with increasing hardness gradient within a material. In addition, the band-type formation of a second phase favors the void nucleation. Larger martensite clusters can lead to significant damage accumulation, since only a small proportion of ferrite may be deformed. Referring to different stress states Tasan et al. could prove that increasing stress triaxiality leads to an earlier onset of void nucleation and equally increases the void density in dual-phase steels [4]. The mentioned triaxiality is often used for the description of the void growth in damage mechanics [5]. Based on this and taking into account the independence of the thickness of the fracture surface from stress state and stress history Martins et al. developed the concept of fracture forming limit curves [6], [7]. Because of the constant critical thickness reduction the critical damage parameter derived from the fracture forming line can be considered as a material parameter. This parameter is used here to quantify the damage tolerance. In addition to the low damage tolerance the edge-crack sensitivity could be a problem by the application of dual-phase steels in industrial practice. By shear cutting operations during manufacturing the formability at the shear-cut-edge is reduced. The reasons for this purpose lie like the causes for damage evolution in the microstructure of the material. A small grain size, a homogeneous structure, the absence of microstructure stringers or bands as well as the minimization of inclusions have a positive effect on the edge-crack sensitivity. To evaluate the edge-crack sensitivity the edge-fracture-tensile-test developed by Feistle et al. shows promising results [8]. With the help of this test, the statements on damage tolerance should be supported in this work.

2. Materials

Dual- and complex-phase steels exhibit a completely different macroscopic material behavior because of their different microstructural characteristics. Dual-phase steels have traditionally a ferritic matrix containing martensite at the grain boundaries, whereas complex-phase steels are characterized by a bainitic microstructure. Other structural phases in a conventional complex-phase steel are ferrite and a smaller amount of martensite. Table 1 shows the volume fraction of phases of the examined materials with a nominal sheet thickness of 1 mm.

Table 1. Volume fraction of phases

	DP1000	CP1000
Ferrite	27.3 %	4.6 %
Bainite/ tempered martensite	59.5 %	93.5 %
Martensite	9.5 %	0.5 %
Retained austenite	2.3 %	0.3 %

The dominant phases of the dual-phase steel (DP1000) are ferrite and bainite. In addition to martensite a small amount of retained austenite is included in the structure. The structure of the complex-phase steel (CP1000) with the volume fractions shown in table 1 corresponds to the classical composition. According to the identified microstructural phases, the dual-phase steel exhibit with 546 HV a significantly greater difference in microhardness than the complex-phase steel with 385 HV. In figure 1, the very fine-grained microstructure of both materials is illustrated by the nital etching in longitudinal section.

For the dual-phase steel a significant amount of stringers and bands could be seen. The phases of the complex-phase steel are very homogeneously distributed, which confirms the right illustration of figure 1. In addition a larger number of titanium carbides can be determined by an energy-dispersive X-ray spectroscopy (EDX) for the complex-phase steel. These brittle inclusions provide a nucleation site to damage development.

IOP Conf. Series: Materials Science and Engineering **159** (2016) 012013 doi:10.1088/1757-899X/159/1/012013

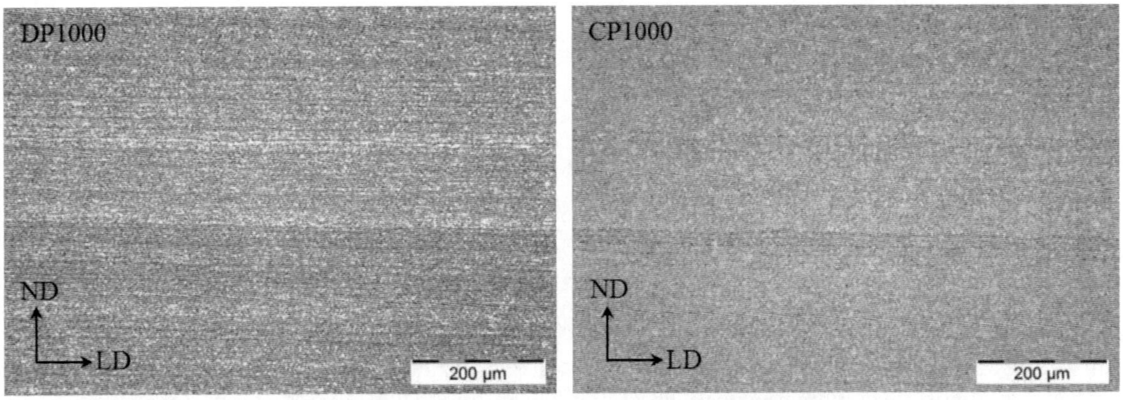

Figure 1. Microstructure after nital etching

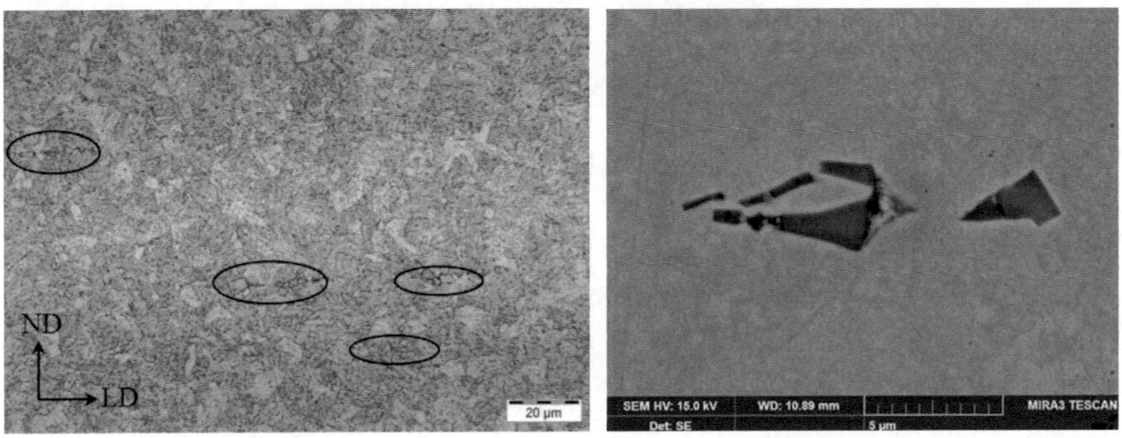

Figure 2. Titanium carbides in the microstructure of CP1000

Besides of the titanium carbides aluminum oxide inclusions can be identified in the structure. These aluminum oxides are likewise present in the dual-phase steel. Moreover softer manganese sulfides can be found here. Table 2 summarizes the mechanical properties of these two steels. The lower yield ratio results from the microstructural composition of the DP1000 and thus results in a relatively high global formability at a strength of 1000 MPa.

Table 2. Mechanical Properties (longitudinal direction)

Material	$R_{p0,2}$ [MPa]	R_m [MPa]	$R_{p0,2}/R_m$ [-]	A_G [%]	A_{80} [%]	n [-]	r_0 [-]	r_{45} [-]	r_{90} [-]	r_m [-]
DP1000	675	1035	0.652	9	13	0.129	0.715	1.088	0.918	0.952
CP1000	903	1012	0.892	5	7	0.053	0.990	1.235	1.080	1.135

The bainitic microstructure of the CP1000 leads, among others due to the smaller difference in hardness between the phases, to an excellent local formability. Both materials have no pronounced normal anisotropy.

3. Experimental methodology

To characterize the damage behavior at different stress states four specimens are used as shown in Figure 3 and 4. For the uniaxial tension (UAT) a classical tensile test specimen according to EN ISO 6892-1 and a smaller one in size is used. The stress state of the plane-strain tension (PST) is obtained by a notched tensile specimen with a notch radius of 4 mm.

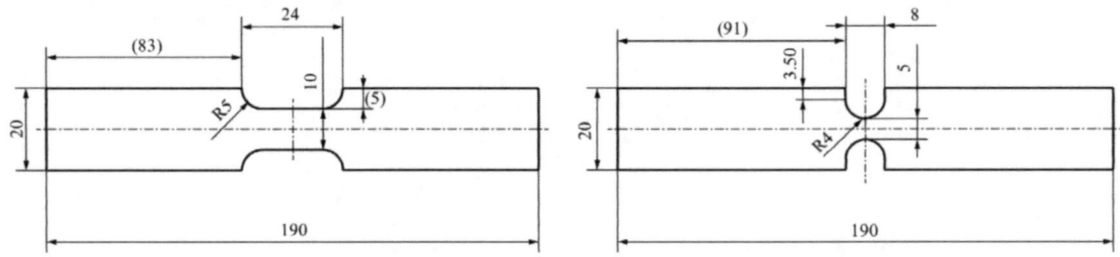

Figure 3. Technical drawing of the smaller in size uniaxial tension specimen (left) and of the plane-strain tension specimen (right)

With regard to damage characterization in the equi-biaxial tensile region (BAT) a cruxiform specimen is unsuitable due to their inhomogeneous strain and stress field. Instead a Nakajima-geometry with a diameter of 200 mm according to EN ISO 12004-2 is used.

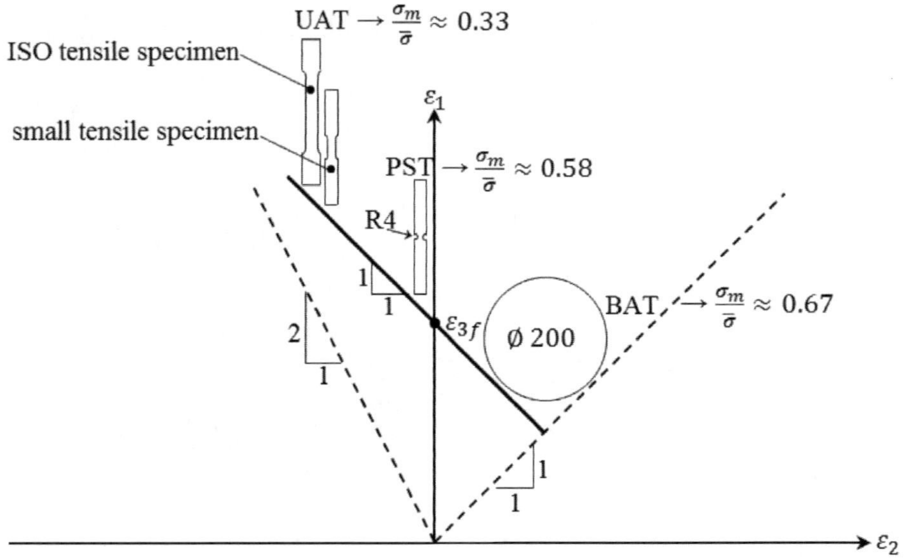

Figure 4. Concept of fracture forming line (FFL) and used specimens [6], [7]

After performing the appropriate quasi-static tensile tests in the longitudinal direction up to fracture, they are stopped at the crosshead displacement of occurrence of maximum force, F_{max}, as well as at 50%, 75%, 85%, 90% and 95% of crosshead displacement at fracture. Instead of test termination at F_{max} the Nakajima-test is stopped at a stamp displacement of 99%. For each stress state and each load level the microstructure is examined with the help of light microscopy, scanning electron microscopy and EDX. To determine the FFL, the thickness reduction is measured at the fractured samples in order to calculate the third principal strain at fracture ε_{3f}. Due to the independence of ε_{3f} of the stress state and the loading history the FFL has a slope of -1 [6], [7]:

$$\varepsilon_{1f} + \varepsilon_{2f} = -\varepsilon_{3f} \approx \text{const.} \tag{1}$$

To quantify the damage tolerance the damage parameter D_{crit} is used [6], [7]:

$$D_{crit} = \int_0^{\bar{\varepsilon}_f} \frac{\sigma_m}{\bar{\sigma}} d\bar{\varepsilon} = -\frac{(1+r_m)}{3}\varepsilon_{3f} \tag{2}$$

This characteristic value is motivated by the damage mechanics in the form of effective strain fracture criteria and its weighting function, the stress triaxiality $\sigma_m/\bar{\sigma}$, which is defined as the ratio of the hydrostatic stress σ_m and the effective stress $\bar{\sigma}$. Further, the anisotropic yield criterion Hill48 was used for the definition by Martins [6], [7]. To support the statement of damage tolerance with edge-crack sensitivity results the edge-fracture-tensile-test developed and executed at the Institute of Metal Forming and Casting (utg) is used for both materials. For this purpose a tensile test (UAT) is performed with both edges milled as wells as one side milled and one shear cut edge. The sample with both edges milled shows a classic ductile failure in the center of the specimen whereas the formability of the shear cut edge is reduced by the stamping process and an edge crack is initiated. This reduced formability is recorded with a digital image correlation system [8].

4. Results and discussion

Firstly the damage initiation mechanisms of both materials are worked out. Based on this the damage evolution is represented comparatively in different stress states. Finally, the damage tolerance of the examined materials is characterized.

4.1. Damage initiation

In the examined DP1000 several damage nucleation sites can be found. In figure 5 the void nucleation by fracture of a manganese sulfide is illustrated.

Figure 5. Damage initiation DP1000

Furthermore, voids are formed by decohesion of the matrix from aluminum oxide inclusions. As can be seen in the right illustration of figure 5, the voids also nucleate at the phase boundaries between martensite and ferrite. This effect could be favored by the difference in hardness between the two phases. Due to the brittleness of the martensite, fracture occurs on the very same. These observations correspond to the work of Tasan et al., who showed for a dual-phase steel damage initiation by particle-matrix decohesion, particle cracking, martensite fracture and phase boundary decohesion [4]. The complex-phase steel hardly shows the latter effects of damage initiation as result of the multiphase nature of high-strength steels.

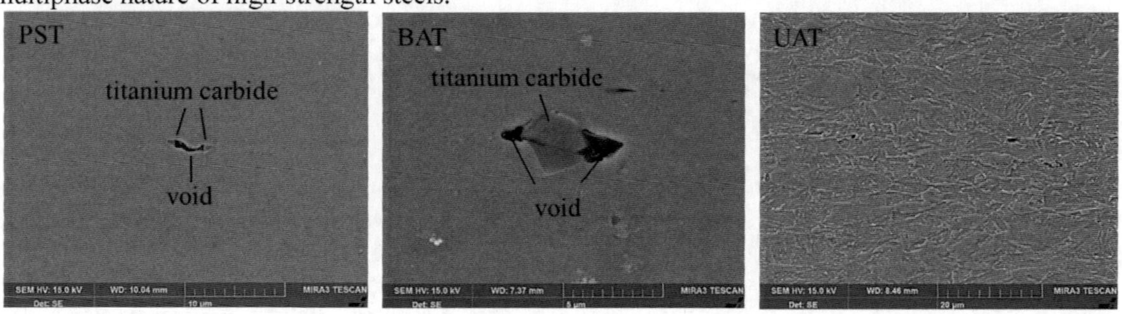

Figure 6. Damage initiation CP1000

The damage initiation is clearly visible here only at inclusions. In figure 6 the void initiation is shown by fractured, brittle titanium carbides. In addition, it comes to decohesion from the matrix. The common occurrence of particle cracking and particle-matrix decohesion could be explained by the high stress triaxiality in the region of BAT. Furthermore, damage initiation for the complex-phase steel is also recorded at aluminum oxide inclusions.

4.2. Damage mechanisms

To assess the damage mechanisms, the porosity, the number of voids and their maximum size are determined by gray-scale analysis in light microscope images of longitudinal cuts of the specimens in a representative region of interest at the respective stress state and load level. Thereby the minimum detectable void size is about 0.305 µm. Table 3 summarizes the results. The dual-phase steel shows an increase in damage accumulation in terms of porosity from UAT through the PST to BAT. Both, the measured quantity as well as the maximum size of the voids increases with rising second principal strain. The complex-phase steel shows a different behavior. No differences between the stress states can be observed. For the UAT the porosity is nearly at a similar level as in the dual-phase steel and therefore lower in the investigated stress states with higher stress triaxiality. In both materials the damage accumulation increases accompanied by a sharply increasing localization.

Table 3. Porosity, size of voids, quantity of voids

		DP1000				CP1000		
		Porosity [%]	Quantity	Max. size [µm]		Porosity [%]	Quantity	Max.size [µm]
	F_{max}	-	-	-	F_{max}	-	-	-
	50 %	-	-	-	50 %	-	-	-
	75 %	< 0.01	19	3.12	75 %	-	-	-
UAT	85 %	0.010	53	5.09	85 %	0.01	19	3.48
	90 %	0.067	147	5.85	90 %	0.034	18	14.09
	95 %	0.095	172	6.78	95 %	0.049	44	14.12
	fracture	0.098	159	7.33	fracture	0.108	43	15.17
	F_{max}	0.017	38	4.3	F_{max}	< 0.01	12	3.7
	50 %	-	-	-	50 %	-	-	-
	75 %	0.030	35	5.09	75 %	< 0.01	16	3.96
PST	85 %	0.053	53	6.14	85 %	0.028	25	8.52
	90 %	0.065	80	6.99	90 %	0.056	35	14.11
	95 %	0.138	142	7.08	95 %	0.090	7	11.66
	fracture	0.172	239	9.2	fracture	0.116	49	15.54
	50 %	0.040	4	7.85	50 %	0.010	7	3.45
	75 %	0.050	5	9.14	75 %	0.030	24	3.96
	85 %	0.120	15	8.13	85 %	0.040	16	4.04
BAT	90 %	0.150	199	6.73	90 %	0.060	12	9.44
	95 %	0.160	175	9.8	95 %	0.080	21	11.87
	99 %	0.320	234	13.28	99 %	0.140	35	15.74
	fracture	0.413	486	14.01	fracture	0.153	41	16.7

Figure 7 shows a fractured Nakajima specimen of DP1000. Compared to CP1000 in Figure 8, one recognizes the significantly larger number of voids. Furthermore, the effect of void coalescence can clearly be seen in Figure 7.

Larger voids can be detected in Figure 8 for the CP steel. Besides a micrograph of notched tensile specimen at the load case of 95 % is shown in Figure 8. The significant localization of the material can be seen here. This is accompanied by the development of a large cavity in the interior of the sample. This cavity develops over several load steps and does not lead to sudden failure. The material behaves in a certain way "tough", hence highly damage tolerant.

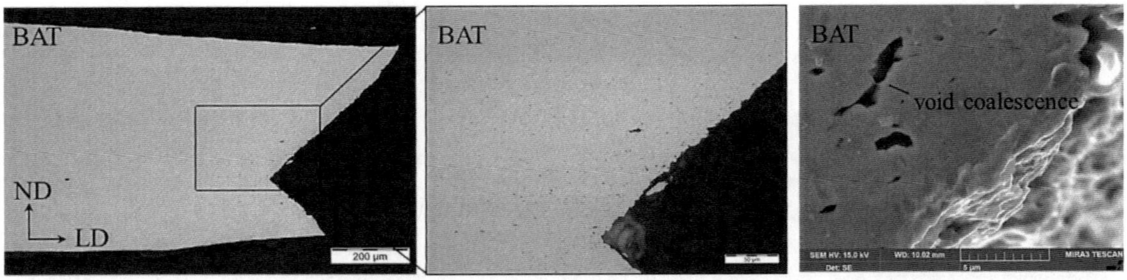

Figure 7. Damage mechanisms DP1000

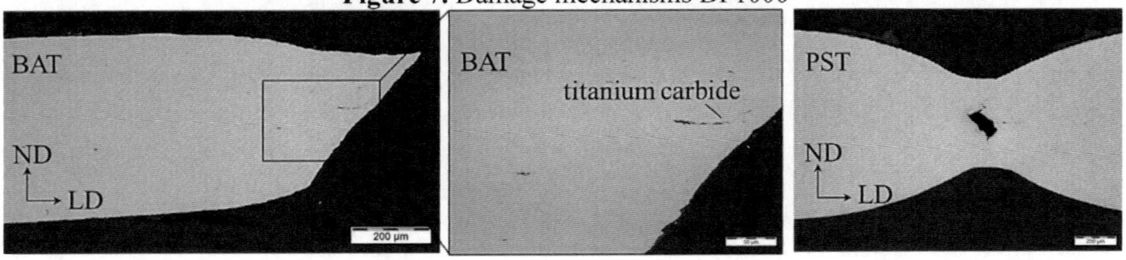

Figure 8. Damage mechanisms CP1000

4.3. Damage tolerance

In the previous sections it was shown that both materials run through the classic stages of ductile damage evolution. Since the concept of FFL is damage mechanically motivated the fracture forming line can be created for both materials according to [6]. Therefore the third principal strain at fracture ε_{3f} must be determined by measuring the fracture thickness for all specimens. The third principal strain at fracture for the DP1000 is -0.514 with a standard deviation of 0.082. ε_{3f} of the CP1000 is amounted -1.114 with a standard deviation of 0.021. The second principal strain is determined via an Aramis 5M digital image correlation system with a strain reference length of 0.5 mm and a frame rate of 2 Hz. Using this second principal strain and the assumption of constant volume, the fracture forming line shown in Figure 9 can be constructed. The slopes of approximately -0.96 for the CP1000 is in accordance with the in [6] developed theory. The slope of the dual-phase steel lies in good agreement with the in [6] and [7] investigated aluminum alloy AA1050-H111.

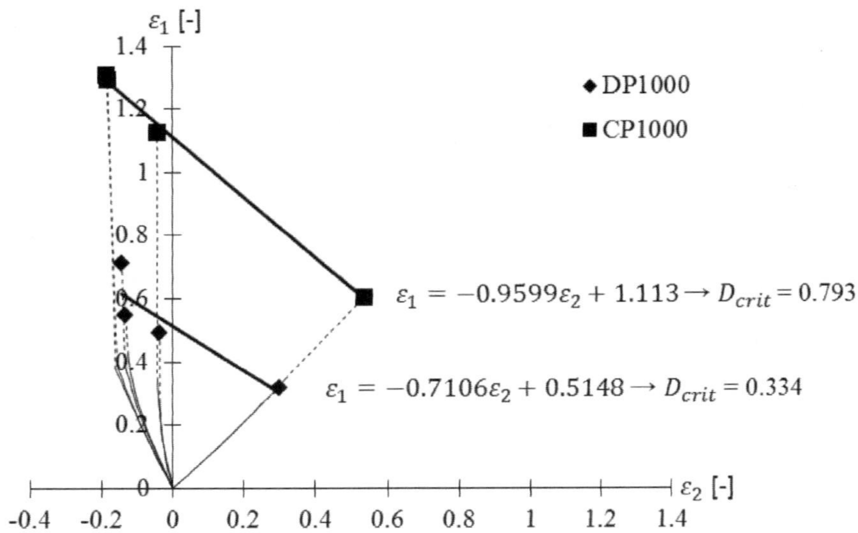

Figure 9. Forming fracture lines with respective loading paths for DP1000 and CP1000

With the third principal strain at fracture and the mean normal anisotropy the critical damage parameter can be determined with equation (2). D_{crit} is determined as 0.793 for the more damage tolerant complex-phase steel more than twice as large as for the dual-phase steel. These results are supported by the results of the edge-fracture-tensile-test [8]. As in table 4 shown for the complex-phase virtually no reduction in the first principal strain can be observed, whereas the dual-phase steel shows significant edge-crack sensitivity with a reduction of first principal strain at fracture of 34%.

Table 4. Results of edge-fracture-tensile-test

	DP1000	CP1000
ε_{1f} at milled cut edge	0.415	0.478
ε_{1f} at milled cut + shear cut edge	0.278	0.453
reduction of ε_{1f}	34 %	5 %

5. Conclusion

The basic damage mechanisms were characterized for both materials. The dual-phase steel as well as the complex-phase steel show regardless of the stress state the mechanisms of ductile damage, namely void initiation, void growth and void coalescence. The voids are nucleated in the dual-phase steel by particle-matrix decohesion, particle cracking, martensite fracture and phase boundary decohesion. Due to the homogeneous, fine-grained bainitic microstructure and the associated low difference in hardness the complex-phase steel shows no damage initiation due to its multiphase nature. Furthermore, the damage accumulation in terms of porosity is independent of the stress state for the complex-phase steel. For the dual-phase steel the damage accumulation increases with increasing second principal strain. For both materials the theory of FFL was confirmed. Using the critical damage parameter, the complex-phase steel could be characterized as damage tolerant and less susceptible to cracking. This statement was supported by the studies on the edge-crack sensitivity for both materials.

Acknowledgements

The authors gratefully acknowledge the experimental support of Martin Kieslich during his employment for the Daimler AG.

References

[1] Till E T, Berger E and Larour P 2008 On an exceptional forming behaviour aspect of ahhs sheets *IDDRG 2008 Int. Conf. (Olofström)* pp 429-440

[2] Gurson A 1977 Continuum theory of ductile rupture by void nucleation and growth: part I-yield criteria and flow rules for porous ductile media *J. Eng. Mat. Tec.* **99** pp 2-15

[3] Pichler A, Tragl E, Kurz T, Werner E, Tsipouridis P, Haglund A, Zajac J, Artimez J M and Ferreno J 2009 *DP grades with improved formability* (Brussels: European Comission Research Fund for Coal and Steels Unit)

[4] Tasan C C, Hoefnagels J P M, ten Horn C H I J and Geers M G D 2009 Experimental analysis of strain path dependent ductile damage mechanics and forming limits *Mech. Mat.* **41** pp 1264-76

[5] McClintock F A 1968 A criterion for ductile fracture by the growth of holes *J. Appl. Mech.* 35 pp 343-71

[6] Martins P A F, Bay N, Tekkaya A E and Atkins A G 2014 Characterization of fracture loci in metal forming *Int. J. Mech. Sci.* **83** pp 112-23

[7] Isik K, Silva M B, Tekkya A E and Martins P A F 2014 Formability limits by fracture in sheet metal forming *J. Mat. Pro. Tec.* **214** pp 1557-65

[8] Feistle M, Krinninger M, Pätzold I and Volk W 2015 Edge-Fracture-Tensile-Test *60 Excellent Inventions in Metal Forming* ed Tekkaya A E, Brosius A and Homberg W (Berlin Heidelberg: Springer-Verlag) pp 193-8

IOP Conf. Series: Materials Science and Engineering **159** (2016) 012014 doi:10.1088/1757-899X/159/1/012014

Evaluation of common tests for fracture characterisation of advanced high-strength sheet steels with the help of the FEA

I Peshekhodov, M Dykiert, M Vucetic and B-A Behrens

Institute of Forming Technology and Machines (IFUM), Leibniz Universität Hannover
An der Universität 2, 30823 Garbsen, Germany

E-mail: peshekhodov@ifum.uni-hannover.de

Abstract. The paper presents results of evaluation of common tests for fracture characterization of advanced high-strength sheet steels with the help of the FEA. The tests include three in-plane shear tests, two uniaxial tension tests, two plane strain tension tests and two equibiaxial tension tests. Three high-strength steels with different yield loci, strain hardening rates and strengths in three different thicknesses each were used. The evaluation was performed based on the spatial distribution of the equivalent plastic strain and damage variable in the specimen at the moment of crack initiation as well as on the time variation of the stress state at the crack initiation location. For in-plane shear, uniaxial tension and plane strain tension, no test can be unconditionally recommended as disadvantages of all studied tests in these groups cannot be neglected. However, in each of these groups, a test can be chosen, which represents an acceptable compromise between its advantages and disadvantages: the shear test on an IFUM butterfly specimen for in-plane shear, the tensile test on a holed specimen for uniaxial tension and the tensile test on a waisted specimen for plane strain tension. On the contrary, the bulge test on a circular specimen with a punch of Ø 100 mm can be unconditionally recommended for equibiaxial tension. In the future, optimisation of the studied tests for in-plane shear, uniaxial tension and plane strain tension appears to be necessary.

1. Introduction

Light-weight materials such as advanced high-strength steels with the ultimate tensile strength above 550 MPa (AHSS), aluminium and magnesium alloys as well as (fibre-reinforced) plastics are gaining more importance in the automotive industry. The main drivers of this trend are high energy costs and stringent legal regulations on CO_2 emissions during production and use of new vehicles. The share of these materials in the global car production accounted for 29 % in 2010 and is expected to grow up to 67 % by 2030 [1] as a recent study of McKinsey & Company states. The share of AHSS is expected to grow in absolute numbers from 15 % to 38 %, which is expected to be the strongest growth among the light-weight materials mentioned above. As opposed to the AHSS, the use of conventional steels with the ultimate tensile strength below 550 MPa is anticipated to fall from 52 % in 2010 to 13 % in 2030 [1]. This is believed to lead to a considerable reduction of the overall steel share in the material mix of the automotive industry from 67 % to 51 % over twenty years [1]. To slow down this decline, the steel manufactures need to work further on improvement of AHSS on one hand and on new solutions for a better exploitation of their advantageous properties in car manufacturing on the other.

FEA-based forming and crash analysis can help to better exploit the advantageous properties of AHSS in car manufacturing. For that, a mathematical formulation of the material yielding, hardening and formability is required. Forming limit curves (FLC) have been long used to describe sheet metal formability [2]. Due to its standardised determination procedure, the FLC is valid for linear strain paths and plane stress states from uniaxial tension to equibiaxial tension. Despite these limitations, failure prediction in a sheet forming process based on the FLC provides an acceptable accuracy for deep-drawing steels and some aluminium alloys [3]. However, failure prediction based on the FLC for AHSS can be insufficient [3]. Especially for shear-dominated stress states or bending, alternative formability descriptions are required [4].

As an alternative approach to predict sheet metal failure, fracture models can be considered. According to these models, fracture initiation is predicted based on a relation between the equivalent plastic strain $\bar{\varepsilon}_{\mathrm{pl}}$ and the equivalent plastic strain at fracture $\bar{\varepsilon}^{\mathrm{f}}_{\mathrm{pl}}$ (or simply fracture strain). The fracture strain is a material property, which depends on the stress state, at which plastic deformation is accumulated. There exist many mathematical models describing the dependence of the fracture strain on the stress state [5, 6, 7]. To quantify a stress state, the stress triaxiality η and normalised Lode angle $\bar{\theta}$ are becoming frequently used in sheet metal forming and crash analysis [8, 9].

To calibrate fracture models, many different experimental procedures exist. Nonetheless, fracture strains of AHSS are scarce in literature and if found, they often differ much from one author to another for similar or even the same steel grades [10, 11, 12]. The differences in fracture strains $\bar{\varepsilon}^{\mathrm{f}}_{\mathrm{pl}}$ obtained by different authors can be explained with the fact that currently there is no consistent requirements nor recommendations concerning the geometry of the material specimens, test procedures and results analysis for their determination [13, 14].

The paper presents an FEA-based evaluation of different tests widely used for fracture characterisation of AHSS, which can be used as a basis for development of such requirements or recommendations. Since sheet steels were considered, the fracture calibration tests were limited to four characteristic plane stress states: in-plane shear, uniaxial tension, plane strain tension and equibiaxial tension. The FE models, evaluation criteria and results are introduced and discussed. At the end, a conclusion and outlook for future work are given.

2. Materials

The evaluation was performed on three different grades and three thicknesses to secure generality of conclusions. A dual-phase (DP) steel, a complex-phase (CP) steel and a press-hardening (PH) steel with the ultimate tensile strength from 600 MPa to 1500 MPa in 0.8 mm, 1.5 mm and 3.0 mm each were chosen. To model the yield behaviour, the Hill'48 yield condition and an isotropic strain-rate dependent hardening formulation were used as shown in figure 1.

To describe the fracture behaviour, the ductile shear fracture and ductile normal fracture of CrahFEM [6] were chosen. Both were combined to describe the fracture behaviour of DP600 and MB-W1500, whereas only the ductile shear fracture model was used for CP1000 since no ductile fracture model data was available for it. The shear fracture model is given by:

$$\bar{\varepsilon}^{\mathrm{f}}_{\mathrm{pl}} = \frac{\bar{\varepsilon}^{\mathrm{f+}}_{\mathrm{pl}} \cdot \sinh\left(f \cdot (\gamma^{-} - \gamma)\right) + \bar{\varepsilon}^{\mathrm{f-}}_{\mathrm{pl}} \cdot \sinh\left(f \cdot (\gamma - \gamma^{+})\right)}{\sinh\left(f \cdot (\gamma^{-} - \gamma^{+})\right)}, \tag{1}$$

where $\gamma = (1 - k_{\mathrm{sf}} \cdot \eta)/(\tau_{\mathrm{max}}/\bar{\sigma})$ is the shear stress parameter with k_{sf} being the stress triaxiality weight factor, τ_{max} being the maximum shear stress and $\bar{\sigma}$ being the von Mises equivalent stress, γ^{-} and γ^{+} are the shear stress parameters at equibiaxial compression and equibiaxial tension, $\bar{\varepsilon}^{\mathrm{f-}}_{\mathrm{pl}}$ and $\bar{\varepsilon}^{\mathrm{f+}}_{\mathrm{pl}}$ are the fracture strains at equibiaxial compresion and equibiaxial tension and f is a model coefficient. The model input parameters are summarized in table 1.

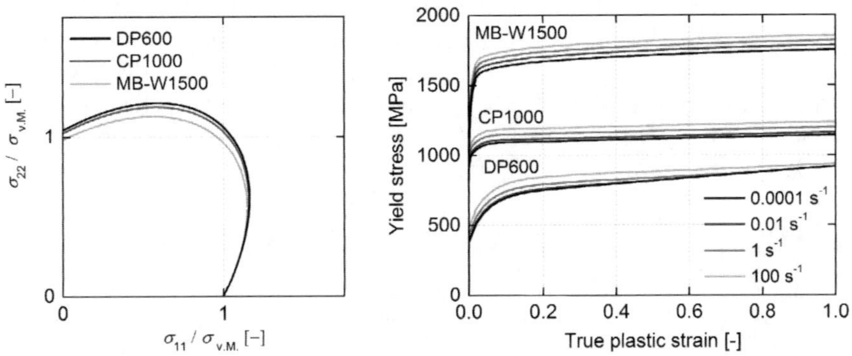

Figure 1. Yield loci (left) and strain-rate-dependent yield curves (right)

Table 1. Coefficients of the CrachFEM ductile shear fracture model [21]

Material	k_{sf}	f	γ^+	γ^-	$\bar{\varepsilon}^{\mathrm{f+}}_{\mathrm{pl}}$	$\bar{\varepsilon}^{\mathrm{f-}}_{\mathrm{pl}}$
DP600	0.120	2.500	1.840	2.160	0.900	1.900
CP1000	0.202	3.800	1.731	2.269	1.190	7.570
MB-W1500	0.210	0.001	1.720	2.280	0.676	1.543

The ductile normal fracture model was defined as follows:

$$\bar{\varepsilon}^{\mathrm{f}}_{\mathrm{pl}} = d \cdot e^{q\beta}, \tag{2}$$

where $\beta = (1 - k_{\mathrm{nf}} \cdot \eta)/(\sigma_{\mathrm{I}}/\bar{\sigma})$ the normal stress parameter with k_{nf} being the stress triaxiality weight factor and σ_{I} being the first major stress, d is the limit strain magnitude and q is the exponent magnitude. The model input parameters are summarised in table 2.

Table 2. Coefficients of the CrachFEM ductile normal fracture model [21]

Material	k_{nf}	d	q
DP600	0	0.0327	3.430
MB-W1500	0.1515	0.0047	5.517

A comparison of the fracture behaviour descriptions given in figure 2 (plane stress states) and figure 3 (plane strain states) reveals that DP600 and MB-W1500 are characterised by similar fracture strains for low and medium stress triaxialities, whereas MB-W1500 behaves less formable than DP600 at high stress triaxialities. CP1000 possesses a considerably higher level of the fracture strain than the other two materials throughout the whole range of the stress states. It should be noted, however, that the fracture behaviours of the investigated materials were described by different testing laboratories with the help of different testing procedures and evaluation approaches and for this reason should not be compared among one another. From a comparison of the shear ductile fracture behaviour and normal ductile fracture behaviours for the DP600 and MB-W1500, it can be seen that fracture initiation for these material takes place mainly due to the ductile shear fracture mode at plane stress states. An exception is the equibiaxial stress state ($\eta \approx 0.67$ and $\bar{\theta} \approx -1$) for MB-W1500, at which the ductile fracture mode dominates. At 3D stress states of higher stress triaxialities, the ductile normal fracture mode dominates fracture initiation in DP600 and MB-W1500.

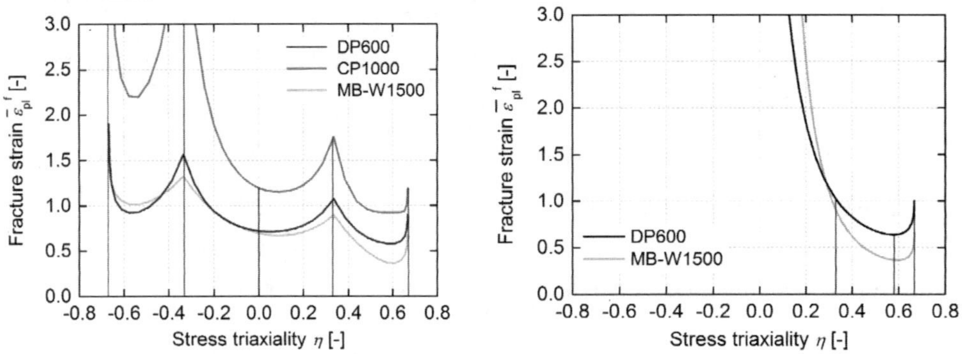

Figure 2. Ductile shear (left) and normal (right) fracture curves for plane stress states

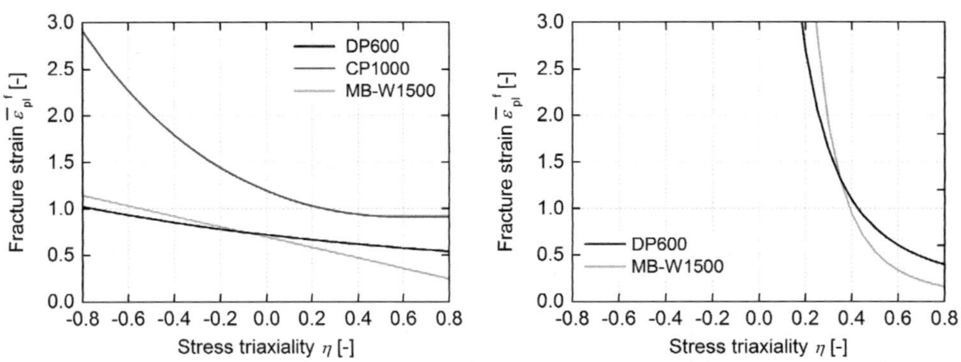

Figure 3. Ductile shear (left) and normal (right) fracture curves for plane strain states

At each increment of the FEA, the incremental equivalent plastic strain $\bar{\varepsilon}_{\mathrm{pl},i}$ in a particular finite element is divided by the fracture strain $\bar{\varepsilon}_{\mathrm{pl}}^{\mathrm{f}}$, which is dependent on the stress state acting in this element during this increment. The division product is accumulated over the increments to obtain the variable D defined by:

$$D = \sum_i \frac{\bar{\varepsilon}_{\mathrm{pl},i}}{\bar{\varepsilon}_{\mathrm{pl}}^{\mathrm{f}}}, \tag{3}$$

which is often referred to as damage variable. At an increment of $D \geq 1$ fracture is assumed. In the present work, the notations D_{s} and D_{n} are used to differentiate between the damage variable according to the shear ductile fracture model and damage variable according to the normal ductile fracture model. From the two, the one is plotted which satisfies $D \geq 1$ first.

IOP Conf. Series: Materials Science and Engineering **159** (2016) 012014 doi:10.1088/1757-899X/159/1/012014

3. Finite element modelling

3D volume elements were used to realistically represent the stress state in the material, especially after the onset of plastic strain localisation. In particular, sufficiently fine (approx. 0.1 mm) continuum eight-node linear reduced-integration elements with an hourglass control as provided in the element library of Abaqus/Standard under the notation C3D8R were used for specimen areas of large deformations. To minimise the computation time, the tools which come into contact with the specimen were modelled as rigid bodies.

The in-plane shear tests included shear tests on an ASTM-type specimen, a modified Miyauchi specimen with a one-sidedely reduced thickness [15] and an IFUM butterfly specimen with a one-sidedely reduced thickness (figure 4). The uniaxial tension tests included a bending test on a flanged specimen (figure 5, left) and a tensile test on a holed specimen of 20 mm width and Ø 10 mm hole diameter. The plane strain tension tests included a bending test on a full specimen (figure 5, right) and a tensile specimen of 20 mm width and 5 mm waist radius (the minimum specimen width of 10 mm). Finally, the equibiaxial tension tests encompassed a bulge test with a hemispherical punch of Ø 100 mm and a circular die of Ø 110 mm and R 10 mm and a bulge test with a hemispherical punch of Ø 20 mm and a circular die of Ø 27 mm and R 0.75 mm.

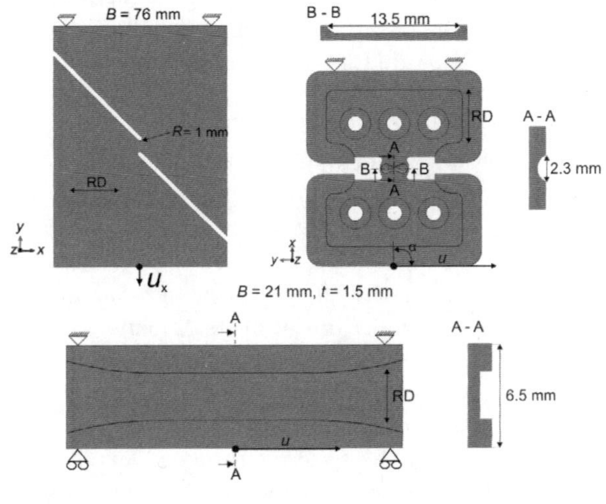

Figure 4. FE models of the in-plane shear tests on the ASTM-type specimen (top left), modified Miyauchi specimen [15] (bottom) and IFUM butterfly specimen with tools (top right)

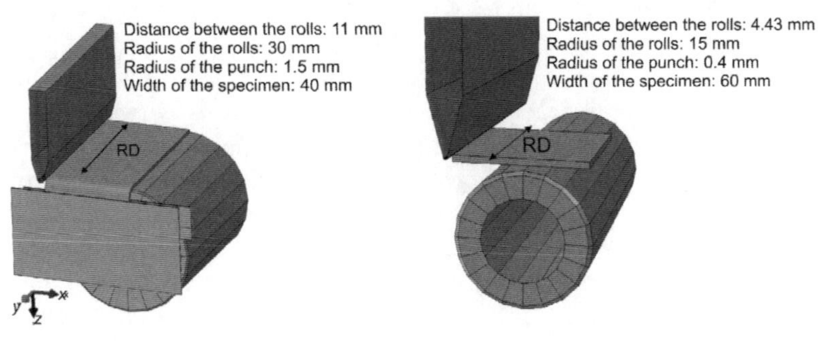

Figure 5. FE models of the bending test on the flanged specimen for uniaxial tension (left) and full specimen for plane strain tension (right)

IOP Conf. Series: Materials Science and Engineering **159** (2016) 012014 doi:10.1088/1757-899X/159/1/012014

4. Test evaluation criteria

The tests were evaluated based on ten evaluation criteria. All criteria are related to the distribution of the plastic strain and damage variable as well as the stress state at the location of the crack initiation and are listed below. Next to each criterion a short explanation is given on why this criterion is important and in what case a test is considered to be advantageous according to this criterion.

Defined crack initiation position independent of the material and thickness. The defined crack initiation position independent of the specimen material and thickness reproducibly sucures a desired stress state at the location of the fracture initiation and facilitates development and subsequent application of standard test evaluation procedures, which is considered to be an advantage of a test.

Guarantee of no crack initiation directly at the specimen edge. Especially some of the shear tests are prone to edge cracking, but also other tests like tensile tests on holed and waisted specimen might exhibit edge fracture. The edge fracture can be considered as a premature failure of the specimen. Therefore, it is classified as a disadvantage of a test.

Guarantee of no crack initiation at the machined specimen surface. To avoid edge cracking many researchers suggested new specimen designs with a specimen analysis area of reduced thickness [16], [17], [18]. In most of the cases, the thickness reduction is performed with the help of machining, which might be accompanied by an influence on the material properties [19] and is therefore considered to be a disadvantage of a test.

Variation of the equivalent plastic strain through thickness. A low variation of the equivalent plastic strain through thickness allows a quick and accurate estimation of the fracture strain via a local measurement of the specimen thickness at the point of crack initiation with the help of a sliding calliper or optical microscope and is regarded to be a test advantage.

Variation of the damage variable through thickness. A low variation of the damage variable through thickness allows accounting for the through-thickness scatter of the material formability. As properties variations of sheet metal in its thickness direction are likely due to its manufacturing process, a homogeneous distribution of the damage variable through thickness is advantageous.

Variation of the stress state at the point of crack initiation. A low variation of the stress state at the fracturing material point of the specimen means less uncertainty in the assignment of a characteristic value of the corresponding stress state parameters and is therefore regarded to be an advantage.

Localisation of the plastic strain in the specimen plane. In most of the cases, tests used for characterisation of the sheet metal formability are evaluated with digital image correlation (DIC). A low localisation of the plastic strain in the specimen plane is an important prerequisite for DIC based on the currently available fineness degree of a typical stochastic black-and-white pattern applied to the surface of the specimen, which is seen as an advantage of a test.

Maximum relative difference between the plastic strain at the location of the crack initiation and most deformed point at the surface. A low relative difference between the plastic strain at the location of crack initiation and most deformed material point at the surface, which can be thought of as a measure of the relative error of the fracture strain determination with DIC, is regarded as a test advantage.

Test dependence on the specimen thickness. This collective criterion is related to the dependence of the location of the fracture initiation *and* stress state at the location of fracture initiation on the specimen thickness.

Test dependence on the specimen material. This collective criterion is related to the dependence of the location of the fracture initiation *and* stress state at the location of fracture initiation on the the specimen material.

5. Results and discussion

5.1. In-plane shear tests

The most important advantages of the shear test on an ASTM-type specimen are its non-reduced thickness and low variation of the equivalent plastic strain and damage variable through thickness (figure 6). But these advantages are outweighed by the disadvantages of a possibility of fracture initiation at the specimen edge and a moderate/high variation of the stress state at the point of crack initiation (figure 9, top), which is due to its close proximity to the specimen edge. Furthermore, this test was found to be highly thickness and material dependent (figure 9, top). Even though this test is currently widely used in industry and academia for characterisation of the fracture behaviour of AHSS, it is not universally suitable for this purpose.

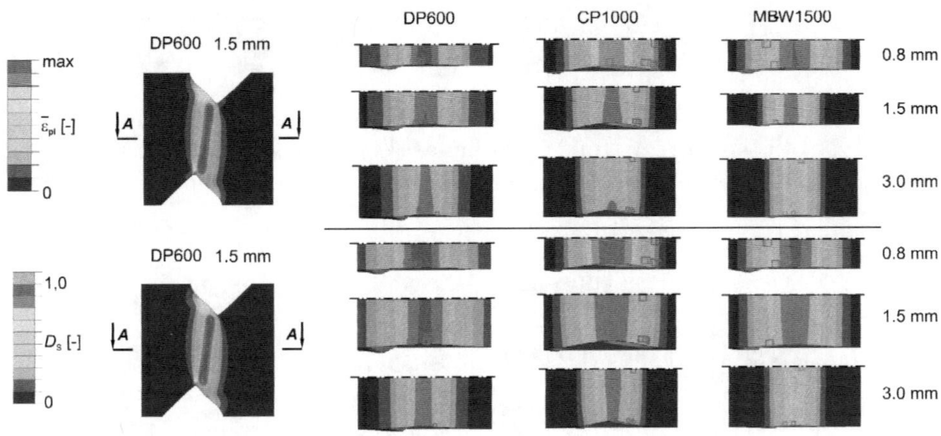

Figure 6. Eq. plastic strain $\bar{\varepsilon}_{\text{pl}}$ (top) and damage variable D_{s} (bottom) in the ASTM-type specimen in the shear test at the moment of crack initiation

In favour of the shear test on the modified Miyauchi specimen [15] speaks the guarantee of no crack initiation at the specimen edge and low maximum relative difference between the plastic strain at the point of crack initiation and most deformed point at the surface (figure 7). The disadvantages include crack initiation at the machined surface and high localisation of the plastic strain in the specimen plane (figure 7). The plastic strain localisation takes place shorty before fracture initiation. Before that, the specimen shows a homogeneous distribution of shear deformation. Therefore, it can be recommended to determine the plastic strain at the onset of this localisation. As the shear test on the ASTM-type specimen, this test was also found to be thickness and material dependent, even though to a substantially lesser degree (figure 9, middle). In summary, this test does not possess the severe disadvantages of the shear test on an ASTM-type specimen and can be recommenden for fracture characterisation of AHSS.

The shear test on an IFUM butterfly specimen with a one-sidedely reduced thickness has an advantage over the other two in-plane shear tests, as its dependency on the specimen thickness and material is negligible (figure 9, bottom). Further advantages include moderate localisation of the plastic strain in the specimen plane and the low maximum relative difference between the plastic strain at the point of crack initiation and most deformed point at the surface (figure 8). The low variation of the stress state at the point of crack initiation and the guarantee of no crack initiation at the specimen edge can be considered as other advantages of this test. To the main disadvantages count the crack initiation at the machined specimen surface and the high variation of the equivalent plastic strain and damage variable through thickness (figure 8). On the whole, this test is not ideal either but it appears to be the best choice among the studied in-plane shear tests.

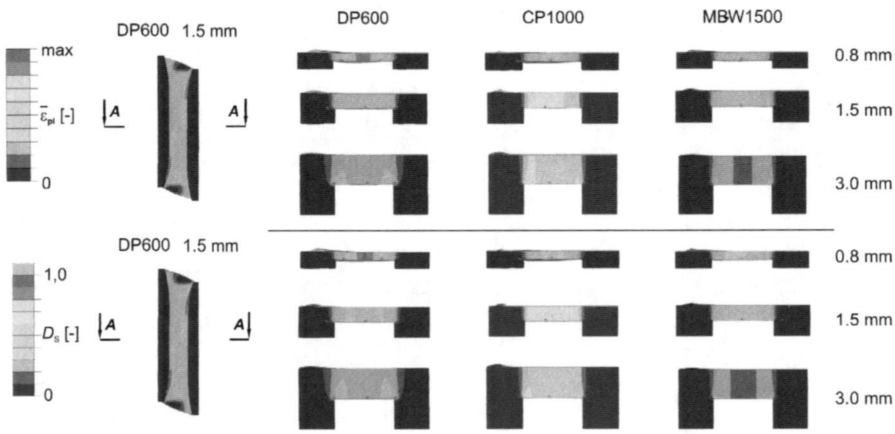

Figure 7. Eq. plastic strain $\bar{\varepsilon}_{\mathrm{pl}}$ (top) and damage variable D_{s} (bottom) in the modified Miyauchi specimen with a one-sidedely reduced thickness [15] in the shear test at the moment of crack initiation

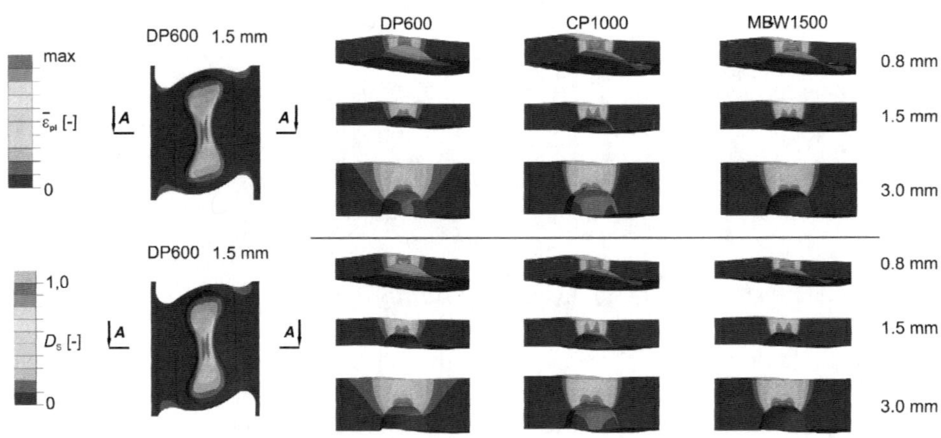

Figure 8. Eq. plastic strain $\bar{\varepsilon}_{\mathrm{pl}}$ (top) and damage variable D_{s} (bottom) in the IFUM butterfly specimen with a one-sidedely reduced thickness in the shear test at the moment of crack initiation

5.2. Uniaxial tension tests

From the two studied tests, the tensile test on a holed specimen has an advantage over the three-point bending test on a flanged specimen, as it guarantees no crack initiation directly at the edge (figure 10 and figure 11). On the contrary, the three-point bending test on a flanged specimen has an advantage over the tensile test on a holed specimen, as it possesses a lower variation of the damage variable through thickness and can therefore better account for possible scatter of the material formability in the thickness direction (figure 10). Furthermore, the three-point bending test on a flanged specimen excels the tensile test on a holed specimen with its slightly lower variation of the stress state at the material point of crack initiation for a given material (figure 12) and a slightly lower localisation of the plastic strain in the specimen plane (figure 10 versus figure 11). Also, the maximum relative difference between the plastic strain at the point of crack initiation and most deformed point at the surface for all materials and thicknesses was found smaller for the three-point bending test on a flanged specimen. The thickness dependency

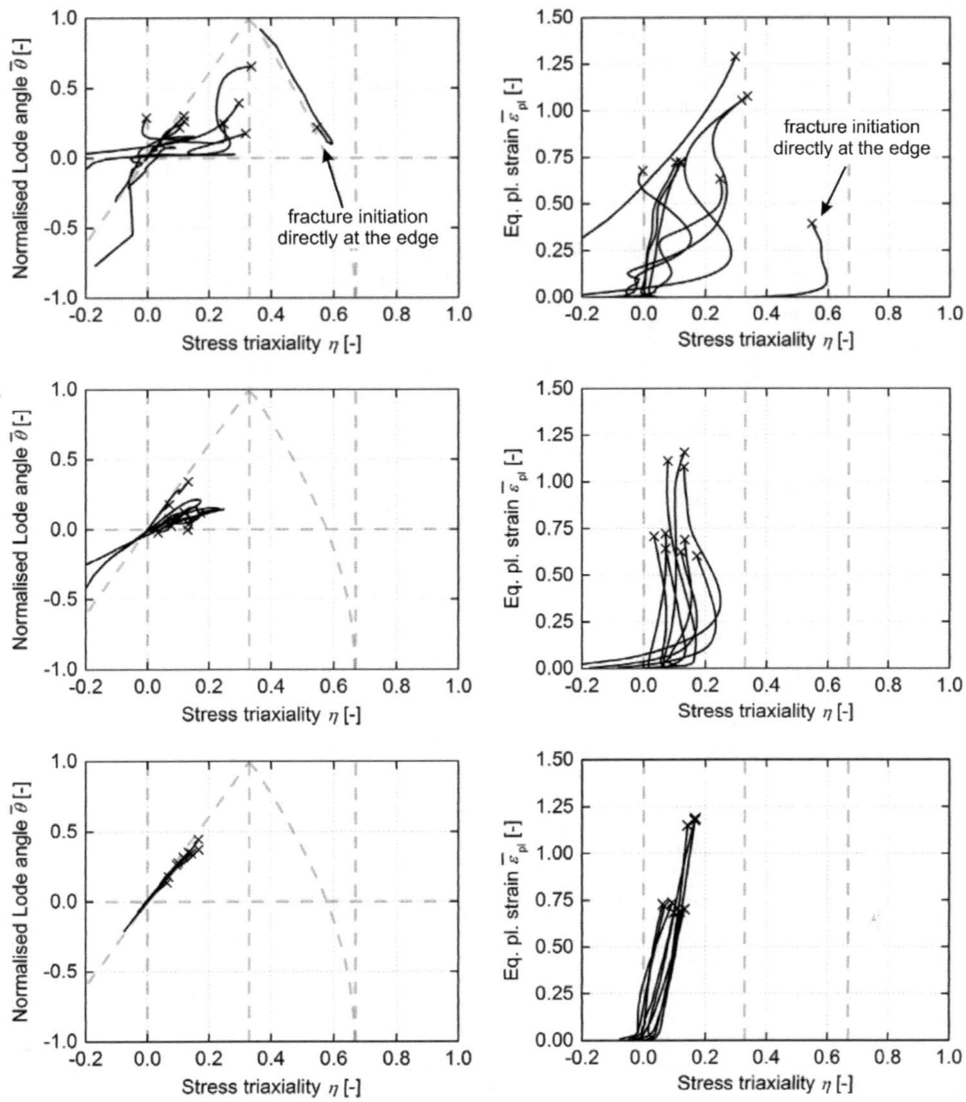

Figure 9. Stress state at the location of fracture initiation in the ASTM-like specimen (top), modified Miyauchi specimen (middle) and IFUM butterfly specimen (bottom) in the shear test for all combinations of the materials and thicknesses each

of the three-point bending test on a flanged specimen was found to be on average lower than for the tensile test on a holed specimen. On the contrary, the latter test performs better what material dependency concerns. To sum up, the tensile test on a holed specimen performs more reproducibly for various materials and thicknesses providing a good safety against edge cracking and a tolerable difference between the equivalent plastic strain at the material point of fracture initiation and most deformed material point at the surface and is considered to be the better choice in the present study even though it does not accurately enough model the stress state of uniaxial tension.

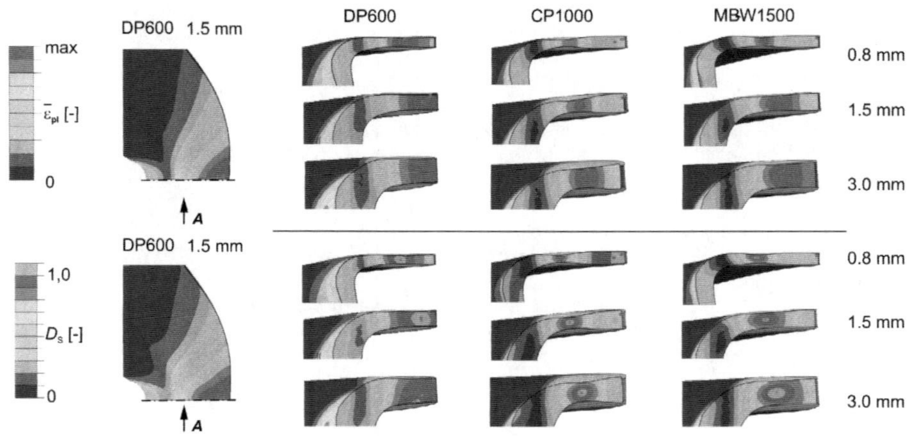

Figure 10. Eq. plastic strain $\bar{\varepsilon}_{\text{pl}}$ (top) and damage variable D_{s} (bottom) in the flanged specimen in the three-point bending test at the moment of crack initiation

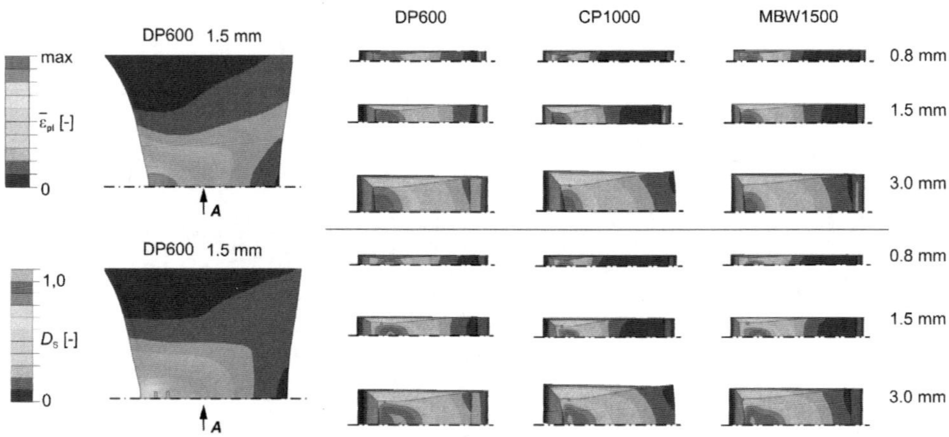

Figure 11. Eq. plastic strain $\bar{\varepsilon}_{\text{pl}}$ (top) and damage variable D_{s} (bottom) in the holed specimen in the tensile test at the moment of crack initiation

5.3. Plane strain tension tests

For thicker materials, the advantages of the three-point bending test on the full specimen outweigh the advantages of the tensile test on the waisted specimen, especially what the stress state concerns (figure 15). However, for ductile materials and low thicknesses, no fracture is achieved at the bending radius, which compromises the test universality. Disregarding the specimen thickness, the three-point bending test on the full specimen possesses a high variation of the damage variable through thickness. In this respect, the tensile test on a waisted specimen delivers better results. But it reveals a high variation of the stress state at fracturing material point. For some materials, like DP600 and MB-W1500 of this study, this may lead to fracture initiation in the ductile normal fracture mode, which is a disadvantage if the test is to be used to characterise ductile shear fracture. On the whole, however, the tensile test on the waisted specimen is found more suitable and more universal for fracture characterisation of AHSS. But for characterisation of shear fracture, its optimisation or an alternative test is required.

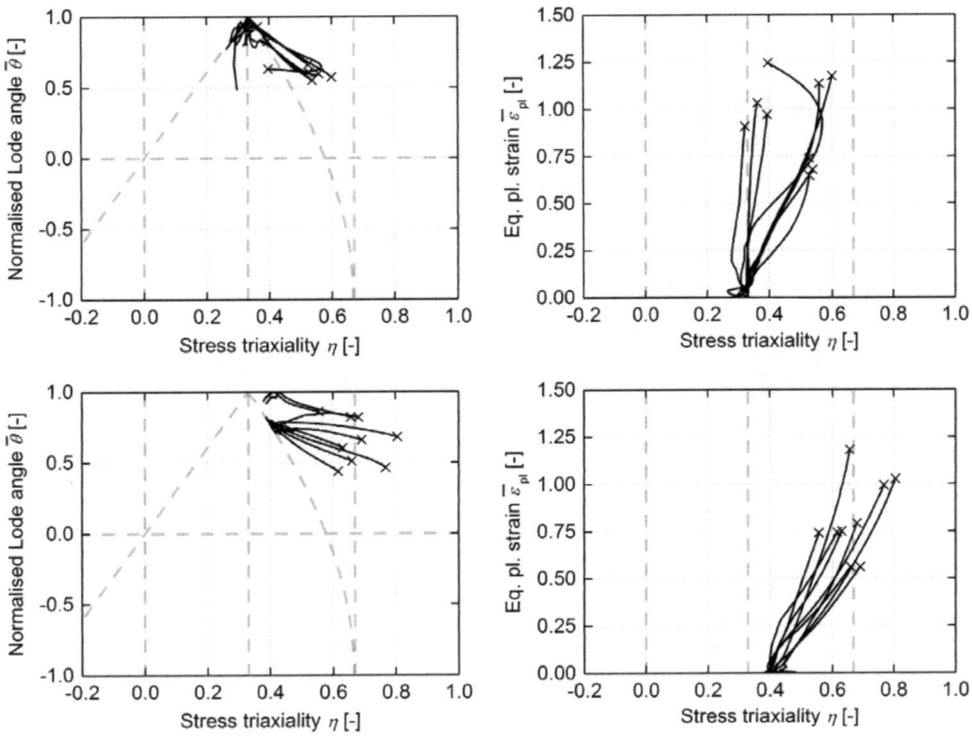

Figure 12. Stress state at the location of fracture initiation in the flanged specimen in the three-point bending test (top) and holed specimen in the tensile test (bottom) for all combinations of the materials and thicknesses each

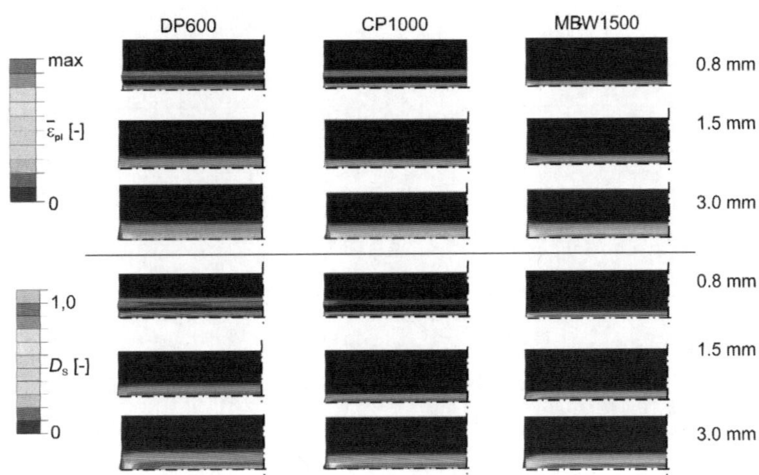

Figure 13. Eq. plastic strain $\bar{\varepsilon}_{pl}$ (top) and damage variable D_s (bottom) in the full specimen in the three-point bending test at the moment of crack initiation

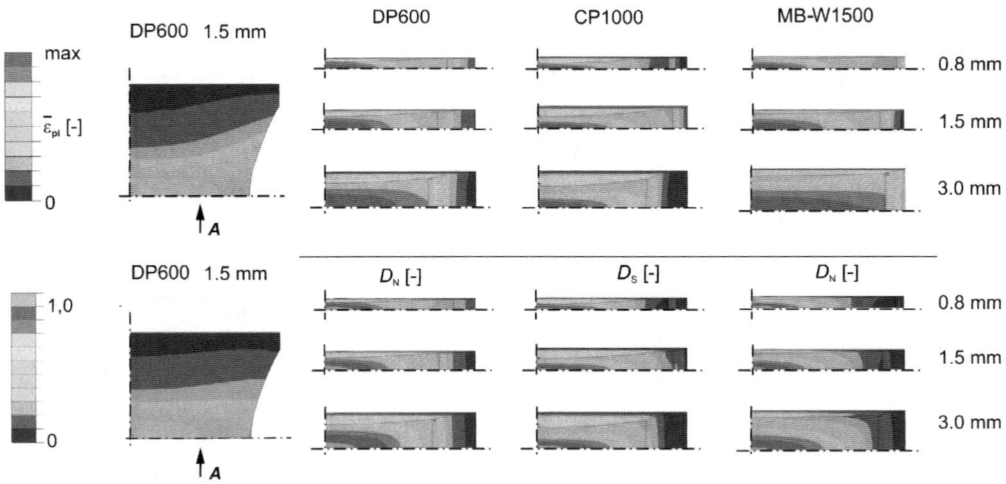

Figure 14. Eq. plastic strain $\bar{\varepsilon}_{\text{pl}}$ (top) and damage variable D_{s} or D_{n} (bottom) in the waisted specimen in the tensile test at the moment of crack initiation

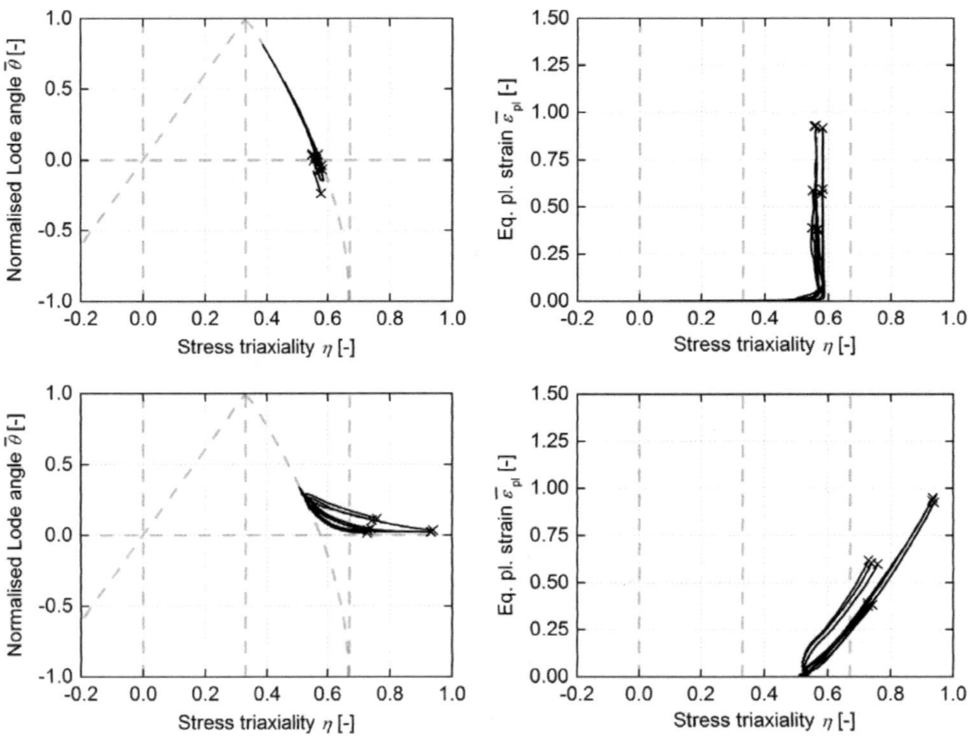

Figure 15. Stress state at the location fracture initiation in the full specimen in the three-point bending test (top) and waisted specimen in the tensile test (bottom) for all combinations of the materials and thicknesses

5.4. Equibiaxial tension tests

Contrary to the previously described test groups, the more favourable test exists in this group, with its advantages overweighing the advantages of its group counterpart: the bulge test with a punch of Ø 100 mm. It is well suitable for the fracture strain determination both with the help of thickness measurement and DIC. Furthermore, the test assures crack initiation at a defined position at the pole of the bulge or in its immediate proximity (figure 16) and delivers a constant stress state independent of the specimen thickness and material (figure 18, top). For DP600 and CP1000 ductile shear fracture occurs. For MB-W1500, ductile normal fracture leads to the specimen failure, though $D_{\mathrm{n}} \approx D_{\mathrm{s}}$. Contrary to that, the bulge test with a punch of Ø 20 mm was found to have a moderate dependency on the specimen thickness, which leads to a disadvantageous variation of the crack initiation position (figure 17) and stress state (figure 18, bottom). Moreover, notable gradients of the equivalent plastic strain and damage variable through thickness for thicker specimens in this test take place (figure 17), which reduce the accuracy of the fracture strain determination via the thickness measurement and hinder accounting for possible scatter of the material formability in the thickness direction.

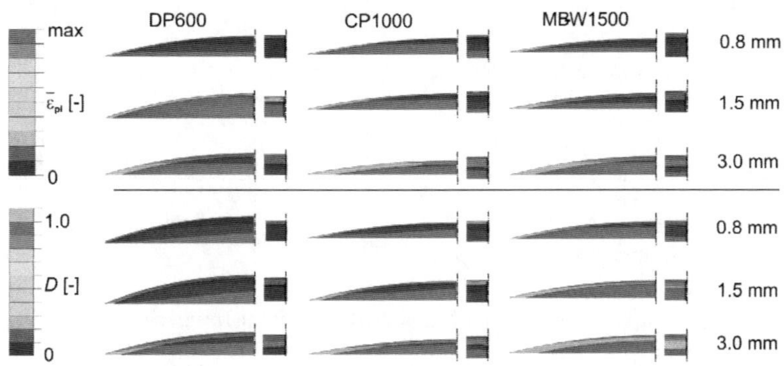

Figure 16. Eq. plastic strain $\bar{\varepsilon}_{\mathrm{pl}}$ (top) and damage variable D_{s} or D_{n} (bottom) in the circular specimen in the bulge test with a punch of Ø 100 mm at the moment of crack initiation

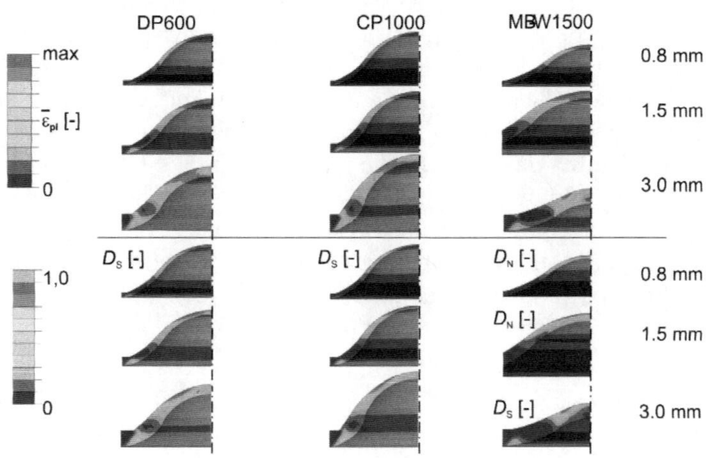

Figure 17. Eq. plastic strain $\bar{\varepsilon}_{\mathrm{pl}}$ (top) and damage variable D_{s} or D_{n} (bottom) in the circular specimen in the bulge test with a punch of Ø 20 mm at the moment of crack initiation

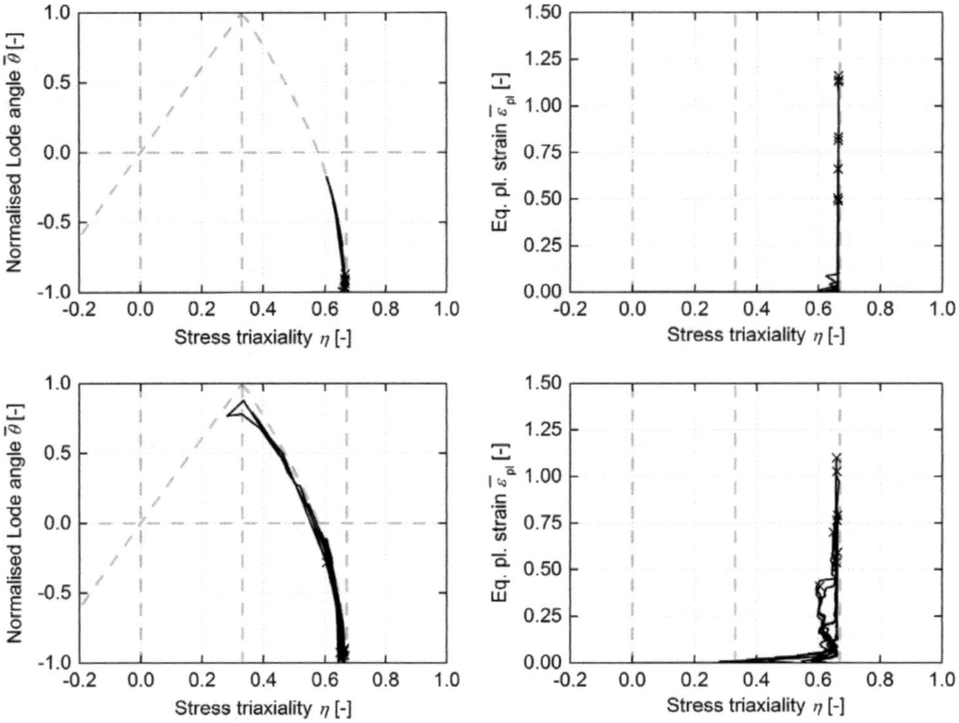

Figure 18. Stress state at the location of fracture initiation in the circular specimen in the bulge test with a punch of Ø 100 mm (top) and circular specimen in the bulge test with a punch of Ø 20 mm (bottom) for all combinations of the materials and thicknesses

5.5. Summary

The test analysis reveals that all the investigated tests have their advantages and disadvantages (table 3). For in-plane shear, uniaxial tension and plane strain tension, no test can be unconditionally recommended as its disadvantages cannot be neglected. However, in each of these groups, a test can be chosen, which represents an acceptable compromise between its advantages and disadvantages: the shear test on an IFUM butterfly specimen for in-plane shear, the tensile test on a holed specimen for uniaxial tension, the tensile test on a waisted specimen for plane strain tension. For equibiaxial tension, the bulge test on a circular specimen with a punch of Ø 100 mm (alternatively, a hydraulic bulge test [20]) can be unconditionally recommended. The results of the shear test on an IFUM butterfly specimen, tensile test on a holed specimen, tensile test on a waisted specimen and bulge test on a circular blank with a punch of Ø 100 mm are summarised in figure 19. It can be seen that these tests provide a basis for characterisation of the fracture behaviour of AHSS in a range of practice-relevant stress states.

6. Conclusion and outlook

The paper presents results of evaluation of different commonly available tests for fracture characterisation of AHSS with the help of the FEA. For in-plane shear, uniaxial tension and plane strain tension, no test can be unconditionally recommended as disadvantages of all studied tests in these groups cannot be neglected. However, in each of these groups, a test can be chosen, which represents an acceptable compromise between its advantages and disadvantages: the shear test on an IFUM butterfly specimen for in-plane shear, the tensile test on a holed

Table 3. Summary of the test results according to the test evaluation criteria

		Defined crack initiation position independent of the material and thickness	Guarantee of no crack initiation directly at the specimen edge	Guarantee of no crack initiation at the machined specimen surface	Variation of the equivalent plastic strain through thickness	Variation of the damage variable through thickness	Variation of the stress state at the point of crack initiation	Localisation of the plastic strain in the specimen plane	Max. relative difference between the plastic strain at the point of crack initiation and most deformed point at the surface	Test dependence on the specimen thickness	Test dependence on the specimen material
in-plane shear	Shear test on the ASTM-type spec.	no	no	yes	low	low	moderate/ high	low/ moderate	high	high	high
	Shear test on the mod. Miyauchi-type spec.	no	yes	no	moderate	moderate	moderate	high	low	moderate	moderate
	Shear test on an IFUM butterfly-type spec.	yes	yes	no	high	high	low	moderate	low	low	low
unia. tension	Three-point bending test on a flanged spec.	no	no	yes	low/ moderate	low/ moderate	low/ moderate	low/ moderate	moderate	low/ moderate	high
	Tensile test on a holed spec.	no	yes	yes	moderate/ high	moderate	moderate	moderate	moderate/ high	high	moderate
plane strain	Three-point bending test on a full spec.	yes	yes	yes	high	high	low	moderate/ high	low	low	low
	Tensile test on a waisted spec.	yes	yes	yes	moderate/ high	moderate/ high	moderate/ high	low/ moderate	high	low/ moderate	low/ moderate
bia. tension	Bulge test with a punch ∅ 100	yes	yes	yes	low/ moderate	low/ moderate	low	low	low	low	low
	Bulge test with a punch ∅ 20	no	yes	yes	moderate/ high	moderate/ high	moderate	low	low	moderate	low

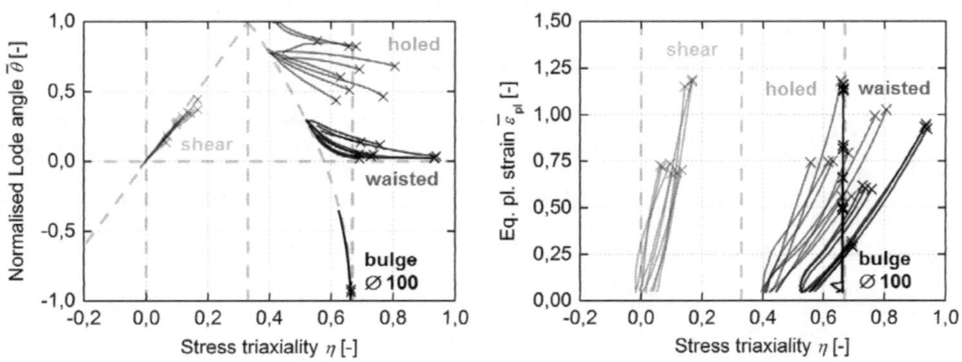

Figure 19. Stress state at the fracture initiation location the preferred specimens

specimen for uniaxial tension, the tensile test on a waisted specimen for plane strain tension. On the contrary, the bulge test on a circular specimen with a punch of ∅ 100 mm can be unconditionally recommended for equibiaxial tension. Together, these four tests provide a basis for fracture characterisation of AHSS. In the future, optimisation of the studied tests for in-plane shear, uniaxial tension and plane strain tension appears to be necessary. In particular, the IFUM butterfly specimen could be optimised for a more homogeneous distribution of plastic deformation through thickness, the holed specimen could be optimised to better model uniaxial tension and the waisted specimen could be optimised to induce shear instead of normal fracture.

IOP Conf. Series: Materials Science and Engineering **159** (2016) 012014 doi:10.1088/1757-899X/159/1/012014

Acknowledgments

The project P 1145 "Test evaluation" [21] of the Research Association for Steel Application (FOSTA), Germany, was initiated by the Working group "Characterisation of damage models" of the Materials committee of the Steel Institute VDEh, Germany. The financial support by the FOSTA is gratefully acknowledged. The authors also thank the members of the project committee Dr. P. Dietsch (ArcelorMittal S. A.), Dr. H. Friebe (Gesellschaft für optische Messtechnik mbH), Dr. H. Gese (MATFEM Partnerschaft Dr. Gese & Oberhofer), Dr. B. Hackl (voestalpine Stahl GmbH), Dr. H. Richter and H. Rösen (ThyssenKrupp Steel Europe AG), M. Schneider (Salzgitter Mannesmann Forschung GmbH) and Dr. C. ten Horn (Tata Steel Europe Ltd.) for the provided material models, test details and valuable comments.

References

[1] Heuss R, Müller N, van Sintern W and Starke A 2012 *Lightweight, heavy impact* (McKinsey & Company)
[2] Banabic D 2010 *Sheet Metal Forming Processes. Const. Modelling and Num. Simulation* (Berlin: Springer)
[3] Roll K 2008 Simulation of sheet metal forming – necessary developments in the future *Proc. Int. Conf. and Workshop on Num. Sim. of 3D Sheet Metal Forming Processes (Interlaken)* (Zurich: IVP, ETH) pp 3-11
[4] Keßler L and Gerlach J 2008 Industrial aspects of material modelling for steel grades in the past, present and future *Proc. Int. Conf. and Workshop on Num. Sim. of 3D Sheet Metal Forming Processes (Interlaken)* (Zurich: IVP, ETH) pp 13-17
[5] Bao Y and Wierzbicki T 2004 A comparative study on various ductile crack formation criteria *J. Eng. Mater. Technol.* **126** pp 314-24
[6] Hooputra H, Gese H, Dell H and Werner H 2004 A comprehensive failure model for crashworthiness simulation of aluminium extrusions *Int. J. Crash.* **9** pp 449-63
[7] Wierzbicki T, Bao Y, Lee Y-W and Bai Y 2005 Calibraton and evaluation of seven fracture models *Int. J. Mech. Sci.* **47** pp 719-43
[8] Till E and Hackl B 2013 Calibration of plasticity and failure models for AHSS sheets *Proc. IDDRG Conf. (Zurich)* (Zurich: Institute of Virtual Manufacturing, ETH) pp 119-24
[9] Bai Y and Wierzbicki T 2013 The concept of damage accumulation for prediction necking and fracture of sheets *Proc. IDDRG Conf. (Zurich)* (Zurich: Institute of Virtual Manufacturing, ETH) pp 9-13
[10] Walters CL 2009 Development of a punching technique for ductile fracture testing over a wide range of stress states and strain rates *PhD Thesis* (Cambridge: MIT)
[11] Chottin J, Hug E and Rachnik M 2011 Damage accumulation in DP1000 sheets submitted to various stress states *Steel Res. Int.* **Spec. Ed. ICTP Conf.** pp 895-900
[12] Behrens B-A, Bouguecha A, Vucetic M and Peshekhodov I 2012 Characterisation of the quasi-static flow and fracture behaviour of DP steel sheets in a wide range of stress states *Arch Civ Mech Eng* **12** pp 397-406
[13] Behrens B-A, Vucetic M and Peshekhodov I 2014 Effiziente Charakterisierung und Modellierung des Fließ- und Versagensverhaltens von Dualphasenstählen *Tagungsband des Umformtechnischen Kolloquiums Hannover* (Hanover: IFUM, Leibniz Universität Hannover) pp 77-97
[14] Keßler L, Richter H and Rösen H 2014 Die Anwendung von erweiterten Bewertungskriterien für die Simulation von Stahlwerkstoffen *Tagungsband des Umformtechnischen Kolloquiums Hannover* (Hanover: IFUM, Leibniz Universität Hannover) pp 99 -109
[15] Partnerschaft Dr. Gese & Oberhofer 2016 *www.matfem.de*
[16] Mohr D and Henn S 2007 Calibration of stress-triaxiality dependent crack formation criteria: a new hybrid experimental-numerical method *Exp. Mech.* **47** pp 805-20
[17] Bai Y 2008 Effect of loading history on necking and fracture *PhD thesis* (Cambridge: MIT)
[18] Behrens B-A, Vucetic M and Peshekhodov I 2014 Calibration of fracture initiation models for advanced high-strength sheet steels in a wide range of stress states using a uniaxial tensile testing machine *Proc. IDDRG Conf. (Paris)* (Senlis Cedex: CETIM) pp 252-57
[19] Behrens B-A, Vucetic M and Peshekhodov I 2015 *Methode zur Charakterisierung des Versagensverhaltens hochfester Stahlblechwerkstoffe in einem breiten Spektrum von Spannungszuständen mithilfe einachsig arbeitender Zugprüfmaschinen* (Dusseldorf: Verlag und Vertriebsgesellschaft mbH)
[20] Vucetic M, Bouguecha A, Peshekhodov I, Götze T, Huinink T, Friebe H, Möller T and Behrens B-A 2011 Num. validation of analytical biaxial true stress - true strain curves from the bulge test *Proc. Int. Conf. and Workshop on Num. Sim. of 3D Sheet Metal Forming Processes (Seoul)* (Melville: AIP) pp 107-14
[21] Behrens B-A and Peshekhodov I 2016 *Evaluation of common tests for fracture characterisation of advanced high-strength sheet steels with the help of the finite element analysis* (Dusseldorf: Verlag und Vertriebsgesellschaft mbH)

IOP Conf. Series: Materials Science and Engineering **159** (2016) 012015 doi:10.1088/1757-899X/159/1/012015

Experimental-numerical evaluation of a new butterfly specimen for fracture characterisation of AHSS in a wide range of stress states

I Peshekhodov, S Jiang, M Vucetic, A Bouguecha and B-A Berhens

Institute of Forming Technology and Machines (IFUM), Leibniz Universität Hannover
An der Universität 2, 30823 Garbsen, Germany

E-mail: peshekhodov@ifum.uni-hannover.de

Abstract. Results of an experimental-numerical evaluation of a new butterfly specimen for fracture characterisation of AHHS sheets in a wide range of stress states are presented. The test on the new butterfly specimen is performed in a uniaxial tensile machine and provides sufficient data for calibration of common fracture models. In the first part, results of a numerical specimen evaluation are presented, which was performed with a material model of a dual-phase steel DP600 taken from literature with plastic flow and fracture descriptions. In the second part, results of an experimental-numerical specimen evaluation are shown, which was conducted on another dual-phase steel DP600, which was available with a description of plastic flow only and whose fracture behaviour was characterised in the frame of this work. The overall performance of the new butterfly specimen at different load cases with regard to characterisation of the fracture behaviour of AHSS was investigated. The dependency of the fracture strain on the stress triaxiality and Lode angle as well as space resolution is quantified. A parametrised CrachFEM ductile shear fracture model and modified Mohr-Coloumb ductile shear fracture model are presented as a result of this quantification. The test procedure and results analysis are believed to contribute to current discussions on requirements to AHSS fracture characterisation.

1. Introduction

Driven by high energy costs and stringent legal regulations on CO_2 emissions during production and use of new vehicles, light-weight materials such as advanced high-strength steels with the ultimate tensile strength above 550 MPa (AHSS), aluminium and magnesium alloys as well as (fibre-reinforced) plastics are becoming extensively used in the automotive industry. According to a recent study of McKinsey & Company, the share of these materials in the global car production accounted for 29 % in 2010 and is expected to grow up to 67 % by 2030 [1]. Thereof, AHSS are expected to show the largest growth in absolute numbers from 15 % to 38 %. As opposed to the AHSS, the use of steels with the ultimate tensile strength below 550 MPa is anticipated to decline drastically from 52 % in 2010 to 13 % in 2030 [1]. As a result of these two expected changes, a noteworthy reduction of the overall steel share in the material mix of this mass-production industry from 67 % to 51 % over the span of twenty years is predicted [1]. To possibly make this decline less pronounced, the steel manufactures are forced to work continuously on improvement of existing AHSS on one hand and on new solutions for a better exploitation of their advantageous properties in car manufacturing on the other.

A better exploitation of AHSS in car manufacturing can be facilitated by FEA-based design of their forming and crash. This requires a mathematical formulation of the yield condition, hardening behaviour and formability. For decades, the forming limit curve (FLC) has been used to describe sheet metal formability [2]. Despite these limitations, such as its validity for linear strain paths and plane stress states ranging from uniaxial tension to equibiaxial tension, the accuracy of failure prediction based on the FLC is acceptable for conventional deep-drawing steels, but not for all AHSS [3]. Especially for shear-dominated stresses or bending loads alternative formability descriptions are required for AHSS [4]. For steel manufacturers it is therefore crucial to provide such alternative formability descriptions along with their AHSS.

As an alternative to the FLC to predict sheet metal failure, fracture models can be considered. In these models, fracture initiation is predicted based on a relation between the equivalent plastic strain $\bar{\varepsilon}_{\mathrm{pl}}$ and the equivalent plastic strain at fracture $\bar{\varepsilon}^{\mathrm{f}}_{\mathrm{pl}}$ (or fracture strain for simplicity). The fracture strain is a material characteristic, which depends on the stress state at which plastic deformation occurs. Many formulations exist for the dependence of the fracture strain on the stress state [5, 6, 7]. To quantify the stress state, the stress triaxiality η and normalised Lode angle $\bar{\theta}$ are becoming often used in the analysis of sheet metal forming and crash [8, 9].

A number of different specimens and tests can be currently used to calibrate fracture models. Still, fracture strains of AHSS are scarce in literature and if found, they often differ much from one author to another for similar or even the same steel grades [10, 11, 12]. The large discrepancies in fracture strains of different authors can be explained with the fact that there currently exist neither consistent requirements nor recommendations to the geometry of the material specimens, test procedures and results analysis for their determination [13, 14].

The paper presents an experimental-numerical evaluation of a new butterfly specimen for fracture characterisation of AHSS. The shape of the specimen is similar to that of the specimens proposed by Mohr and Henn [15] and Bai [16] but has geometry modifications, which make the position of the crack initiation more predefined, reduce the test sensitivity on the specimen holder displacement and act against undesired material thinning. The test is performed in a uniaxial tensile testing machine and provides sufficient data for calibration of common fracture models. In the first part of the paper, results of a numerical specimen evaluation are presented, which was performed with a material model of a dual-phase steel DP600 taken from literature [17] with both plastic flow and fracture behaviour descriptions. With the help of this numerical evaluation, spatial distribution of the equivalent plastic strain and damage variable in the specimen as well as time variation of the stress state at the location of fracture initiation were investigated. In the second part of the paper, results of an experimental-numerical test evaluation are shown, which was conducted on another dual-phase steel DP600 of 1.4 mm thickness, which was available with a description of plastic flow only and whose fracture behaviour of AHSS was characterised in the frame of this work. With the help of this experimental-numerical evaluation, the performance of the new butterfly specimen with regard to characterisation of the fracture behaviour was investigated. Finally, modelling of the DP600 fracture behaviour with the help of the CrachFEM ductile shear fracture model and modified Mohr-Coloumb fracture model is discussed.

2. Numerical evaluation

2.1. Material

The evaluation was performed on a dual-phase steel DP600 of 1.4 mm. To model the yield behaviour, the Hill'48 yield condition and an isotropic hardening formulation were taken from literature [17] and are graphically presented in figure 1. To describe the fracture behaviour, the CrachFEM ductile shear fracture model [6] was used:

$$\bar{\varepsilon}^{\mathrm{f}}_{\mathrm{pl}} = \frac{\bar{\varepsilon}^{\mathrm{f+}}_{\mathrm{pl}} \cdot \sinh\left(f \cdot (\gamma^{-} - \gamma)\right) + \bar{\varepsilon}^{\mathrm{f-}}_{\mathrm{pl}} \cdot \sinh\left(f \cdot (\gamma - \gamma^{+})\right)}{\sinh\left(f \cdot (\gamma^{-} - \gamma^{+})\right)}, \tag{1}$$

where $\gamma = (1 - k_{sf} \cdot \eta)/(\tau_{max}/\bar{\sigma})$ is the shear stress parameter with k_{sf} being the stress triaxiality weight factor, τ_{max} being the maximum shear stress and $\bar{\sigma}$ being the von Mises equivalent stress, γ^- and γ^+ are the shear stress parameters at equibiaxial compression and equibiaxial tension respectively, $\bar{\varepsilon}_{pl}^{f-}$ and $\bar{\varepsilon}_{pl}^{f+}$ are the fracture strains at equibiaxial compresion and equibiaxial tension and f is a model coefficient. The model input parameters are summarized in table 1 [17]. A graphical representation of the model is given in figure 2 [17].

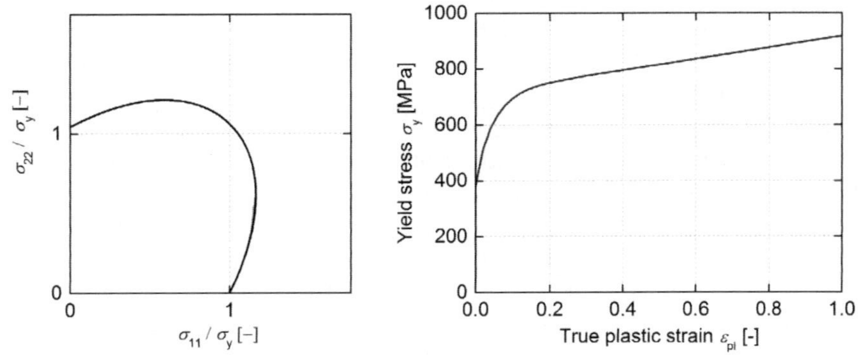

Figure 1. Yield locus (left) and yield curve (right) of DP600 [17]

Table 1. Coefficients of the CrachFEM shear fracture model of DP600 [17]

k_{sf} [-]	f [-]	γ^+ [-]	γ^- [-]	$\bar{\varepsilon}_{pl}^{f+}$ [-]	$\bar{\varepsilon}_{pl}^{f-}$ [-]
0.120	2.500	1.840	2.160	0.900	1.900

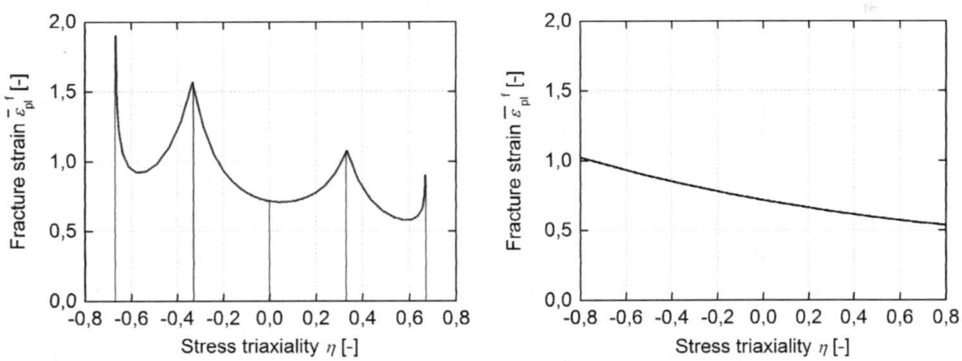

Figure 2. Fracture curve of DP600 for plane stress states (left), plane strain states (right) [17]

At each increment of the FEA, the incremental equivalent plastic strain $\bar{\varepsilon}_{pl,i}$ in a particular finite element is divided by the fracture strain $\bar{\varepsilon}_{pl}^{f}$, which is characteristic for the stress state acting in this element during this increment. The division product is then to be accumulated over the increments to obtain the variable D defined by:

$$D = \sum_i \frac{\bar{\varepsilon}_{pl,i}}{\bar{\varepsilon}_{pl}^{f}}, \tag{2}$$

which is referred to as damage variable. At an increment of $D \geq 1$ fracture initiation is assumed.

2.2. Finite element model

The new specimen, the geometry of which is similar to that of the specimens proposed by Mohr and Henn [15] and Bai [16], is shown in figure 3 and can be divided into the following areas:

- two load introduction areas over which the external load is introduced from the specimen holder into the specimen via friction forces (1.1 and 1.2),
- one load transfer area with a material cutout on the left and on the right side of the specimen, which follows a smooth concave-convex curve (2),
- one analysis area with reduced material thickness by 25 % of the initial sheet thickness from each of the two sheet sides, which is entirely enclosed within the load transfer area along a smooth concave-convex loop (3).

The interaction of the load transfer area and analysis area lead to localisation of plastic strain in the middle of the analysis area independent of the direction of the load applied in the sheet plane. Thus, the first crack initiates in the material interior and not at the specimen edge. This ensures that the material formability at a desired stress state and not the specimen edge formability at a uniaxial stress state is determined. The specimen is extracted from the sheet with the help of wire erosion so that the longer side of the specimen is oriented parallel to the rolling direction of the sheets. The thickness reduction in the analysis area is carried out by means of machining, whereby the roughness of $R_a < 0.8$ and $R_z < 6.3$ or finer is ensured [18].

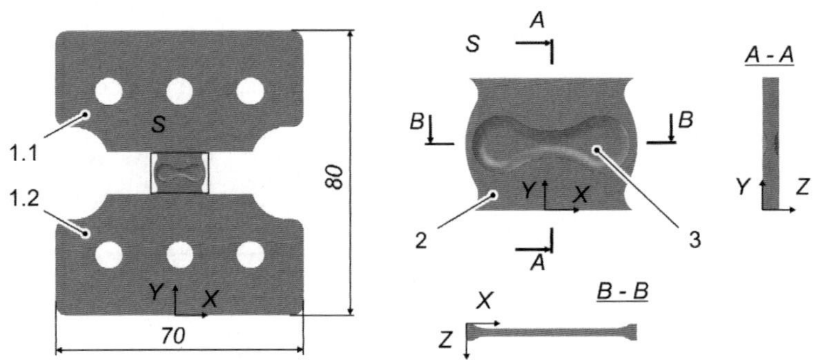

Figure 3. Geometry of the new butterfly specimen with two load introduction areas (1.1 and 1.2), one load transfer area (2) and one analysis area (3)

The stress and strain fields in the new specimen and specimen of Bai [16] due to a load arbitrary oriented in the specimen plane were investigated with a FE model shown in figure 4. Parts of the specimen holder and the fastening plates, which come into contact with the specimen were modelled as elastic bodies. The specimen was clamped between the specimen holder and the fastening plates with the help of pressure of $p_z = 204\,\text{MPa}$ applied to six circle areas on the fastening plates, which imitate the contact areas beneath the disks of the six screws. The magnitude of the pressure was chosen to correspond to a fastening moment on a M8 screw of 35 Nm. 3D volume elements were used to realistically represent the stress state in the material, especially beyond the onset of plastic strain localisation. In particular, sufficiently fine (0.1 mm) continuum eight-node linear reduced-integration elements with an hourglass control as provided in the element library of Abaqus/Standard under the notation C3D8R were used for specimen areas of large deformations. Furthermore, a thin layer of complaint shell elements was applied to one of the specimen surfaces for evaluation of strain paths. The specimen deformation was induced by applying a displacement u_{xy} on all the outer surface nodes of one of the specimen holders. The direction of the displacement u_{xy} was defined by the load application angle α. The following α were investigated: $-3.0°$, $12.5°$, $28.0°$, $43.5°$, $59.0°$, $74.5°$ and $90.0°$, each of them stayed constant during the corresponding test.

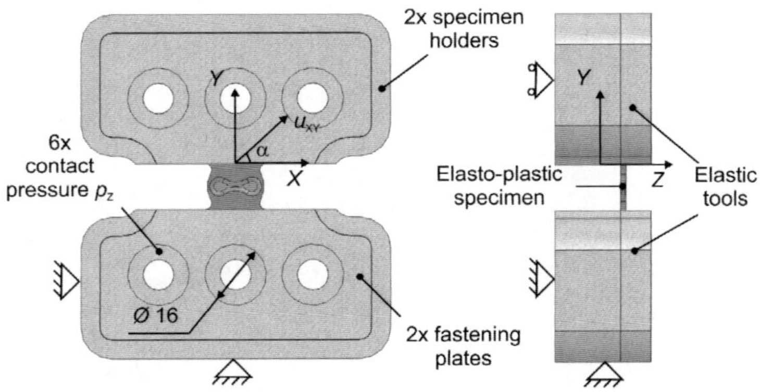

Figure 4. Components and boundary conditions of the finite element model of a tensile test on butterfly specimen according to Bai [16] and new butterfly specimen

2.3. Results and discussion

The objectives of the performed geometry optimisation of the new butterfly specimen as compared to the butterfly specimen according to Bai [16] were:

- contraction of the zone of high plastic deformation along the shorter side of the specimen,
- extension of the zone of high plastic deformation along the longer side of the specimen,
- suppression of material thinning in the zone of high plastic deformation.

To reach the objectives, the curvature radius of the waist of the analysis area both in the specimen plane and in the plane orthogonal to it was decreased. Furthermore, the analysis area of reduced material thickness was completely enclosed into the load transfer area of the initial material thickness. The latter leads to two outwardly convex bridges of the initial material thickness connecting the two opposite sides of the specimen, which support the stress state and act against material thinning. An exemplary result of the optimisation is shown in figure 5.

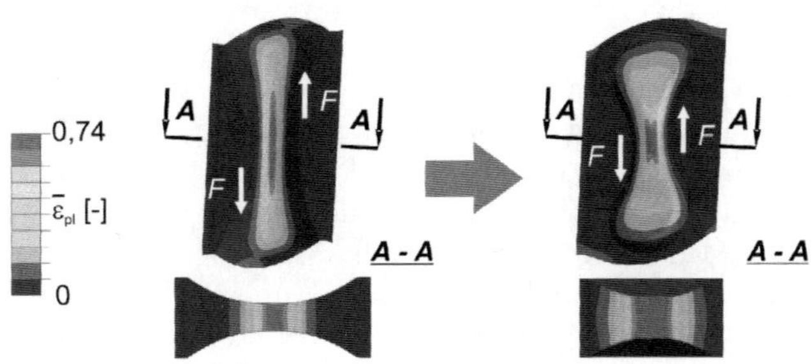

Figure 5. Equivalent plastic strain in the butterfly specimen according to Bai [16] (left) and new butterfly specimen (right) at the moment of fracture initiation for $\alpha = -3.0°$

Figure 5 illustrates the contraction of the zone of high plastic deformation along the shorter side of the specimen and its extension along the longer side of the specimen. This increases the safety of fracture initiation at a desired stress state as demonstrated in figure 6. For the diagrams in figure 6, it was assumed that due to inhomogeneity of the real material formability fracture initiation may take place at any finite element in which the condition $D \in [0.98 \cdot D_{\max}; D_{\max}]$ with $D_{\max} \geq 1$ is satisfied. Under this assumption, figure 6 shows that the differences in the stress triaxialities among possible locations of fracture initiation is lower for the new butterfly specimen.

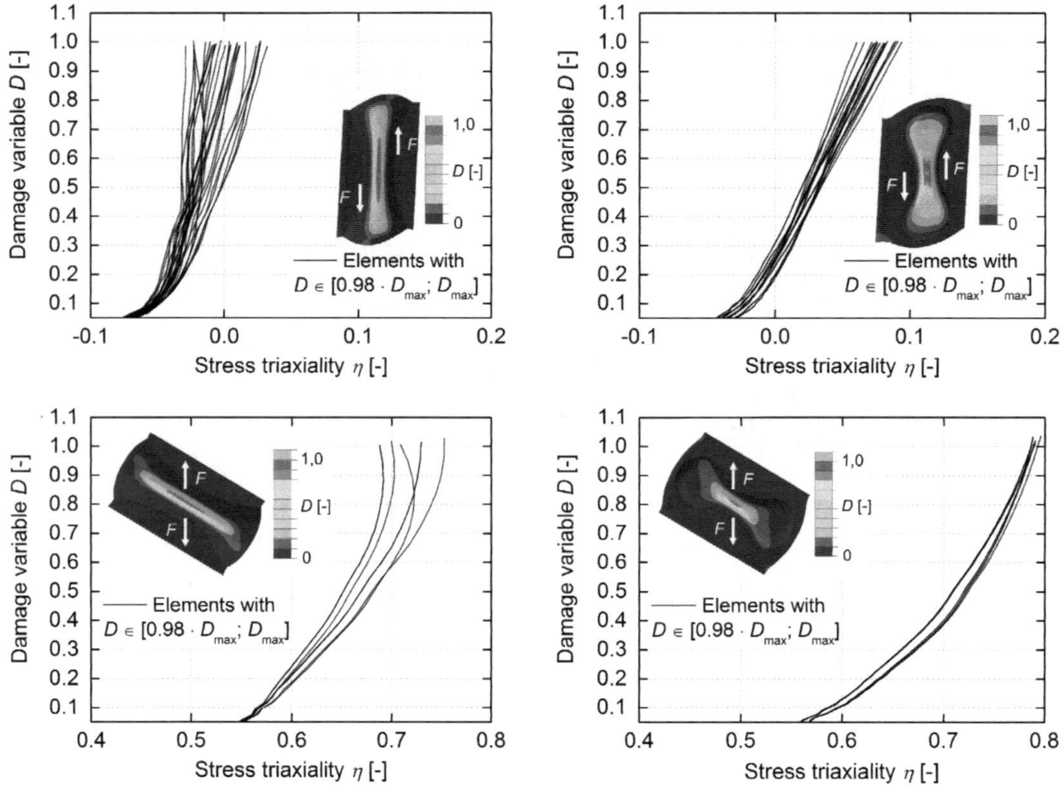

Figure 6. Stress triaxiality in the elements of potential fracture initiation for the butterfly specimen according to Bai [16] (left) and new butterfly specimen (right) for $\alpha = -3°$ (top) and $\alpha = 59°$ (bottom)

Figure 5 also illustrates the extension of the zone of high plastic deformation along the longer side of the specimen. This favourably decreases the sensitivity of the equivalent plastic strain at the location of fracture initiation to the relative displacement of the opposite specimen holder parts as presented in figure 7. This decreased sensitivity allows a more accurate fracture characterisation with the help of conventional uniaxial tensile testing machines.

Finally, the effect of the enclosure of the analysis area of reduced material thickness with the load transfer area of initial material thickness on the stress state is illustrated in figure 8. The stress state in the new butterfly specimen is slightly less sensitive to the variation of the load application angle. In other words, a more accurate definition of the desired stress state via definition of the load application angle is possible. For stress triaxialities $\eta > 0.33$ at the location of fracture initiation, the new butterfly specimen exhibits less local material thinning. This becomes manifest in lower variation of the stress triaxiality at the fracture initiation location.

The analysis of damage variable distribution in the analysis area of the new butterfly specimen obtained for the studied seven load application angles reveals that the fracture initiation is expected at the surface of the analysis area for $\alpha = -3.0°$ and $\alpha = 12.5°$ and is likely to occur at the surface of the analysis area for $\alpha = 28.0°$ as shown in figure 9, top. For these load application angles, fracture initiation takes place due to plastic deformation at plane stress states and the fracture strain can be determined with the help of digital image correlation of sufficiently fine resolution (specimen surface). On the contrary, the maximum damage variable is reached in the middle of the specimen for $\alpha = 43.5°$, $\alpha = 59.0°$, $\alpha = 74.5°$ and $\alpha = 90.0°$

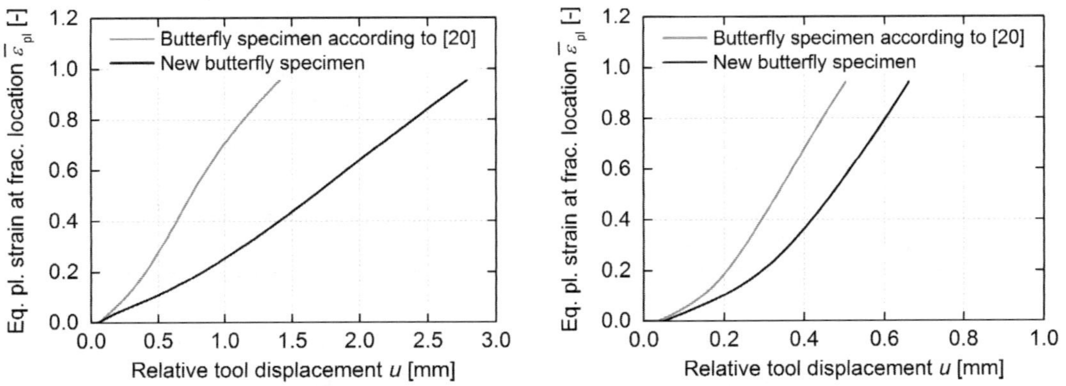

Figure 7. Equivalent plastic strain at the location of fracture initiation versus tool displacement for the butterfly specimen according to [16] and proposed butterfly specimen for the load application angle $\alpha = -3°$ (left) and $\alpha = 59°$ (right)

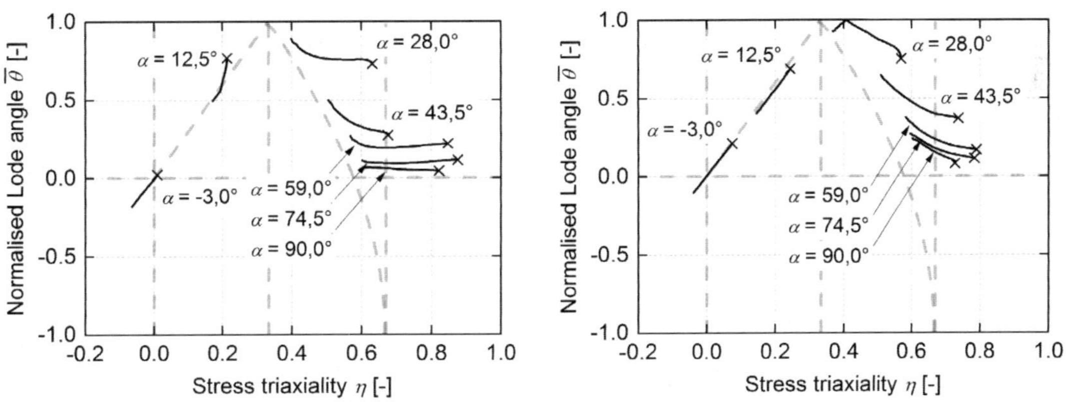

Figure 8. Development of the stress state at the location of fracture initiation for the butterfly specimen according to Bai [16] (left) and the new butterfly specimen (right) for α

as shown in figure 9 bottom. For these load application angles, fracture initiation takes place due to plastic deformation at three-dimensional stress states and other methods like FEA are required to determine the fracture strain (specimen middle). The transition between these two characteristic locations of fracture initiation in the tensile tests on the new butterfly specimen occurs at the stress triaxiality of $\eta \approx 0.33$.

As a last result, development of the stress states at the location of fracture initiation in the space of the stress triaxiality and equivalent plastic strain as well as development of the strain state at the most deformed material point at the surface in the space of the second major strain and first major strain are presented in figure 10. From figure 10 it can be seen that the change of the load application angle from $\alpha = -3.0°$ to $\alpha = 43.5°$ goes along with a change of the stress triaxiality, whereas the change of the load application angle from $\alpha = 43.5°$ to $\alpha = 90.0°$ lead only to minor changes of the stress triaxiality and is rather characterised by a change of the normalised Lode angle (compare figure 8, right and figure 10, left). The strain paths at the most deformed material point at the surface for different load application angles are presented in figure 10, right and cover the range of strain states from pure shear to plane strain.

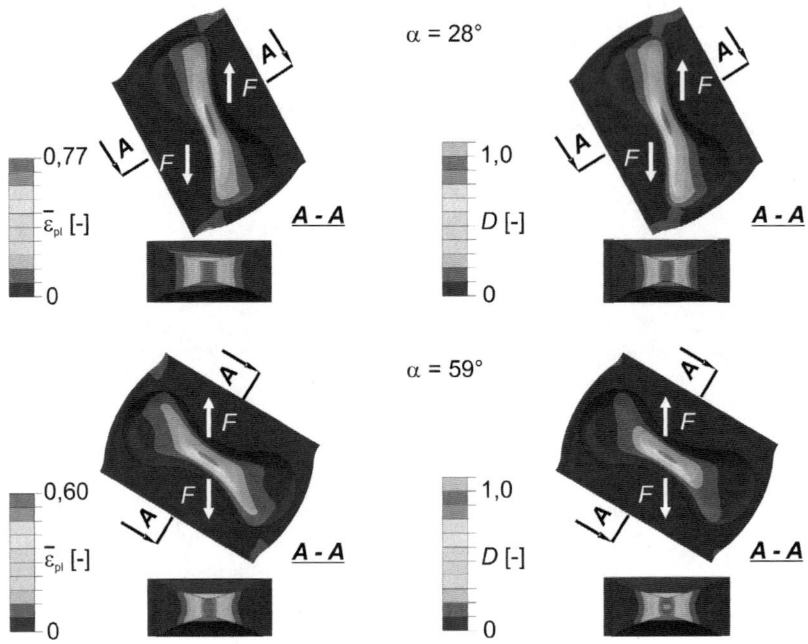

Figure 9. Equivalent plastic strain (left) and damage variable (right) in the proposed butterfly specimen at the moment of fracture initiation for $\alpha = 28.0°$ (top) and $\alpha = 59.0°$ (bottom)

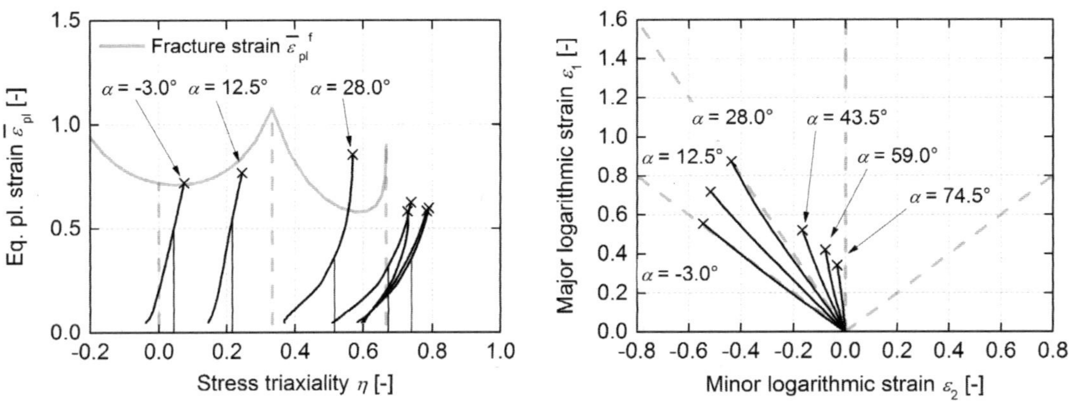

Figure 10. Stress state at the location of fracture initiation (left) and strain path at the location of the highest equivalent plastic strain on the specimen surface (right) of the new specimen

3. Experimental-numerical evaluation

3.1. Experimental set-up

The tests on the new butterfly specimen were realised in a uniaxial tensile testing machine with the help of a tool shown in figure 11, left. The tool consists of two fixtures which are firmly attached to the pistons of the machine and have a slot to accommodate two half-disk-shaped specimen holders. The specimen holders can be adjusted in this slot allowing a desired specimen orientation relative to the test load direction. Each fixture and the corresponding specimen holder have seven holes through which pins are inserted to lock the specimen holders in a particular orientation and prevent them from rotating during the test. The tool allows the following load application angles of $-3.0°$, $12.5°$, $28.0°$, $43.5°$, $59.0°$, $74.5°$ and $90.0°$, which remain constant during the test. The tests were carried out with the piston speed of 0.05 mm/s to ensure quasistatic deformation. During the tests, deformation of the analysis area of the specimen was continuously recorded with the help of the optical measurement system Aramis (GOM mbH) as shown in figure 11, right. Furthermore, the relative displacement of two reference points on the opposite specimen holders close to the analysis area of the specimen were optically recorded. The relative displacement of these two reference points at the moment of fracture initiation provided the boundary conditions for the FEA of the tests.

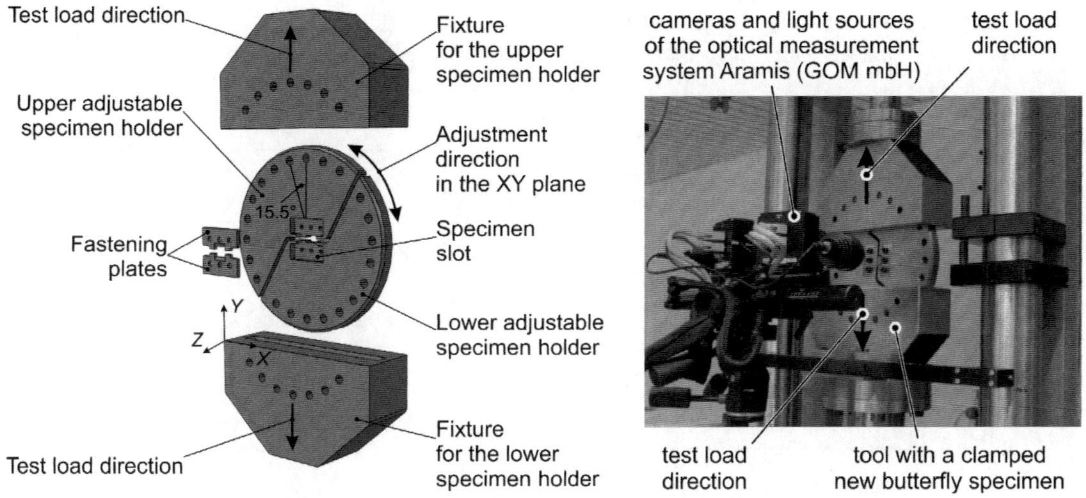

Figure 11. Tool (left) and set-up (right) for tensile tests on the new butterfly specimen

3.2. Material

A DP600, 1.4 mm, produced by Salzgitter Flachstahl GmbH was used. The elastic behaviour of the material was described with the elastic modulus of 210 GPa and the Poisson ratio of 0.3. The plastic behaviour was modelled with the Hill'48 yield condition with $F = 0.460$, $G = 0.534$, $H = 0.466$, $N = 1.366$, $L = 1.500$ and $M = 1.500$ [18] and a combined Swift-Hocket-Sherby model determined with the help of uniaxial tensile and hydraulic bulge tests [19] given by

$$\sigma = 0.479 \cdot \left(642 \cdot \left(0.00596 + \bar{\varepsilon}_{\mathrm{pl}}\right)^{0.660}\right) + 0.521 \cdot \left(1224 - (1224 - 757) \cdot e^{-15.5 \cdot \bar{\varepsilon}_{\mathrm{pl}}^{0.820}}\right) \quad (3)$$

3.3. Finite element model

The numerical analysis of the tests was performed to determine the development of stress state and equivalent plastic strain at the location of fracture initiation over time. The location of fracture initiation was assumed based on the results of the numerical evaluation of the test from the previous chapter, which was performed with both the plasticity model and fracture model for another DP600 from [17]. The same finite element model schematically shown in figure 4 was used. The relative tool displacement for each of the seven load application angles was set to be equal to the relative tool displacement obtained in the experiment for the corresponding load application angle at the moment of fracture initiation. Three element sizes were used to investigate the dependence of the fracture strain on the space resolution: 0.025 mm, 0.100 mm and 0.180 mm. It should be noted here that element with the edge length of 0.2 mm are the largest elements, which can be used to accurately model geometry of the new butterfly specimen.

3.4. Results and discussion

The finite element model used in the study was validated based on a comparison of the experimentally and numerically determined test forces, which are exemplary shown for the two extreme load application angles in figure 12, left. It can be concluded that the quality of the used material description and boundary conditions is acceptable. The high scatter of the test force at the load application angle $\alpha = 90.0°$ is mainly explained with the scatter of the material thickness in the analysis area of the specimen and not with the scatter of the material properties. Figure 12, right shows representative strain paths determined with the help of digital image correlation (DIC) for the load application angles, which lead to fracture initiation at the specimen surface (plane stress states). The strain paths in figure 12 are deliberately shown for small and medium strains only as at higher strains the data determined with DIC went corrupt due to excessive shear deformation of the paint pattern applied to the surface.

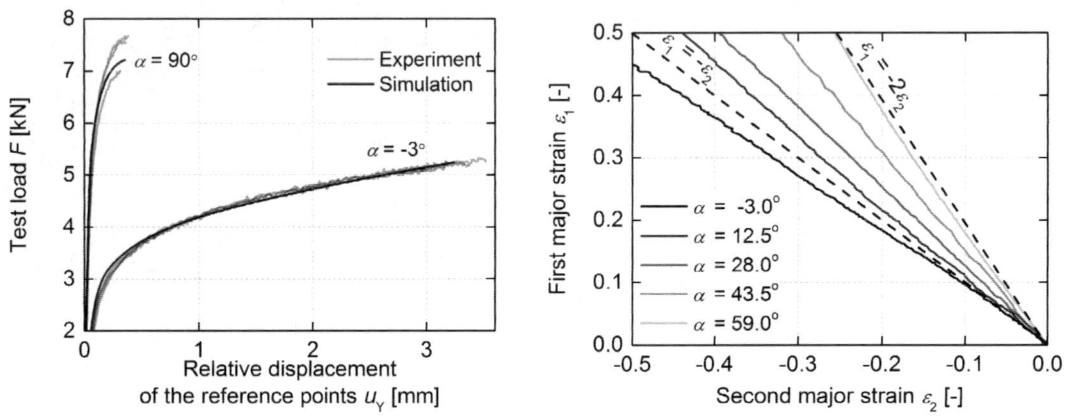

Figure 12. Exemplary experimentally and numerically determined test forces (left) and optically determined strain paths at the location of the highest deformation (right)

The development of the stress state at the location of fracture initiation is presented in figure 13. Based on the results of the numerical test evaluation given in the previous chapter, it is assumed that fracture initiation takes place at the surface of the analysis area due to plastic deformation at plane stress states for the load application angles α of $-3.0°$, $12.5°$, $28.0°$, $43.5°$ and $59.0°$ and in the middle of it due to plastic deformation at 3D stress states for the load application angles α of $74.5°$ and $90.0°$. Due to the horizontal movement of the specimen holders, a stress state shift towards lower stress triaxialities is observed (figure 10, left and figure 13, right). In figure 13, right the characteristic stress states for each load application angle are denoted by the thin black lines. The characteristic stress state parameters were determined as a weight centre of the area below the corresponding curve as used by e. g. Wierzbicki et al. [7].

The equivalent plastic strain at the location of fracture initiation numerically determined for the tool displacement, at which fracture initiation took place in the experiment was defined to be the fracture strain. The so determined fracture strain was found to be dependent on the size of the finite elements or, in other words, on the space resolution. This dependency was found to be linear in the studied range of the virtual strain gauge lengths l_0 and is in its turn stress state dependent as shown in figure 14, left. The space resolution dependency of the equivalent plastic strain is intrinsic for ductile materials exhibiting strain localisation and is more pronounced for triaxial tensile stress states. In the present work, it is considered with the help of equation 4, which is constructed based on the obtained numerical results for three different finite element sizes (or virtual strain gauge lengths l_0) mentioned above. The dependency of the virtual strain

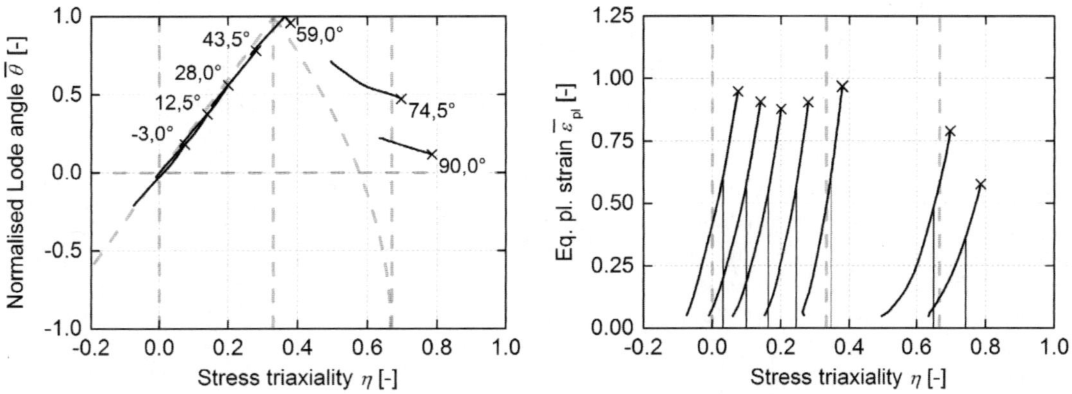

Figure 13. Stress state at the location of fracture initiation in the tensile tests on the new butterfly specimen with an indication of the characteristic stress state of DP600

gauge length factor b on the stress triaxiality is shown in figure 14, right.

$$\bar{\varepsilon}^{\text{f}}_{\text{pl}}(l_0) = \bar{\varepsilon}^{\text{f}}_{\text{pl}}(l_0 = 0) + b \cdot l_0, \tag{4}$$

where $\bar{\varepsilon}^{\text{f}}_{\text{pl}}(l_0)$ is the fracture strain for the strain gauge length l_0 and b is the strain gauge length factor from table 2. It should be noted that equation 4 is valid for $l_0 \in [0.02\,\text{mm}; 0.2\,\text{mm}]$. The results of the performed determination of the input data for parametrisation of fracture models for three different space resolutions are summarised in table 2. A comparison of the fracture data with the fracture data obtained by Basaran [20] for a DP600 of 1.0 mm and $l_0 = 0.075$ mm is presented in figure 15, right and reveals a good accordance.

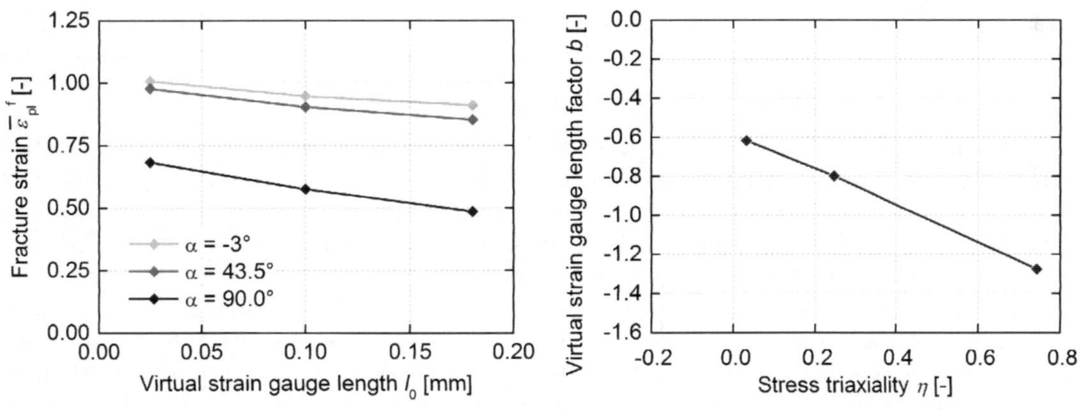

Figure 14. Fracture strain versus virtual strain gauge length for different α (left) and dependence of the virtual strain gauge length factor on the stress triaxiality (right) for DP600

The data from table 2 was subsequently approximated with the CrachFEM ductile shear fracture model given by equation 1. The model parameters are in table 3. A comparison of the graphical representation of the CrachFEM model from [17] and graphical representation of the CrachFEM model determined in this work is given in figure 16. From the comparison it can be seen that the mean positions of the two fracture surfaces with regard to the fracture strain

Table 2. Exp.-numerical results of the tensile tests on the new butterfly specimen of DP600

Load appl. angle α [°]	Stress triaxiality η [-]	Normalised Lode angle $\bar{\theta}$ [-]	Fracture strain $\bar{\varepsilon}^{f}_{pl}$ for the strain gauge length $l = 0.01\,\mathrm{mm}$ [-]	Strain gauge length factor b acc. to equation 4 [-]
-3.0	0.032	0.047	0.947	0.610
12.5	0.100	0.243	0.907	0.673
28.0	0.162	0.436	0.878	0.731
43.5	0.246	0.683	0.904	0.809
59.0	0.348	0.960	0.941	0.990
74.5	0.649	0.514	0.790	1.183
90.0	0.743	0.144	0.576	1.272

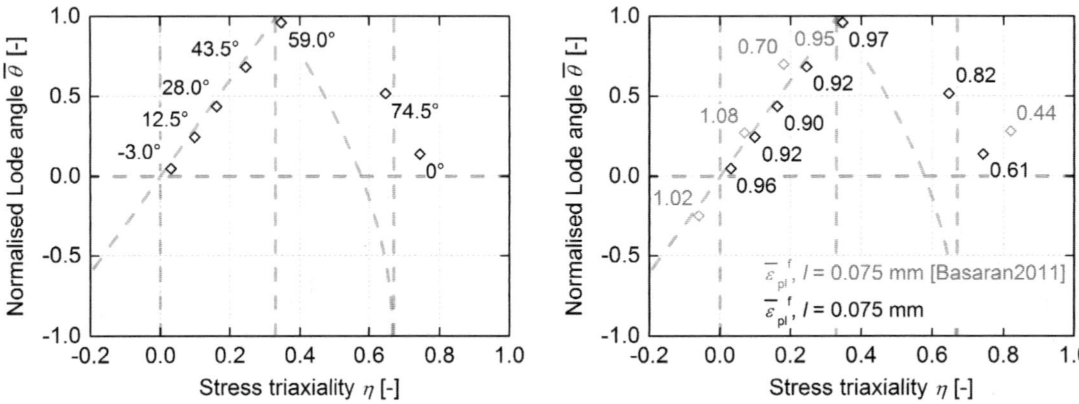

Figure 15. Characteristic stress states at the location of fracture initiation (left) and comparison of the test results with those of Basaran [20] (right) for $l = 0.075\,\mathrm{mm}$ and DP600

axis are similar. The difference between the two lies in the dependence of the fracture strain on the normalised Lode angle. This dependence was found to be less pronounced for the fracture surface determined in this work as compared to the fracture surface from [17]. The shape of the fracture surface of DP600 determined in this work correlates well with the results of Basaran [20]. It can be concluded that the dependence of the fracture strain of DP600 on $\bar{\theta}$ is not that pronounced as it is often assumed to be as e. g. in [17]. A reasons for the higher curvature of the fracture surface along the normalised Lode angle axis in [17] may be inaccurate determination of the fracture strain in the shear test. As shown by Behrens et al. [17], the use of an ASTM-similar shear specimen in the shear test can lead to underestimated values of the fracture strain and, hence, to a more curved fracture surface along the normalised Lode angle axis.

The dependence of the fracture strain on the virtual strain gauge length l_0 is illustrated in figure 17. From figure 17, it can be clearly seen that the dependence of the fracture strain on the virtual strain gauge length l_0 is more pronounced for higher stress triaxialities and can be neglected for compression-dominated stress states characterised by $\eta < -0.33$. This behaviour of the material in the FEA can be explained with plastic strain localisation prior to fracture, which is better resolved by finer finite elements leading to higher values of computed strains. Naturally, the degree of plastic strain localisation is defined by the stress state.

IOP Conf. Series: Materials Science and Engineering **159** (2016) 012015 doi:10.1088/1757-899X/159/1/012015

Table 3. Coefficients of the CrachFEM shear fracture model of DP600

l [mm]	k_{sf} [-]	f [-]	γ^+ [-]	γ^- [-]	$\bar{\varepsilon}_{\mathrm{pl}}^{\mathrm{f+}}$ [-]	$\bar{\varepsilon}_{\mathrm{pl}}^{\mathrm{f-}}$ [-]
0.020	0.259	0.1490	1.655	2.346	0.910	1.360
0.100	0.340	0.0213	1.546	2.454	0.802	1.360
0.200	0.436	0.0490	1.418	2.582	0.671	1.360

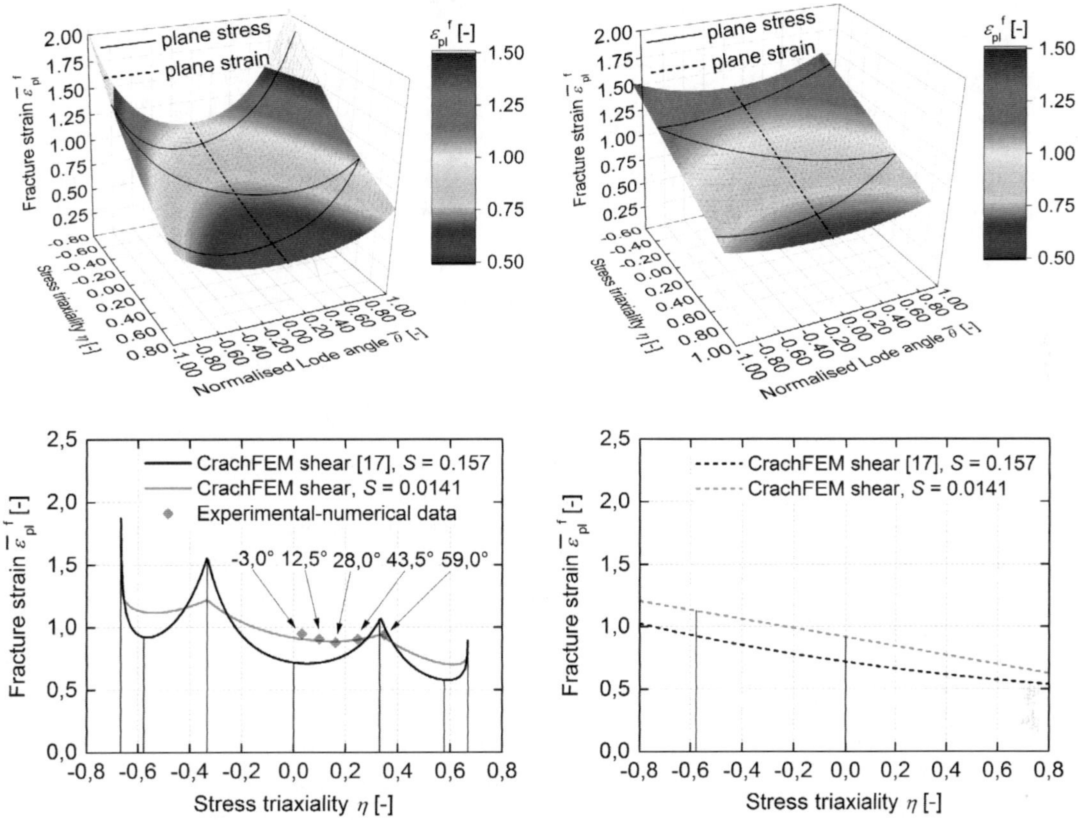

Figure 16. Fracture surface of DP600 according to the CrachFEM model from [17] (top left) and CrachFEM model from this work for the virtual gauge length of 0.1 mm (top right) and fracture curves for plane stress states (bottom left) and plane strain states (bottom right) with S being the sum of squared residuals

Finally, the fracture data from table 2 for the virtual strain gauge $l_0 = 0.2$ mm was approximated with the modified Mohr-Coloumb shear fracture model [21], which can also be used to predict fracture initiation in AHSS [22]. The determined model parameters are summarised in table 4. It should be noted here, however, that not the original MMC model [21], which corresponds to a combination of the Mohr-Coulomb fracture criterion with a Lode-angle-dependent plasticity model, is used. Instead, an inconsistent version of the MMC model, as discussed e. g. by Mohr and Marcadet [23], which is just a fitting function for the fracture data is made use of. A comparison of the fracture surfaces is presented in figure 18. The deviations of the model predictions from the seven data points used for their parametrisation can be seen from table 5. From figure 18 and table 5 it can be seen that both models approximate the

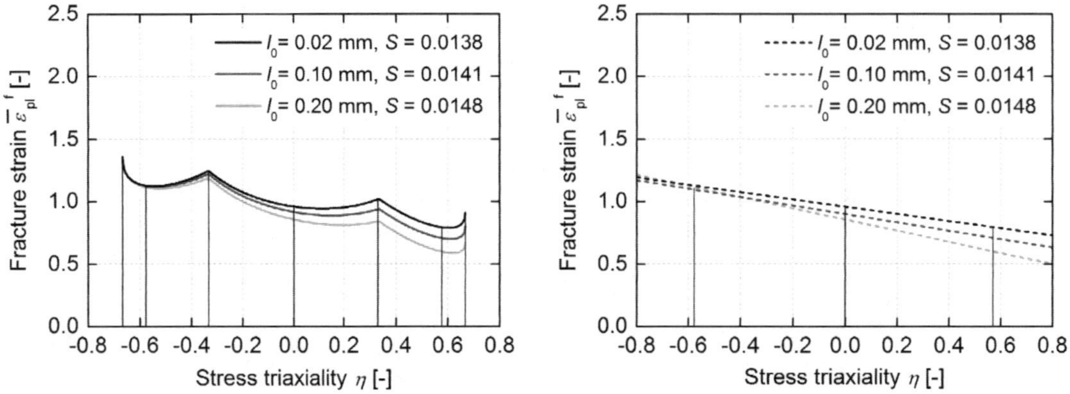

Figure 17. Fracture curves of DP600 according to the CrachFEM shear fracture model for different virtual strain gauges l_0 at plane stress states (left) and plane strain states (right)

fracture data similarly and accurately enough. In the present study, the focus is put on fracture characterisation at shear-dominated stress states of $\eta \in [0; 0.33]$. In this range both models perform equally well. The difference between the two becomes notable for $\eta < 0$. In the case of the symmetric shape of the fracture surface given by the CrachFEM model with regard to $\bar{\theta} = 0$, the increase of the fracture strain at plane stress states with a decreasing stress triaxiality in the range of $\eta \in [-0.33; 0]$ is defined by the overall slope of the fracture surface along the stress triaxiality axis. On the contrary, the asymmetry of the fracture surface given by the MMC model with regard to $\bar{\theta} = 0$allows a steeper increase of the fracture strain at plane stress states with a decreasing stress triaxiality in the range of $\eta \in [-0.33; 0]$. Therefore, the MMC fracture model parameterised with the fracture data from table 2 may give slightly better predictions of fracture initiation due to plastic deformation at compression-superimposed shear stress states characterised by $\eta \in [-0.33; 0]$. Plastic deformation at such stress states may occur, for example, in the flange area of a sheet in a deep-drawing process.

Table 4. Coefficients of the MMC shear fracture model of DP600

l [mm]	A [-]	n [-]	C_1 [-]	C_2 [-]	C_3 [-]
0.200	1500	0.635	0.291	900.0	1.070

Table 5. Exp.-numerically determined data and model predictions for DP600 ($l_0 = 0.2\,\mathrm{mm}$)

η [-]	0.032	0.100	0.162	0.246	0.348	0.649	0.743
$\bar{\theta}$ [-]	0.047	0.243	0.436	0.683	0.960	0.514	0.144
$\bar{\varepsilon}_{pl}^{f}$ (experiment) [-]	0.886	0.839	0.805	0.823	0.842	0.671	0.449
$\bar{\varepsilon}_{pl}^{f}$ (CrachFEM) [-]	0.843	0.820	0.810	0.809	0.824	0.594	0.527
$\bar{\varepsilon}_{pl}^{f}$ (MMC) [-]	0.870	0.821	0.797	0.793	0.823	0.593	0.547

IOP Conf. Series: Materials Science and Engineering **159** (2016) 012015 doi:10.1088/1757-899X/159/1/012015

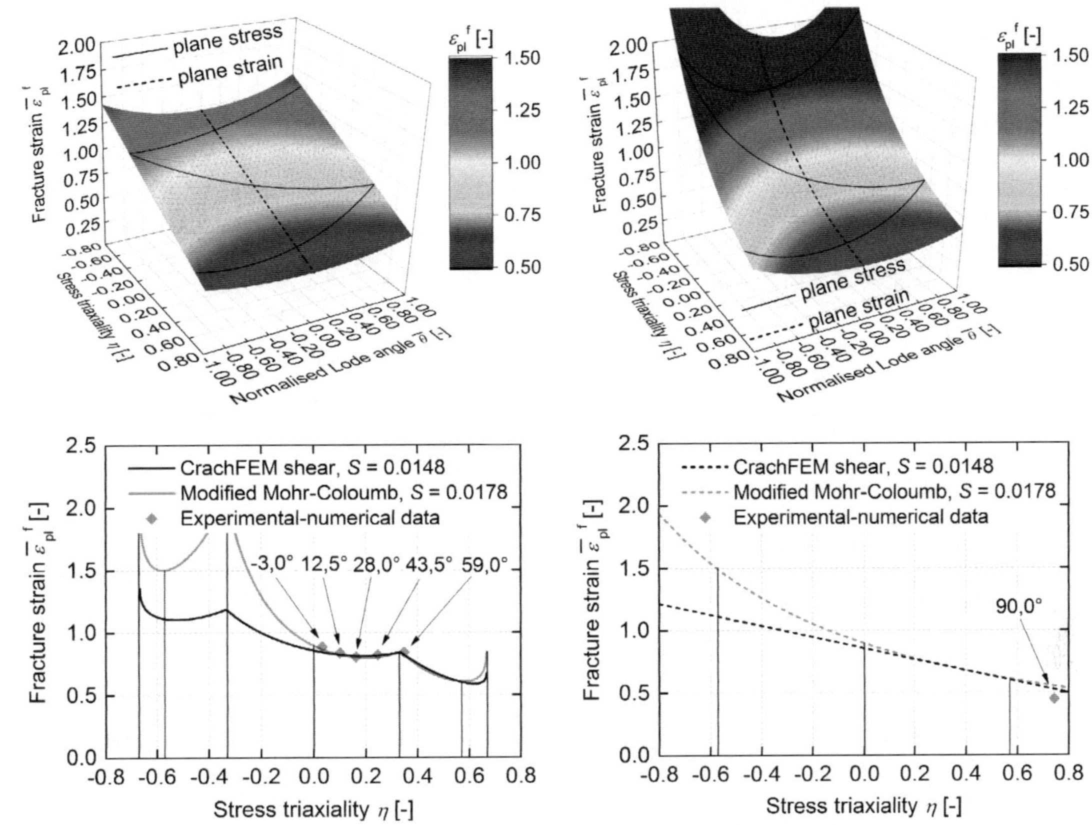

Figure 18. Fracture surfaces of DP600 according to the CrachFEM model (top left) and MMC model (top right) for the virtual gauge length of $l_0 = 0.2$ mm and fracture curves for plane stress (bottom left) and plane strain states (bottom right)

4. Conclusion and outlook

The work proved applicability of the new test on the new butterfly specimen to fracture characterisation of AHSS sheets. The advantages of the new specimen include an improved safety of fracture initiation at a desired stress state and decreased sensitivity of the plastic strain to the relative tool displacement. Furthermore, the new specimen exhibits less local material thinning during the test. The disadvantages of the new specimen remain in high production costs and fracture initiation at the machined surface at some load cases. The test allows material testing at two shear-tensile, two tensile-shear and one uniaxial tension plane stress states as well as two three-dimensional tensile stress states. For the five shear-tensile/tensile-shear plane stress states, determination of the fracture strain with digital image correlation is possible. For the other two stress states, other methods like FEA need to be used. In the future, material testing at five out of seven stress states seems to be more appropriate. The dependence of the fracture strain on the normalised Lode angle for DP600 was found to be less pronounced as it is often assumed. The dependence of the fracture strain on the space resolution of the FEA was found to be more pronounced for three-dimensional tensile stress states and negligible for compression-dominated stress states. Both the CrachFEM fracture model and MMC fracture model perform equally well in the studied range of stress states. The difference between the two models becomes notable for negative stress triaxialities only. The test procedure and results analysis are believed to contribute to current discussions on requirements to AHSS fracture characterisation.

Acknowledgments
Most of presented results were obtained in the frame of the projects P 972, P 1007 and P 1145 of the Research Association for Steel Application (FOSTA), Dusseldorf, Germany. The financial support of the FOSTA and technical support of the industrial project committees are gratefully acknowledged. From the industrial project committees, the authors especially thank Ford Forschungszentrum Aachen GmbH (A. Bach) for the provided material and Volkswagen AG (Dr. C. Sunderkötter) for funding of the tool manufacturing. Valuable support of Dr. B. Hackl (voestalpine Stahl GmbH) with the approximation of the fracture data is much appreciated.

References

[1] Heuss R, Müller N, van Sintern W and Starke A 2012 *Lightweight, heavy impact* (McKinsey & Company)
[2] Banabic D 2010 *Sheet Metal Forming Processes. Const. Modelling and Num. Simulation* (Berlin: Springer)
[3] Roll K 2008 Simulation of sheet metal forming - necessary developments in the future *Proc. NUMIFORM (Interlaken)* (Zurich: IVP, ETH) pp 3-11
[4] Keßler L and Gerlach J 2008 Industrial aspects of material modelling for steel grades in the past, present and future *Proc. NUMIFORM (Interlaken)* (Zurich: IVP, ETH) pp 13-17
[5] Bao Y and Wierzbicki T 2004 A comparative study on various ductile crack formation criteria *J. Eng. Mater. Technol.* **126** pp 314-24
[6] Hooputra H, Gese H, Dell H and Werner H 2004 A comprehensive failure model for crashworthiness simulation of aluminium extrusions *Int. J. Crash.* **9** pp 449-63
[7] Wierzbicki T, Bao Y, Lee Y-W and Bai Y 2005 Calibraton and evaluation of seven fracture models *Int. J. Mech. Sci.* **47** pp 719-43
[8] Till E and Hackl B 2013 Calibration of plasticity and failure models for AHSS sheets *Proc. IDDRG (Zurich)* (Zurich: Institute of Virtual Manufacturing, ETH) pp 119-24
[9] Bai Y and Wierzbicki T 2013 The concept of damage accumulation for prediction necking and fracture of sheets *Proc. IDDRG (Zurich)* (Zurich: Institute of Virtual Manufacturing, ETH) pp 9-13
[10] Walters CL 2009 Development of a punching technique for ductile fracture testing over a wide range of stress states and strain rates *PhD Thesis* (Cambridge: MIT)
[11] Chottin J, Hug E and Rachnik M 2011 Damage accumulation in DP1000 sheets submitted to various stress states *Steel Res. Int.* **Proc ICTP** pp 895-900
[12] Behrens B-A, Bouguecha A, Vucetic M and Peshekhodov I 2012 Characterisation of the quasi-static flow and fracture behaviour of DP steel sheets in a wide range of stress states *Arch Civ Mech Eng* **12** pp 397-406
[13] Behrens B-A, Vucetic M and Peshekhodov I 2014 Effiziente Charakterisierung und Modellierung des Fließ- und Versagensverhaltens von DP-Stählen *Tagungsband des UKH* (Hanover: IFUM, LUH) pp 77-97
[14] Keßler L, Richter H and Rösen H 2014 Die Anwendung von erweiterten Bewertungskriterien für die Simulation von Stahlwerkstoffen *Tagungsband des UKH* (Hanover: IFUM, LUH) pp 99 -109
[15] Mohr D, Henn S 2007 Calibration of stress-triaxiality dependent crack formation criteria: A new hybrid experimental-numerical method *Exp Mech* **47** pp 805-820
[16] Bai Y 2008 Effect of loading history on necking and fracture *PhD thesis* (Cambridge: MIT)
[17] Peshekhodov I, Dykiert M, Vucetic M, Behrens B-A 2016 Evaluation of common tests for fracture characterisation of advanced high-strength sheet steels with the help of the FEA *IOP Conference Series: Materials Science and Engineering* **Proc. IDDRG (Linz)** (Bristol: IOP Publishing)
[18] Behrens B-A, Vucetic M and Peshekhodov I 2014 Calibration of fracture initiation models for advanced high-strength sheet steels in a wide range of stress states using a uniaxial tensile testing machine *Proc. IDDRG (Paris)* (Senlis Cedex: CETIM) pp 252-57
[19] Vucetic M, Bouguecha A, Peshekhodov I, Götze T, Huinink T, Friebe H, Möller T and Behrens B-A 2011 Num. validation of analytical biaxial true stress - true strain curves from the bulge test *Proc. NUMIFORM (Seoul)* (Melville: AIP) pp 107-14
[20] Basaran M 2011 Stress state dependent damage modelling with a focus on the Lode angle influence *Dissertation* (Aachen: RWTH)
[21] Bai Y, Wierzbicki T 2010 Application of extended MC criterion to ductile fracture *Int J Frac* **161** pp 1-20
[22] Li Y, Luo M, Gerlach J, Wierzbicki T 2010 Prediction of shear-induced fracture in sheet metal forming *J Mat Proc Tech* **210** pp 1858-69
[23] Mohr D, Marcadet SJ 2015 Micromechanically-motivated phenomenological Hosford-Coulomb model for predicting ductile fracture initiation at low stress triaxialities *Int J Sol Struc* **67-68** pp 40-55

Influence of specimen size and sheet thickness on the material behavior of AZ31B under uniaxial tension

S Suttner[1] **and M Merklein**[1]

[1] Institute of Manufacturing Technology, Department Mechanical Engineering, Friedrich-Alexander-Universität Erlangen-Nürnberg (FAU), Egerlandstraße 13, 91058 Erlangen, Germany

E-mail: sebastian.suttner@fau.de

Abstract. Concerning a low specific density, magnesium alloys offer a great potential to reduce the part weight. Due to the conditional formability of magnesium alloys at room temperature, forming processes are mostly realized at elevated temperatures. Within this contribution, uniaxial tensile tests for two different specimen geometries with a gauge length of 50 mm and 2 mm and a cross section of 12.5x1 mm² and 2x1 mm² are carried out for the magnesium alloy AZ31B at an elevated temperature of 200 °C. Since the specimen size is reduced, the influence of the sheet thickness needs to be analyzed. The analysis of the sheet thickness results in a change of the determined flow curve above a strain level of 0.1 and a much earlier failure of specimen with an initial sheet thickness of 1.5 mm. The changed material behavior can be explained by a different microstructure, e.g. the influence of the grain size. In summary, a characterization of the magnesium alloy AZ31B at elevated temperatures can be realized with a miniaturized test setup.

1. Introduction

The realization of lightweight design is driven by the demand on reducing greenhouse gas emission. One possibility to reach this aim is the replacement of conventional materials like steel with lighter materials, as aluminum or magnesium. Another chance of reducing the part weight is the maximum utilization of the material's performance by an exact characterization of the material behavior. Therefore, the knowledge of local mechanical properties enables an accurate validation of process- and part design to extend the process limits [1]. Especially, the variation of local material behavior of a part can be characterized for the validation of a numerical simulations [2]. For testing local material properties, a miniaturization of well-established test setups, as the uniaxial tensile test is needed. Regarding tension and compression tests, the miniaturization of the measuring area to a measuring length of 2.0 mm permits a detection of the material behavior under uniaxial compression without an early buckling [3]. Another method to prevent the specimen from buckling is the usage of additional stabilization plates in the relevant area with side forces, which includes an influence of friction [4].

In case of magnesium alloys, the material consists of a hexagonal close-packed microstructure with a low number of slip systems, which leads to an asymmetric behavior of the beginning of plastic yielding between tension and compression testing [5]. Disregarding this material behavior during numerical process design leads to significant differences of the predicted results compared with experimental results [6]. Another challenge for the usage of magnesium alloys is the low formability at room temperature. Due to the low number of slip systems at room temperature, forming processes of magnesium alloys are realized at temperatures above 150 °C [7]. In this context, an increase of the forming temperature from 20 °C to 200 °C improves the ductility of AZ31 around 38% [8]. Thus, experimental setups with miniaturized specimen design for compressive load but also for testing at elevated temperatures up to 200 °C are needed for the numerical process design of magnesium alloys. However, the miniaturization of the specimen size begs the risk of losing the transferability of the outcome to results of standardized test methods.

For this reason, the investigation of the influence of miniaturized specimen size for the characterization result under uniaxial tension of the magnesium alloy AZ31B at an elevated temperature of 200 °C is superior aim of this research work. Therefore, the test results are compared to results of a test setup with a standardized specimen geometry. Concerning a small specimen size, the influence of sheet thickness can be important and needs to be investigated. Therefore, AZ31B magnesium alloy sheets with an initial sheet thickness of 1.0 mm and 1.5 mm are investigated.

2. Material, experimental setups and specimen geometry

2.1. Material

In this research work, the wrought magnesium alloy AZ31B with initial sheet thicknesses t_0 of 1.0 mm and 1.5 mm is investigated. The main alloying elements are aluminum (3%) and zinc (1%), which increase the tensile ductility compared to pure magnesium. The chemical composition of AZ31B according to ASTM B90/B90M is shown in Table 1.

Table 1. Chemical composition [wt%] of AZ31B according to ASTM B90/B90M

	Al [wt%]	Zn [wt%]	Mn [wt%]	Si [wt%]	Cu [wt%]	Fe [wt%]	Ni [wt%]
AZ31B	2.5 - 3.5	0.6 - 1.4	> 0.2	> 0.1	> 0.05	> 0.05	> 0.005

The grain size of both investigated magnesium sheets is measured by EBSD (Merlin Gemini II, Carl Zeiss AG) and averaged for approximately 500 grains. In case of the sheet thickness 1.0 mm, the average grain size is detected to 5.3 µm. For a sheet thickness of 1.5 mm, the average grain size is almost twice as large with 10.8 µm.

2.2. Specimen geometry

Tensile test specimens are cut out via laser cutting (Tru Laser Cell 7020, Trumpf GmbH + Co. KG). Due to an observed influence of strain hardening, which falsifies the obtained yield stress (see figure 1), no further milling operation is done in the measuring area of the examined specimen.

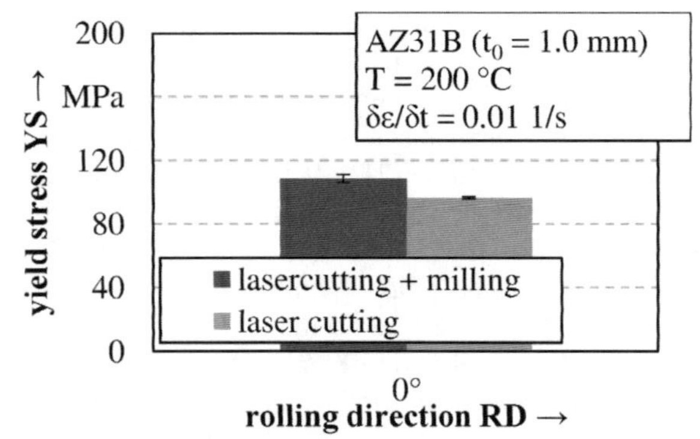

Figure 1. Influence of milling on the beginning of plastic yielding YS of AZ31B at 200 °C test temperature

Additionally, the results without the influence of strain hardening are rather comparable to literature values [9]. In order to qualify a miniaturized specimen geometry (Type B) for uniaxial tensile tests, a standardized specimen geometry (Type A) according to DIN 50125 is chosen as reference (see figure 2).

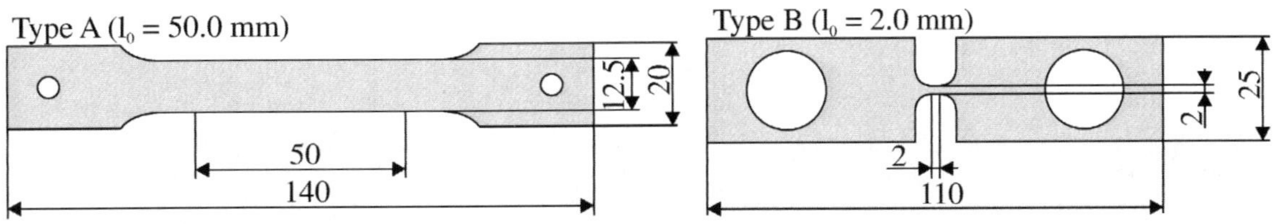

Figure 2. Principal of the investigated specimen geometries

The referencing geometry Type A consists of a measuring length of 50.0 mm and a width of 12.5 mm in the measuring area. Since buckling occurs during compression of large specimen [10], a specimen geometry according to Staud et al. [11] is used for a miniaturization of the measuring area. The miniaturized specimen (Type B) includes a measuring length and width of 2.0 mm. In contrast to specimen A with a length to width ratio of 4.0, the miniaturized specimen B contains of a length to width ratio of 1.0.

2.3. Experimental setup

For the experimental observations, two different testing machines are used. In case of the reference tests with standardized specimen geometry, a Gleeble 3500 (DSI) testing machine is used with a 98 kN load cell and an optical DIC system (Aramis, GOM mbH) with a two-dimension camera including a resolution of 4 megapixels (see figure 3). The temperature in the relevant area of the specimen is measured via thermal couple type K. Miniaturized tensile tests are realized with a built-in device (see figure 4) developed by Schaub et al. [3].

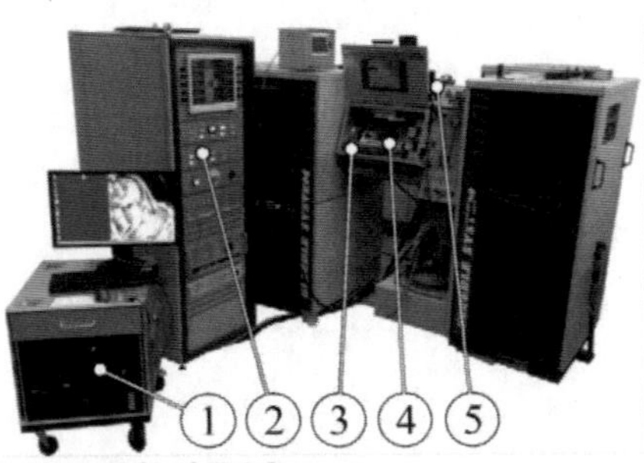

(1) optical DIC system
(2) control unit
(3) specimen chamber
(4) specimen and mechanical clamps
(5) CCD camera with light

Figure 3. Gleeble 3500 (DSI) testing machine with optical DIC system for standardized specimen size (Type A)

(1) load cell
(2) hydraulic clamping
(3) conductive heating
(4) measuring area
(5) pyrometer
(6) optical DIC system
(7) stabilization plate
(8) external light

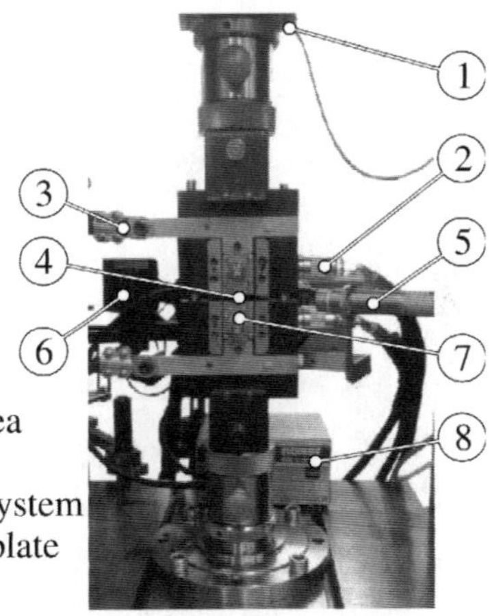

Figure 4. Test device for miniaturized specimen (Type B) with optical DIC system built-in a universal testing machine Z100 (Zwick AG)

The test device in figure 4 is mounted in a universal testing machine Z100 (Zwick AG) with a 100 kN load cell, while the measuring zone is heated via conductive heating. Concerning the good thermal conductivity of magnesium, the electrical linkage is modified to reduce the electrical resistance of the system. The temperature is measured by pyrometer and controlled by an external device. Local strains are measured by an optical DIC system with a two-dimensional CCD camera with a resolution of 4 megapixels. The specimen is mounted between stabilizer plates to prevent an early buckling of the measuring zone. Therefore, a hydraulic clamping system is used with a maximum clamping pressure of 40 MPa. All tests are performed at an elevated temperature of 200 °C and a nominal strain rate of 0.01 1/s. Tests with three trails (n = 3) are carried out for each configuration and the relevant values and curves are averaged. For the determination of average strains, an evaluation area for each specimen geometry is chosen at an average equivalent strain of 0.2 according to von Mises (see figure 5 and 6).

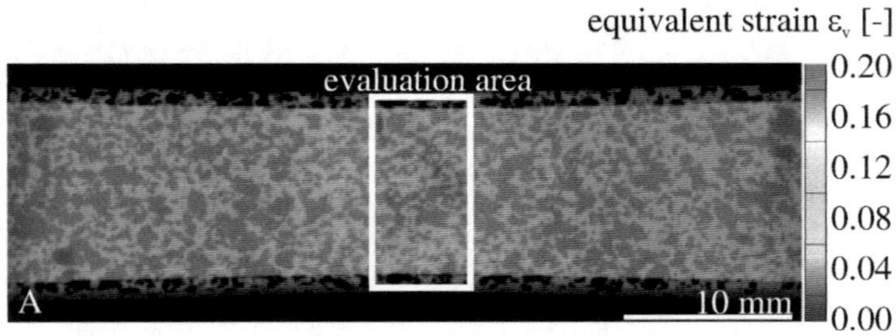

Figure 5. Strain distribution of the reference specimen (Type A) and evaluation area in the center of the specimen according to an inhomogeneous strain distribution

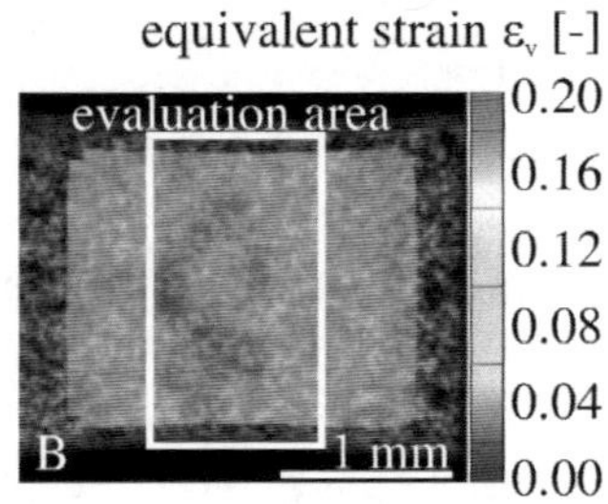

Figure 6. Strain distribution and evaluation area of miniaturized specimen (Type B)

Since an inhomogeneous strain distribution exists in the reference specimen due to the parabolic temperature propagation in a specimen with large length, an evaluation area in the center of the specimen is chosen with a maximum deviation of the strain around 5 %. In contrast to this, the evaluation area of the miniaturized specimen is selected to 50 % of the initial length due to an almost homogeneous strain distribution (see figure 6).

3. Experimental results

3.1. Flow curve

The flow curves of the reference tests (l = 50.0 mm) and the miniaturized tensile tests (l = 2.0 mm) are presented in figure 7 for a sheet thickness of 1.0 mm and at an elevated temperature of 200 °C. Due to the small evaluation area around the maximum column, flow curves beyond the tensile strength investigated. The test specimens are extracted in 0 ° to the rolling direction (RD) of the sheet.

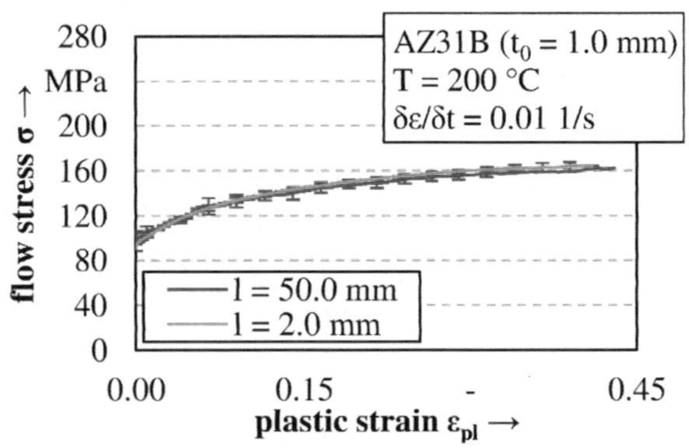

Figure 7. Comparison of flow curves of AZ31B at an elevated temperature of 200 °C

The flow curves are almost congruent regarding the determined standard deviation (sd). Moreover, plastic strain values until 0.41 are measured in case of miniaturized tensile tests, while maximum plastic strains of 0.45 are observed for reference tests. In addition, a similar hardening behavior of AZ31B at an elevated temperature of 200 °C is noticed, while the selected area around the maximum column of the strain level allows the detection of flow curves until higher strains.

3.2. Mechanical properties

In order to qualify the miniaturized tensile test for the determination of mechanical properties, the beginning of plastic yielding YS at a plastic strain of 0.002 and the tensile strength TS for the reference geometry (l = 50.0 mm) and the miniaturized specimen (l = 2.0 mm) are compared in figure 8 and figure 9. To investigate the influence of anisotropy, the mechanical properties are obtained for tests in 0°, 45° and 90° to the rolling direction (RD).

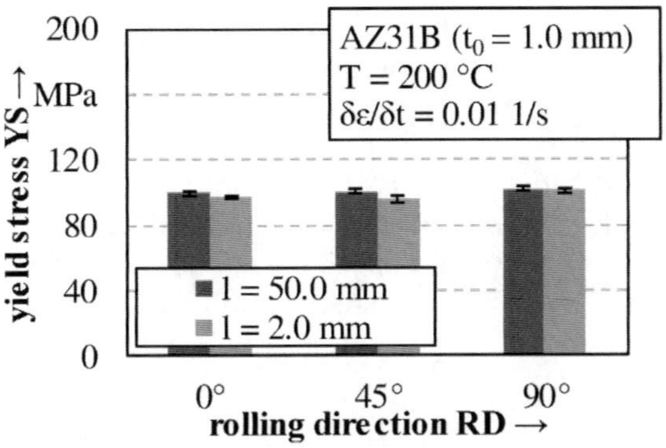

Figure 8. Beginning of plastic yielding YS at an elevated temperature of 200 °C

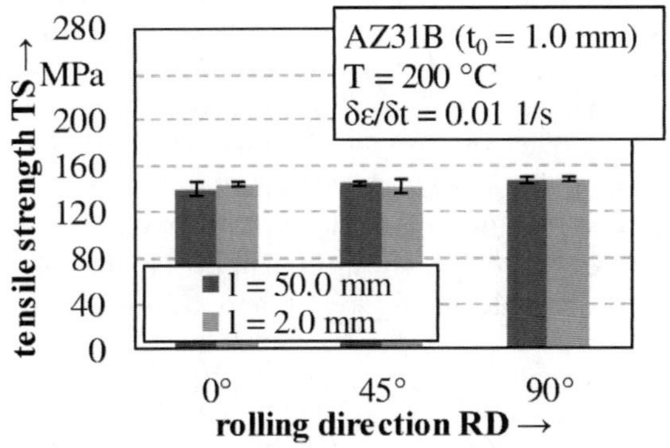

Figure 9. Tensile strength TS at an elevated temperature of 200 °C

An earlier beginning of plastic yielding is detected for miniaturized specimen extracted in 0° and 45 ° to the rolling direction, while no significant difference is observed for tests in 90° to RD. With a yield stress of 99.4 MPa ± 1.16 MPa in case of the reference tests and a yield stress of 96.4 MPa ± 0.63 MPa for miniaturized tests a deviation smaller than 3.1 % is determined. Contrariwise, the comparison of tensile strength TS leads to an opposite behavior in case of tests in 0 ° to RD.

With a tensile strength TS of 139.8 MPa ± 5.4 MPa for the referencing tests in 0 ° to RD (l = 50.0 mm) and a tensile strength of 143.6 MPa ± 1.92 MPa in case of miniaturized tensile tests (l = 2.0 mm) in 0 ° to RD, a deviation smaller than 2.7 % exists. Regarding the standard deviation, no significant influence of the specimen size on the beginning of plastic yielding and the tensile strength exists. In addition, no dependency on anisotropy is detected for the yield stress and tensile strength at an elevated temperature of 200 °C. This means, that the results of mechanical properties of both specimen geometries are comparable and no influence of miniaturization is recognizable.

3.3. Strain path and Lankford coefficient r

Since the reduced specimen size leads to a change of the length to width ratio from 4.0 in case of the reference geometry to 1.0 for the miniaturized specimen geometry, the evolution of the major and minor strain is observed (see figure 10). Therefore, the strain path for both specimen geometries is presented. Moreover, the average standard deviation (sd) for both test setups is listed.

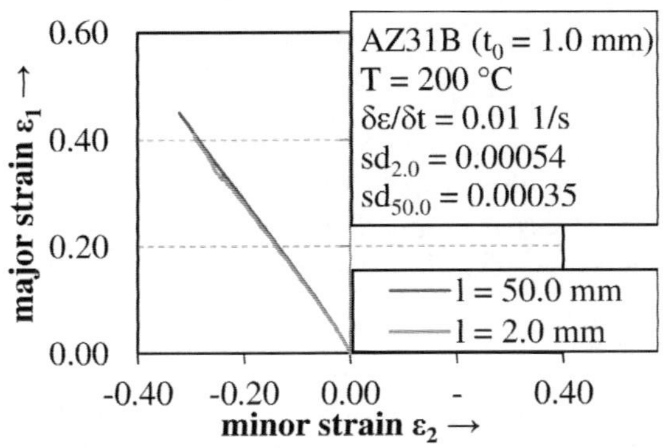

Figure 10. Strain path represented by major and minor strain evolution for the reference geometry Type A (l = 50.0 mm) and the miniaturized geometry Type B (l = 2.0 mm)

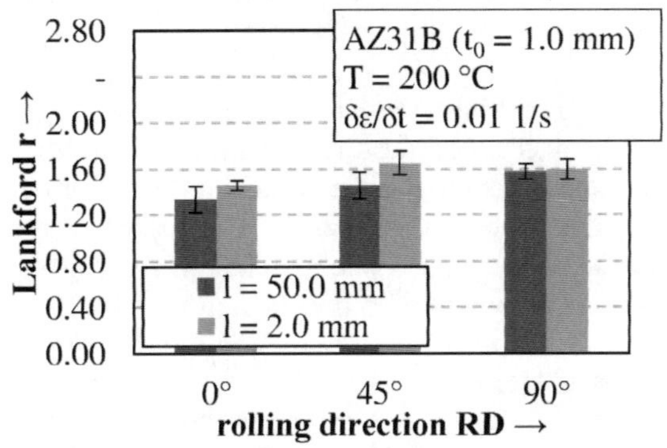

Figure 11. Comparison of the Lankford coefficient of AZ31B at an elevated temperature of 200 °C for the reference geometry and the miniaturized specimen

Furthermore, the influence of the aspect ratio on the anisotropic behavior is determined regarding the Lankford coefficient r (see figure 11). Relating to the evolution of the strain path in figure 10, a similar behavior for the miniaturized specimen geometry (l = 2.0 mm) compared to the reference geometry (l = 50.0 mm) is observed, while uniaxial tensile tests with miniaturized specimen lead to lower reachable strains. Additionally, higher Lankford coefficients r are observed for the miniaturized tensile tests (see figure 11). Considering the standard deviation, no significant difference exists, while miniaturized specimen in 45 ° to RD indicate a slightly higher anisotropic behavior than the reference tests. With a Lankford coefficient of 1.46 ± 0.04 in 0° to RD, the results of the miniaturized specimen are comparable to Fukuda et al. [9] with a Lankford coefficient of 1.52 at an elevated temperature of 200 °C. Although, no significant differences of the miniaturization of the specimen on the evolution of the strain path (see figure 10) and the Lankford coefficients (see figure 11) are observable.

3.4. Influence of sheet thickness

Since the miniaturized specimen width is 2.0 mm, the influence of the sheet thickness on the material behavior of AZ31B at an elevated temperature of 200 °C is analyzed. Therefore, the determined flow curves of test specimen with an initial sheet thickness of 1.0 mm are compared to the results of miniaturized tests with an initial sheet thickness of 1.5 mm (see figure 12). Additionally, the influence of sheet thickness on the yield stress YS at a plastic strain of 0.002 is observed (see figure 13).

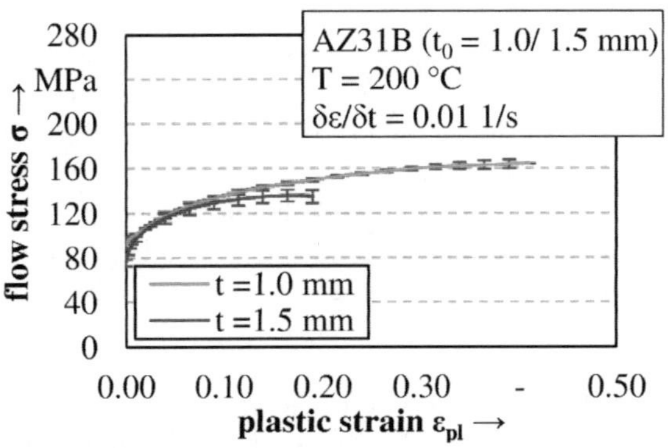

Figure 12. Flow curves of AZ31B for miniaturized tensile specimen with an initial sheet thickness of 1.0 mm and 1.5 mm at an elevated temperature of 200 °C

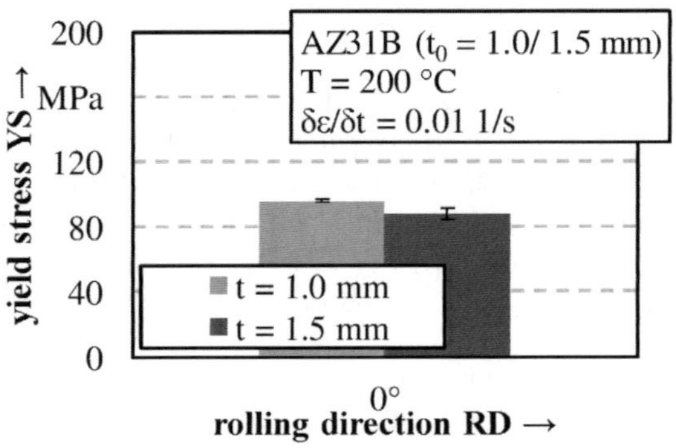

Figure 13. Beginning of plastic yielding at a plastic strain of 0.002 for specimen with an initial sheet thickness of 1.0 mm and 1.5 mm at an elevated temperature of 200 °C

The resulting flow curves are almost congruent to a plastic strain of 0.1, while stresses for tests with a sheet thickness of 1.5 mm are slightly smaller. At a plastic strain level higher than 0.1, a relevant difference of the hardening behavior occurs with a smaller stress level for miniaturized tensile tests at a sheet thickness of 1.5 mm. Moreover, a significant earlier failure due to cracking is observed in case of the thicker sheet metal. Accordingly, an earlier beginning of plastic yielding is obtained for a sheet thickness of 1.5 mm (see figure 13).

4. Discussion

The influence of a miniaturization of the specimen size on the mechanical behavior of AZ31B under uniaxial tension at an elevated temperature of 200 °C is observed within this research work. The resulting flow curves (see figure 7) show no significant difference of stress gain between a standardized specimen geometry (reference geometry) and the investigated miniaturized specimen, while miniaturized tensile tests result in a lower reachable plastic strain. In this context, the flow curves are determined for an area around the maximum column of the strain distribution, where the selected area influences the maximum reachable strain due to the region of averaging the strain values. A direct relationship of the maximum reachable strain level and the miniaturization of the specimen size cannot be observed.

The mechanical properties, the beginning of plastic yielding (see figure 8) and the tensile strength (see figure 9) clarify an assimilable material behavior in case of miniaturized specimen compared to the results of the reference geometry with a measuring length of 50.0 mm. Additionally, the results of both test setups illustrate that no significant influence of the anisotropy on the yield stress and the tensile strength exists at an elevated temperature of 200 °C. Fukuda et al. [9] explained the material behavior at elevated temperature above 473 K (= 200 °C) by the activation of additional slip systems at elevated temperatures and dynamic recovery, which reduces the texture induced anisotropic behavior of the initial sheet metal.

Concerning the evolution of the strain path (see figure 10), no significant difference of major and minor strain development is observed for miniaturized specimen. However, a fractional lower strain level can be achieved with miniaturized specimen, which is caused by the size of the selected evaluation area.

Otherwise, the determined Lankford coefficients (see figure 11) are slightly higher in case of miniaturized specimen with a length to width ratio of 1.0 compared to the reference geometry with an aspect ratio of 4.0. Regarding the estimated standard deviation, this influence is not significant. Comparing the Lankford coefficient in 0 ° to RD with results from literature [9], a smaller deviation in case of miniaturized specimen is identified compared to the results of the reference geometry. Additionally, no steady increase of the Lankford coefficient with higher angle to the rolling direction is ascertained for miniaturized tensile tests as well as for the reference tests with larger specimen. Only a trend is observed in case of the reference tests, which is not significant regarding the standard deviation. As already mentioned for the stress values, no significant anisotropic behavior is examined for the Lankford coefficient, which can be explained by the activation of additional slip systems and dynamic recrystallization, where a predominant forming direction is attenuated.

Finally, the influence of sheet thickness is analyzed. Since the specimen width of the miniaturized geometry is 2.0 mm, an increase of the sheet thickness from 1.0 mm to 1.5 mm leads to a significant difference in the development of the hardening behavior above a strain level of 0.1 (see figure 12) and a decrease of the obtained yield stress at a plastic strain of 0.002 (see figure 13). Additionally, a much earlier failure due to cracking is detected in case of a sheet thickness of 1.5 mm. Regarding the determined grain size, which is twice as large in case of a sheet thickness of 1.5 mm, the higher reachable strains and the higher beginning of plastic yielding for a thickness of 1.0 mm are explainable by a better dislocation motion due to smaller grains. A similar influence of the grain size on the elongation behavior at 200 °C is presented by Mabuchi et al. [12]. Zhu et al. explained the increase of the yield stress by the effect of twinning, where the critical resolved shear stress (CRSS) increases significantly with a decrease of grain size [13]. Thus, additional observations on the microstructure are needed to analyze the material behavior for different sheet thicknesses.

Hence, the discrepancy in the mechanical properties and the material behavior can be explained by the differences of the grain size, while no direct influence of the width to thickness ratio is detected for the investigated sheet thicknesses. This can be seen in case of small strains less than 0.1, where the flow behavior is almost congruent, but with a smaller beginning of plastic yielding for AZ31B magnesium alloy sheets with a thickness of 1.5 mm.

5. Conclusion

Uniaxial tensile tests are carried out for the wrought magnesium alloy AZ31B at an elevated temperature of 200 °C and two different test setups to analyze the influence of specimen miniaturization on the resulting mechanical behavior. Moreover, the influence of sheet thickness on the characterization results is investigated. Although, the results confirm no significant difference between the standardized specimen and the miniaturized specimen geometry, while a lower strain level can be detected in case of the small-sized specimen.

Additionally, no influence of anisotropy on the yield stress and tensile strength is observed for both test setups, which can be explained by the activation of additional slip systems and dynamic recrystallization at an elevated temperature of 200 °C. The analysis of the sheet thickness results in a change of the determined flow curve above a strain level of 0.1 and a much earlier failure of specimen with an initial sheet thickness of 1.5 mm compared to sheets with a thickness of 1.0 mm. The changed material behavior can be explained by a different microstructure, e.g. the influence of the grain size, which is almost twice as high for a sheet thickness of 1.5 mm compared to the sheets with a thickness of 1.0 mm.

In summary, a characterization of the magnesium alloy AZ31B at elevated temperatures can be realized with a miniaturized test setup. Especially for the determination of mechanical properties under compressive loading due to an asymmetric material behavior of magnesium between tension and compression, a miniaturization of the measuring area is needed to prevent an early buckling of the specimen. Therefore, a transferability of the results of small-sized specimen to results with standardized specimen is required regarding the identification of complex yield criteria.

Acknowledgments
The authors are grateful to the German Research Foundation (DFG) for funding the research project "Contribution to an efficient FE-based design of magnesium sheet metal parts" (ME 2043/40-1). Supplementary, a special thank belongs to Mr. Luigi De Marinis for the input during his bachelor of science thesis "Contribution to a mechanical and microstructural investigation of the magnesium alloy AZ31 under uniaxial tension at elevated temperature".

References
[1] Olofsson J and Svensson I L 2013 The effects of local variations in mechanical behaviour – Numerical investigation of a ductile iron component *Materials & Design* **43** 264-271
[2] Ambriz R, Chicot D, Benseddig N, Mesmacque G and de la Torre S D 2011 Local mechanical properties of the 6061-T6 aluminium weld using micro-traction and instrumented indentation *European Journal of Mechancis –A/Solids* **30** 307-315
[3] Schaub A, Lechner M and Merklein M 2013 Druckversuch bei erhöhten Temperaturen *Tagungsband Werkstoffprüfung* ed H J Christ (Düsseldorf: Stahleisen GmbH) pp 85-90
[4] Lou X Y, Li M, Boger R K, Agnew S R and Wagoner R H 2007 Hardening evolution of AZ31B Mg sheet *Int J Plasticity* **23** 44-86
[5] Nobre J P, Noster U, Kornmeier M, Dias A M and Scholtes B 2002 *Key Eng. Mater.* **230-232** 267-270
[6] Ghaffari Tari D, Worswick M J, Ali U and Gharghouri M A 2014 Mechanical response of AZ31B magnesium alloy: experimental characterization and material modeling considering proportional loading at room temperature *Int. J. Plasticity* **55** 247-267
[7] Juchmann P and Wolff S 2004 New perspectives with magnesium sheet *6th Proc. Int. Conf. on Magnesium Alloys and Their Applications* ed K U Kainer (Weinheim: Wiley-VCH Verlag GmbH & Co. KGaA) pp 1006-1012
[8] Ullmann M, Berge F, Neh K, Kawalla R and Vogt H-P 2014 Application of Magnesium Sheets and Strips in Vehicle Construction *AutoMetForm - SFU* ed R Kawalla (Freiberg: ACATRAIN e.V.) pp 37-45
[9] Fukuda I, Harada Y and Ohtsuka S 2014 Effect of Temperature on Stretchability of Anisotropic AZ31 Magnesium Alloy Sheet *Procedia Eng.* **81** 742-747
[10] Nguyen N-T, Seo O S, Lee C A, Lee M-G, Kim J and Kim H Y 2014 Mechanical Behavior of AZ31B Mg Alloy Sheets under Monotonic and Cyclic Loadings at Room and Moderately Elevated Temperatures *Materials* **7** 1271-1295
[11] Staud D and Merklein M 2009 Zug-Druck-Versuche an Miniaturproben zur Erfassung von Parametern für kinematische Verfestigungsmodelle. *Tagungsband Werkstoffprüfung* ed M Borsutzki and S Geisler (Düsseldorf: Stahleisen GmbH) pp 211-218
[12] Mabuchi M, Chino Y and Iwasaki H 2003 Influence of Grain Size on Elongation at Elevated Temperatures in AZ31 Mg Alloy *Mater. Trans.* **44** 490-495
[13] Zhu S Q, Yan H G, Liao X Z, Moody S J, Sha G, Wu Y Z and Ringer S P 2015 Mechanisms for enhanced plasticity in magnesium alloys *Acta Mater.* **82** 344-355

Edge crack sensitivity of lightweight materials under different load conditions

I Tsoupis[1] **and M Merklein**[1]

[1] Institute of Manufacturing Technology, Department Mechanical Engineering, Friedrich-Alexander-Universität Erlangen-Nürnberg (FAU), Egerlandstraße 13, 91058 Erlangen, Germany

E-mail: ioannis.tsoupis@fau.de

Abstract. This study addresses the analysis of edge crack sensitivity of DP800 steel and AA5182 aluminum alloy in dependency of punching and machining operation as well as load case of subsequent forming. The inserting of a round hole by punching with defined punch-to-die-clearance, milling and drilling is compared. Subsequent forming is performed by standardized hole expansion test and by Nakajima-tests with three different specimen geometries. Local strain distribution at the surface for Nakajima-tests is measured by optical strain measurement technique and investigated in order to evaluate local deformation before failure. Additionally, resulting hole expansion ratio λ is determined. Significant higher λ as well as local strain values ε_{max} are achieved by machined holes. This is directly coupled to higher local formability and stretchability for both materials. Furthermore, the load condition has a strong impact on the edge crack sensitivity of the material. Prior failure is observed with changing stress conditions using different specimen geometries also influencing the reachable maximum failure strain. Higher edge crack sensitivity is observed for DP800, which is in good accordance to the material properties in terms of ductility and strength. These data in dependency of the process parameter can be used for the design of automotive components.

1. Introduction

Within the fabrication of automotive components multi-stage process sequences with punching and subsequent forming operations are commonly used. In this context holes in blanks are required for assembly in case of safety relevant structural components or as relief holes for deep drawing operations of body parts, for instance. According to economic reasons the holes are commonly punched in industry. This leads to a significant reduction of materials formability. More particularly, pre-punched areas have a negative effect on forming operations, which primarily occur during component production, concerning crack initiation. Especially, when modern lightweight materials are processed the formability of the sheared edges is restricted. First cracks occur in these regions as a result of work hardening and pre-damage. Consequently the quality of the hole is crucial to the arising damage intensity. Nowadays, the impact of shearing edge quality on subsequent forming is mainly assessed by the standardized hole expansion test [1]. However, the loading condition on the material and the strain evolution differs clearly to processes like deep drawing, as within the hole expansion test tangential tensile stresses and strains represent the dominant load leading to final failure [2]. Furthermore, higher local strains occur by hole expansion compared to uniaxial tensile loads [3]. Relating to material properties, the achievable hole expansions ratio λ, defined as the ratio of the initial and the resulting diameter of the hole, is generally higher with lower material strength [4] and higher

Lankford coefficient [5]. Also the type of cutting process [6], the punch-to-die clearance [7] as well as the position of the burr out of the shearing process and anisotropy of material [8] have a significant influence on the achievable hole expansion ratio and consequently, on edge crack sensitivity. Other investigations showed that test speed and clamping force of the blank holder have no significant influence on the hole expansion ratio [9]. However, the evaluation method by manual abortion of the test due to crack occurrence causes a high scatter in the results and is stated to be strongly subjective [10]. Additionally, other parameters like for example the tool affect the results [11]. First semi-automatic evaluation methods were published recently [12]. In consideration of the described aspects alternative approaches need to be developed [13] that can efficiently reflect the characteristics and influences of pre-punched blanks to subsequent forming processes [14]. In the scope of this work the formability of differently produced edges by punching and machining is analysed for various load cases for DP800 steel and AA5182 aluminum. A new evaluation method is presented in order to determine achievable hole expansion ratios and maximum strain levels.

2. Materials and methods

2.1. Materials

The materials used in this investigation are 5000 series aluminum alloy AA5182 as well as fine grained and cold rolled advanced high strength steel (AHSS) DP800, both with a sheet thickness of $t_0 = 1.00$ mm. These materials are typically used in automotive industry for structural components and body parts. Tensile tests at room temperature were carried out according to SEP 1240 with specimen geometry after DIN EN 10002-1 appendix B shape 2, in order to determine the mechanical properties of the steel sheets. The influence of the material orientation is taken into account by testing the specimen 0°, 45° and 90° to the rolling direction (RD). Figure 1 (a) shows representative true stress true strain curves of the materials and their extrapolation by analytical models, in particular by Hockett-Sherby in case of DP800 and a combination of Swift-Hockett-Sherby for AA5182. Independent of the rolling direction DP800 shows a stronger hardening in the beginning of the forming for strain levels $\varepsilon < 0.1$ compared to the aluminum alloy, whereas the hardening of AA5182 is constantly increasing. In sum this leads to higher hardening exponent for AA5182 ($n_A = 0.33$) than for DP800 ($n_S = 0.13$). The mechanical properties ultimate tensile strength and total elongation reveal an isotropic character of both materials as there are no significant differences in the strength values. The higher strength levels and lower ductility of DP800 steel have to be considered in the experimental analysis.

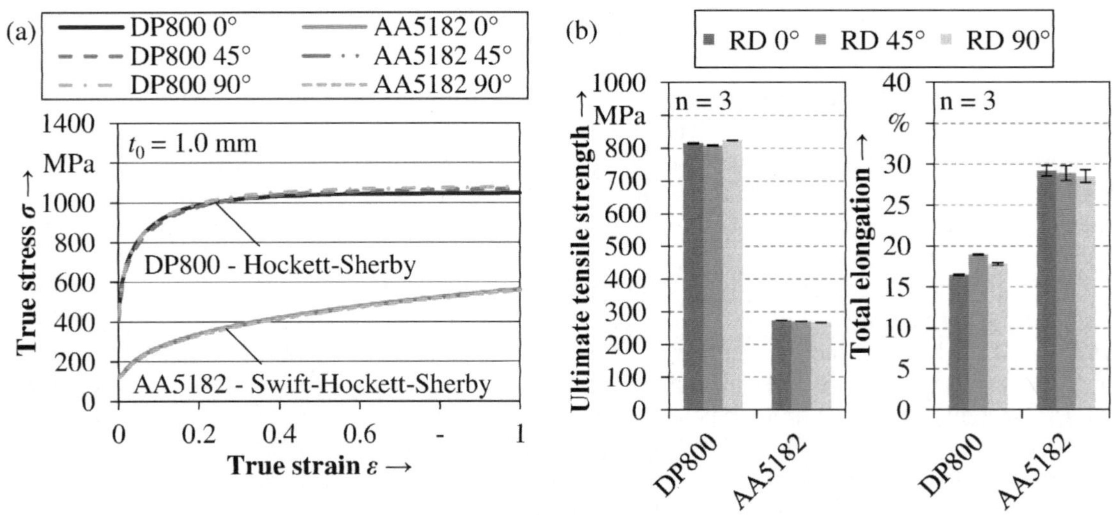

Figure 1. (a) Approximated and extrapolated true stress true strain curves
(b) Ultimate tensile strength and total elongation

2.2. Methods for the investigation of edge crack sensitivity

The edge crack sensitivity is evaluated by a combined analysis of achievable hole expansion ratios and the maximum strain levels in the regions of crack initiation. In order to respect different load conditions on the material, two test setups are incorporated within this work. On the one hand the hole expansion test according to ISO 16330 [1] with rectangular specimen is performed to study the stretchability of punched edges related to present industrial standards. On the other hand a stretch forming test in a modified Nakajima [15] test set up is used to stretch the hole in plane. Three geometries of specimen are tested to realize different strain and stress conditions. In particular, a full specimen (N-R0) and two notched specimen with different notch radii (N-R45 and N-R75) are used. The optical strain measurements system ARAMIS (GOM mbH, Braunschweig) is applied for both setups. This offers the advantage of investigating the deformation field at the outer surface of the specimen. Additionally, the system is used for determining the hole expansion ratio λ on basis of CCD-camera recorded pictures. In this context two of the main drawbacks of existing evaluation method can be overcome. By using the camera data the time of crack initiation can be exactly contained user independently and the measuring results are not influenced by springback effects of the material, which is the case by manual abortion criteria and following evaluation after unloading the specimen. Consequently, the hole expansion ratio is determined more accurately, reproducibly and practice-oriented.

The hole expansion ratio λ for the specimen of hole expansion test as well as for the Nakajima specimen is determined by the methodology shown in Figure *2* (a). Initial (D_0) and resulting (D_h) diameter of the hole are measured on basis of recorded pictures of the camera. By setting six pixel-points at the edge of the hole two isosceles triangles are defined, which are oriented perpendicular to each other to ensure an orthogonal measuring according to ISO 16330 [1]. By the points 1-3 and 4-6 two measure circuits are defined, which diameters are the basis for the calculation of the averaged diameter of the hole D_h. In relation to standard specification the occurrence of first crack through the sheet thickness is the criteria for measuring the resulting diameter D_h.

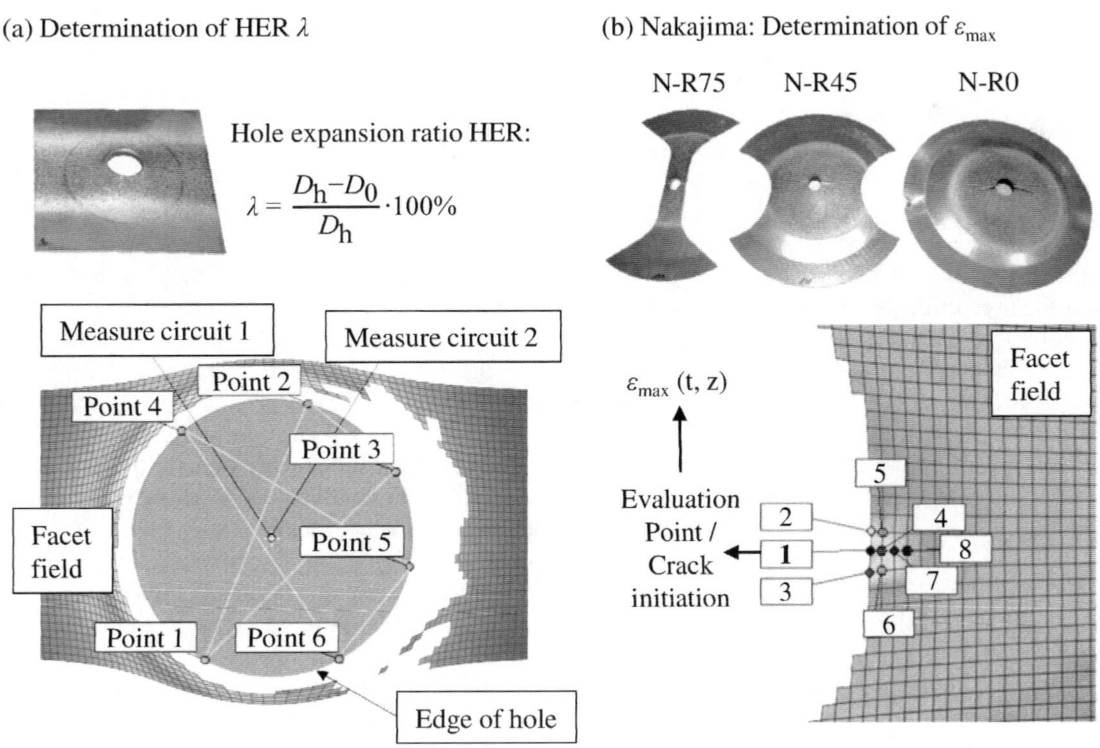

Figure 2. Procedure for the determination of hole expansion ratio (a) and maximum strain level (b)

IOP Conf. Series: Materials Science and Engineering **159** (2016) 012017 doi:10.1088/1757-899X/159/1/012017

The evaluation procedure for the Nakajima specimen N-R0 is analogue to the described one. As different load conditions for N-R45 and N-R75 conduct to increasing tensile stresses the hole is not stretched evenly. This leads to oval shaped holes with increasing notch radii. Thus, the determination of λ is based on the calculation of the major and minor diameter of the elliptical holes. This has to be considered by comparative evaluation of resulting hole expansion ratios.

The examination of the maximum strain levels for the Nakajima tests is carried out by the analysis of the deformation field at the outer surface of the specimen determined by ARAMIS system. The surrounding region of the crack is analysed by the strategy shown in Figure 2 (b). 8 measuring points in the facet field, where the crack initiates, are defined to capture the immediate surrounding within the deformation field. In this study point 1 is considered as it represents the maximum strain at the crack. The evolution of maximum strain value ε_{max} is analysed over the punch travel. The frame rate of image capturing is 10 Hz, the square and constant size of the facets is 0.25 mm x 0.25 mm. All experiments are evaluated until 0.1 s before first crack through the sheet thickness occurs. Consequently, the maximum strain evolution in critical regions can be opposed to each other comparatively for different specimen geometries and in dependency of the cutting process.

In order to investigate the influence of the cutting processes on the edge crack sensitivity three different industrial relevant processes are incorporated within this work. Considering machining as a process, milling and drilling are used to process a hole. In addition punching under a variation of the punch-to-die-clearance on two levels (c_1 = 0.10 mm, c_2 = 0.15 mm) is performed. In all processes the initial diameter of the hole is D_0 = 10 mm. Furthermore the influence of the punching process on the sheared edge is taken into account by varying the position of the burnish and consequently the fracture zone with regard to the punch in the subsequent forming process. On the one hand the fracture zone is located in the tensile stress dominated area averted to the punch (burr up) and on the other hand it is located facing the punch (burr down). Summarised, the parameters cutting process, punch-to-die-clearance c and the position of the burnish / fracture zone in case of punching, subsequent forming process and the material are varied and investigated in the next chapter.

3. Results

3.1. Hole expansion ratio

The hole expansion ratio λ determined by the described evaluation method is summarized in Figure 3 in dependency of the investigated parameters. Three valid tests are performed for the evaluation. The highest hole expansion ratios are reached in the hole expansion test, because the load condition is different to the Nakajima test. As within the hole expansion test a first flanging of the hole leads to further tangential stresses and strains, the hole is expanded in plane within the Nakjima test, which leads to more critical stresses directly at the edge of the hole. Also, significant higher hole expansion ratios for both materials are achievable by the cutting processes drilling and milling when HET and N-R0 are considered. The reason for this is the higher edge quality of the hole for these processes as there is no visible pre-damage and barely a hardening of the edge. For DP800 steel this is also the case for N-R45 and N-R75 as higher λ are measured for drilling and milling. The aluminum seems not to be that sensitive as there is no significant difference in the resulting hole expansion ratio. This can be explained by the high total elongation values of AA5182 in uniaxial tension. With increasing notch radii the predominant strain condition is more equal to uniaxial tension and thus failure is mainly caused by critical tensile stresses than by the edge condition. This is also an explanation for the small disparity of λ for this load cases regarding DP800. Comparing milling and drilling, higher values of λ are reached with milling in the HET, but vice versa in the N-R0. For the other Nakajima tests no significant difference can be identified, what makes it difficult to state an expressive conclusion.

However, concerning the process punching a trend to higher hole expansion ratios with increasing the punch-to-die-clearance from c_1 = 0.10 mm to c_2 = 0.15 mm is detected. This seems not to be logical as a measuring of the characteristic zones of the cutting edges shows, that higher punch-to-die-clearances lead to higher fracture zones and thus to higher pre-damage [16]. Conversely, further investigations by hardness measurements of the sheared edges reveal significant higher hardening of the edge within the smaller punch-to-die-clearance of c_1 = 0.10 mm. Consequently, the remaining formability for subsequent forming operation is reduced, which might be the more dominant factor within hole expansion when just considering the punch-to-die-clearance in the investigated range. Nevertheless, the hole expansion ratio is reduced, when changing the position of the fracture zone from down to up regarding the punch. In this context the influence of shear induced pre-damage becomes apparent. Within the fracture zone first cracks initiate due to higher roughness of the surface compared to the burnish zone, which can be seen in exemplary pictures of the sheared edge in Figure 3. Many small cracks occur circumferential the hole when fracture zone is located up. Again this trend is not that distinctive for the investigated aluminum confirming the higher sensitivity of dualphase steel on the formability of holed edges in sum. In comparison of both materials AA5182 shows higher hole expansion rations for all analysed parameter combinations by trend. This can be lead back to the higher ductility and lower strength values compared to DP800. A more sophisticated analysis of occurring strains and reachable punch stroke seems to be constructive to understand the present findings more in detail.

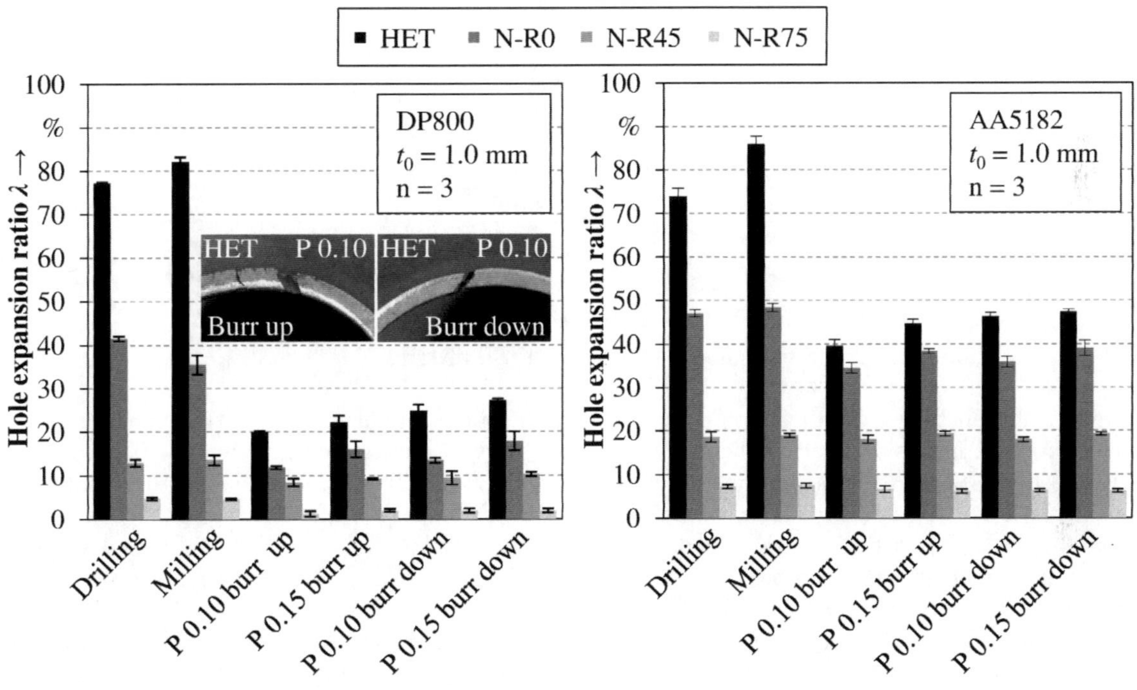

Figure 3. Hole expansion ratios for DP800 and AA5182 material

3.2. Evolution of maximum major strain

The evolution of the maximum major strain ε_{max} according to the presented evaluation method for Nakajima test specimens is summarized in Figure 4 for DP800 steel and in Figure 5 for AA5182 aluminum. For every parameter combination one representative curve is shown out of three experiments. The contact of punch and blank starts after a punch travel of 5 mm. After another 5 mm the maximum major strain starts to increase with ongoing plastification of the material. The slope of the curve is strongly dependent on the considered geometry of the specimen and thus on the stress and strain condition. With increasing notch radii also the slope of the curve is increasing independently of the material. Considering drilling and milling higher values of maximum major strain ε_{max} in comparable levels of $0.5 < \varepsilon_{max} < 0.55$ are reached for all specimen geometries in case of DP800 steel. However, maximum punch travel differs widely, which is in direct relation with the achievable hole expansion ratio λ. The main reason for the deviation is the obtained stress conditions on the material within the specimen geometries. For increasing notch radii the stress condition at the hole changes starting from stretching within N-R0 to more critical conditions. From N-R45 to N-R75 the percentage of uniaxial tension and stress state is raising leading to prior failure. Simultaneously, the mentioned fact that maximum strain levels are equal shows that sustainable strains of the material seem to have a major influence on the failure behaviour, which is directly dependent on the load condition.

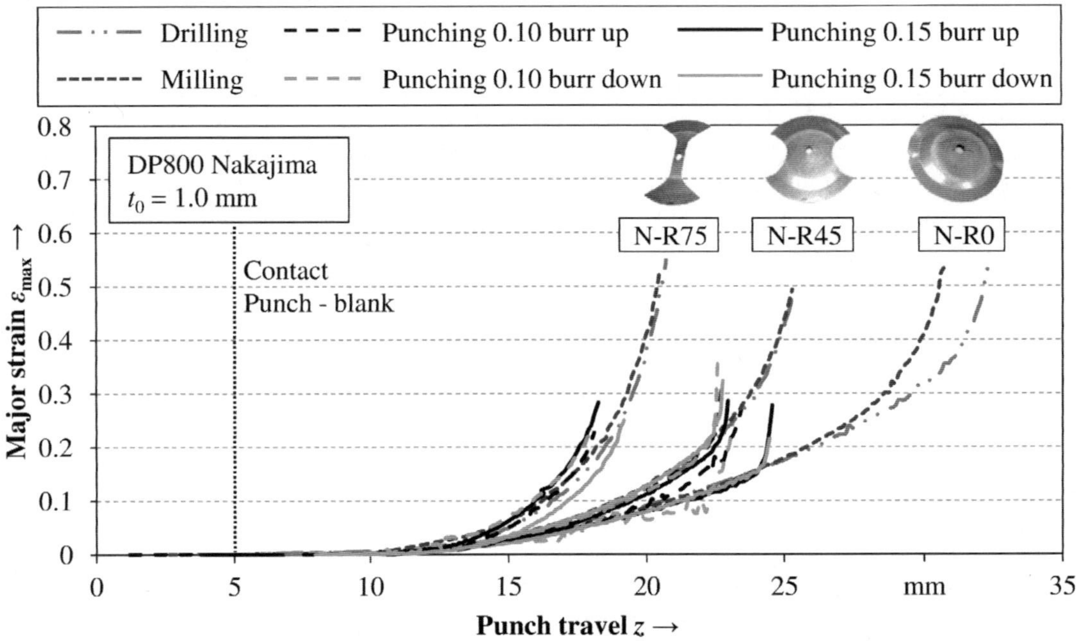

Figure 4. Evolution of the maximum major strain for DP800 material

In case of punching a significant reduction for both the punch stroke and the maximum major strain is recognizable. In this context the slightly better formability of punched edges with c_2 = 0.15 mm for N-R0 becomes clear as the punch travel is higher than for c_1 = 0.10 mm, what is consistent with the results of HET. This trend is also visible for the other specimen geometries, even though not that distinctive, because critical tension load is the dominant factor for failure as discussed before. Summarized, punching-induced pre-damage clearly affects the formability of the edge and decreases significantly the strain to failure for DP800.

In general, the results concerning milling and drilling are transferable to AA5182 aluminum. In contrast to the investigated steel the influence of specimen geometry is more pronounced respecting obtained maximum strain values. Strain is more concentrated for notched specimen due to resulting load condition and within the aluminum alloy the higher hardening exponent leads to higher strains. However, the influence of cutting process on the reachable strains is not as significant as for the DP800 steel. The strain-stroke-curves follow a similar trend for the considered specimen geometries, see Figure 5. The strong increase of the strains in the end of forming before fracture is a direct result of beginning necking, which simultaneously initiates shear fracture. As for N-R0 stretching of the hole is the load case the deformation of the material is distributed more evenly over a larger area leading to better formability and also higher hole expansion ratios. Apart from that sustainable strains before fracture are lower compared to the other geometries. Another effect that can be seen in the curves for AA5182, especially for the ones of N-R0, is the stepwise run of the curve resulting in a rapid slope in the end. This can be lead back to the Portevin-Le Chatelier effect, which is a result of dynamic strain aging or related to plastic strain inhomogeneity in solid solutions [17]. The effect is rated not to have a significant influence on the resulting formability itself and consequently is not considered within this work.

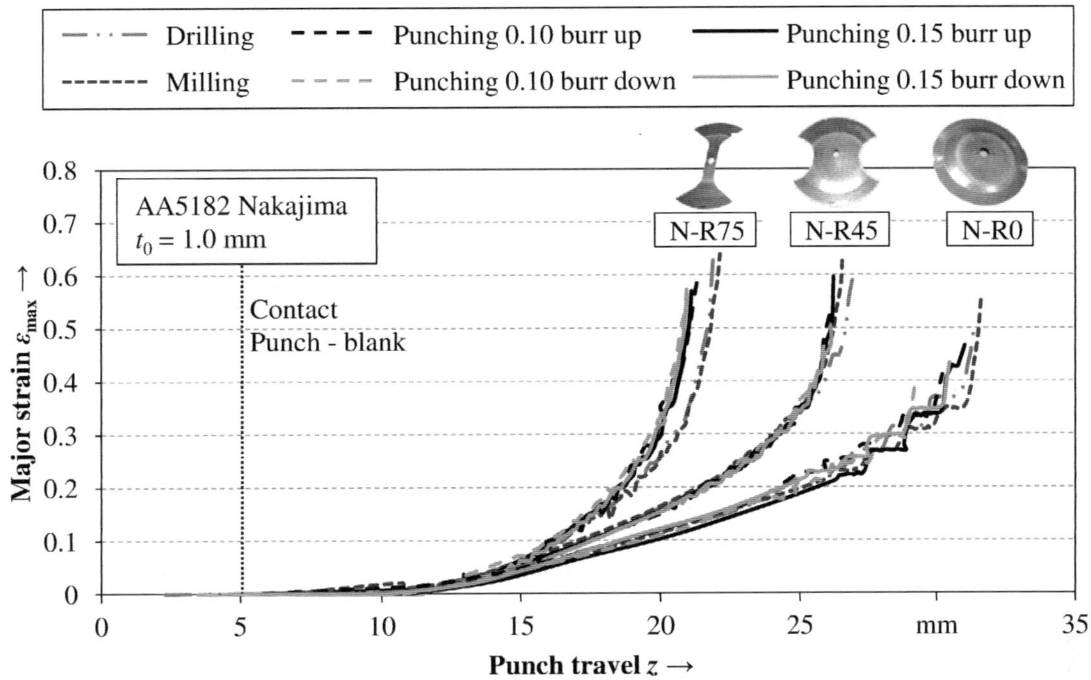

Figure 5. Evolution of the maximum major strain for AA5182 material

Comparing these results with the ones from the hole expansion test, no direct correlation can be seen. Based on the results of the hole expansion test the significant influence of the cutting process and the loading condition is more distinctive in the absolute value λ. However, no information on sustainable strains of the material and the formability around the edge can be provided by HET. This leads to the conclusion that a combined investigation using the presented methods to determine the hole expansion ratio and maximum major strain is necessary to evaluate the formability concerning the edge crack sensitivity in dependency of the considered parameters.

4. Conclusion and outlook

Within this research the edge crack sensitivity of AHSS steel DP800 and aluminum alloy AA5182 is investigated in dependency of the cutting processes milling, drilling and punching as well as the subsequent forming process. A novel evaluation method based on an optical measurement system is presented in order to determine the hole expansion ratio and the maximum strain levels at the region of crack occurrence. Edge crack sensitivity is significantly affected by the cutting process. Machining processes drilling and milling provide comparable results concerning the hole expansion ratio and the reachable maximum stain levels in dependency of the load condition and the material. Edges produced by these processes are more stable against crack sensitivity. Punching affects the formability of the edges significantly as the hole expansion ratio is reduced for both materials. However, the sensitivity is more pronounced for DP800 steel, which in sum also shows lower formability of the edge due to higher strength and lower ductility in comparison to AA5182. In case of punching the position of the fracture zone up within subsequent forming decreases the formability according to induced pre-damage and rougher surface in this zone. Also a decrease of the punch-to-die-clearance from 0.15 mm to 0.10 mm reduces the edge formability due to greater hardening of the material in the region of the edge. The results can be used in the design process or for the deduction of characteristic values in the numerical simulation for the estimation of the edge crack sensitivity in dependency of the investigated parameters. In further investigations the influence of material microstructure is investigated as the current results do not give information on the damage and failure modes in particular. Additionally, a study of the deformation field around the crack seems to be promising to understand these mechanisms more in detail.

Acknowledgements
The Authors would like to thank the European Research Association for Sheet Metal Working e.V. (EFB) and the German Federation of Industrial Research Associations e. V. (AiF) for the financial support of the research project EFB 14/213 (AiF 17737N). Furthermore, the authors would like to thank Almut Töpperwien and Sascha Gick for encouraging the development of the digital evaluation method with ARAMIS.

References

[1] ISO 16630 2009 Metallic materials Method of hole expanding test *International Organization for Standardization* (Berlin: Beuth)
[2] Shi M F, Chen X 2007 Prediction of Stretch Flangeability Limits of Advanced High Strength Steels using the Hole Expansion Test *SAE Transactions* **116** 742-747
[3] Karelova A, Krempaszky C, Dünckelmeyer M, Werner E, Hebesberger T, Pichler A 2009 Formability of Advanced High Strength Steels Determined by Instrumented Hole Expansion *Materials Science and Technology* 1358-1368
[4] Chen X, Haoming J, Chu Z, Lian C, Lu C 2014 Hole expansion characteristics of ultra high strength steels *Procedia Engineering* **81** 718-723
[5] Takahashi M 2003 Development of High Strength Steel for Automobile *Nippon Steel Technical Report* **88** 2-7
[6] Pathak N, Butcher C, Worswick M 2013 Influence of the sheared edge condition on the hole expansion of dual phase steel *Proc. IDDRG2013 (Zürich)* 213-218
[7] Mori K I, Abe Y, Suzui Y 2010 Improvement of stretch flangeability of ultra high strength steel sheet by smoothing of sheared edge *Journal of Materials Processing Technology* **210** 653-659
[8] Tsoupis I, Hildering S, Merklein M 2014 Bending of high-strength low-alloyed steel with respect to edge crack sensitivity caused by shearing operations *Procedia Engineering* **81** 712-717
[9] Dünckelmeyer M, Karelova A, Krempaszky C, Werner E 2009 Instrumented Hole Expansion Test *Proc. International Doctoral Seminar 2009 (Smolenice)* 1-8
[10] Braun M, Borsutzki M, Bülter M, Carlsson B, Larour P, Richter A, Atzema E, Geisler S 2010 Ergebnisse eines europäischen Ringversuchs zum Lochaufweitungsversuch nach ISO 16630 *Proc. Werkstoffprüfung 2011 (Berlin)* 317-326
[11] Schneider M, Eggers U 2011 Investigation on punched edge formability *Proc. IDDRG 2011 (Bilbao)*1-12
[12] Goncalves J, Fouques D, Bellut X, Zhang L, Huang M, Deventer J V, Farrand B 2014 More Fruitful and Robust Hole Expansion Testing Conditions *Proc. IDDRG 2014 (Paris)* 216-221
[13] Westhäuser S, Schneider M 2015 Edge cracking sensitivity - valuation methods and material related solutions *Automotive Engineering Congress 2015 (Nürnberg)*
[14] Larour P, Freudenthaler J, Grünsteidl A, Wang K 2014 Evaluation of alternative stretch flangeability testing methods to ISO 16630 standard *Proc. IDDRG 2014 (Paris)* 188-193
[15] Nakajima K, Kikuma K, Hasuka K 1971 Study on the formability of steel sheets *Yawara Technical Report* **284** 678-680
[16] Tsoupis I, Hildering S, Merklein M 2012 Methodology for the analysis of the process behaviour of advanced high strength steels in bending and shearing operations. *Key Engineering Materials* **504-506** 895-900
[17] Abbadi M, Hähner P, Zeghloul A 2002 On the characteristics of Portevin-Le Chatelier bands in aluminum alloy 5182 under stress-controlled and strain controlled tensile testing. *Materials Science and Engineering* A **337** 194-201

IOP Conf. Series: Materials Science and Engineering **159** (2016) 012018 doi:10.1088/1757-899X/159/1/012018

Tribological behavior of a new green industrial lubricant for stamping operations – Application to Stainless Steels

A Buteri[1], M Borgeot[1, 2], X Roizard[2], F Lallemand[3], J-M Melot[3], L Morand[4]

[1] APERAM Research Center, F-62330 Isbergues – France
[2] FEMTO-ST Institute, F-25009 Besançon – France
[3] UTINAM Institute, F-25009 Besançon – France
4 CRISTEL, F-25490 Fesches-le-Châtel – France

E-mail: aurelien.buteri@aperam.com

Abstract. Thanks to a fruitful collaboration between different academic and industrial research entities, a new green lubricant solution AFULudine has been developed for stamping operations in substitution to the usual mineral oils, with the idea and the wish to combine different essential properties such as a low environmental impact and ease-of-use according to the market needs and the more and more restrictive environmental legislation (e.g. REACH). Based on the Self-Assembled-Monolayer (SAM) principle and so, on a perfect control of the chemical reaction between the solution and the substrate (grafting of molecules onto the surface), AFULudine offers an efficient technical answer for improving stamping processes.
The present study, focused on stainless steel stampings (comparison between a 441-1.4509 grade and a 304-1.4301 grade), investigates the performances of this new green industrial lubricant at different levels: from laboratory tests to industrial conditions through the production of hundreds of parts. Additional results coming from tests made on a rotational pin-on-disk tribometer will allow us to appreciate and retrieve more local information about the tribofilm creation during sliding. Moreover, the comparison with different mineral oils currently used at industrial scale, will strengthen the AFULudine performances. Indeed, this new solution usually outperforms a majority of such oils whatever their viscosity and their own composition: formulation, content of extreme pressure additives (Cl, S, P)...

1. Introduction

Lubrication of stamping operations is today widely assumed as absolutely necessary to allow part feasibility and preserve surface integrity of both stamped parts and tools. This finding appears to be particularly important concerning the forming operations of stainless steel grades, whatever the mechanical properties, for which well-known issues of adhesion or transfer of work material to the tool can easily occur if no precautions are taken [1, 2]. The figure 1 illustrates the current lubrication offer dedicated to stainless steels. Mainly based on petrochemical processes and widely dominated by few industrial companies, a compromise between performance, correlated to friction coefficient, and the easy-of-use, linked to the viscosity, can be done. Today, no concrete alternative solutions exist to substitute the most efficient lubricants, containing additives such as sulphur, phosphorous or chlorine. These compounds are designed to chemically react with the metal surface during mechanical solicitations, forming sheared layers of sulfides, phosphides and chlorides according to the induced working temperatures [3]. In fact, the higher efficiency of the chlorine additive with stainless steels is mainly due to a good adequacy between the working temperatures inherent to drawing processes (around 100°C and perhaps more if you consider the temperature reached at the extreme surface) and the activation temperature of this additive (between 100°C and 150°C according to the type of chlorine – C chain length, etc…).

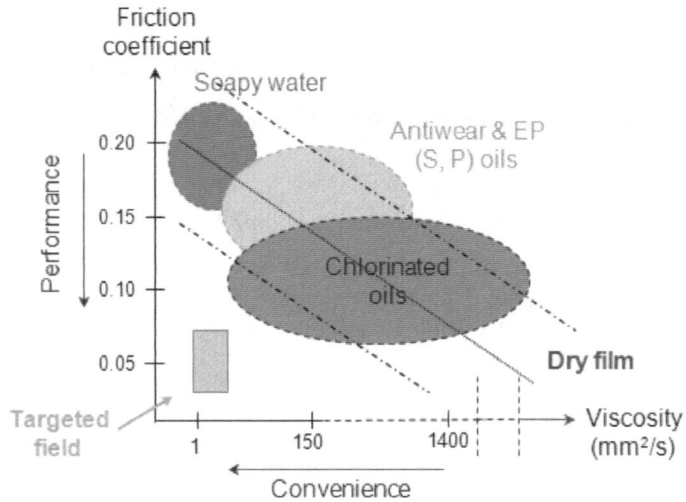

Figure 1. Schematic view of the current offer in terms of lubricants dedicated to stainless steels – a daily compromise has to be done between convenience (easy-of-use) and product performance. The AFULudine target is highlighted by the green box – Viscosity at 40°C – The convenience aspect takes into account the oil application (difficult for viscosity levels > 500 mm^2/s) but also the degreasing step.

The development of the AFULudine solution is based on the wish to perfectly control the chemical reaction between specific active molecules and metallic oxidized substrates such as stainless steels. With the idea to voluntarily control chemical reactions instead of hoping that these ones will occur during sliding, the developed solution ensures surface functionalisation by using organic molecular assemblies like Ulman et al. previously did [4], and more especially by using alkylphosphonic acid molecules. These molecules are well-known to act as a protective layers formed by chemical reaction with the metal surface due to a fairly high level of chemical activity [5]. Raman et al [6] studied their ability to create a Self-Assembled-Monolayer (SAM) on the surface of an oxidized austenitic stainless steel (316L). In more recent studies [7, 8, 9, 10], significant improvement of the tribological behaviour of different substrates (copper, titanium, stainless steels) was demonstrated and assigned to alkylphosphonic acids molecules used as nano-size liquid additives in lubricant, highlighting more especially the formation of low-friction tribofilm during sliding, responsible for the surface protection. Such active molecules are the very essence of the AFULudine solution, which constitutes today a more environmentally friendly lubrication solution than the standard petrochemical lubricants.

The present study investigates at different experimental scales until industrial production the full potential of the AFULudine solution for two different stainless steel grades: a 1.4301-189ED austenitic grade and a 1.4509-K41 ferritic grade, according to mechanical solicitations representative of deep drawing operations. The lubrication strength for such shrink drawing deformation path is widely validated today. Verma et al. [12] concluded for example that the limiting drawing ratio (LDR) is decreasing with increase of friction. In addition to the performance comparison with some industrial mineral oils dedicated to stainless steels, the influence of the AFULudine processing conditions (solution concentrations + lubrication process) on the product performance will be highlighted. Direct correlations between a tribometer especially designed by Roizard et al. [11] to be highly representative of stamping operations [11] and two stamping presses (a laboratory press and a single effect production press) will be furthermore established.

2. Experimental

2.1. Material

Several industrial austenitic and ferritic stainless steel grades were used as received with different industrial finish surface states: 2R (bright annealed surface + Skin-Pass operation) or 2B (surface after a pickling step followed by a Skin-Pass operation), without any additional preparation steps whatever the mechanical characterizations. The intrinsic roughness variations between these different surface states will not be took into account in the present study. Recent works on ferritic stainless steels [9] showed a limited contribution of the roughness in the tribofilm formation. The table 1 and table 2 sum up the standard chemical compositions of these grades as well as their main tensile properties.

Table 1. Chemical composition of the Stainless Steel grades

Grade (EN-ASTM-Commercial codes)	C	Si	Mn	Cr	Ni	Others
1.4509 – 441 – K41	0.015	0.60	0.30	17.80	-	Ti+Nb=0.65
1.4301 – 304 – 189ED	0.040	0.40	1.20	18.20	8.10	-
1.4301 – 304 – 189DDQ	0.045	0.40	1.10	18.20	9.10	-

Table 2. Tensile properties of the Stainless Steel grades

Grade (EN-ASTM-Commercial codes)	Thickness (mm)	UTS (MPa)	YS (MPa)	A%	r_N
1.4509 – 441 – K41	1	480	310	30	1,285
1.4301 – 304 – 189ED	1	630	285	57	0,970
1.4301 – 304 – 189DDQ	0,8	610	270	57	0,936

UTS = Ultimate Tensile Stress, YS = Yield Strength, A% = Maximal elongation, r_N *= Normal anisotropy*

2.2. AFULudine solution & other mineral oils

Based on the use of alkylphosphonic acids as organic friction modifiers, the AFULudine solution has been thought to ensure an industrial easy of use thanks to a hydro-alcoholic solvent. Several AFULudine solutions were investigated in the present study all based on a single carbon chain length. These ones will not be described in detail in the present paper in terms of composition and process in order to preserve as much as possible the internal know-how. They will be identified mainly thanks to their relative concentration in active alkylphosphonic molecules compared to a normalized reference: from C1 (2%) to C9 (300%), with C7 (100%) defined as the current reference. Likewise, the different industrial processes used to lubricate the substrates will not be fully described either, but simply identified by the following notation: from P1 to P5. The contact time with the surface (decreasing from P1 to P4, leading to a productivity improvement), the quantity of lubricant and the type of lubrication system (pulverization, …) are the changing parameters between the different processes, which can be after all considered as perfectly adapted to standard industrial production equipments (except the P5 process corresponding to a specific lab treatment). The table 4 sums up the concentration and process conditions used according to the experimental devices used.

Other industrial lubricants from the petrochemical industry were characterized on this study and confronted to the AFULudine solution. Highly representative of the market, more than a dozen of the mineral oils, containing different chemical additives such as extreme-pressure (S, P, Cl) or other anti-wear elements were tested. The different oils will only be distinguished from one to another by their

IOP Conf. Series: Materials Science and Engineering **159** (2016) 012018 doi:10.1088/1757-899X/159/1/012018

respective Chlorine content: with or without Cl. No distinction in terms of viscosity will be done. The grammage was maintained quite constant, between 5 and 10g/m^2 according the oil viscosity.

2.3. Tribological measurements & Stamping characterizations

The AFULudine performances were tested and quantified at different scales: from lab tests, with a rotational pin-on-disk tribometer and a mechanical Erichsen stamping press (20T), to industrial equipments dedicated amongst others to the daily production of pans (Cristel®). The figure 2 illustrates the schematic view of the single effect hydraulic press as used by Cristel® [13].

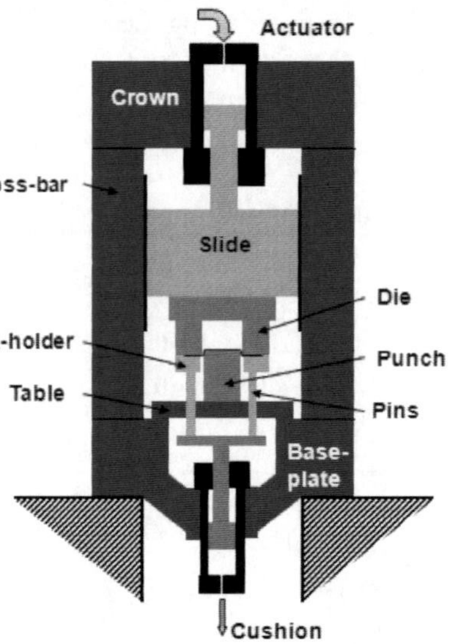

Figure 2. Schematic view of a single effect press [13]

The rotational pin-on disk tribometer (with a circular track of 3.5 mm radius) has been used in this study for quantifying the intrinsic tribofilm resistance [7, 9] and its ability to support very long solicitation distance according to the experimental settings used (Table 3).

As a complement to these tribological data, stamping tests were carried out with an Erichsen press (20T capacity – 142.20 model), with the wish once again to quantify the intrinsic impact of the lubricant on the Limit Drawing Ratio (LDR). Indeed, when changing only the lubricant, LDR value is directly correlated to the lubricant performance due to the different friction phenomena inherent to the forming operation. Each characterization has been performed at room temperature, with equivalent experimental settings (Table 3) and replicated 3 times to ensure the repeatability. With a constant blank-holder load equal to 6kN (leading to different pressures according to the disk's size), the experimental parameters have been optimally chosen regarding the deep drawing field of each grade/thickness couple. Moreover, for each test, a recording system allowed us to measure the drawing force applied during to the punch stroke.

To finally validate the AFULudine performance at an industrial scale, collaboration was set up with a faithful customer of Aperam: Cristel, producer of high-end cooking utensils, which currently use an efficient chlorinated mineral oil for its stamping operations. Different production campaigns were carried out for positioning the AFULudine performance in series production: small batches of 20 pans (∅240mm) were firstly done for studying the impact of the surface preparation (P1, P2, P3 and P4) before opting for a 200 parts set in order to integrate potential mass production effects (production rate: 4parts/min). All tests were conducted with a constant C7 concentration on a single effect press as illustrated by the figure 3, with instrumented tools continuously recording for example the maximal drawing force applied by the slide during the punch stroke. As a result of the device conception, the only measurement of the drawing force applied by the slide informs directly of the lubrication contribution to the forming operation, provided that other parameters (material, blank-holder force, tools kinematic) remain unchanged.

Table 3 and 4 synthesize the experimental devices and conditions used in this study to compare the AFULudine solution with the different industrial oils.

Table 3. Experimental settings used with the different equipments.

Equipments	Experimental settings	Substrate
Rotational pin-on-disk tribometer	V = 770mm/min – F = 5N (Hertzian mean contact pressure = 860MPa) – Sliding distance = 22mm Steel ball 100C6 (∅5mm)	K41 (1mm)
Erichsen Press (20T)	V = 20mm/min – BH Force = 6kN – ∅Punch = 33mm Z160CDV12 quenched tools ∅ variation between 2 successive discs = 2mm	K41 & 189ED (1mm)
Industrial Press (500T)	V = 2880mm/min – BH Pressure = 5.5MPa ∅Punch = 214mm – TiCN coating - ∅discs = 415mm (β=1.73)	189DDQ (0.8mm)

Table 4. Lubrication and process conditions according to the experimental devices – The C7 concentration can be considered as a reference.

Equipments		Lubricant concentration									Process				
		C1	C2	C3	C4	C5	C6	*C7*	C8	C9	P1	P2	P3	P4	P5
Rotational pin-on-disk		X	X		X			*X*			X				
Erichsen Press	*Concentration effect*	X	X	X	X	X	X	*X*	X	X	X				
	Oil's comparison							*X*			X				X
Industrial Press								*X*			X	X	X	X	

3. Results and discussion

3.1. Lab. tests

3.1.1. Influence of the concentration

Before comparing the AFULudine solution with other current mineral oils, the optimal working concentration of alkylphosphonic acid molecules has been firstly defined thanks to the LDR determination of a K41 1mm thick grade (Figure 3). In this figure, concentration variations are reminded by the different blue points. Instead of one specific optimal concentration, the existence of a large concentration range is highlighted, between C4 (40%) and C7 (100%) within which maximal LDR is reached (2.24±0.01). Out of this range, the performance is drastically decreasing, more especially for lower concentrations for which a galling phenomenon can also be observed (C1 concentration). For higher concentration levels, the performance decrease is less critical with a ceiling value around 2.21 (±0.01), apparently constant. This performance deterioration has no evident cause but must be due to variation in terms of surface functionalisation and/or tribofilm formation. Although the LDR values are the same for the concentrations from C4 to C7, a ranking based on the maximal force applied by the punch can be established (orange points). Thus, for the same deep drawing operation, C7 concentration induces a reduction of 2,8% of the maximal force required to perform part stamping compared to C4 concentration. More broadly, a decrease of the maximal punch force appears to be directly correlated to the increase of concentration.

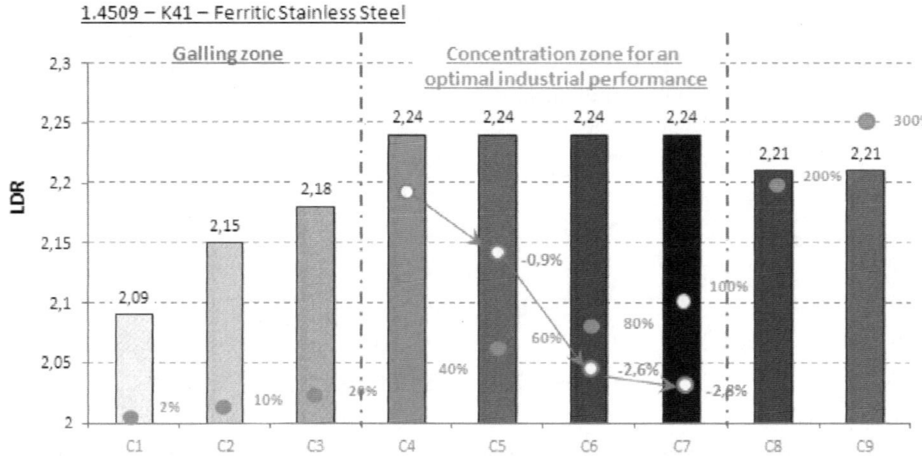

Figure 3. LDR evolution according to the AFULudine concentration: from C1 (2%) to C9 (300%) with C7 defined as reference value (100%) – Samples have been prepared thanks to the P1 process.

One of the reasons which could explain this concentration influence concerns the efficiency and the resistance of the induced tribofilm during the stamping operation according to the friction solicitations. Roizard and al [9] has recently studied the intrinsic mechanisms relative to the formation of an efficient low-friction tribofilm on the surface of a ferritic stainless steel (1.4509). A healing effect was thus highlighted and mainly correlated to the ability for new active molecules to graft at the metal surface in order to continuously replenish the tribofilm. Without internal source in terms of new molecules, a constant but irremediable deterioration of the friction coefficient has been observed until galling occurred. According to these findings, even if the test conditions were different compared to the present study, the tribofilm resistance will be defined by the quantity of active molecules available which depends on the initial concentration and the process application. These two parameters are responsible for a grafted layer having its intrinsic properties.

The figure 4 explores the friction response of samples prepared with the same AFULudine solutions than previously studied, with variations in terms of concentration between C1 and C7 (P1 preparation). Since the early stages of the test, variable performance of the tribofilm can be noticed, depending on the concentration levels in correlation with the previous LDR results. Indeed, the lower concentration, the higher friction response, even if the values are quite acceptable, between 0.08 and 0.13. Nevertheless, significant differences are quickly observed with a constant and chaotic increase of the friction coefficient for the lower C1 concentration, finally leading to a galling phenomenon. SEM views of the figure 5 illustrate the track topology after 200 cycles performed with the C1 solution (yellow point of the figure 4), bringing to light a significant degradation of the surface state. The complete absence of phosphorous on the surface after sliding as revealed by the EDS analysis strengthens the hypothesis of a complete deterioration of the tribofilm during sliding.

Increasing the concentration level of active molecules is finally leading to improve the low-friction tribofilm efficiency and its time stability until galling. A good correlation exists between this results and the ranking presented in figure 4 in terms of LDR for equivalent concentrations, mainly explained by the intrinsic tribofilm formation whatever the mechanical solicitations.

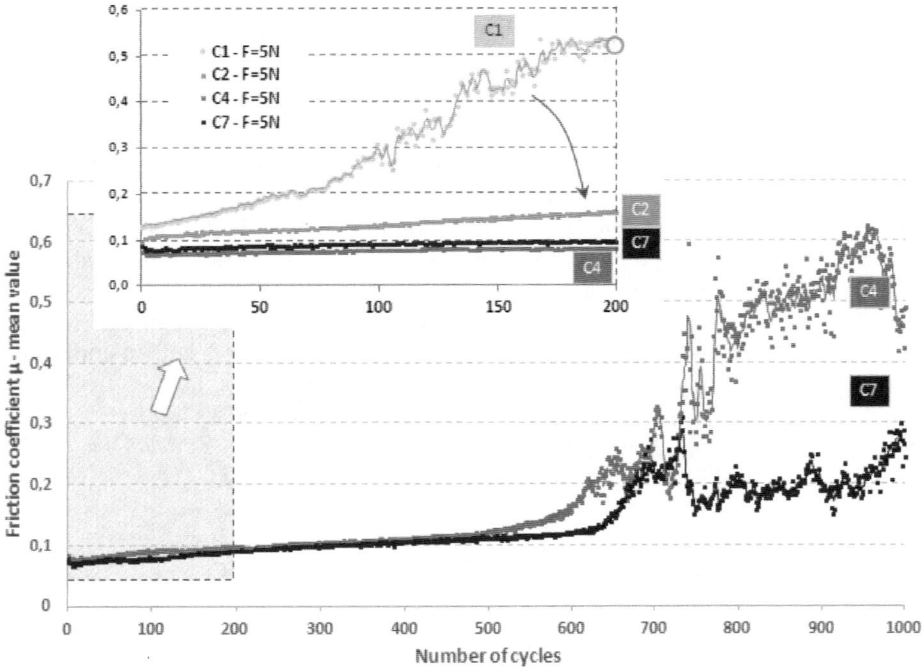

Figure 4. Evolution of mean friction coefficient versus number of cycles (Rotational pin on disk tribometer) for four different AFULudine concentrations: C1, C2, C4 and C7, from light to dark curves respectively. Samples have been prepared thanks to the P1 process

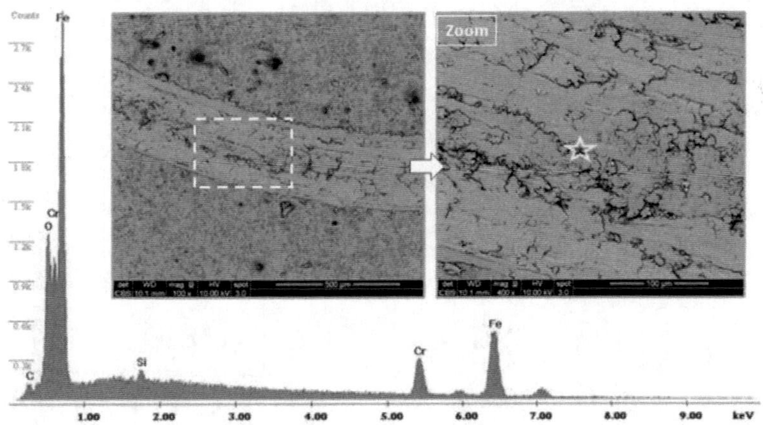

Figure 5. SEM images (back scattered electrons) of the track after sliding test on rotational pin on disk tribometer where 200 cycles were performed with the C1 AFULudine solution, prepared thanks to the P1 process. EDS analysis was performed on the yellow star location.

3.1.2. Comparison with other industrial lubricants

The figure 6 compares LDR values obtained for more than a dozen of industrial lubricants, only distinguished from one to another by their chlorine content, with the AFULudine performance at the optimal concentration (C7). A laboratory grease (based on graphite) and a new non-oil lubricant suggested as a green alternative to mineral oils were also tested and compared. The exercise was done for a ferritic (1.4509-K41) and an austenitic (1.4301-189ED) stainless steel, presented in figure 7 a) and b) respectively.

IOP Conf. Series: Materials Science and Engineering **159** (2016) 012018 doi:10.1088/1757-899X/159/1/012018

Even if the ferritic stainless steel grade highlights on average logically higher LDR values than with the austenitic grade, thanks to a direct effect of the material anisotropy, it is interesting to remark an equivalent ranking of the different lubricants according to their chlorine content: higher drawability is allowed by using this extreme pressure additive, whatever the oil viscosity and the substrate nature. It is also interesting to underline the remarkable performance of the AFULudine solution, at least equivalent or even better than a majority of chlorinated oils. Its performance even matches with the laboratory grease one for the 1.4301-189ED austenitic grade. Unlike the direct competitor non-oil lubricant which appears to be sensitive to the substrate nature, the operating principle of AFULudine is relevant for all the stainless steels grades. Moreover, the scope of progress is still great as can be seen on the figure 6 a) with a 2.35 LDR value reached for the AFULudine solution prepared in laboratory conditions (C7-P5). The main difference between the P1 and P5 processes lies in the time required for preparing the samples. New developments are currently engaged to achieve equivalent performance in industrial conditions.

It may also be noticed that oil 2 used with both stainless steel grades refers to the chlorinated lubricant daily used by Cristel. With LDR values equal to 2.25 (1.4509-K41) and 2.11 (1.4301-189ED), this one can be considered as appropriate.

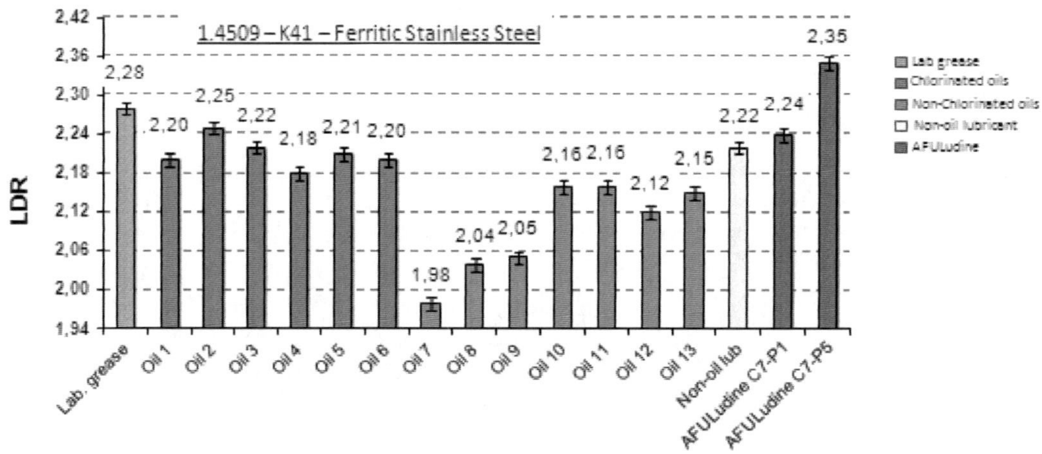

a)

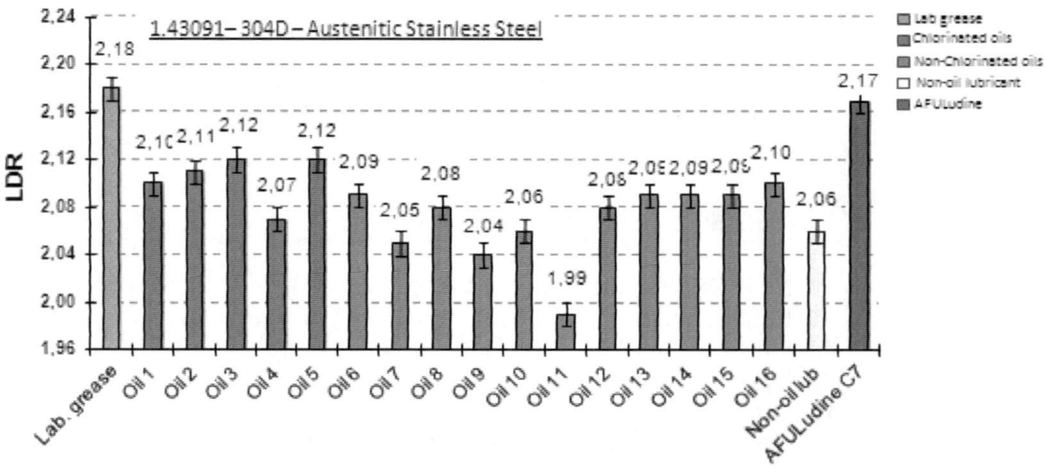

b)

Figure 6. LDR comparison between different industrial oils with (orange bars) or without (blue bars) chlorine additive, a lab grease (green bars), a non-oil industrial lubricant (yellow bars) and the AFULudine solution (pink bars - C7 concentration – P1 preparation) for a) a 1.4509 ferritic stainless steel and b) a 1.4301 austenitic stainless steel. Notice that for the ferritic grade, the specific laboratory preparation P5 of the AFULudine solution (C7) was also tested.

3.2. *Industrial validation*

3.2.1. *Small batch production – Influence of the lubrication conditions*

The figure 7 illustrates the evolution of the maximal force applied by the machine slide during stamping according to the number of parts produced. Whatever surface preparations (from P1 to P4), AFULudine solution induces more efficient lubrication conditions than the ones obtained with the current lubricant daily used by Cristel® (Figure 7). Indeed, regardless of the lubrication method, the use of AFULudine is systematically associated with a significant decrease of the maximal force required to perform a part stamping, with a saving between 5,3% and 9,8% compared to the standard Cristel chlorinated oil, highlighting a high level of performance.

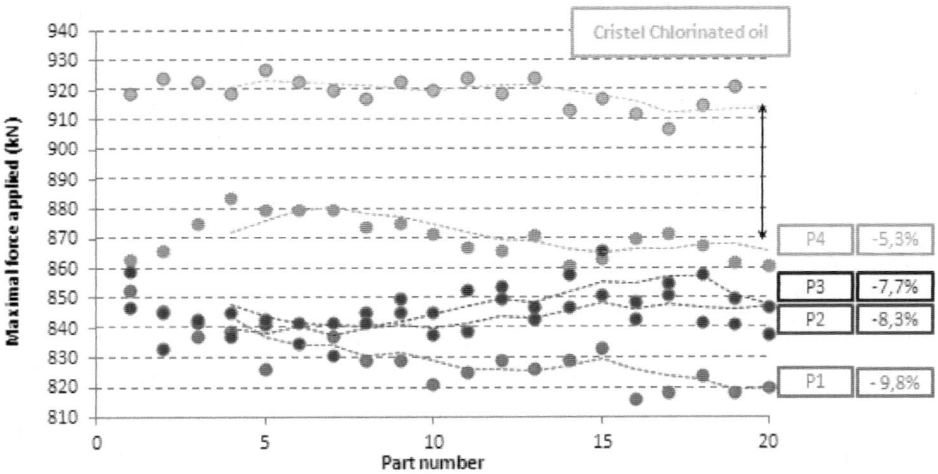

Figure 7. Evolution of the maximal force applied by the slide according to the number of pieces - Comparison between different AFULudine preparations, P1 (pink), P2 (purple), P3 (dark blue) and P4 (light blue), and the chlorinated oil used by Cristel (green). Doted lines correspond to 4 points sliding average curves – C7 concentration for all tests.

The differences observed between the P1 to P4 conditions are directly correlated to the quantity of alkylphosphonic acids molecules grafted on the metal surface and their organization, mainly depending on the associated process. As explained on the experimental section (2.3) the four preparation conditions differ, amongst others, by the contact time between the AFULudine solution and the metal surface, ranging from minutes to seconds. According to a previous study made on an equivalent 304 stainless steel [14], even if it may be reasonable to suppose that molecules immediately graft at the surface due to a spontaneous reaction, the contact time has been highlighted as having a significant influence on the layer quality. Indeed, Roy and al. [14] reported that increasing the grafting time induces an increase of the recovery rate, up to a critical point (for few hours of grafting time). It is not surprising therefore that performance variation occurs depending on the lubrication process used. Both concentration and contact time are so influencing the tribofilm efficiency.

Moreover, it is important to remind that, with a constant blank-holder pressure equal to 5.5MPa (Table 2), the production conditions are more severe than the ones used on the 20t press at the laboratory scale (equal to 3.53MPa for a ∅57mm disk - β ratio = 1.73).

3.2.2. *Series production – Influence of the temperature*

While the previous small batch productions was not able to underline the inherent thermal effect of series production (increase of the temperature with the number of pieces), the bigger series (200 pans) carried out with the AFULudine solution (C7-P3) brought to light a transitional regime before reaching from the eightieth part a more stable working domain (Figure 8). During this transition period, different stop (from few seconds to few minutes) were carried out, mainly motivated by quality tracking. Each one of them is responsible for an immediate increase of the maximal force applied by the machine after restarting. The more symptomatic example is the force increase observed after the main stop done (5 minutes) at the piece 27. Without changing any process parameter, the only variable between the piece 27 and 28 is the tools temperature. For information only, an increase of the punch temperature (thermocouple located 2mm below the punch radius surface for practicality reason) equal to 20°C has been measured during the 200 pans production. This one can be considered as relevant and significant although it is not the most critical area in terms of thermal heating during stamping operations (e.g. the die radius). Indeed, the thermograph highlights a visible increase of the temperature induced by each stamping operation. A positive sensitivity of the AFULudine solution to the working temperature has been thus identified.

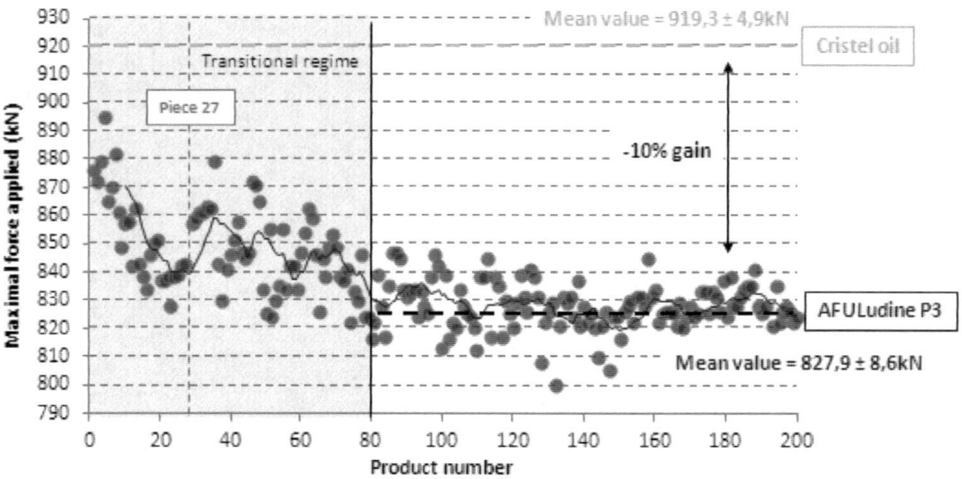

Figure 8. Evolution of the maximal force applied by the slide according to the number of product for a 200 pieces series production – AFULudine solution used with the C7 concentration and the P3 lubrication process. The doted line corresponds to a 10 points sliding average curve.

More significantly, AFULudine use is leading once again to a significant gain in terms of working energy required to perform a piece: 10% compared to the current Cristel oil, with a mean maximal force applied by the slide equal to 827,9kN, which appears to be constant after 80 pans. The system has so reached a necessary stable regime domain for insuring the process robustness. Moreover, the 200 produced pans are perfectly acceptable in terms of quality, without generating any deterioration of the coated tools.

4. Conclusions

Several previous studies showed the effectiveness of alkylphosphonic acids to improve friction behaviour of different metallic substrates (Cu, Ti, Stainless steels...). Mainly representative of laboratory conditions (using different tribometers), the transition to an industrial scale, targeting more especially stamping operations, had not been investigated previously.

The present study goes further into the industrialization of the concept of the Self-Assembled-Monolayer through the AFULudine solution by providing results obtained on both laboratory and industrial presses. With a lower friction coefficient and/or force applied for carrying out stamped parts compared to other lubricant solutions, AFULudine appears to be relevant to offer an alternative to mineral oils and more especially to the chlorinated ones that highlight better performances than most non-chlorinated lubricants. Of course, AFULudine will not systematically replace these daily used lubricants which can be efficient or irreplaceable for specific technical or economical reasons. But for once, a more eco-friendly lubricant solution is conceivable. By the way, this new lubricant should be

IOP Conf. Series: Materials Science and Engineering **159** (2016) 012018 doi:10.1088/1757-899X/159/1/012018

integrated in industrial production line before the end of the year. Indeed, as illustrated by the collaboration with Cristel, AFULudine is well-adapted to mass production of stainless steel stamped parts.

References

[1] Andreasen JL, Bay N, Andersen M, Christensen E, Bjerrum N. Screening the performance of lubricants for the ironing of stainless steel with a strip reduction test. Wear (1997) 1-5.

[2] Bay N. Olsson DD, Andreasen JL. Lubricant test methods for sheet metal forming. Tribology International 41 (2008) 844-853.

[3] Huang W, Tan Y, Dong J, Chen B. Tribological properties of the film formed by borated dioctyl dithiocarbamate as an additive in liquid paraffin. Tribology International. 35 (2002) 787-791.

[4] Ulman A. An introduction to ultrathin organic films: Langmuir-Blodgett to self-assembly. New-York Academic Press (1991).

[5] Rudnick L.R. Lubricant additives – Chemistry and applications. CRC Press, 2nd edition 2009.

[6] Raman A, Dubey M, Gouzman I, Gawalt ES. Formation of Self-Assembled Monolayers of alkylphosphonic acid on the native oxide surface of SS316L. Langmuir 22 (2006) 6469-6472.

[7] Roizard X, Heinrichs J, Et Taouil A, Jacobson S, Olsson M, Melot JM. Insights into sliding wear and friction behavior of copper in ethanol containing alkylphosphonic acid molecules. Tribology International 96 (2016) 141-148.

[8] Zhang C, Liu Y, Wen S. Excellent tribological behavior of hexadecylphosphonic acid films formed on titanium alloy. Science China – Technological Sciences 57 (2014) 1816-1823.

[9] Roizard X, Heinrichs J, Buteri A, Jacobson S, Borgeot M, Carpentier L, Melot JM, Lallemand F. Global rethinking of the lubrication of ferritic stainless steel sheets – Low friction and wear using strongly diluted alcohol solution of alkylphosphonic acid. NordTrib Congress 2016.

[10] Moine MM, Roizard X, Melot JM, Carpentier L, Cornuault PH, Lallemand F, Rauch JM, Heintz O, Lallemand S. Grafting and characterization of dodecyl-phosphonic acid on copper. Surface and Coatings technology 232 (2013) 567-574.

[11] Roizard X, Pothier JM, Hihn JY, Monteil G. Experimental device for tribological measurement aspects in deep drawing process. Journal of Materials Processing Technology 209 (2009) 1220-1230.

[12] Verma RK, Chandra S. An improved model for predicting limiting drawing ratio. Journal of Materials Processing technology 172 (2006) 218-224.

[13] Col A. L'emboutissage des aciers. Ed. Dunod – L'Usine Nouvelle – 2010.

[14] Roy F, Et Taouil A, Lallemand F, Melot JM, Roizard X, Heintz O, Moutarlier V, Hihn JY. Influence of modification time and high frequency ultrasound irradiation on self-assembled of alkylphosphonic acids on stainless steel: Electrochemical and spectroscopic studies. Ultrasonics Sonochemistry 28 (2016) 268-275.

IOP Conf. Series: Materials Science and Engineering **159** (2016) 012019 doi:10.1088/1757-899X/159/1/012019

Influence of the tool temperature increment on the coefficient of friction behavior on the deep drawing process of HSS

I Gil[1], L Galdos[1], E Mugarra[1], J Mendiguren[1], E Saenz de Argandoña[1]

[1] Mechanical and Manufacturing Department, Mondragon University
Loramendi 4, 20500 Arrasate-Mondragon, SPAIN

e-mail: igil@mondragon.edu, Web page: www.mondragon.edu/eps

Abstract. The use of High Strength Steels (HSS) in the deep drawing processes has an impact on the temperature that is achieved on the die surfaces. Due to the heat that is created through the deformation of the material and the friction itself, the tools temperature increase considerably up to approximately 100°C. This temperature increment has an effect on both the wear of the surface and also the coefficient of friction (COF). In this work the influence of the tool temperature on the coefficient of friction is studied. For that, Strip Drawing Tests have been carried out at different tool temperatures with a DP780, High Strength Steel. Moreover, different contact pressures have been considered in the study to analyse the combined effect of the contact pressure with the temperature increment. It has been proved that the temperature increment has to be taken into account to predict accurately the behavior of the coefficient of friction between the sheet and the tool. This change in the coefficient of friction has a high impact on the prediction of the deep drawing process.

1. Introduction

The increasing use of High Strength Steels (HSS) has become common practice in the automotive industry. This growth is due to the global trend to reduce the CO_2 emissions to the atmosphere. Nevertheless, the processing of these materials has induced some drawbacks that have to be taken into account. The forming of these materials requires increased work and contact pressures [1], which is translated to high friction forces producing significant temperatures on the tool and blank surfaces. Moreover, due to the complexity of the final geometry of the majority of the automotive components, significant deformation levels have to be achieved. In order to achieve these levels of deformation, a remarkable plastic deformation energy is produced and as consequence high amount of heat is generated. In addition, the automotive industry requires high productivity of around 32 strokes per minute, which prevents the surfaces of the tools to dissipate the absorbed heat. The generated heat can increment the temperature of the surfaces up to 181°C at some punctual moments [2].

Kim et al. [3] analyzed the temperature distribution of the sheet and the tools in a deep drawing process of a cylindrical cup and observed a maximum temperature of 85.5°C on the sheet. Nevertheless, the study focusses on an isolated stroke regardless of the temperature rise effect during serial production. Hazra et al. [4] analyzed the temperature increment in big serial production of real components and concluded that modest temperature changes did not affect material properties but altered lubricant behavior. Pereira et al. [2] studied separately and coupled the influence of the frictional heating and the deformation-induced heating on the temperature rise of the tool surfaces. It was observed that the frictional heating is primarily responsible for the peak temperatures on the die surfaces.

Several researchers have observed the effect of the tool surfaces temperature increment on the deep drawing process. Kim et al. [5] showed that the formability of the High Strength Steels is affected by these temperature changes, particularly when shear fracture failure occurs. Sachnik [6] analyzed the zinc abrasion of the galvanized sheets that is produced in the deep drawing due to the achieved temperatures on the tool surfaces. Merklein et al. [7] tested a DX56-D+Z at different contact conditions, including different temperatures, but did not achieve temperatures over 60°C. Even though, as the tool temperature increased from 20 to 40 °C, the coefficient of friction experienced an increment of 6.5%. However, there has been noticed that a deeper research is needed in terms of the effect that the temperature changing has on the coefficient of friction.

The modelling of complex tribological conditions is one of the biggest goals to improve the accuracy in sheet metal forming simulations. Together with the blankholder force and the drawbeads restriction force, the coefficient of friction represents the restriction that the material has to overcome to flow through the surfaces of the tool. Furthermore, the deformation and stress distribution depends directly on the applied restriction to the material flow. The complexity of a precise definition of the friction coefficient lies on the number of variables which affect on it. The coefficient of friction is already known to be influenced by several factors within the forming processes as the sliding velocity and the applied contact pressure. But the influence of the temperature increasing at the tool surfaces caused by the forming of High Strength Steels on the coefficient of friction has not been analyzed yet.

In this work, the coupled effect of the applied contact pressure and the tool surface temperature on the tribological behavior has been experimentally investigated. A novel Strip Drawing test facility was developed to provide a temperature control on the tool. This equipment has been used to carry out Strip Drawing tests at different tool temperatures and contact pressures to show the influence of each parameter on the coefficient of friction of a DP780. The obtained results clarify the remarkable effect of the temperature increment on the tribological behavior in deep drawing processes of High Strength Steel components.

2. Methodology

The developed temperature controlled Strip Drawing Test consist of two blocks which have been modified to be able to be heated up to a defined temperature. The block temperatures were maintained at the commanded temperature by means of a cartridge die heater (Hasco). In each block two cartridges are placed to heat the material while a thermocouple is measuring the current temperature, see figure 1. The temperature control was realized by a closed-loop which received the current temperature from the thermocouple. As the thermocouple was located at a certain distance from the surface, there was a difference between the measured temperature at the thermocouple and the one at the block surface. So first, these differences were measured to know which consign had to be defined to reach each block surface temperature.

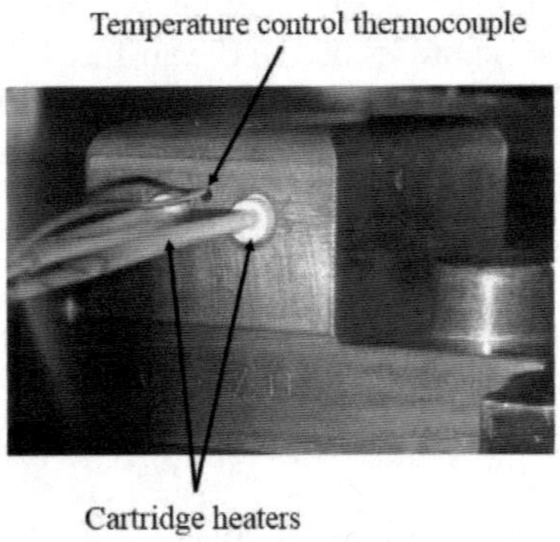

Figure 1. Location of the cartridges and the temperature control thermocouple

These blocks are located in a biaxial machine installation at the Mondragon University. Once that the temperature of the blocks surfaces is stabilized at the commanded value, a cylinder applies a determined contact pressure to the sheet and then another cylinder pulls the sheet in the transversal direction to the contact pressure overcoming the friction force that is produced between the surfaces. Both the applied contact pressure and the sheet pulling force are measured during the whole test for the posterior calculation of the coefficient of friction. The schematic illustration of the test is presented in the figure 2. The material of the blocks was GGG70 and it was followed exactly the same procedure to get the surfaces which were in contact with the sheet as the die manufacturers standards. There was no extra lubrication applied to the sheet apart from the lubrication that the sheet brings itself.

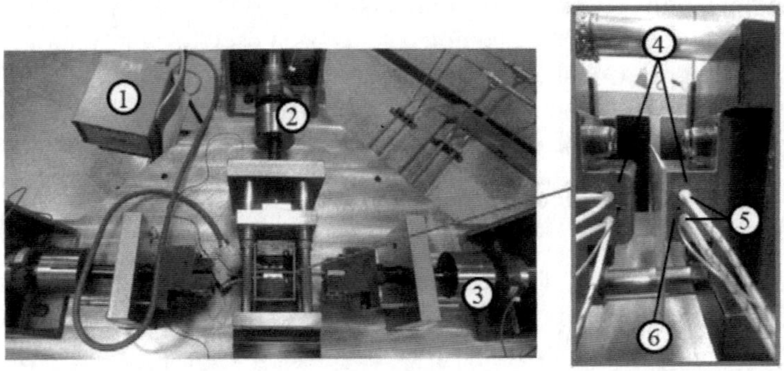

1	Cartridge die heater (Hasco)
2	Contact pressure application cylinder
3	Sheet pulling cylinder
4	Strip Drawing test blocks
5	Cartridges
6	Temperature control thermocouple

Figure 2. Strip Drawing Test installed at the biaxial machine and temperature controlled

The biaxial machine provides the possibility to vary the contact pressure in a wide range. In this study, a contact pressure range between 1 and 16 MPa has been defined. The contact pressure values have been obtained from a numerical simulation of a B-Pillar reinforcement from a DP780 sheet where the above presented range was shown. In addition, four different temperatures have been defined: 20°C (room temperature), 50°C, 80°C and finally 100°C.

3. Results and discussion

As it has been mentioned before, the sheet material is a DP780 High Strength Steel. The analyzed material was covered by 50 microns of zinc coating in each sheet surface. The obtained experimental results are shown in figure 3.

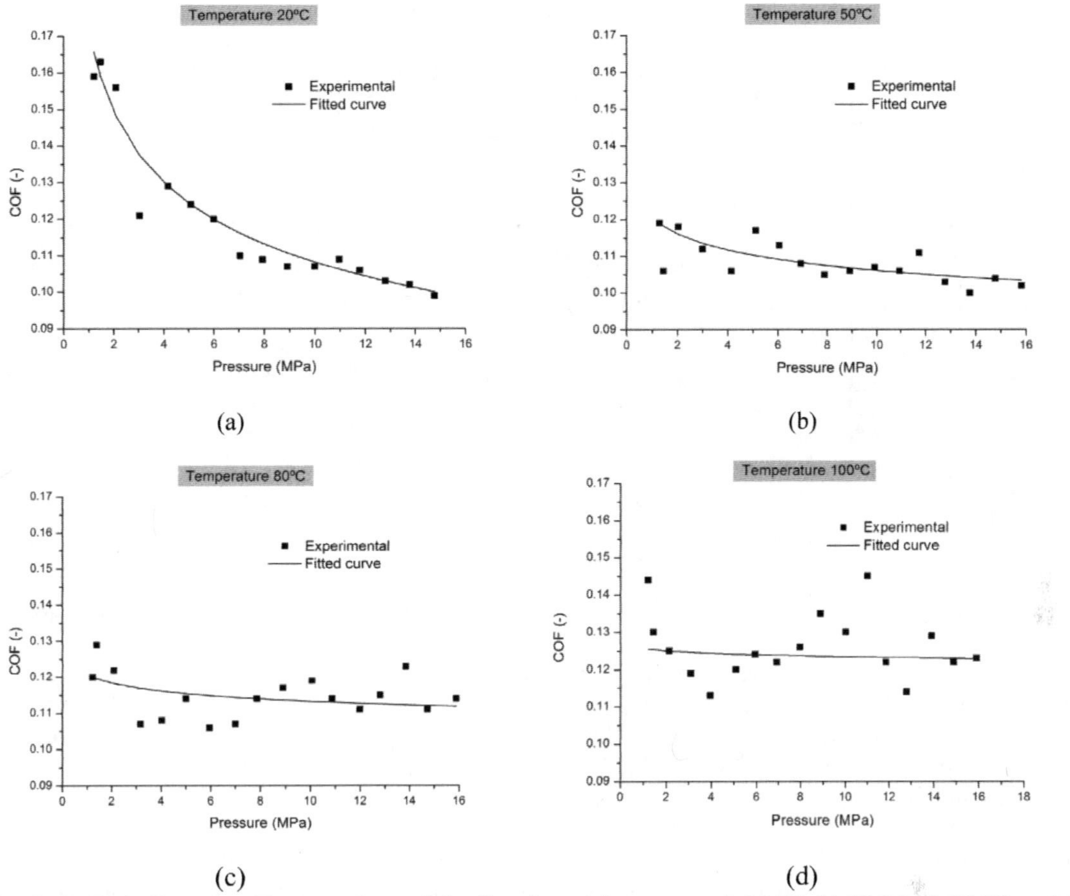

Figure 3. Strip Drawing Test results and the fitted model curve at a) 20 °C, b) 50 °C, c) 80 °C and d) 100 °C. One test was carried out at each condition

The experimental results have been fitted by the model proposed by Filzek et al. [8], where the coefficient of friction is represented in function of the applied contact pressure. This model represents the tribological behavior of the sheet by the equation (1),

$$\mu = \mu_0 \left(\frac{p}{p_o} \right)^{n-1} \tag{1}$$

where μ_0 and p_0 are reference values and n is an exponent defined within the range $0 < n < 1$. The results have shown that the tool temperature affect in the stability of the coefficient of friction. As the temperature increases, the coefficient of friction has a greater variation. Except from the experimental results carried out at 100 °C, the general tendency of the coefficient of friction is to decrease as the contact pressure increases. However, this tendency seems to be reduced as the tool temperature increases. The most evident case is at 100 °C where there is not a clear tendency of the coupled influence between the temperature and the contact pressure. The author suggest that as Merklein et al. stated [7], the change of the tribological behavior comes from the alteration that the viscosity of the lubricant suffers due to the temperature increment on the tool surfaces.

In figure 4 it is presented the comparison between the different tendencies that the experimental results showed in each thermal condition.

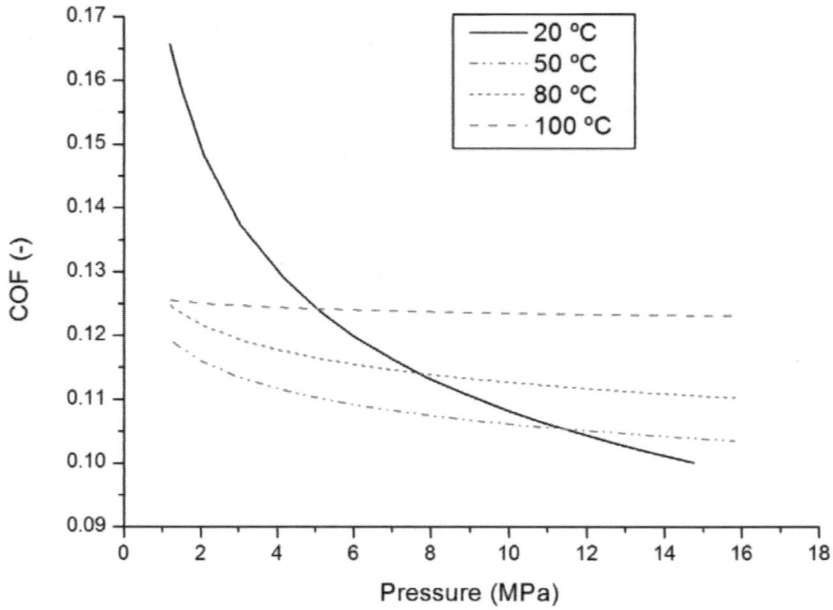

Figure 4. Coefficient of friction tendency at different contact pressures and tool temperatures

These results showed that the temperature increment has the major effect on the tendency of the coefficient of friction when different contact pressures are applied. As the temperature increases, the tribological behavior showed a lower dependency on the contact pressure variations. In fact, at 100°C the value of the coefficient of friction maintains almost constant independently to the contact pressure that was applied.

4. Conclusions

These results add another extra deal to take care of when forming High Strength Steels components. The heat that is generated due to the high energy required to deform the material plus the high friction forces is transferred to the tool surfaces altering the viscosity of the lubricant and as consequence the tribological behavior between the tool and sheet surfaces. As it is known, the definition of the coefficient of friction is a key factor to predict accurately the deformation and stress distribution along the sheet and its direct impact on the springback and final geometry of the part. As consequence the temperature increment of the tools caused by the continuous production of High Strength Steels components is not a negligible parameter from the point of view of the coefficient of friction response.

The most remarkable conclusions obtained from this work are listed as follows:

- It has to be taken into account that the major temperature increment happens in tool areas where high deformations and contact pressures appear as die shoulders, punch radius or drawbeads zones. The results show at higher pressures than 11 MPa that the coefficient of friction is higher as the temperature increases. This effect can result in galling problems at those areas apart from inaccurate predictions in terms of deformation and stress distribution in the sheet.
- The inhomogeneus distribution of the temperature of the surfaces of the tools hinders the correct definition of an equal coefficient of friction model for all the different areas.
- A cooling of the tools which would be able to maintain the tool surfaces at the same temperature may facilitate the adequate establishment of a unique coefficient of friction model for the whole contact along the sheet.
- The obtained results demonstrate the necessity of using a friction model which considers the coupled effect of the temperature and the applied contact pressure. However, this model would

require a numerical simulation software able to predict the heating up of the tool surfaces along the deep drawing processes.

Acknowledgements
The work presented in this paper has been carried out in cooperation with MATRICI SCOOP and has been financially supported by the project HRD: High Reliability Dies under the program RETOS with reference number RTC-2015-3643-4 and financed by Ministry of Economy and Competitiveness of the Spanish Government.

References
[1] Pereira M P, Yan W and Rolfe B F 2008 Contact pressure evolution and its relation to wear in sheet metal forming *Wear* **265** 1687–99
[2] Pereira M P and Rolfe B F 2014 Temperature conditions during "cold" sheet metal stamping *J. Mater. Process. Technol.* **214** 1749–58
[3] Kim H, Altan T and Yan Q 2009 Evaluation of stamping lubricants in forming advanced high strength steels (AHSS) using deep drawing and ironing tests *J. Mater. Process. Technol.* **209** 4122–33
[4] Hazra S, Williams D, Roy R, Aylmore R and Smith A 2011 Effect of material and process variability on the formability of aluminium alloys *J. Mater. Process. Technol.* **211** 1516–26
[5] Kim J H, Sung J H, Piao K and Wagoner R H 2011 The shear fracture of dual-phase steel *Int. J. Plast.* **27** 1658–76
[6] Sachnik P, Volk W, Golle R and Hoffmann H The Influence of Temperature on Zinc Abrasion in Deep Drawing Processes *Key Eng. Mater.* **611-612** 1039–46
[7] Merklein M, Zoeller F and Sturm V 2014 Experimental and numerical investigations on frictional behaviour under consideration of varying tribological conditions *Adv. Mater. Res.* **966-967** 270–8
[8] Filzek J, Ludwig M and Groche P 2011 Improved FEM Simulation of Sheet Metal Forming with Friction Modelling using Laboratory Tests (IDDRG)

IOP Conf. Series: Materials Science and Engineering **159** (2016) 012020 doi:10.1088/1757-899X/159/1/012020

In-Situ-measurement of restraining forces during forming of rectangular cups

M Singer[1, 2] and M Liewald[2]

[1]GSaME, University of Stuttgart, Nobelstr. 12, 70569 Stuttgart, Germany
[2]IFU, University of Stuttgart, Holzgartenstr. 17, 70174 Stuttgart, Germany

E-mail: markus.singer@gsame.uni-stuttgart.de

Abstract. This contribution introduces a new method for evaluating the restraining forces during forming of rectangular cups with the goal of eliminating the disadvantages of the currently used scientifically established measurement procedures. With this method forming forces are measured indirectly by the elastic deformation of die structure caused by locally varying tribological system. Therefore, two sensors were integrated into the punch, which measure the restraining forces during the forming process. Furthermore, it was possible to evaluate the effects of different lubricants showing the time dependent trend as a function of stroke during the forming of the materials DP600 and DC04. A main advantage of this testing method is to get real friction corresponding data out of the physical deep drawing process as well as the measurement of real acting restraining forces at different areas of the deep drawing part by one single test. Measurement results gained by both sensors have been integrated into LS-Dyna simulation in which the coefficient of friction was regarded as a function of time. The simulated and deep drawn parts afterwards are analysed and compared to specific areas with regard to locally measured thickness of part. Results show an improvement of simulation quality when using locally varying, time dependent coefficients of friction compared to commonly used constant values.

1. Introduction

Models simplifying physical effects are still necessary to measure effects of friction in sheet metal forming processes. The disadvantages of all model tests performed are the simplifications of the deep drawing process that are necessary to measure the restraining forces. The majorities of the known tests for evaluating restraining forces in deep drawing are limited to a specific part of the flange or die radius and do neither take into account friction impact of entire part nor the occurring interdependencies. Commonly performed strip drawing tests do not take into account surface enlargement of material nor include the measurements of tangential compression stresses during deep drawing processes. These factors indeed do change the tribological system and to date it was impossible to describe the entire system using only one single test.

Increasing demands on weight saving in car body industries result into manufacturing and processing of high strength steels because less material is required for the same level of strength [1]. Additional to that, the process window has to be extended by modifying the friction [2] to ensure proper manufacturing of such components made of sheet metal. To model the occurring tribological system, a broad knowledge base of physical interdependencies of the influencing factors is necessary. Therefore, many simple model tests were performed in the past in the field of sheet metal forming to gain knowledge about friction and corresponding modelling for simulation objectives. Commonly used tests are mentioned in [3] and represent a single part of the tool, e.g. the flange. The tests are designed in a way in which they can be controlled easily and single parameters can be determined. Thus, they present substantial deviations from the deep drawing processes [4]. The simplifications of the process are necessary to measure the occurring restraining forces directly, but lead to a loss of information and decrease the quality of the simulation. To classify the model tests, [5] defines six categories, in which the load condition of the test is closer to the real component with decreasing number of category. The load condition in the experiment has to be as close to the real component as possible to improve the prediction accuracy of the experiment. To measure restraining forces during a real deep drawing process [6] uses 3 component load cells that are integrated into the die of a deep drawing tool. [7] used same 3 component load cells and a segmented blank holder, where the restraining force of the complete segment was recorded. Unfortunately these methods for in-situ measurement of restraining forces are not capable of being integrated to series production tools.

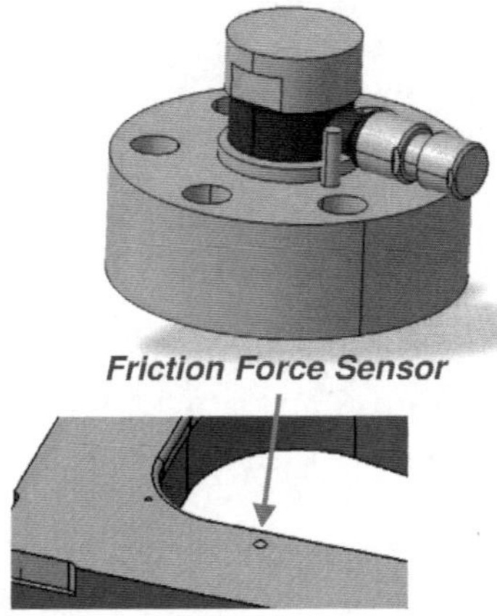

Figure 1. 3D-friction force sensor to measure local restraining forces [6].

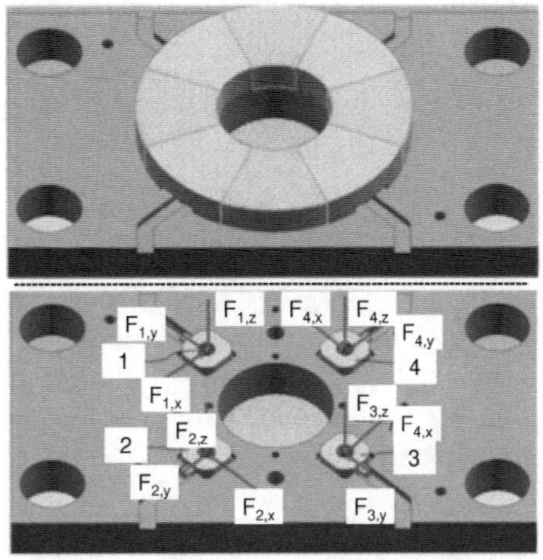

Figure 2. Segmented tool with four integrated 3D-force load cells to measure restraining forces [7].

Due to high cost for performing friction tests and high engineering effort in terms of proper transfer of measured friction values into FEA codes of today, the automotive industry mainly uses generalised and constant coefficients of friction. Objective of this paper for that reason is to present a new method for in-situ-measurement of restraining forces during forming. Lubricants having different viscosities, drawing tests performed at different drawing speeds, and different materials are measured in a real deep drawing process and modelled in FEA within the scope of this work. Additionally, two different drawn sheet metal components were selected and analysed. To evaluate the quality of the simulation, sheet thickness reduction occurring in both parts were compared between simulation and experiment.

2. Experimental procedure and measurement method

2.1. Experimental procedure and setup

Deep drawing experiments for determining occurring restraining forces were performed without any part failures based on using rectangular cup geometry. AIDA servo press machine at IFU with a maximum ram capacity of 6300 kN was used for this. The tool can be used for single strokes or continuous test objectives. Used material (deep drawing steel DC04) was chosen with a thickness of 0.7 mm and a high strength steel (DP600) was chosen with a thickness of 1.2 mm. Amount of lubrication was given by 1.5 g/m^2 of CLF65E (Raziol) with a viscosity of 65 mm^2/s at 40°C and CLF400E (Raziol) with a viscosity of 400 mm^2/s at 40°C. Drawing tests were processed with two different ram velocities (3 strokes/min; 9 strokes/min). The sensors were integrated into the corner and in middle of straight area of the punch. All tests were performed at room temperature.

IOP Conf. Series: Materials Science and Engineering **159** (2016) 012020 doi:10.1088/1757-899X/159/1/012020

2.2. Introduction of measurement method

The changeable restraining forces are measured indirectly by the elastic deformation of the tool caused by different tribological systems. Therefore, the piezoelectric sensor (Kistler Typ 9247A), a longitudinally sensitive pin, was used. The pin is mounted by an M5 thread and the end of the sensor transfers the strain of the surrounding material to the quartz sensing element. The electrical charge output is proportional to the change in mechanical strain [8]. The elastic deformation of the punch is dependent on the occurring drawing force. According to [9] this force can be determined by Siebel`s formula

$$F_{total} = F_{id} + F_{fric,BH} + F_{fric,DR} + F_{bnd,bckbnd} \quad (1)$$

with F_{id} = ideal forming force,
$F_{fric,BH}$ = restraining force resulting from friction by blank holder,
$F_{fric,DR}$ = cable friction force at die radius,
$F_{bnd,bckbnd}$ = force from bending and back bending.

In this equation, F_{id} and $F_{bnd,bckbnd}$ are dependent on geometrical and material properties, $F_{fric,DR}$ and $F_{fric,BH}$ are dependent on geometry, blank holder force, material properties and coefficient of friction. When using same testing parameter, except variations in speed or lubrication, measured changes are only dependent on changes in the tribological system. The main advantage of this measurement method is that no direct contact of sensor and blank occurs and as a result of that the restraining forces are not influenced and changed by the measuring method. Figure 3 shows the integrated sensor positioned with a 45° degree orientation to drawing direction. Figure 4 shows the graph of the measured sensor signals for different tribological systems. To calibrate these sensor signals and use them as coefficient of friction in simulation, a measured curve was modified with a factor to fit sheet thickness from simulation to the experimental measured. The same factor was used for all variations in tribological system.

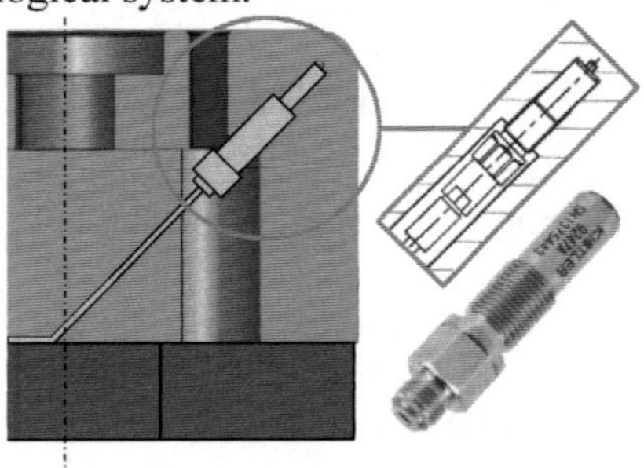

Figure 3. Punch with integrated sensor to measure restraining forces

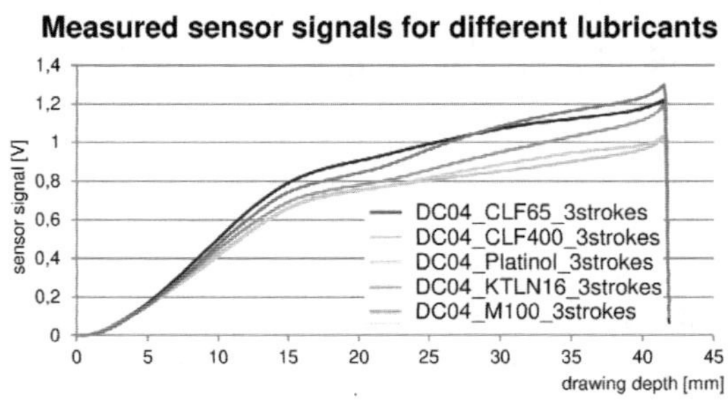

Figure 4. Measured sensor signals for different lubricants.

3. Numerical Procedures

The deep drawing process of rectangular cups was modelled using the FE-software LS-Dyna to compare any changes of the sheet thickness between experiment and simulation. Flow curves were determined in a uniaxial tensile test using a Zwick tensile test device. Figure 5 shows the extrapolation of the flow curves using the Hockett-Sherby criterion. Additional information on recording of such flow curves can be found in [10].

Different geometrical areas of the tool, punch, die, and blank holder are modelled in segments in order to integrate different coefficients of friction into simulation model (Figure 6). LS-Dyna was used as FE-Solver, FE-Models first were built up in DynaForm 5.9 and modified with LS-Prepost 4.1. Drawing speed, sheet metal thickness and blank holder pressures were recorded with measurements in the real deep drawing process and were set equivalent in simulation. All time steps are multiplied with the factor 0.01, which results into shorter simulated drawing process and reduces simulation time.

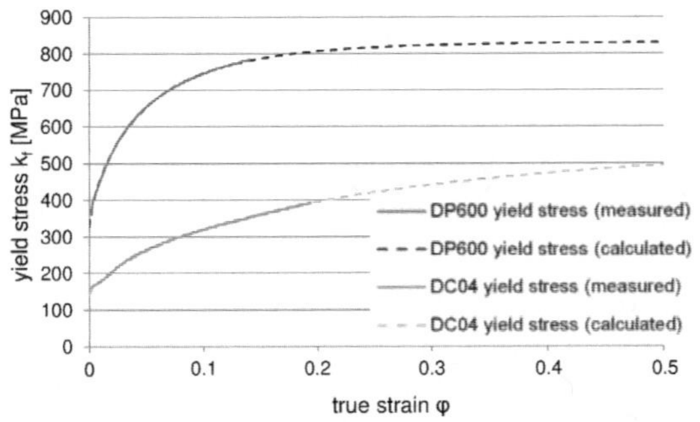

Figure 5. Measured and extrapolated flow curves of used DC04 and DP600

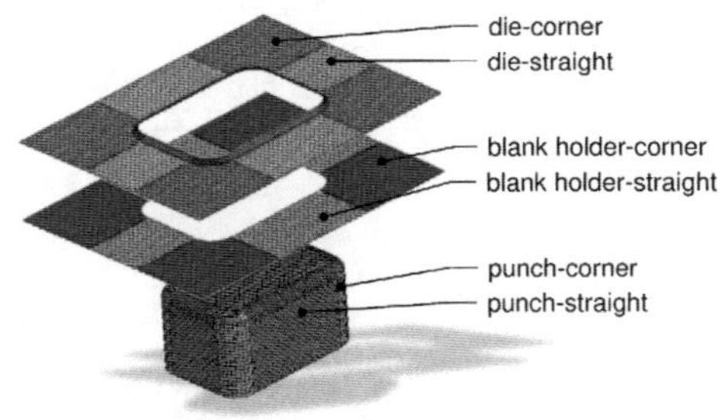

Figure 6. Simulation model of segmented die, blank holder, and punch

LS-Dyna does not allow the integration of time dependent coefficients of friction, but the program supports a thermal friction option which uses temperature dependent coefficients of friction. Therefore, a coupled thermal-mechanical simulation was built up, in which the tool and blank are heated continuously, and thermal friction coefficients were defined ($\mu(\vartheta) \rightarrow \vartheta(t) = \mu(\vartheta(t))$). The speed of ram and punch was set according to the real deep drawing tests of press machine in order to consider the speed and movement of punch realistically and to assign measured sensor signals to the right time step of the simulation. The measured sensor signal at corner and straight side of the punch were used as temperature dependent coefficient of friction for the corresponding part of the tool.

The validation of integrating time dependent coefficients of friction is shown in Figure 7. An abstracted strip drawing test was modelled in LS-Dyna with the same speed profile of AIDA servo press ram during drawing. The temperature is increasing continuously from 20°C to 40°C to assign a specific coefficient of friction to each temperature. The force in x-direction was evaluated to prove that the coefficient of friction is changing in the required way.

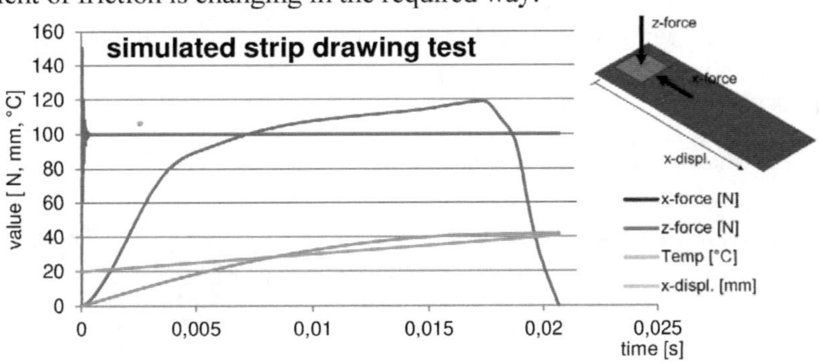

Figure 7. Simulated strip drawing test with time dependent coefficient of friction and resulting restraining force to validate the used simulation model.

4. Experimental and numerical results

Ten measuring points were chosen along two lines to compare simulation with experiment. First line diagonal from the part corner of the rectangular cup; second is set along the straight side. The sheet thickness was measured tactile, initial sheet thickness of DC04 was 0.70 mm; DP600 had a thickness of 1.2 mm. To show the performance of a simulation with integrated coefficients of friction that were measured in a deep drawing process, distribution of sheet thickness of cups deep drawn with different amounts of lubricant was compared to corresponding simulation being performed with the generalised coefficient of friction as well as to executed, real experiments (Figure 8). A satisfactory correlation could be achieved especially for the straight area. Lubrication reveals a higher affection on thinning in the straight than in the corner area. The different lubrication and the resulting tribological system is not that explicit due to the occurring tangential compression stresses.

IOP Conf. Series: Materials Science and Engineering **159** (2016) 012020 doi:10.1088/1757-899X/159/1/012020

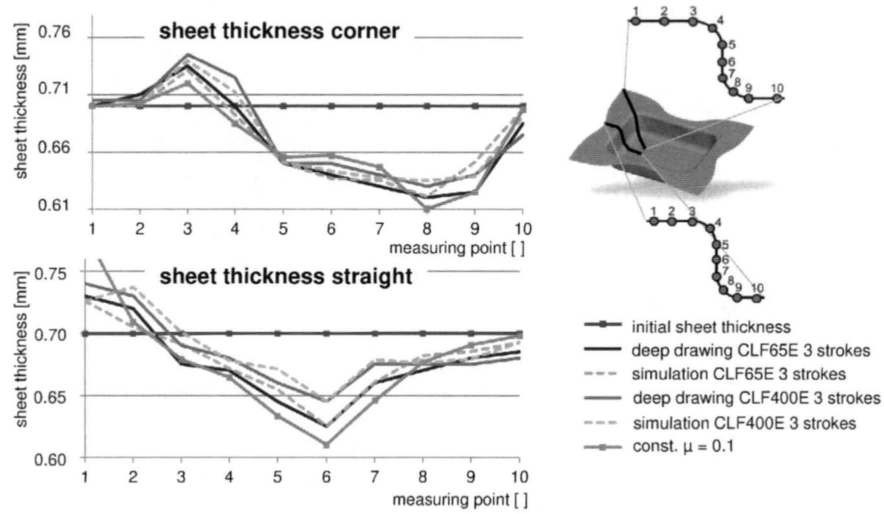

Figure 8. Simulated and experimentally measured distribution of sheet thickness for different lubricants at different part areas (solid lines: real deep drawing, interrupted lines: simulation).

Figure 9 shows the influence of variation of drawing velocities on sheet thicknesses. The velocities used were measured in the real forming process at 3 and 9 strokes per minute, which results into an average speed of v1 = 45 mm/s and v2 = 135 mm/s. Distribution of sheet thickness when simulated with measured sensor signals shows a good correlation to deep drawn experiment in comparison to a generalised coefficient of friction. Especially along the straight area of part perimeter, sheet thickness was predicted exactly. The reduction of thinning, when raising the drawing speed, is caused by the hydrodynamic effects that reduce the coefficient of friction. Occurring deviation at measuring points 8 and 9 of the corner area develops during the bending of the sheet material and is mostly affected by the material model. Finally coefficient of friction is not the most dominant factor due to the small relative movement between sheet material and tool.

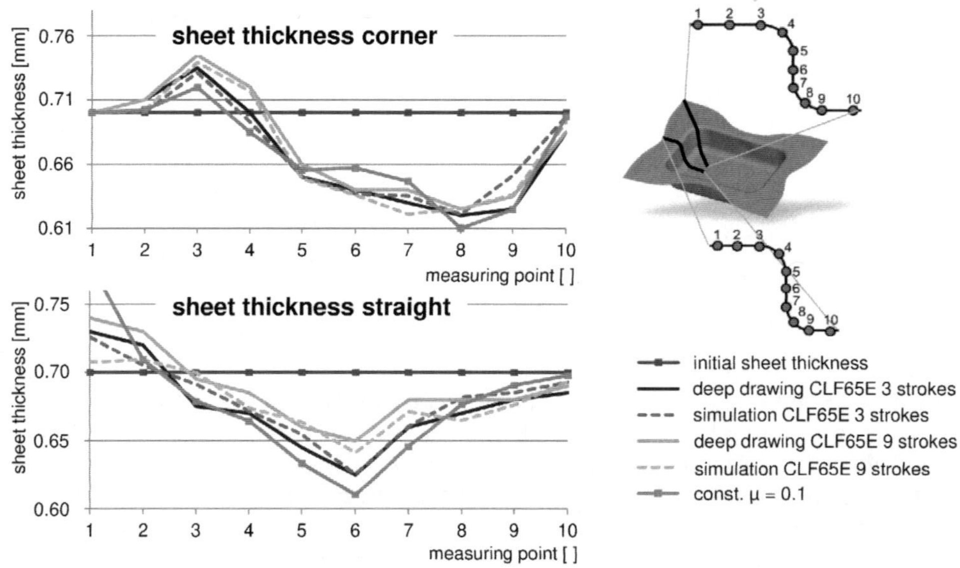

Figure 9. Simulated and experimentally measured distribution of sheet metal thickness for different drawing velocities at different part areas (solid lines: real deep drawing, interrupted lines: simulation).

In order to ensure that the sensor signals are performing and working independently from the material, the high strength steel DP600 also was analysed in a real deep drawing process as well as in simulation. Figure 10 shows the deviation from initial sheet thickness in percent to compare the two different materials with different initial sheet thicknesses. The deviation corresponds very well for the two areas, corner and straight, as well as for both materials. The percentage thinning of points 7 to 9 is lower for DP600 in comparison to DC04. This is caused by the higher yield strength of the DP600 sheet material. The deep drawing experiments were performed for both materials with the same blank holder force of 275 kN. Due to the higher strength, higher loads can be carried and a smaller thinning occurs.

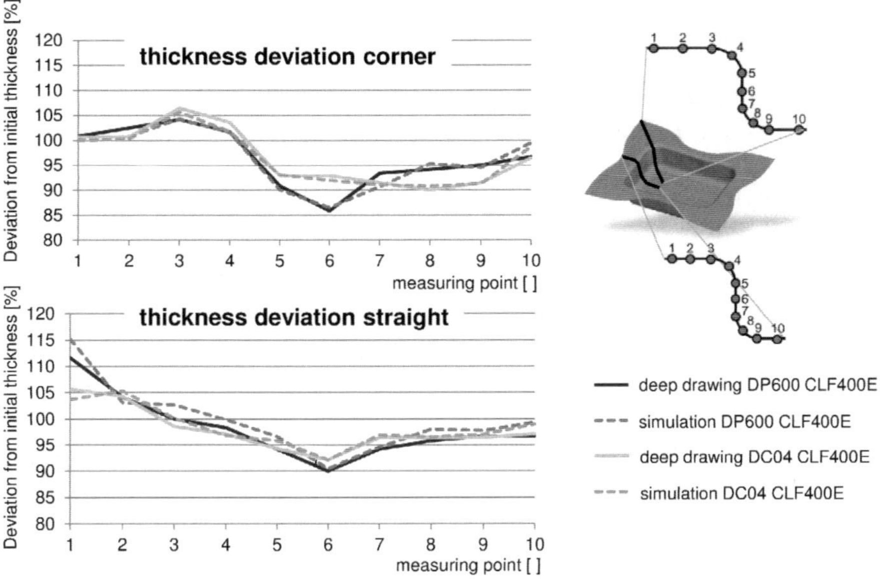

Figure 10. Simulated and experimentally measured change of initial sheet thickness for DP600 and DC04 at different part areas (solid lines: real deep drawing, interrupted lines: simulation).

5. Conclusions

Deep drawing tests were performed for different tribological systems with rectangular cup geometry. Restraining forces were measured in-situ for two geometric areas and integrated in a LS-Dyna simulation. The integration was feasible with LS-Dyna using a thermo-mechanical simulation with defined heated tool and blank as well as the thermal friction option where a specific coefficient of friction is attached to each temperature. This procedure allows the integration of time dependant coefficients of friction. Die, punch and blank holder were modelled segmented and measured signals were assigned to the according area. The feasibility of the novel in-situ measuring method of restraining forces was proved by a comparison of material thinning with rectangular cups. Results show that effects of lubrication, drawing velocity and sheet material are reproducible in FE-simulation when using the mentioned method. Results show, that the use of varying coefficients of friction improves the prediction quality of simulation in matters of thinning compared to the use of constant values.

Further investigations will deal with additional part areas as well as sensors in the die to evaluate the friction forces for specific geometric areas. Furthermore, a model to calculate the coefficient of friction dependant on the measured forces will be developed. The comparatively easy integration of the sensor into a deep drawing tool offers an immense potential for in-situ-monitoring of the friction forces and to detect deviations during a production process.

Acknowledgments
This work was supported by the "Graduate School of Excellence advanced Manufacturing Engineering" (GSaME) and the "Deutsche Forschungsgemeinschaft" (DFG). The research was performed at the Institute for Metal Forming Technology (IFU) at the University of Stuttgart.

References
[1] Kuziak R, Krawalla R and Waengler S 2008 Advanced high strength steel for automotive industry *Archives of Civil and Mechanical Engineering* **VIII-2** 103–17.
[2] Meiler M, Pfestorf M, Geiger M and Merklein M 2003 The use of dry film lubricants in aluminium sheet metal forming *Wear* **255** 1455–62.
[3] Bay N, Olsen D D and Andreasen J L 2008 Lubricant test methods for sheet metal forming *Tribology International* **41** 844–53
[4] De Souza J 2008 *New Design Approaches for Sheet Metal Forming Dies using Polymeric Materials* Liewald M (Stuttgarrt) Beiträge zur Umformtechnik **61** (Frankfurt: DGM Informationsgesellschaft).
[5] Gesellschaft für Tribologie 2002 *Arbeitsblatt 7 - Tribologie Definition, Begriffe, Prüfung.*
[6] Hengelhaupt J and Vulcan M 2006 Robuste Prozesse beim Ziehen großflächiger Karosserieteile *Neuere Entwicklungen in der Blechumformung* **2006** 291-318
[7] Dietrich F, Recklin V, Filzek J and Hoerning M 2016 FricOn – Reibungsberücksichtigung in der Umformsimulation *9. Forum Tribologische Entwicklungen in der Blechumformung* 77-89
[8] Kistler – Datasheet 2006 *Miniature Longitudinal Measuring Pin M5 for Indirect Measurement of Forces in Machinery and Tools Type 9247A*
[9] Grüner M and Merklein M Determination of friction coefficients in deep drawing by modification of Siebel's formula for calculation of ideal drawing force *Production Engineering* **8-5** 577-84
[10] Hockett J E and Sherby O D 1975 Large strain deformation of polychystalline metals at low homologous temperatures *Journal of Mechanics and Physics of Solids* **23** 87-98

IOP Conf. Series: Materials Science and Engineering **159** (2016) 012021 doi:10.1088/1757-899X/159/1/012021

Friction and lubrication modeling in sheet metal forming simulations of a Volvo XC90 inner door

M Sigvant[1, 2], J Pilthammar [1,2], J Hol[3], J H Wiebenga[3], T Chezan[4], B Carleer[5] and A H van den Boogaard [6]

[1] Volvo Cars, Dept 81153 Stamping CAE & Die Development, Olofström, Sweden
[2] Blekinge Institute of Technology, Valhallavägen, 371 41 Karlskrona, Sweden
[3] TriboForm Engineering, Hengelosestraat 500, 7500 AM Enschede, the Netherlands
[4] Tata Steel, P.O. Box 10.000, 1970 CA IJmuiden, the Netherlands
[5] AutoForm Engineering, Emil-Figge-Str. 76-80, D-44227 Dortmund, Germany
[6] University of Twente, P.O. Box 217, 7500 AE, Enschede, The Netherlands

E-mail: mats.sigvant@volvocars.com

Abstract. The quality of sheet metal formed parts is strongly dependent on the tribology, friction and lubrication conditions that are acting in the actual production process. Although friction is of key importance, it is currently not considered in detail in stamping simulations. This paper presents a selection of results considering friction and lubrication modeling in sheet metal forming simulations of the Volvo XC90 right rear door inner. For this purpose, the TriboForm software is used in combination with the AutoForm software. Validation of the simulation results is performed using door inner parts taken from the press line in a full-scale production run. The results demonstrate the improved prediction accuracy of stamping simulations by accounting for accurate friction and lubrication conditions, and the strong influence of friction conditions on both the part quality and the overall production stability.

1. Introduction

The quality of sheet metal formed parts is strongly dependent on the tribology, friction and lubrication conditions that are acting in the actual production process. These friction conditions are dependent on the tribology system, i.e. the applied sheet material, coating, tooling material, lubrication- and process conditions. Although friction is of key importance, it is currently not considered in detail in stamping simulations. The current industrial standard is to use a constant (Coulomb) coefficient of friction. This limits the overall simulation accuracy as also demonstrated in an earlier work of the authors for a U-bend application [1].

At the Stamping CAE & Die Development Department at Volvo Cars, it is concluded that friction and lubrication modeling is the way forward for improving stamping simulation accuracy. This paper presents a selection of results considering friction and lubrication modeling in stamping simulations of the Volvo XC90 right rear door inner, demonstrating the strong influence of tribology and friction conditions on both the quality of the door inner part and the overall production stability.

First the overall project approach will be outlined. Next, the production process of the door inner part and the corresponding stamping simulation models will be introduced. A description of the project results including validation of the simulation results based on door inner parts taken from the press line is provided next. Finally, the conclusions and points of future work are described.

2. Approach

The approach followed in this work is visualized in Figure 1 whereby the TriboForm software is used in combination with the AutoForm software.

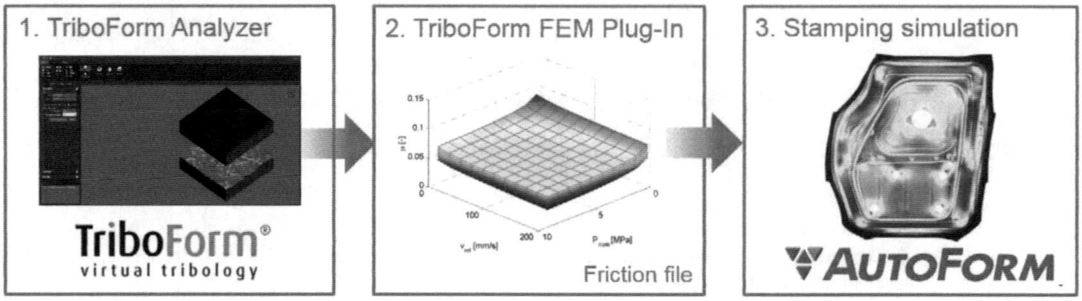

Figure 1. Approach for friction and lubrication modeling in sheet metal forming simulations

The modelling approach comprises three steps. In step 1, a TriboForm friction analysis is performed on a user-defined metal-lubricant combination (see Section 2.1). A friction model is generated for the selected metal-lubrication combination, which can be exported to a friction file (see Section 2.2). Finally in step 3, the generated friction file can be used in AutoForm by making use of a TriboForm FEM Plug-In as described in Section 2.3.

2.1. Simulation of friction and lubrication conditions

Tribological conditions in metal forming processes are dependent on local process and lubrication conditions, loading and local strain state of the sheet material as demonstrated in [2, 3]. Before starting a TriboForm friction analysis, information of the tribology system is required as a user input, i.e. the applied sheet material, coating and tooling material, lubrication type, lubrication amount and process conditions. This information can either be entered by the user or extracted from a database, i.e. the TriboForm Library, as further described in Chapter 3.

The TriboForm software allows for multi-scale modeling of a time and locally varying friction coefficient under a wide range of process conditions. The physically-based models included in TriboForm enable friction modeling in the mixed lubrication regime. This is achieved by coupling a boundary lubrication friction model [4] and a hydrodynamic friction model [5].

The boundary lubrication model includes models to describe the change in tribological properties during forming due to normal loading, deformation of the underlying bulk material and sliding. The models provide an expression for the fractional real contact area, used as an input for the friction calculation. Shear stresses at the interface are obtained by accounting for the influence of ploughing and adhesion during sliding. For this purpose, the plateaus of the deformed sheet asperities are assumed to be perfectly flat, in which tool contact patches (a collection of neighboring tool asperities that are in contact) are penetrating. The shear stress acting on individual contact patches is calculated based on the theory described by Challen and Oxley [6]. By adding the individual contributions of all contacting tool patches, the boundary shear stress τ_{asp} can be obtained.

The calculated deformation of sheet asperities is used to calculate the volume of the lubricant entrapped into non-contacting surface pockets. This information is subsequently used to calculate the fluid film thickness h and the hydrodynamic pressure distribution p_{lub} (i.e. the load carrying capacity of the lubricant). To solve p_{lub}, an FE approach has been adopted as described in [5], introducing hydrodynamic contact elements with additional pressure degrees of freedom. The viscous shear stress at the fluid–solid interface τ_{lub} is calculated based on the obtained hydrodynamic pressure distribution. The summation of the shear stress acting between contacting surface asperities τ_{asp} and the shear stress acting at the fluid-solid interface τ_{lub} is used to obtain the desired coefficient of friction μ, see Equation 1.

$$\mu = \frac{\tau_{asp}+\tau_{lub}}{p_{nom}} \quad (1)$$

2.2. Friction model

The TriboForm software calculate friction coefficients for a predefined range of process settings, i.e. local contact pressures, relative sliding velocities, plastic strains in the sheet material and interface temperatures. A four dimensional matrix is constructed containing friction coefficients for all possible combinations of process settings. To use this data-set within large scale FE simulations, and to guarantee computational efficiency, a four dimensional model is adopted to describe the calculated data points which is stored in a friction file (see Figure 1). Using the TriboForm software, a friction file can be created per tribology system, i.e. the sheet and tooling materials, coatings and lubricants used in actual metal forming production. A friction file has to be constructed only once for a specific tribology system, after which it can be used in different FE forming simulations where the same combination is used.

2.3. Stamping simulations

Next, the friction file is easily imported into the commercial FE software AutoForm using the TriboForm FEM Plug-In. That is, if an AutoForm simulation is started, the FEM Plug-In reads the friction file from TriboForm and enables the usage of the friction model within the AutoForm simulation. As a result, the constant coefficient of friction in AutoForm is replaced by a friction model. Instead, a local- and time-dependent (nodal) friction coefficient is computed each increment and used in the computation of the equilibrium of the finite element model for the materials, coatings and lubricants used in actual metal forming production of the door inner part.

3. Door inner production

An impression of the door inner part highlighted in the all-new Volvo XC90 is shown in Figure 2. The quality of the door inner part is strongly dependent on the friction and lubrication conditions in the actual production process. For varying production conditions, like stroke rate, cushioning force or lubrication conditions, the door inner part can either show wrinkling or fracture as indicated in the right images in Figure 2. Moreover, in the course of a production run, these quality issues can either appear or disappear based on the drift of the actual process conditions like tooling temperature during the production run.

It is known in production that by changing the process- or lubrication conditions, these quality issues can be prevented. The aim of this work is to account for realistic friction and lubrication conditions in stamping simulations to improve the simulation accuracy and enable the detection and prevention of tribology related quality issues in the virtual design process.

Figure 2. Right rear door inner highlighted in the Volvo XC90 (left) and potential areas of quality issues like wrinkling and splits in production (right).

3.1. Tribology system: sheet – lubricant – tooling

The door inner part is produced using a VDA239 CR4 GI sheet material with a thickness of 0.7 mm and an EDT surface finish. The material tests and data for the BBC2005 material model is determined according to the methodology as described in [6]. The data used in the study are presented in Table 1 and Figure 3.

Table 1. Material data for the BBC2005 model.

σ_0 [MPa]	σ_{45} [MPa]	σ_{90} [MPa]	σ_b [MPa]	R_0	R_{45}	R_{90}	R_b	M
156.6	160.0	156.0	187.0	1.81	1.34	1.88	0.98	4.5

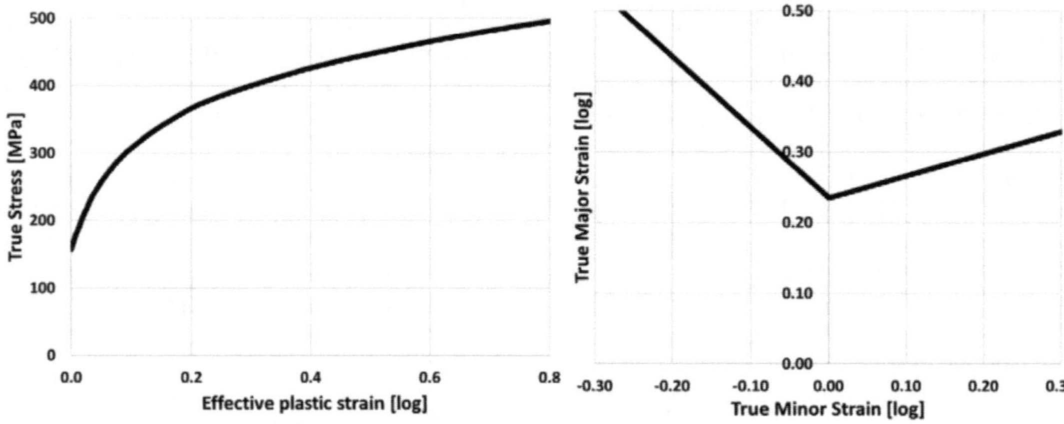

Figure 3. Hardening curve (left) and Forming Limit Curve at onset of localization (right) used in the forming simulations.

The sheet material is delivered with a Fuchs Anticorit RP4107S lubricant. Measurements in production have shown that the lubrication amount ranges between 0.7 g/m^2 and 2.2 g/m^2 at different locations on both sides of the sheet. An average value of 2.0 g/m^2 will be used in the following numerical studies. Future work will include numerical studies on the influence of (variation of) lubrication amount, distribution and type. A temperature dependent relation for the viscosity of the lubricant is provided by Fuchs Schmierstoffe GmbH and implemented in the TriboForm software.

The tooling material type is nodular iron GGG70L. The tools are hardened at the positive tool radii and chrome plated at selected areas. The actual tooling geometries have been determined by 3D scanning and implemented in the forming simulations. The scanned data of the blank holder and addendum of the die has been morphed with the deformed surfaces from a structural FE-analysis as a guide. This FE-analysis studies the complete die and press structure loaded with the blank holder force used in production.

3.2. Full-scale production run

A full-scale production run of 1700 parts is performed at a mechanical transfer press-line at Volvo Cars in Olofström, Sweden. The corresponding velocity profile is recorded and implemented in the forming simulations. The blank is contour cut from a 1700 mm wide coil and the pitch is 1553 mm. The stroke rate is set (and limited) to 8 strokes/min. Increasing the stroke rate for the current tribology system results in wrinkles in the part as shown in Figure 2.

3.3. Friction simulations

The friction and lubrication conditions corresponding to the materials and lubricants used in actual door inner production are simulated using the TriboForm software R1.0. As an input for the friction calculations, the real 3D surface topographies of the sheet and die surfaces are required. This information can be imported by the user or extracted from the TriboForm Library.

In this work, the virgin CR4 GI sheet surface topographies have been measured by 3D confocal microscopy, see Figure 4 (left) for an impression. These measurements have been performed for the sheet material by Tata Steel and imported in the TriboForm software. The die surfaces have been measured by the Volvo Cars at varying locations using epoxy replicas. A single representative chrome plated die surface measurement is taken and used in the TriboForm software for the friction calculations, see Figure 4 (right). Figure 5 (left) shows the projection of the measured die surface topography and the sheet surface topography with a predefined amount of lubricant in the TriboForm software.

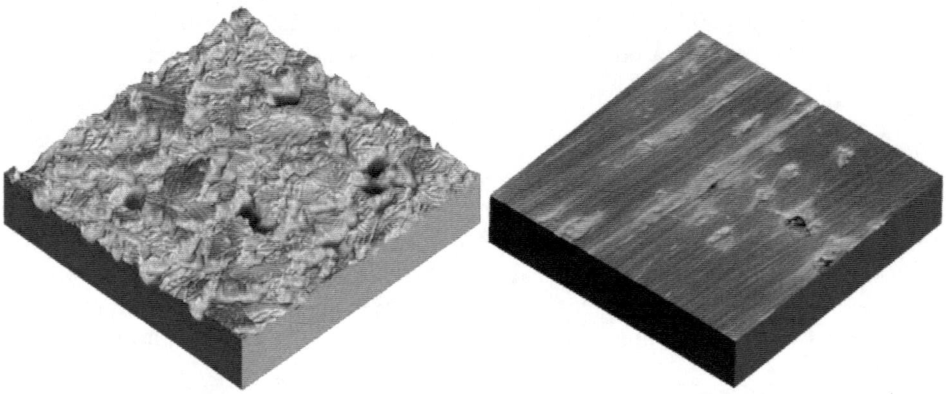

Figure 4. Impression 3D surface texture sheet (left) and tooling (right).

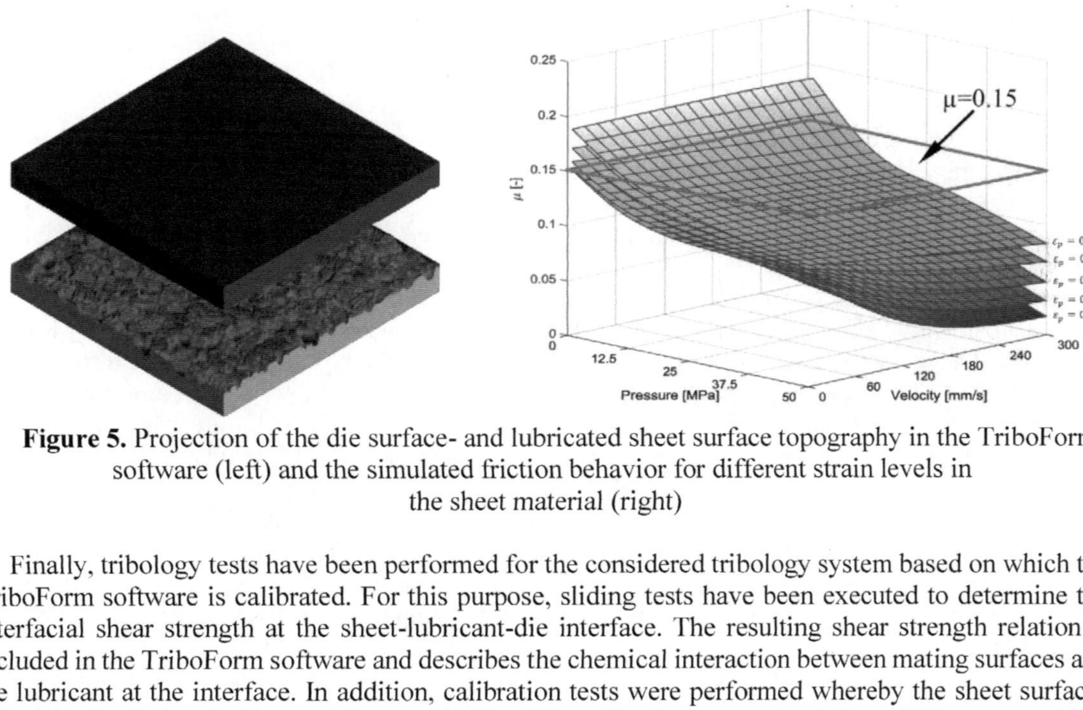

Figure 5. Projection of the die surface- and lubricated sheet surface topography in the TriboForm software (left) and the simulated friction behavior for different strain levels in the sheet material (right)

Finally, tribology tests have been performed for the considered tribology system based on which the TriboForm software is calibrated. For this purpose, sliding tests have been executed to determine the interfacial shear strength at the sheet-lubricant-die interface. The resulting shear strength relation is included in the TriboForm software and describes the chemical interaction between mating surfaces and the lubricant at the interface. In addition, calibration tests were performed whereby the sheet surfaces have been loaded and subsequently measured by 3D confocal microscopy at three different occasions: as received, after normal loading and after normal loading and sliding. The resulting relation of the real area of contact of the sheet surface topography for varying loading conditions is entered in the TriboForm software and used for calibration purposes.

3.4. Sheet metal forming simulations
The door inner forming process is simulated with AutoForm^{plus} R6.0. The scanned and morphed tooling surfaces have been implemented, thus also including geometrical draw-beads. The BBC2005 material model is used for all simulations. The ram speed in the simulation is taken identical to the ram speed of the mechanical press-line.

4. Results
Following the approach presented in Figure 1, the first step is to execute the friction calculations. The resulting friction behavior will be described in Section 4.1. The friction results are then exported from the TriboForm software and imported in AutoForm. Section 4.2 will described the forming simulation- and experimental validation results.

4.1. Friction and lubrication modeling in sheet metal forming simulations
The projection of the measured die surface topography onto the measured sheet surface topography as shown in Figure 5 (left) is the basis for the simulation model used for the friction calculations in TriboForm. The friction conditions are calculated by loading and sliding the die surface over the sheet surface for a lubrication amount of 2.0 g/m^2 and pre-defined ranges of process conditions, i.e. contact pressure, relative sliding velocity, plastic strain in the sheet material and interface temperature. The calculation times of the friction analyzes range between 5 and 10 minutes on a standard quad-core desktop computer.

Figure 5 (right) displays the simulated friction behavior. It shows that the friction behavior is dependent on both contact pressure, relative sliding velocity and plastic strain in the sheet material. Each friction surface is valid for a certain plastic strain in the sheet material. As the plastic strain increases from 0 to 0.4, with an interval of 0.1, the friction coefficient decreases.

4.2. Forming simulations and experimental validation
Forming simulations are performed whereby only the description of the friction conditions is changed. Simulations are performed using the friction model as presented in Section 4.1 and the Coulomb friction model. Regarding the latter, sheet metal forming simulations are generally performed at Volvo Cars using a value for the coefficient of friction of 0.15 for steel sheet which is therefore taken as a reference in this work.

The forming simulation results are validated using strain, draw-in and geometry measurements on door inner parts picked out from the press line at different times in a full-scale production run. This work will present the validation results for a part picked at the start of the production run. The difference in true major strain from forming simulations and measurements on the upper surface are presented in Figure 6. In red areas, the simulation overestimates the major strain, while in blue areas is the simulation underestimates the strain. Using a constant friction coefficient of $\mu = 0.15$ results in too large major strains in several areas, especially in vertical walls. Using the TriboForm friction file will results in lower friction coefficients in areas with high contact pressures, e.g. in radii and draw beads, and also in areas with high relative velocity between the sheet and the die surfaces. This results in a better agreement between simulated and measured major strains. The comparison for the minor strains show the same trend, i.e. that the accuracy of the sheet metal forming simulation results increase using the TriboForm friction file.

The validation of the draw-in results are presented in Figure 7 represented as the 3D position of the edge of the part projected on a horizontal plane. The simulated draw-in results are strongly dependent on both an accurate description of friction and geometrical description of the actual tools and draw-beads. A deviation between production draw-in results and simulation results using a coefficient of friction of $\mu = 0.15$ is observed. The simulation results using the TriboForm friction file still show a deviation in some areas with the experimental draw-in results, but generally show an improved simulation accuracy.

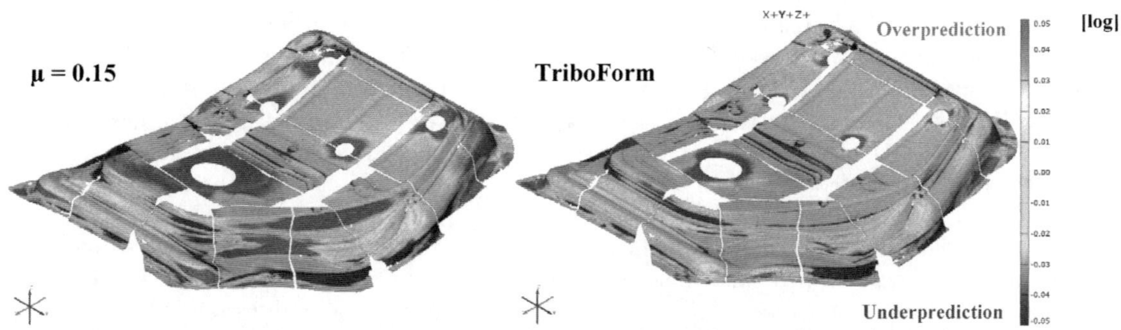

Figure 6. Difference in true major strain between simulations and experimental measurements.

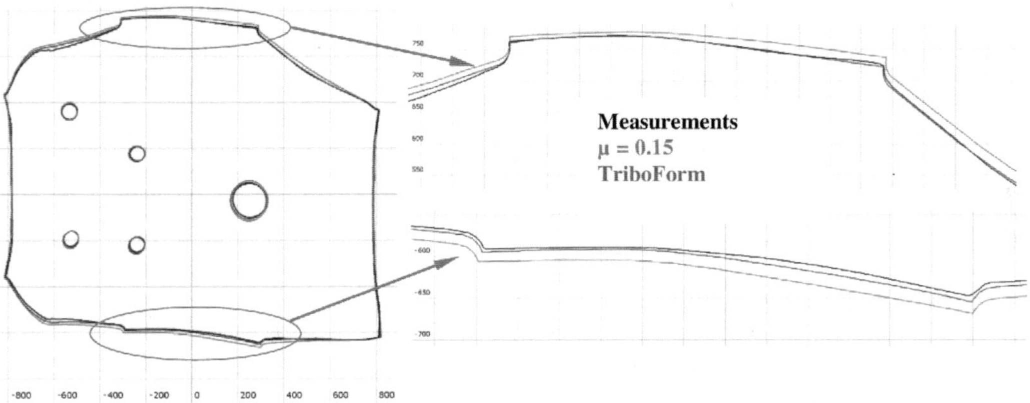

Figure 7. Experimentally measured and simulated draw-in results for the door inner part

The part shape after forming is displayed in Figure 8. With a constant friction coefficient of $\mu = 0.15$ there are no wrinkles on the part predicted. However, with the TriboForm friction file there are two wrinkles on the addendum predicted. The parts that were picked out at the production line had these wrinkles as well. Once again this demonstrates that forming simulations using the TriboForm friction file show a better agreement with measurements and observations made in production.

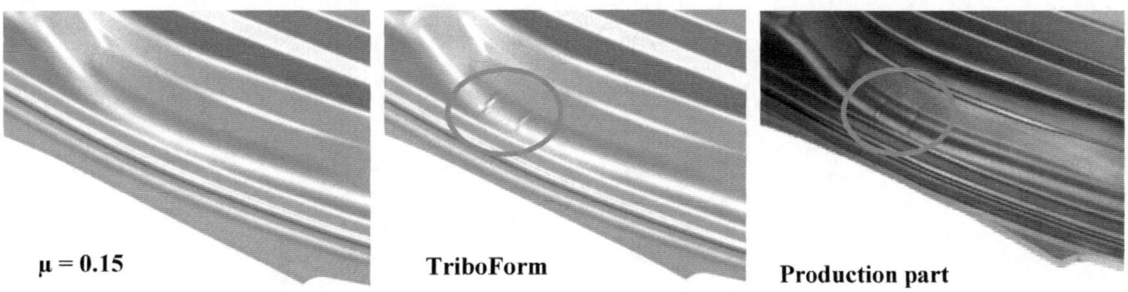

Figure 8. Simulated and actual part shape after forming showing wrinkles on the addendum

5. Conclusions and future work

The results presented in this paper demonstrate the strong influence of tribology and friction conditions on both the quality of the door inner part and on the overall production stability. Moreover, it demonstrates that accounting for realistic and accurate friction and lubrication conditions bring metal forming simulations to a higher level and improve the prediction accuracy of stamping simulations.

Major benefits of the presented approach for Volvo Cars are the following. First of all, the TriboForm software is based on physical models with input parameters that can be efficiently collected from a database or measured with minimal effort. This enables to accurately predict the results of sheet metal forming operations before manufacturing the dies. Secondly, it enables the simulation of friction for the materials and lubricants used in actual production of automotive parts and reduces the demand for experimental testing and try out. Overall, it enables Volvo Cars to further reduce lead time and development cost through the use of more accurate stamping simulations with enriched simulation functionalities.

Future work will include more detailed simulation studies on the effect of varying tool coatings and sheet coatings on the quality of the door inner part. Moreover, future work will focus on numerically studying the influence of temperature dependent friction and lubrication conditions and the resulting transient effects in sheet metal forming production and its effect on the overall production stability. Also the strain rate effects of sheet material will be included in future studies.

Acknowledgement
The authors are grateful for the financial support from Volvo Cars.

References

[1] Sigvant M, Hol J, Chezan T and van den Boogaard A H, 2015, Friction modeling in sheet metal forming simulations: Application and validation on an U-bend product, *Proc. of FTF 2015 (Zürich, Switzerland).*

[2] Grueebler R and Hora P, 2009, Temperature dependent friction modeling for sheet metal forming. *International Journal of Material Forming, **2**:251–254.*

[3] Ludwig M, Müller C and Groche P, 2010, Simulation of dynamic lubricant effects in sheet metal forming processes, *Key Engineering Materials, **438**: 171–178.*

[4] Hol J, Meinders V T, de Rooij M B and van den Boogaard A H, 2014, Multi-scale friction modeling for sheet metal forming: The boundary lubrication regime, *Tribology Int., **81**: 112-128.*

[5] Hol J, Meinders V T, Geijselaers H J M, and van den Boogaard A H, 2015, Multi-scale friction modeling for sheet metal forming: the mixed lubrication regime. *Tribology Int., **85**: 10–25.*

[6] Banabic D, Carleer B, Comsa D S, Kam E, Krasivskyv A, Mattiasson K, Sester M, Sigvant M and Zhang X, 2010, Sheet Metal Forming Processes, Constitutive Modelling and Numerical Simulation, *Springer*.

Investigation of surface finishing of carbon based coated tools for dry deep drawing of aluminium alloys

J Steiner[1], K Andreas[1] and M Merklein[1]

[1] Institute of Manufacturing Technology, Department Mechanical Engineering, Friedrich-Alexander-Universität Erlangen-Nürnberg (FAU), Egerlandstraße 13, 91058 Erlangen, Germany

E-mail: jennifer.steiner@fau.de

Abstract. Global trends like growing environmental awareness and demand for resource efficiency motivate an abandonment of lubricants in metal forming. However, dry forming evokes increased friction and wear. Especially, dry deep drawing of aluminum alloys leads to intensive interaction between tool and workpiece due to its high adhesion tendency. One approach to improve the tribological behavior is the application of carbon based coatings. These coatings are characterized by high wear resistance. In order to investigate the potential of carbon based coatings for dry deep drawing, friction and wear behavior of different coating compositions are evaluated in strip drawing tests. This setup is used to model the tribological conditions in the flange area of deep drawing operations. The tribological behavior of tetrahedral amorphous (ta-C) and hydrogenated amorphous carbon coatings with and without tungsten modification (a-C:H:W, a-C:H) is investigated. The influence of tool topography is analyzed by applying different surface finishing. The results show reduced friction with decreased roughness for coated tools. Besides tool topography the coating type determines the tribological conditions. Smooth tools with ta-C and a-C:H coatings reveal low friction and prevent adhesive wear. In contrast, smooth a-C:H:W coated tools only lead to slight improvement compared to rough, uncoated specimen.

1. Introduction

Sheet metal forming operations like deep drawing represent very energy efficient processes [1]. However, the consumption of lubricants and detergents for cleaning procedures in conventional deep drawing processes has a negative impact on the environment. Dry forming aims at the abandonment of lubricants or utilization of volatile lubricants to eliminate cleaning and drying steps [2]. This reduces the number of production steps and time. Furthermore, the realization of lubricant-free forming would improve the sustainability and further increase the resource efficiency of manufacturing of sheet metal parts. Without lubrication, tool and workpiece surface have direct contact which leads to intensive interaction. This causes distinctive friction and wear. Especially, while forming aluminum alloys, adhesive wear needs to be considered in dry forming due to its high reactivity towards tool steel. Former investigations have shown that under laboratory and dry conditions the friction coefficient increases by factor of 10 to 20 depending on the aluminum alloy [3]. High friction leads to wear. In this regard, adhesion occurs as main wear mechanism at the initial tool-workpiece contact. Due to the adhesion and friction coefficients of 0.3 and higher lubricant-free deep drawing will lead to insufficient surface quality and tool wear. Thus, measures need to be taken to reduce adhesion tendency of aluminum alloys towards tool steel. A promising approach for wear reduction under lubricant-free conditions is the application of coatings as a separating layer between tool and workpiece. Especially carbon based coatings are frequently used when friction reduction and increased wear resistance are necessary [4]. Horiuchi et al. [5] figured out that applying diamond like carbon based coatings could improve tribological conditions in terms of reduced friction coefficients and punch load for deep drawing of 5xxx aluminum alloys. However, this investigation focused on blank-sided coatings which seem not suitable for industrial applications. Few studies analyzed the influence of substrate and coating topography on coating properties itself and tribological performance. Singh et al. have analyzed the influence of varying steel substrate roughness values on the coating characteristics [6]. Their results reveal that a higher substrate roughness leads to growing coating roughness which causes a greater risk of cracks and surface damages during indentation tests. Thus, it seems that smoother substrate and coating topographies are beneficial for the coating durability. However, no studies have been conducted regarding the tribological performance of varying coating roughness values. Numerical investigations from Lindholm et al. revealed that shape and distribution of coating asperities influence the distribution of tensile stresses. Thereby, small and sharp asperities cause spots of extreme load conditions [7]. Jiang and Arnell analyzed the wear behavior of different coating roughness in ball-on-disc tests [8]. Their study showed that above an arithmetic mean roughness value Ra of 0.93 µm wear rate and variation of test results increase rapidly. Due to its closed tribological system, the results of a ball-on-disc tribometer cannot be transformed directly to sheet metal forming operations. Sheet metal forming is characterized by an open tribological system with planar contact conditions where constantly new sheet material is drawn through the contact zone. Additionally, the ball was made out of tungsten carbide which means totally different material and surface properties compared to sheet metal forming. Thus, the following investigation aims to evaluate the tribological behavior under process relevant conditions for commonly used sheet materials like aluminum alloys. By comparing ta-C, a-C:H:W and a-C:H coated with uncoated tools the effect of coating type on friction and wear is evaluated. The influence of surface treatment before coating deposition is investigated for a-C:H:W and a-C:H coatings compared to uncoated tools with different surface roughness. Furthermore, for ta-C coated tools, the impact of tool topography is analyzed by varying surface finishing after coating deposition. Overall goal of this study is to increase the understanding of correlation between coating type, tool topography and resulting tribological conditions.

2. Materials and coatings

In the present study the aluminum alloy AA5182 is selected as workpiece material with a sheet thickness of 1 mm and an electrical discharge texture (EDT). AA5182 is commonly used in automotive industry for inner car body parts [9]. The cold working steel 1.2379 (X155CrVMo12) is chosen as tool material with a hardness of 60 HRC. To ensure that the hardness endures the coating deposition a special hardening process suitable for post treatment at higher temperatures was applied. All tools are machined with a combined lapping and polishing process with 9 µm grain size - in the following called rough polishing - to achieve a plane surface without preferential direction. As second surface finishing strategy polishing with a grain size of 1 µm – called smooth polishing – is applied. The process chains are varied to investigate the influence of different finishing operations. The details of surface finishing are explained in chapter 3.2.

Carbon based coatings are known for their low friction and high wear resistance. Thus, to prevent tool sided wear under dry conditions three different carbon based coating types are investigated. Coatings based on carbon with diamond like characteristics (DLC) are classified according to their ratio of graphite (sp^2), diamond (sp^3) and hydrogen (H). In order to obtain a broad range of coating properties and analyze their differing tribological behavior, coatings with a high sp^3 ratio (ta-C) and a low sp^3 ratio containing hydrogen (a-C:H and a-C:H:W) are applied. The coating designs are schematically shown in Figure 1. The ta-C coatings are deposited by using vacuum arc evaporation (Laser-Arc-Technology). As adhesive film a Cr layer is deposited by sputtering on friction jaw surfaces to guarantee a smooth ta-C coating. The functional ta-C layer has an average thickness of 1.2 µm and an average hardness of 4000 HV 0.001 was measured. Besides ta-C coatings, two kinds of hydrogenated amorphous carbon coatings are analyzed. Beneath a-C:H:W and a-C:H coating an adhesive Cr layer and a WC interlayer is deposited. The chromium layer is deposited by using the arc evaporation method. The coating deposition is performed with the coating machine TT-300 from H-O-T by physical vapor deposition/plasma assisted chemical vapor deposition (PVD/PACVD). The a-C:H:W coating is deposited in an argon-acetylene atmosphere by reactive unbalanced magnetron sputtering of a binder-free WC target. Finally, for a-C:H coating a top layer without tungsten is deposited. The functional a-C:H:W coating shown in Figure 1 b) has an average thickness of 4 μm and an average hardness of 1000 HV 0.001. In comparison, the a-C:H coating with a thickness of 1.8 μm has a hardness of 2200 HV 0.001.

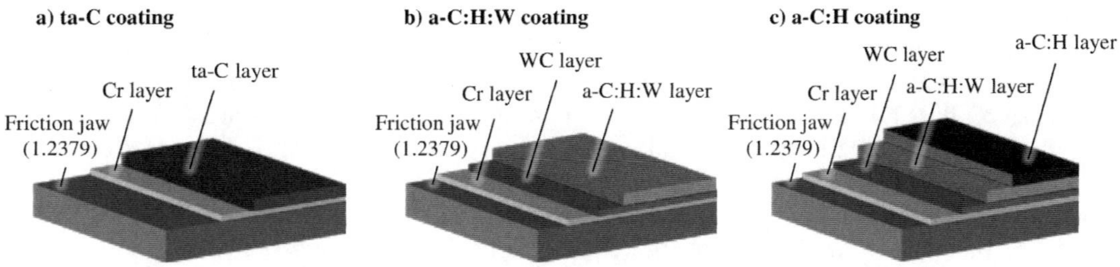

Figure 1. Coating design of a) ta-C, b) a-C:H:W and c) a-C:H coating

3. Experimental procedure

3.1. Test setup

In order to evaluate the tribological behavior of the different coating types and surface finishing strategies flat strip drawing tests are performed. This test setup models the tribological conditions of the area between blank holder and die in a conventional deep drawing process [10]. Thus, the strip drawing test is commonly used for tribological investigations in sheet metal forming. Furthermore, this setup enables a direct determination of friction coefficients and represents an open tribological system with a planar contact zone. In contrast to closed tribological systems, new sheet material is transferred constantly into the contact area like in real deep drawing processes. The schematic test setup is shown in Figure 2. A sheet metal strip is positioned between an upper fixed and a lower movable friction jaw. A defined normal force F_N is applied by moving the lower friction jaw upwards. The strip is clamped on one side and afterwards drawn through the friction jaws with a constant relative velocity v_{rel}.

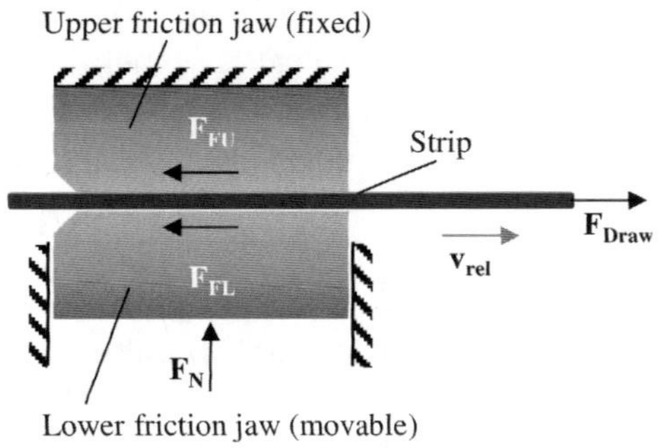

Figure 2. Schematic test setup of flat strip drawing test

Over a drawing length of 190 mm the drawing force F_{Draw} is recorded. This force corresponds to the sum of the upper and lower friction forces F_{FU} and F_{FL}. The friction coefficient μ can be calculated according to (1) by using the Coulomb friction law. Due to tool contact on both sides of the strip, only half of the friction force needs to be considered for determining μ.

$$\mu = \frac{F_{FU}+F_{FL}}{2 \cdot F_N} \qquad (1)$$

IOP Conf. Series: Materials Science and Engineering **159** (2016) 012022 doi:10.1088/1757-899X/159/1/012022

3.2. Experimental design

The experimental design is summarized in Table 1. A constant drawing velocity of 100 mm/s is used for all tests. This relative velocity represents a commonly applied relative velocity in sheet metal forming operations [11]. For each parameter combination, three strips are drawn to ensure a statistical certainty. In general, the level of contact pressure in the flange area of a deep drawing process varies between 1 and 10 MPa [11]. Former investigations revealed that under dry conditions with metallic bright tools contact pressures of up to 1.5 MPa can be realized without macroscopic deformation or failure of aluminum strips. Therefore, in a first run the contact pressure was set to this level. To proof the transferability of the experimental results to other process parameters, promising combinations of coating type and finishing operation which reveal low friction and no wear, are additionally tested at a pressure level of 3.0 MPa. In order to analyze the influence of varying surface topographies, for each coating type two different roughness levels are analyzed. As a reference, two uncoated friction jaws are investigated with a rougher and a smoother surface. All friction jaws were rough polished in a first step with an oil based diamond suspension with a grain size of 9 µm. Afterwards, the friction jaws 1-s to 4-s were polished in three steps with diamond suspension up to a minimum grain size of 1 µm. In the following, this process is called smooth polishing. Both pairs of ta-C coated friction jaws where smooth polished before applying the coating. The investigation of the ta-C coatings focuses on the influence of the surface treatment after the coating deposition. Therefore, the friction jaws 2-r were coated but not polished whereas 2-s were polished after the application of ta-C coating. The impact of surface treatment after coating deposition is not analyzed for a-C:H:W and a-C:H because previous studies showed that strip drawing is not possible without final surface treatment. Without polishing after coating deposition high friction causes fracture of the strips. Thus, merely the influence of a rough (3-r and 4-r) and smooth (3-s and 4-s) polishing before the coating deposition was analyzed. Before the strip drawing tests, all strips are cleaned in an aceton bath to remove the basic lubrication which is conventionally applied in the rolling mill as anticorrosive measure.

Table 1. Experimental design and parameters for strip drawing (velocity v_{rel} = 100 mm/s; n_{AA5182} = 3)

Friction jaw	*Coating type*	*Final surface treatment*	*Contact pressure (MPa)*
1-r	None	Rough polishing	1.5
1-s	None	Smooth polishing	1.5
2-r	ta-C	Smooth polishing -> coating	1.5
2-s	ta-C	Smooth polishing -> coating -> polishing	1.5 / 3.0
3-r	a-C:H:W	Rough polishing -> coating -> polishing	1.5
3-s	a-C:H:W	Smooth polishing -> coating -> polishing	1.5
4-r	a-C:H	Rough polishing -> coating -> polishing	1.5 / 3.0
4-s	a-C:H	Smooth polishing -> coating -> polishing	1.5 / 3.0

4. Results

4.1. Surface characterization before strip drawing tests

Before performing strip drawing tests, tactile stylus and optical surface measurements of the tool surfaces are conducted with the “Mar Surf GD 120” and the nanofocus “µSurf” microscope, respectively. The topographies of the friction jaws after the final surface treatment are summarized in Figure 3. This figure gives a qualitative impression about the initial surface structure and the comparability of the surface properties between different friction jaws. The rough and smooth specimens reveal distinct differences in appearance of the surface structure. The rough polished tools without coating and with a-C:H:W and a-C:H coating show similar topographies whereas the rough configuration of the ta-C coating looks smoother. Reason is the smooth polishing step before the coating deposition. All smooth surfaces in Figure 3 reveal a similar and smooth topography. Small surface defects occurring during the growing of the coatings can be detected at the coated friction jaws. Especially, the smooth configuration of a-C:H depicts some distinct surface defects. However, all samples in the second row appear much smoother compared to the topographies above.

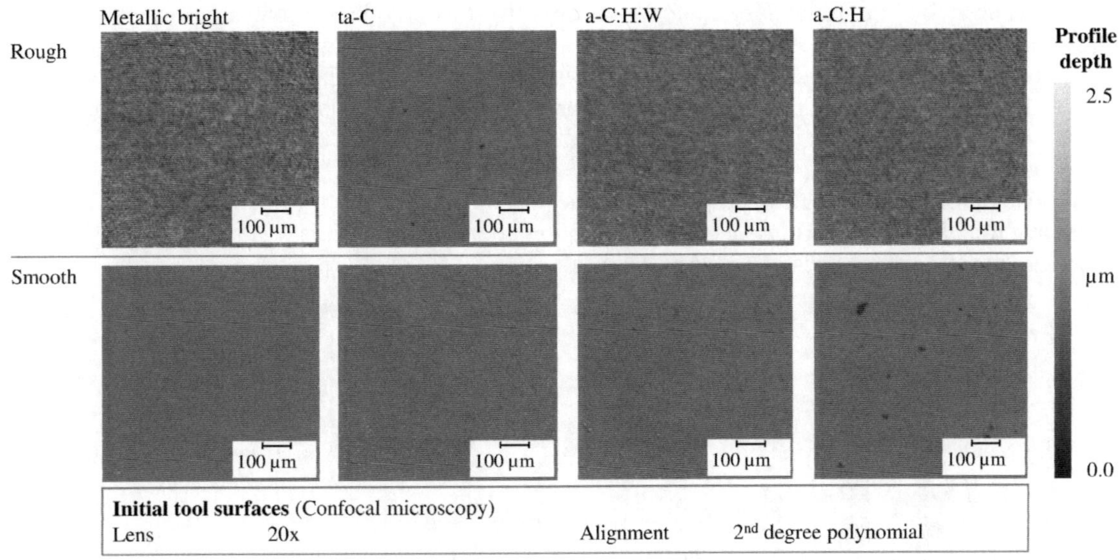

Figure 3. Initial tool topography of varied coating types and surface treatments

In addition to image-based measurements, two dimensional roughness measurements are performed for a quantitative evaluation. The arithmetic mean roughness Ra, the reduced peak height Rpk and the reduced valley depth Rvk were analyzed in Figure 4 for all friction jaws to get an impression of average roughness as well as of height and depth of asperities.

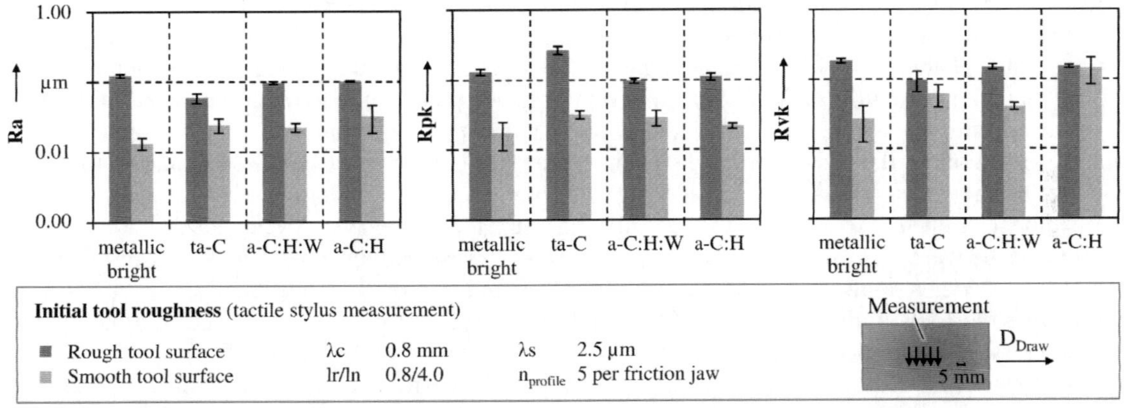

Figure 4. Initial tool roughness of varied coating types and surface treatments

Compared to the rough configurations, the values for Ra and Rpk are reduced significantly for the smooth variants. The development of Rvk shows a significant roughness reduction only achieved for the smooth surfaces without coating and with a-C:H:W coating. For ta-C coated jaws only a slight decrease of Rvk is accomplished with the polishing step after the coating deposition. The rough and smooth polished a-C:H coated tools reveal nearly the same level for the reduced valley depths. Reason for high Rvk values of smooth a-C:H surface is the occasional occurrence of approximately 1 µm deep holes in the top layer which could already be seen in the topography in Figure 3. These holes are caused by dirt particles or dust which could not be removed during the cleaning process which was performed before coating deposition. However, the height and distribution of roughness peaks is more important than those of valleys because the peaks are in direct contact with the sheet surface. To assure the comparability for all surface finishing strategies, the Rpk values vary in a small range between 0.02 and 0.03 µm for the smooth tools and between 0.10 and 0.13 µm for the rough tools. Only the rough configuration of ta-C coated tools reaches a much higher level of Rpk around 0.27 µm which is caused by the roughness increase during the coating deposition. Reasons for surface roughening are elevated temperatures during deposition, the formation of carbon clusters and the higher growing rate at asperities compared to valleys. Thus, the roughening during dynamic growth of the films depends on substrate roughness, temperature, gas mixture, ion energy and bias voltage.

4.2. Determination of friction coefficients

For the evaluation of tribological properties of the investigated tools, the recorded drawing forces during strip drawing are used to determine the friction coefficients. Figure 5 summarizes the friction coefficients for all topographies and coating types.

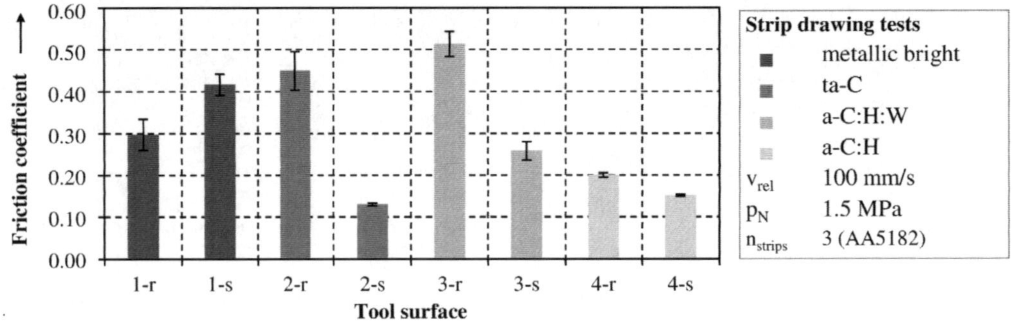

Figure 5. Friction coefficients according to strip drawing tests for varied tool surfaces

The friction coefficients vary in a broad range between 0.13 and 0.54. Under dry conditions with uncoated and rough tools a friction coefficient of 0.3 is determined. Smooth polishing of the metallic bright tool surface leads to further increase of friction up to a level of 0.42. In general, the coated tool surfaces could not decrease the friction coefficients compared to uncoated tools. The rough tools coated with ta-C and a-C:H:W cause even higher friction. A friction coefficient of 0.45 is reached for the rough ta-C coated jaws (2-r) whereas rough a-C:H:W coated tools (3-r) lead to a maximum friction of 0.54. Beneficial tribological conditions in terms of lower friction compared to uncoated tools could be achieved with all smooth polished, coated jaws and with rough polished a-C:H coated tools. A friction reduction of 0.04 is accomplished with smooth a-C:H:W coated tools compared to rough, metallic bright tools. However, considering the high standard deviation this is only a slight improvement. Depending on the surface finishing a-C:H coated jaws reveal general low friction coefficients between 0.20 and 0.15 (4-r and 4-s). The lowest friction coefficient is determined for smooth ta-C coated tools (2-r) with a friction coefficient of 0.13. Additionally, the tribological behavior of the tools with the lowest friction coefficients (2-s, 4-r and 4-s) is analyzed at a higher contact pressure of 3.0 MPa. Before these tests, the jaws where prepared by cleaning with acetone without further mechanical treatment. The resulting friction coefficients for both levels of pressure are compared in Figure 6. The tests with smooth ta-C and a-C:H tool surfaces reveal almost the same level of friction for both contact pressures. A slight increase of friction coefficient from 0.20 to 0.23 is determined for rough a-C:H jaws at the higher pressure of 3.0 MPa. Thus, at the relevant pressure levels in the flange area of deep drawing processes the investigated tribological systems (2-s, 4-r and 4-s) assure stable friction conditions.

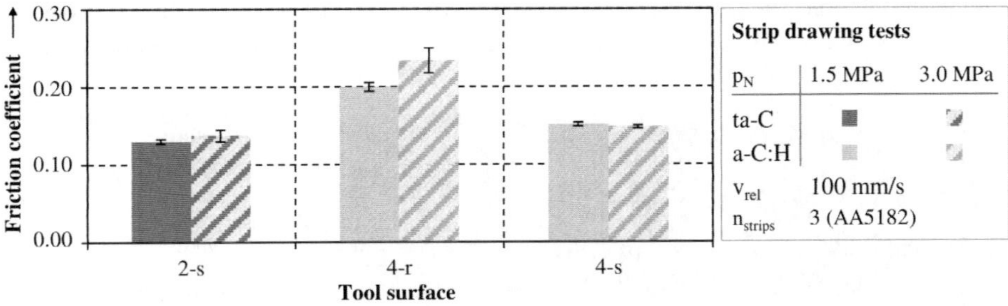

Figure 6. Friction coefficient for 1.5 and 3.0 MPa for smooth ta-C coated jaws and a-C:H coated jaws

4.3. Surface characterization after strip drawing tests

In order to evaluate tool sided wear a surface characterization was performed after strip drawing tests at 1.5 MPa contact pressure. Photographs and topography measurements need to be considered for a comprehensive wear evaluation, because topography measurements with a size of 0.8 mm x 0.8 mm are not representative for the whole contact area with a size of 100 mm x 55 mm. As the determined friction coefficients vary in a wide range between 0.13 and 0.54 the wear distribution varies as well for each jaw. Pictures are taken from each tool to assess the complete contact area of 100 mm x 55 mm. The contact surface of the friction jaws after strip drawing is represented in Figure 7.

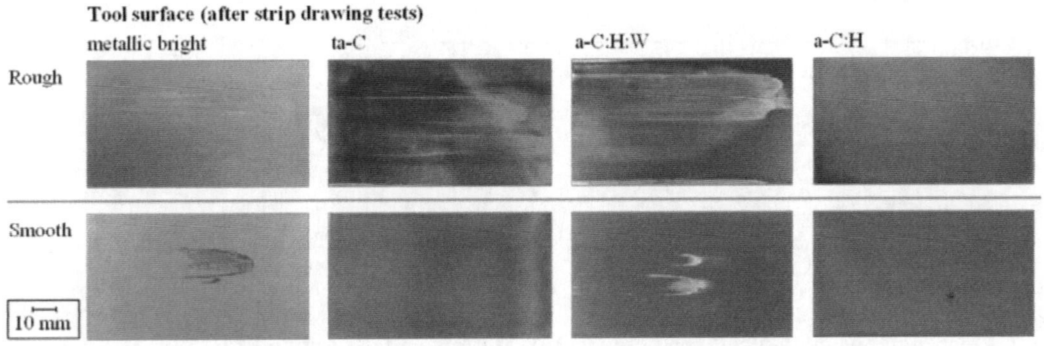

Figure 7. Photographs of tool surfaces after strip drawing tests

Adhesion could be identified as main wear mechanism. Both configurations of metallic bright jaws reveal adhesive wear. The smooth and uncoated tool surface reveals a higher amount of adhesion. Adhesion causes local deformation of the strip which increases the resistance against relative movement and thus leads to higher friction coefficients like for the smooth uncoated tools. For ta-C coated jaws a high amount of adhesive aluminum particles is detected for the rough topography whereas no wear occurs on the smooth surface. Distinctive wear arises for tools with tungsten modified coatings. A smaller amount of adhesion concentrated in the middle of the friction jaw can be found at the smooth tool whereas almost the complete contact area is covered with aluminum adhesion for rough a-C:H:W coated tools. Friction jaws coated with a-C:H display no visible signs of wear, regardless of surface finishing. Overall, the experiments which resulted in the highest friction levels (1-s, 2-r and 3-r) reveal a high amount of adhesion. The configurations ta-C with smooth surface and both a-C:H variants which resulted in low friction between 0.13 and 0.23 show no visible signs of wear. Besides photographs, the topographies of the tool surfaces are characterized after the tests using confocal microscopy. These measurements are performed to get information about the height of wear marks. Furthermore, the jaws which revealed no macroscopic adhesion are analyzed for microscopic signs of wear. For each pair of friction jaws one topography measurement is exemplarily shown in Figure 8. It could be assured that smooth, ta-C (2-s) and a-C:H coated jaws (4-r and 4-s) prevent initial adhesion also at a µm-scale. An increased profile depth is measured for metallic bright (1-r and 1-s) and a-C:H:W coated (3-r and 3-s) as well as rough, ta-C coated jaws (2-r). The adhesion height on metallic bright, smooth jaws reaches a level of 20 µm. Adhesive wear tracks for rough, uncoated and rough, ta-C and both a-C:H:W coated jaws reveal a maximum profile depth of around 10 µm.

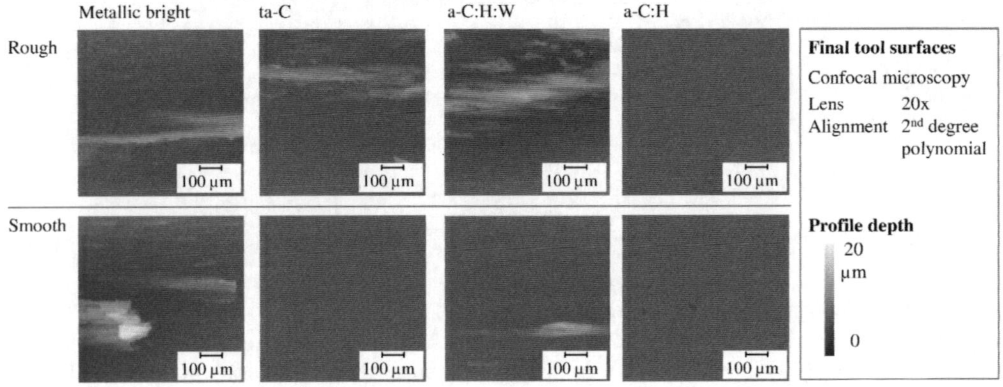

Figure 8. Topographies of tool surfaces after strip drawing tests

5. Discussion

By varying coating type and final surface treatment, friction coefficients can be varied in a broad range under dry conditions. The general relation between surface roughness and friction development is summarized in Figure 9.

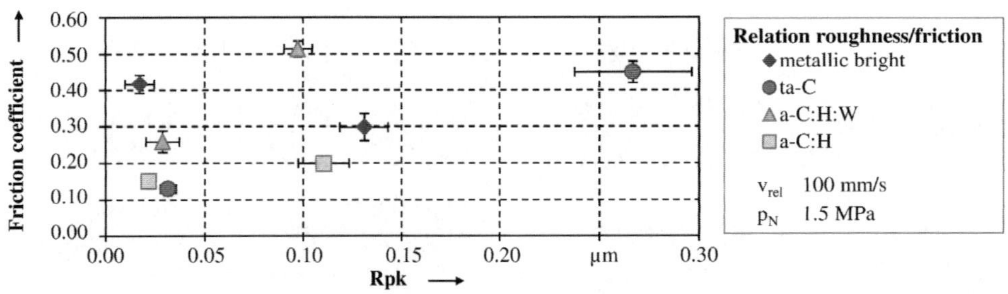

Figure 9. Relation between reduced peak height Rpk of tools and determined friction coefficients

The coating application was performed to reduce friction and adhesion, because the carbon based coatings act as separating layer between tool steel and sheet material. Thus, the coatings substitute the detaching function which is conventionally fulfilled by lubricants. The successful prevention of adhesive wear depends on coating type and roughness of the coated surface. Besides chemical properties like surface free energy and electrochemical potentials, adhesion tendency is related to material hardness [12]. Materials with higher hardness evoke higher internal metal-metal bond energies which lead to increased resistance towards adhesion [13]. This correlation might explain the friction development for smooth surfaces. The metallic bright specimen reveals the lowest hardness of about 700 HV causing the highest adhesion and friction coefficient. In contrast, ta-C coated tools which have the highest hardness result in lowest friction and no adhesive wear. Additionally, for the smooth variants the friction increases with growing amount of metallic material on the tool surface. Thus, the highest friction is measured for uncoated jaws and the lowest for ta-C and a-C:H coatings without any metallic phase. Looking at the specimens with higher roughness this correlation is not entirely valid anymore. Here, the influence of mechanical interaction of contacting asperities seems predominant for coated surfaces. The higher hardness of coatings in combination with increased roughness leads to high local stresses which cause break out of microscopic particles of the brittle surface. Evidence for this effect might be the growing arithmetic mean roughness from 16% to 26% comparing the initial surface with the surface after strip drawing tests in areas without visible adhesion. An increased roughness could not be measured for uncoated specimens. Additionally, higher roughness peaks cause a rising interlocking of roughness asperities. Thus, the resistance against relative movement increases which leads to higher friction coefficients for rough, coated specimens. Besides mechanical and topographical properties, the determined friction depends as well on chemical properties of the coating. Tungsten modified coatings result in an insufficient tribological behavior. For rough and smooth configurations approximately the same values of Rpk were measured for a-C:H:W and a-C:H coatings. However, the friction coefficient is more than 2.5 times higher for rough a-C:H:W coatings and 1.7 times higher for smooth a-C:H:W compared to a-C:H coatings. Reason for these differences might be a higher affinity of aluminum towards the metallic components in tungsten modified coatings. Additionally, a-C:H:W coatings reveal only half the hardness of a-C:H which promote growing adhesion tendency. As a reference, experiments with rough and smooth uncoated jaws were conducted. Contrary to the results of coated tools, a smoother, metallic bright surface lead to an increase of friction of 39% which is caused by growing adhesive wear due to increased real contact area. When tool roughness decreases, the real contact area increases as shown in Figure 10. Hence, an increased contact area leads to a growing target area for adhesion occurrence and thus results in higher friction in combination with increased wear. In contrast, smooth, coated surfaces did not result in higher friction because the coating itself functions as wear-resistant, separating layer between tool steel and sheet material. This reduces adhesion for a-C:H:W and prevents adhesion in case of ta-C and a-C:H coatings. At similar peak heights, smooth ta-C and a-C:H coatings reveal low friction and no wear. However, with ta-C coatings a 16% lower friction coefficient is achieved. This example and the results for a-C:H:W coatings show that not only low roughness but also a suitable coating type has to be considered to improve the tribological conditions in dry contacts.

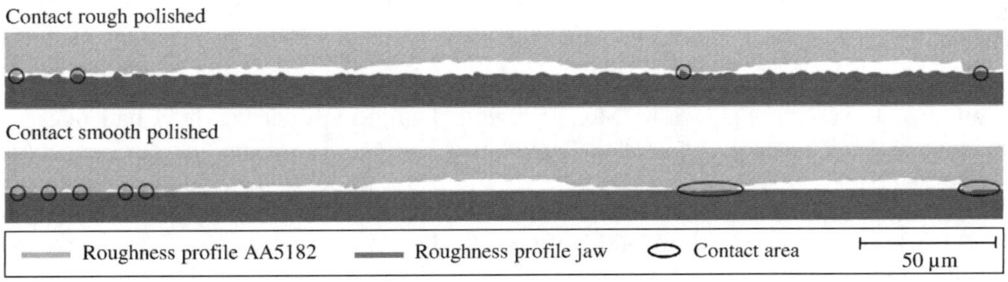

Figure 10. Contact area between AA5182 and metallic bright jaws for rough and smooth surfaces

6. Conclusion and outlook

In this paper the tribological properties in terms of sliding friction coefficients and wear behavior were characterized for different coating types. Moreover, the influence of varied surface topographies was investigated. As a reference, metallic bright tools were analyzed. To determine friction coefficients under close to forming conditions, flat strip drawing tests were performed. In general, the properties of a tribological system depend on chemical bonds and physical forces [12]. Thus, the tribological behavior revealed a strong dependency on surface roughness and coating type. Whether chemical properties like bonding energies and reactivity or mechanical properties like interlocking asperities dominate the frictional behavior depends on their specific characteristics. The relation of reduced peak height Rpk and friction coefficient is reverse for coated and uncoated tools. The investigated aluminum alloy shows a high affinity towards metallic bright jaws due to high chemical bonding forces between aluminum and steel. A reduction of surface roughness increased the real contact area and consequently the target area for adhesion occurrence. Thus, applying smoother tool surfaces leads to increased adhesive wear and friction in case of uncoated jaws. Coated specimens revealed lower friction at a decreased level of surface roughness. Increased friction and partly occurrence of adhesion is measured for rough coated tools. Here, adhesion is mainly caused by interlocking of roughness asperities in relation with high local stresses at contacting roughness peaks. For tungsten modified coatings, both, chemical and mechanical bonding occurs. Therefore, the friction coefficient is significantly higher for smooth and rough surfaces compared to the other coating types. A high chemical inertness occurs for a-C:H and ta-C which resulted in beneficial tribological behavior for smooth surfaces. In case of a-C:H coatings even the rough configuration showed low friction and no visible adhesion. In contrast, the more than two times rougher ta-C coatings reveal relatively high friction and adhesion. Thus, applying ta-C coatings without a surface treatment after coating deposition could not be recommended. There must be a critical value of Rpk between 0.11 and 0.27 μm at which the interlocking and the local contact stresses are too high to further prevent adhesion even with low reactive coatings. Further tests need to be conducted with surface qualities between these roughness levels to identify a promising compromise between duration and costs for surface finishing and tribological behavior. In order to achieve the overall goal of lubricant-free deep drawing operations with aluminum alloys suitable surface modifications need to be developed. Experiments with uncoated tools showed that even at low pressure and with low and high surface roughness severe galling occurs. Carbon based coatings could improve the tribological properties in direct contact with aluminum alloys depending on their chemical, mechanical and topographical properties. First of all, coating types with low reactivity and without metallic elements revealed better tribological conditions. Besides coating structure and doping elements, a surface treatment after coating deposition and general low coating roughness is beneficial to further reduce friction under dry conditions. Moreover, high coating hardness has a positive impact on friction and wear behavior. Further investigations with non-metallic doping elements in ta-C and a-C:H coatings need to be performed to evaluate if a further friction reduction is possible.

Acknowledgements
The authors thank the German Research Foundation (DFG) for supporting this investigation by funding the SPP 1676 project ME 2043/43-1. Furthermore the authors thank Mrs. Zhao from institute of engineering design (KTmfk) and H-O-T GmbH & Co.KG for coating deposition of the tools.

References

[1] Altan T and Tekkaya E 2012 *Sheet metal forming fundamentals* (Ohio: ASM International) p 5
[2] Vollertsen F, Schmidt F 2014 Dry Metal Forming: Definition, Chances and Challenges, *International Journal of Precision Engineering and Manufacturing – Green Technology* **1** 59–62
[3] Merklein M, Schmidt M, Tremmel S, Wartzack S, Andreas K, Häfner T, Zhao R and Steiner J 2015 Investigation of Tribological Systems for Dry Deep Drawing by Tailored Surfaces *Dry Metal Forming - OAJ FMT* **1** 42-56
[4] Association of German Engineers (VDI) 2012 *VDI Guideline 2840* (Berlin: Beuth)
[5] Horiuchi T, Yoshihara S and Iriyama Y 2012 Dry deep drawability of A5052 aluminum alloy sheet with DLC-coating *Wear* **286-287** 79–83
[6] Singh R K, Xie Z H, Bendavid A, Martin P J, Munroe P and Hoffman M 2008 Effect of substrate roughness on the contact damage of DLC coatings *Diamond and Related Materials* **17** 975–9
[7] Lindholm P, Björklund S and Svahn F 2006 Method and surface roughness aspects for the design of DLC coatings *Wear* **261** 107–11
[8] Jiang J and Arnell R D 2000 On the running-in behaviour of diamond-like carbon coatings under the ball-on-disk contact geometry *Wear* **239** 1–9
[9] Romhanji E, Popović M, Glišić D, Stefanović M and Milovanović M 2004 On the Al-Mg alloy sheets for automotive application: Problems and solutions *Metallurgija* **10** 205-16
[10] Severo V, Vilhena L, Silva P N, Dias J P, Becker D, Wagner S and Cavaleiro A 2009 Tribological Behaviour of W-Ti-N Coatings in Semi-Industrial Strip-Drawing Tests *Journal of Materials Processing Technology* **209** 4662–7
[11] Birkert A, Haage S and Straub M 2013 *Umformtechnische Herstellung komplexer Karosserieteile* (Berlin, Heidelberg: Springer) pp 179–209
[12] Menezes P, Nosonovsky M, Ingole S P, Kailas S V, Lovell M R 2013 *Tribology for Scientists and Engineers: From Basics to Advanced Concepts* (New York: Springer Science and Business Media) pp 108-110
[13] Vijh A K 1975 The influence of metal-metal bond energies on the adhesion, hardness, friction and wear of metals *Journal of Materials Science* **10** 998–1004

IOP Conf. Series: Materials Science and Engineering **159** (2016) 012023 doi:10.1088/1757-899X/159/1/012023

Tailored boron steel sheet component properties by selective laser heat treatment

N Asnafi[1], R Andersson[2], M Persson[3] and M Liljengren[4]

[1] VA Automotive, Box 34, SE-281 21 Hässleholm, Sweden, and School of Science & Technology, Örebro University, SE-701 82 Örebro, Sweden
[2] Duroc Special Steel, SE-971 88 Luleå, Sweden
[3] Duroc Laser Coating, SE-971 88 Luleå, Sweden
[4] Industrial Development Center, Vällaregatan 30, SE-293 38 Olofström, Sweden

E-mail: nader.asnafi@oru.se

Abstract. This investigation is focused on the stamping behaviour of boron steel, the properties of which are modified by selective laser heat treatment. Both CO_2 and fibre lasers are tested. By using different laser processing parameters, the hardening depth in the 1 mm thick boron steel sheet Boloc 02 is varied. Four routes are tested and verified. The forming operation (in which a so-called flexrail beam is produced) in all four routes is conducted at ambient (room) temperature. The *Reference* route comprises stamping of the sheet. The *GridBlank* route starts with selective laser heat treatment of the blank, after which the blank is allowed to cool down, moved to a hydraulic press and stamped. In the *GridTube* route, the blank is first stamped, after which the part is moved to a laser cell and selectively laser heat treated. The fourth route, the *RapidLaser* route, is similar to the *GridBlank* route, but a higher laser speed is used to promote higher total productivity. The *GridBlank* route results in the highest hardness values and the best shape accuracy. The initial sheet material exhibits a hardness of 200 HV, while the parts produced in the *GridBlank* route exhibit a hardness of 700 HV.

1. Introduction

The automotive industry has constantly sought new technology solutions to meet increasingly stringent emission norms, provide greater passenger safety & incessantly drive down the cost per vehicle produced. In Sweden, too, there is a consensus that the transportation sector must continue to deliver value to society & industry, with minimal environmental impact & fewer traffic injuries/fatalities, while remaining globally competitive. To achieve the above, the industry must enhance productivity, augment part performance & durability, accommodate greater volume flexibility with existing production systems as well as continue vehicle light-weighting efforts by inducting high-strength steels & light alloys (of Al, Mg & Ti). This demands use of new, cost-effective, adaptable & flexible process technologies capable of handling new materials & higher customer adaption/product individualization.

The present investigation is driven by the established ability of a laser to selectively heat treat materials to locally modify properties. When applied to sheet materials, the above approach can provide a rapid & cost effective pathway to locally tailor properties in a pre-determined manner to either aid forming operations during component manufacture or enhance overall performance of formed sheet metal parts by imparting superior properties in select locations as demanded by the operating environment. Due to the immense flexibility of present day lasers, this approach will be of significant interest to adopt in production and will generate, in turn, higher customer adaption/product individualization both in supplier-OEM relationship & OEM-market segments' differentiation. Compared to for instance hot stamping, the present approach requires much smaller investments, is much less process and type bound, and is more flexible both in in-house production and in cases where supplier integration should be considered.

1.1. The technical concept

High power lasers find widespread use in the automotive industry for many manufacturing operations spanning cutting, welding & surface modification. However, their application for sheet metal parts has been restricted largely to cutting & welding (e.g., tailor welded blanks) & their utility in sheet metal part fabrication steps such as forming or in augmenting performance of sheet components remains largely unexplored. Selective heat treatment with a laser involves heating the sheet material according to a pre-determined specific pattern, in the form of discrete localized areas of certain shape/size, continuous "lines" of a particular width, pattern of such "lines" to generate a grid pattern or other desired contours. The heat treatment may also be followed by cooling. An appropriate heat treatment protocol, & an optimum treatment pattern, promises to yield local sheet metal formability improvements &/or realize local strengthening/reinforcement in sheet material.

As an example, consider a sheet of boron steel. During selective heat treatment, the boron steel can be converted into austenite by heat treatment, after which cooling can be employed to promote conversion of austenite into martensite selectively in treated regions, for the martensite to act as the reinforcing structure & impart strength to the sheet material component either before or after forming. Similarly, selective heat treatment can also be utilized to improve formability during stamping, although the window of properties required to be realized in the sheet metal can be entirely different in this case. Illustrative examples of grid patterns or portions of grid patterns that can be investigated to assist forming or to tailor the component properties are depicted in Figure 1.

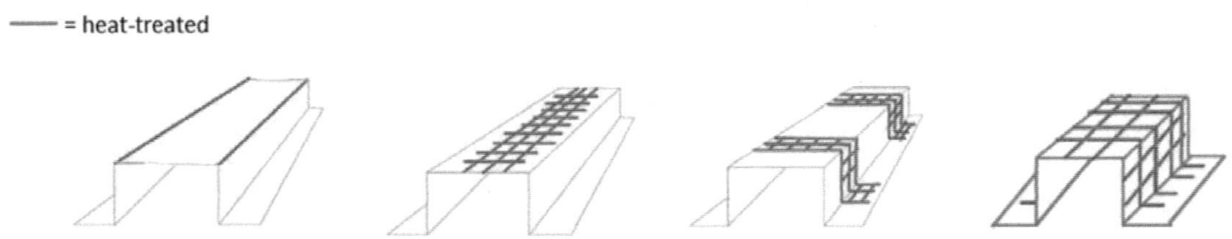

Figure 1. Illustrative examples of some selective heating grid patterns.

It may be appreciated that the specific heat treatment pattern to be followed, including its exact shape & precise location, needs to be designed (based on reverse engineering or detailed simulations) & later realized based on identified requirements. One such potential selective heat treatment grid is illustrated in Figure 2. Here, the forming may be executed immediately after selective heat treatment, such that parts of the sheet material, in particular the martensite grid pattern, are already at a temperature higher than the ambient. On the other hand, the objective might be to improve the component properties/performance by reinforcement, the selective heat treatment being conducted either before or after sheet material shaping.

1.2. Objective

Given the above, the overall objective is to investigate selective laser heat treatment of sheet metals from a new knowledge generation perspective by comprehensively determining processing-structure-property relationships, ascertaining influence of selective treatment on subsequent forming operation, testing/optimizing part behaviour for different applications/requirements & eventually demonstrating proof of tailored sheet metal formability concept on a couple of actual components. The present paper

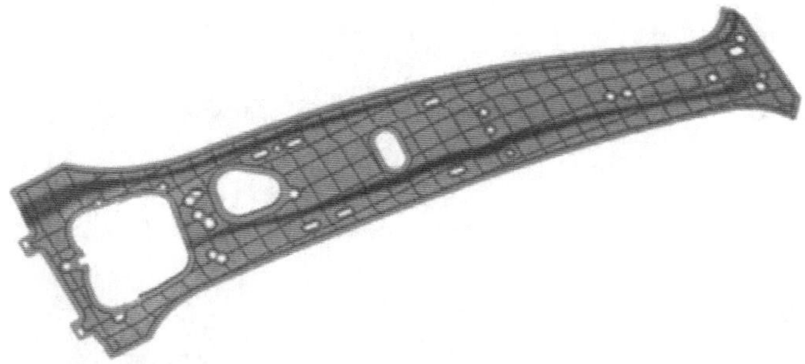

Figure 2. Illustration of shapes & positions of a heat treated grid on a true component.

is focused on the stamping behaviour of boron steel sheet selectively laser heat-treated in accordance to four (4) different routes.

1.3. State of the art

Various laser-based operations have been considered for sheet metal processing, including shock processing, shearing, bending, forming etc. & also widely employed in the manufacture of tailor welded blanks [1-5]. Incidentally, the term "tailored blanks" initially referred to joining of sheets of similar or dissimilar materials &/or with identical or different thickness. More recently, "tailor heat treated blanks" involving material properties adjusted by local heat treatment, have also been pursued [6]. The use of high strength steels or various light alloys for body-in-white applications has been limited by their inherent low formability [7]. Selective heat treatment can potentially improve the forming behaviour of these materials by modifying sheet material properties at specific locations in a predetermined manner to facilitate high drawing depth, sharp edges etc. A laser provides a promising pathway for the above, although deciding an appropriate heat treatment layout to accomplish the above can be challenging [8]. Vollertsen & Hofmann used a CO_2-Laser to demonstrate enhancement of formability in several 6000 series aluminium alloys [9]. Later, Geiger et al. used a lower wavelength Nd:YAG laser to allow for more efficient coupling, thereby eliminating need for a graphite layer [10]. There have been subsequent studies devoted to investigating the efficacy of local laser heat treatment on sheet metal forming [11]. These have dealt with Aluminium alloys [12-14], Magnesium alloys [15] as well as ultra-high strength steels widely used in the automotive industry [16]. However, notwithstanding its acknowledged potential, comprehensive investigations on the subject remain scarce. Particularly in Sweden, research efforts in this direction are yet to be initiated despite the fact that the technology, once developed, has the capability to be relevant to various other industries beyond the automotive.

1.4. Pilot study

A pilot study was conducted to demonstrate that sheet material properties can be modified in a controlled manner by manipulating the laser processing parameters. These have involved the following:

- CO_2 & fiber laser processing
- Fixed laser spot size (ϕ6.24 mm for CO_2 & ϕ5.50 mm for fiber)
- Different laser powers & scan speeds
- Heat treatment of 1.0 mm thick boron steel Boloc 02 sheet material

Figure 3 reveals cross-sectional micrographs of a 1 mm thick Boloc 02 steel sheet CO_2 laser treated under two different conditions (laser power = 650 W, scan speed = 500 mm/min and case depth = 0.6 mm in Figure 3 (a) and laser power = 2300 W, scan speed = 5000 mm/min and case depth = 0.17 mm in Figure 3 (b)) & the results are both educative & relevant. At the outset, they reveal that it is possible to manipulate & optimize the laser heat treatment process to achieve the desired impact (in this case, an enhanced hardness of 470 HV compared to initial sheet hardness of 220 HV). Furthermore, they also demonstrate that the desired surface hardness can be obtained under varied operating conditions & with controllable case depths, thereby suggesting that the process can be optimized to achieve faster processing speeds, offering an opportunity to generate varied properties.

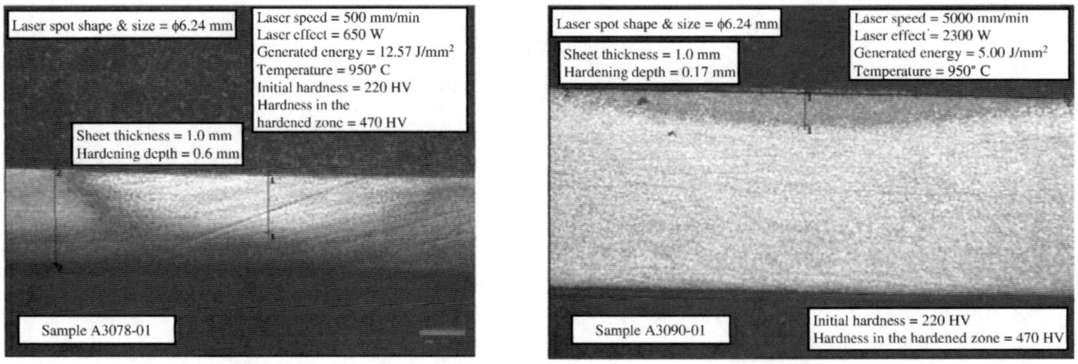

Figure 3. Typical cross-sectional micrographs of a 1.0 mm thick boron steel Boloc 02 laser heat treated with a CO_2 laser at (a) power = 650 W & (b) power = 2300 W.

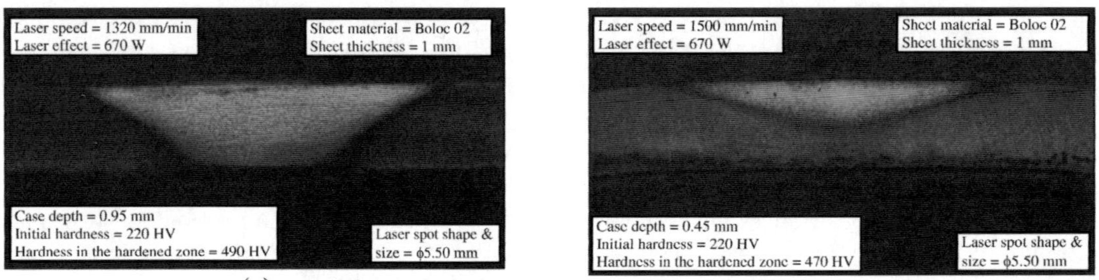

Figure 4. Typical cross-sectional micrographs of a 1.0 mm thick boron steel Boloc 02 laser heat treated with a fiber laser at 670 W and with a scan speed of (a) 1320 mm /min & (b) 1500 mm /min.

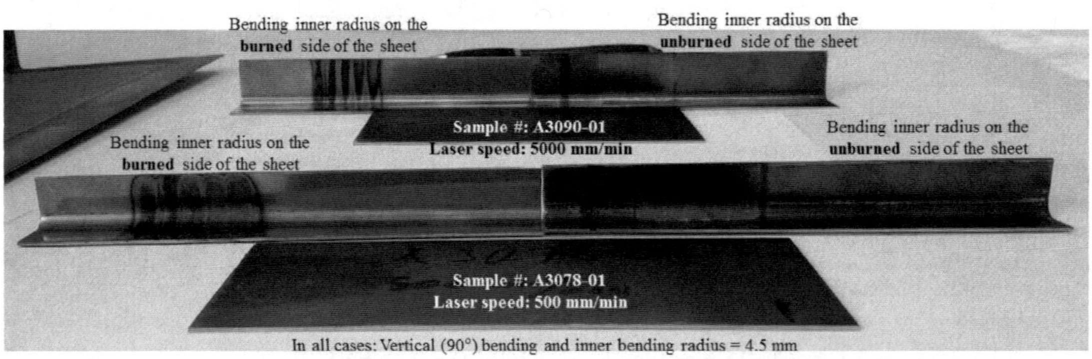

Figure 5. Vertical (90°) bending tests of selectively laser heat treated blanks (Figure 3).

Results of similar trials were conducted using a fiber laser, with the Boloc 02 sheet material graphitized for better coupling, are depicted in Figure 4.

Vertical (90°) bending tests were conducted on the selectively laser heat treated blanks of 1 mm thick boron sheet steel Boloc 02 shown in Figure 3. Bending was conducted so that the inner bend radius was located on both the burned side and the unburned side of the sheet. Inner bending radius was 4.5 mm in all cases and all tests were successful (no cracks or other defects were detected), Figure 5.

2. Materials

In this investigation, 1.0 mm thick boron steel sheet Boloc 02 was used. Table 1 displays the chemical composition of this material. The mechanical properties of 1 mm thick Boloc 02 are shown in Table 2,

Table 1. Chemical composition of Boloc 02.

Material	C (%) min-max	Si (%) min-max	Mn (%) min-max	P (%) max	S (%) max	Cr (%) min-max	B (%) min-max
Boloc 02	0.20-0.25	0.20-0.35	1.00-1.30	0.030	0.010	0.140-0.260	0.0015-0.0050

in which R_e is the yield strength, R_m the ultimate tensile strength and A_{80} is the fracture elongation in tensile testing.

Table 2. Mechanical properties of 1 mm thick Boloc 02.

Material	Thickness (mm)	R_e (MPa)	R_m (MPa)	A_{80} (%)
Boloc 02	1.0	ca 340	ca 480	28

3. Experimental procedure
To prove the concept, the flexrail shown in Figure 6 was selected as the demonstration component. The stamping tool set used to form this flexrail is displayed in Figure 7.

Figure 6. The flexrail produced in accordance to the four production routes (shown in Figure 8) in this investigation,

Figure 7. The stamping tool set used to form the flexrails.

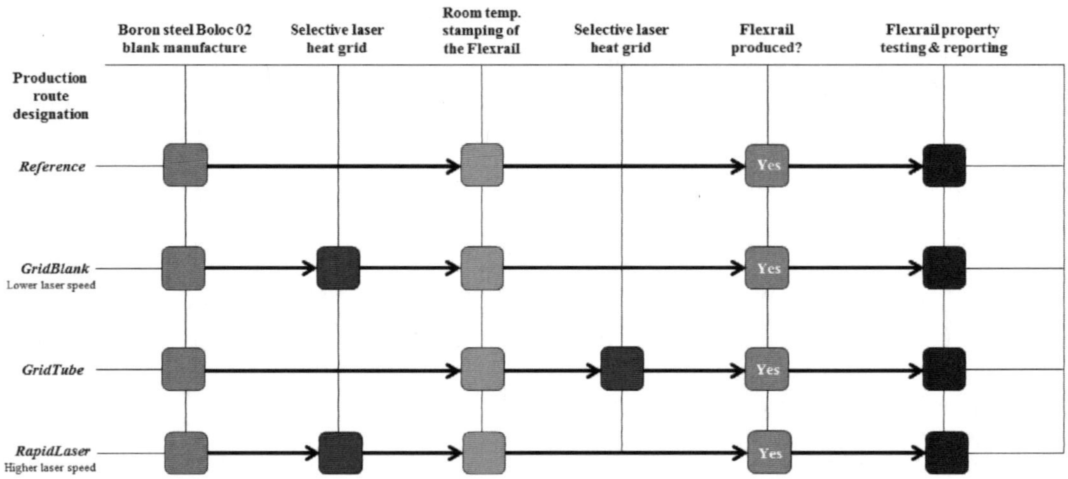

Figure 8. These four (4) production routes were followed in this investigation.

Four (4) different production routes were used to make the flexrail mentioned above. These four production routes are depicted in Figure 8. The blank size and shape used in all four routes are displayed in Figure 9.

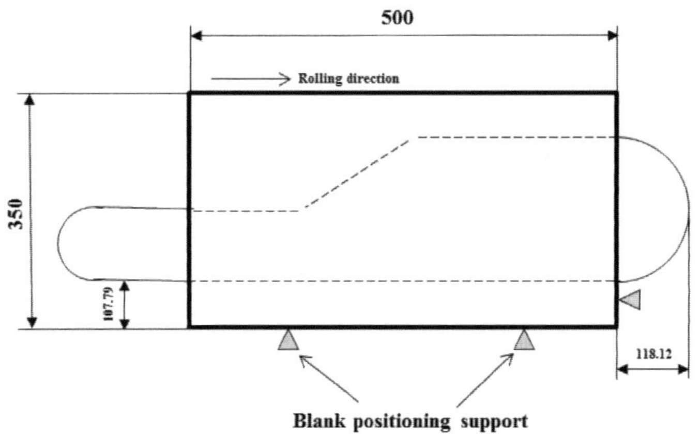

Figure 9. Blank size and shape versus the stamping die cavity.

The four production routes exhibited in Figure 8 comprise:

Reference route: In this route, 1 mm thick blanks of Boloc 02 (Figure 9) were produced in City 1, transported to City 2 and stamped at ambient (room) temperature in the tool set displayed in Figure 7 in City 2.

GridBlank and *RapidLaser* routes: In these routes, 1 mm thick blanks of Boloc 02 (Figure 9) were produced in City 1, selectively laser heat treated in City 1 in such a fashion that the heat treatment grid pattern constituted the network depicted in Figure 10, allowed to cool down (still air cooling) in City 1, transported to City 2 where they were stamped at ambient (room) temperature in the tool set displayed in Figure 7. In the *GridBlank* route, the laser processing parameters in Figure 4(a) were used. For *RapidLaser* blanks, the laser processing parameters in Figure 4(b) were applied.

GridTube route: In this route, 1 mm thick blanks of Boloc 02 (Figure 9) were produced in City 1, transported to City 2, where they were stamped at ambient (room) temperature in the tool set displayed in Figure 7 in City 2. After stamping, the stamped flexrails were moved to a fibre laser cell in City 2, where they were selectively laser heat treated in such a fashion that the grid pattern constituted a network with a quadratic side length of 20 mm (such as that shown in Figure 10).

To selectively laser heat treat the flexrails stamped in the *GridTube* route (Figure 8), the laser robot path must be generated based on the scanned shape of the *GridTube* flexrail (Figure 11). Since another laser cell (located in City 2) was used and the selective laser heat treatment must be conducted on stamped flexrails, the laser processing parameters needed to be found and optimized. After 12 test run, the following parameters yielding a hardness of 550 HV were selected: spot size & shape = ϕ5 mm, laser speed = 25 mm/s, temperature = 1500-1530 °C, the flexrail was graphitized before selective laser heat treatment, no cooling used at any time (except still air). Figure 12 depicts the cross-sectional micrograph of this twelfth (12th) test run.

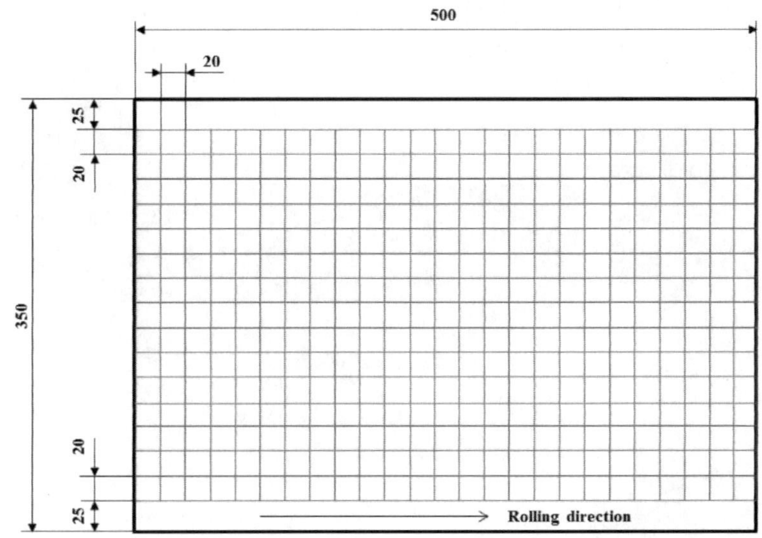

Figure 10. The selective laser heat treatment pattern used in the production routes *GridBlank* and *RapidLaser*.

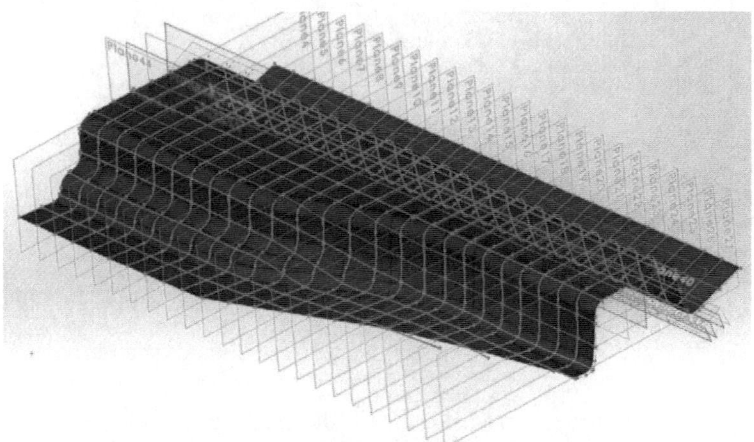

Figure 11. The scanned shape of the *GridTube* flexrail was used to generate the laser robot path.

Figure 12. The cross-sectional micrograph of the sheet laser heat treated in accordance to the parameters set after 12 test runs.

To produce comparable flexrails, the process parameters during stamping needed to be kept constant regardless of the production route. From a formability perspective, the *GridBlank* option (Figure 8) is the toughest case. The *GridBlank* blanks are selectively hardened in such a fashion that the hardening case is almost as large (0.95 mm) as the sheet thickness (1.00 mm) and the laser heat treatment pattern constitutes a network, Figures 4(a) and 10. Therefore, the stamping try-outs started with the *GridBlank* option lubricated with 1.5 g/m^2 Aral Ropa 4332. Stamping with a binder pressure of 100 bar resulted in a crack. Reducing the binder pressure in combination with trimming a portion of the initial blank and finally shimming corresponding to the sheet thickness (1 mm) resulted in a approved flexrail, Figure 13.

These process parameters, i.e. trimmed initial blank, a binder pressure of 90 bar, shimming corresponding to 1 mm, and using 1.5 g/m^2 Aral Ropa 4332 as lubricant, were used in all stamping operations regardless of the selected production route (Figure 8). In other words, all flexrails in this study were stamped with these parameters regardless of the production route.

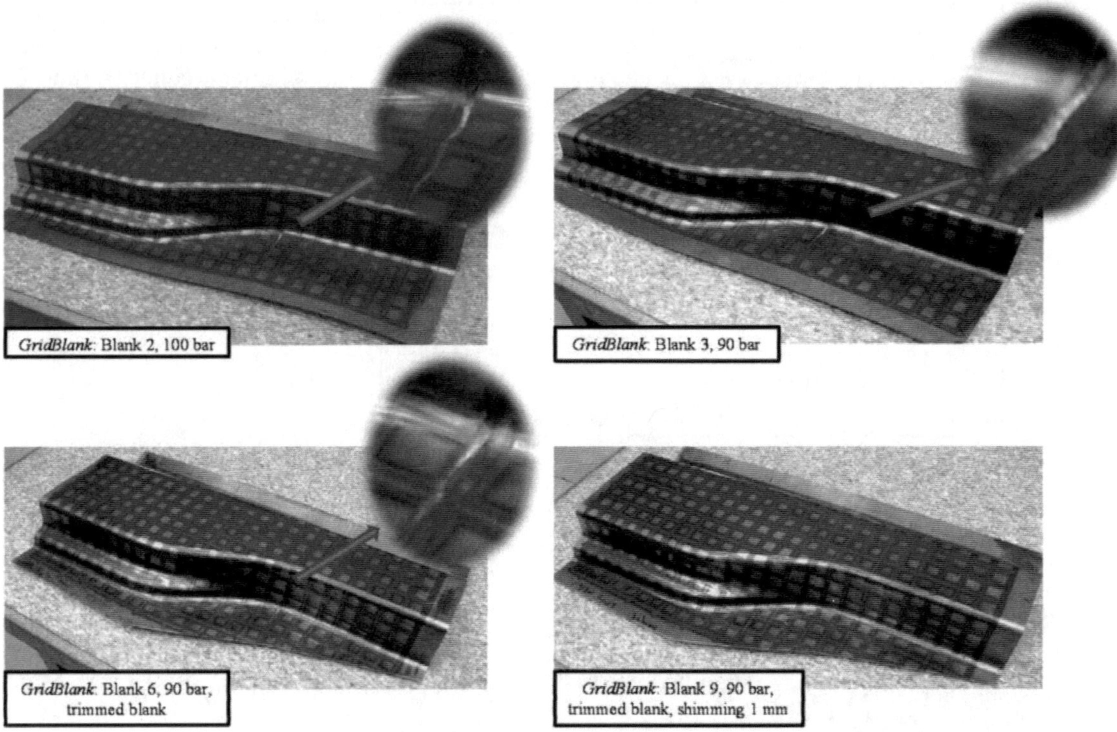

Figure 13. Stamping try-outs with the *GridBlank* option to find the stamping process parameters that should be and were kept constant regardless of the production route.

4. Results

The flexrails, the side members or beams, manufactured in accordance to the production routes described in Figure 8, *Reference*, *GridBlank*, *GridTube* and *RapidLaser*, are shown in Figure 14. Note in this figure that the *GridTube* flexrail is displayed during selective laser heat treatment.

The blanks made for the *GridBlank* and *RapidLaser* routes, Figure 8, were selectively laser heat treated and allowed to cool down in City 1, and transported to City 2, where they were stamped at ambient (room) temperature in the tool set displayed in Figure 7. These blanks, particularly the select-

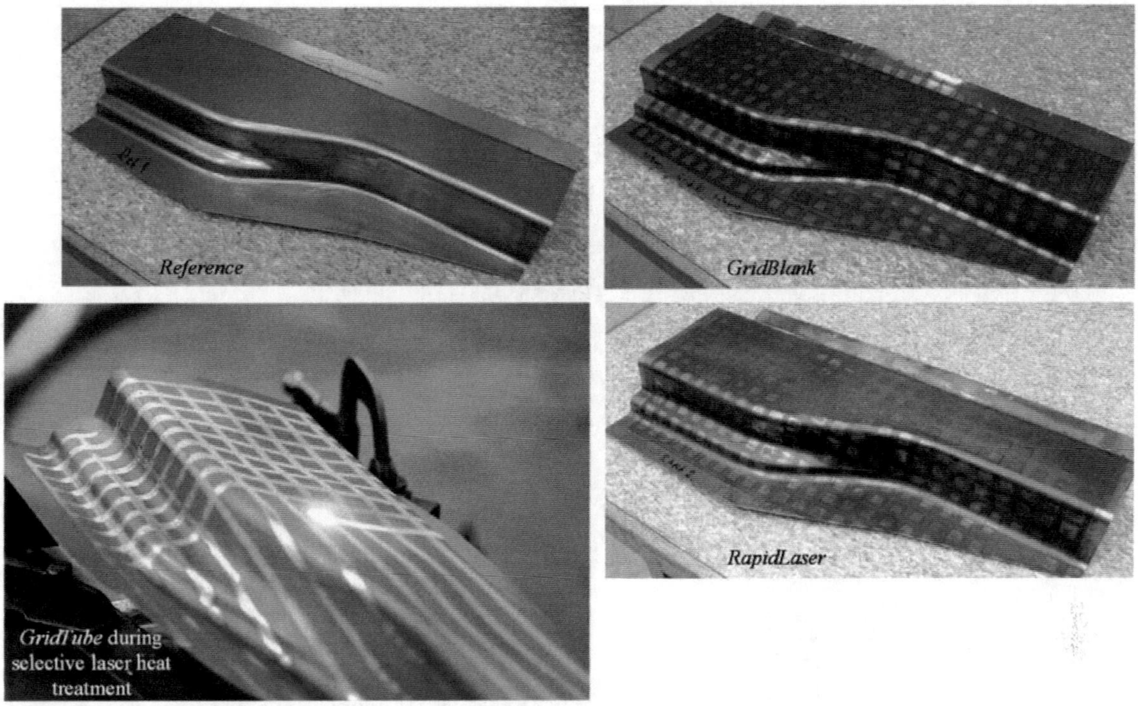

Figure 14. The flexrails manufactured in accordance to the production routes described in Figure 8: *Reference*, *GridBlank*, *GridTube* and *RapidLaser*. Note that the *GridTube* flexrail is displayed during selective laser heat treatment.

ively heat treated blanks in the *GridBlank* route, were not flat/not plane before stamping. This type of blank positioned on the die prior to stamping is displayed in Figure 15. The blank is however stretched out during the stamping and the stamped flexrail is not "buckled". See the *GridBlank* flexrail, top right photograph in Figure 14.

The punch force was registered, as the blanks were stamped (or the flexrails were formed). Figure 16 displays the maximum punch force (during stamping) for the different production routes (Figure 8). As shown in Figure 16, 20% and 41% higher maximum punch forces were required for the *RapidLaser* and *GridBlank* routes respectively compared to the reference route.

Figure. 15. The selectively laser heat treated blank positioned on the die prior to stamping a flexrail according to the *GridBlank* route in Figure 8.

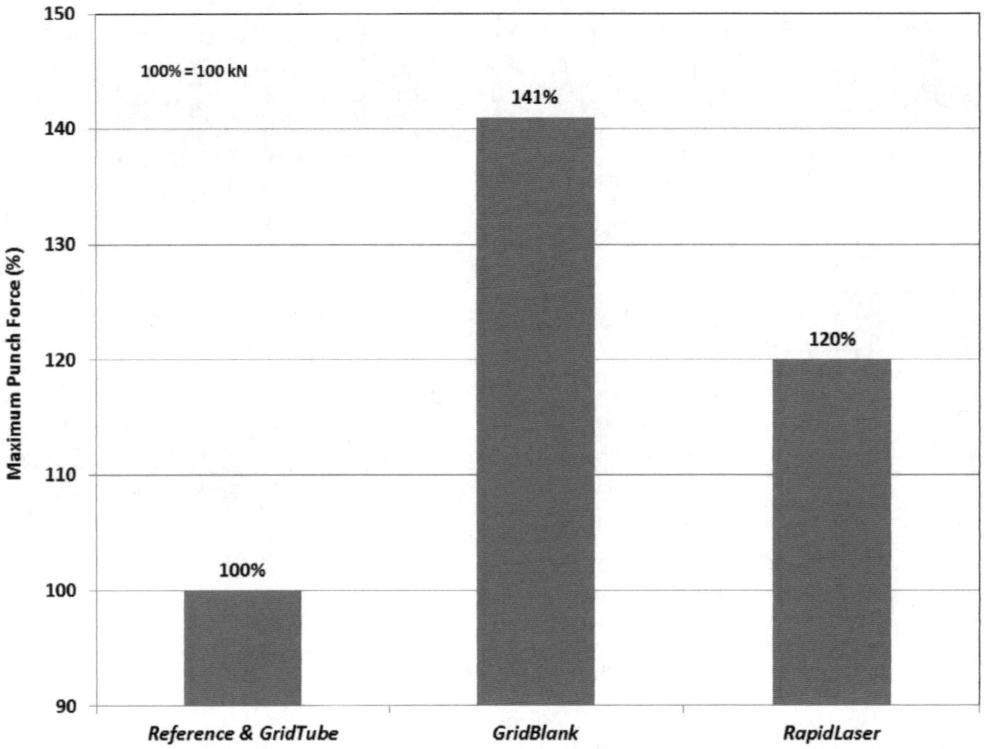

Figure 16. The maximum punch force (during stamping) for different production routes (Figure 8).

The flexrails produced in accordance with the production routes shown in Figure 8 were scanned. These flexrails were then compared in three sections, A-A, B-B and C-C, with respect to shape accuracy, Figure 17. The scanned shapes of the produced flexrails are displayed at the above-mentioned 3 sections in Figure 18. As shown in Figure 18, the *GridBlank* route displays the smallest springback (or the best shape accuracy) in all three studied sections.

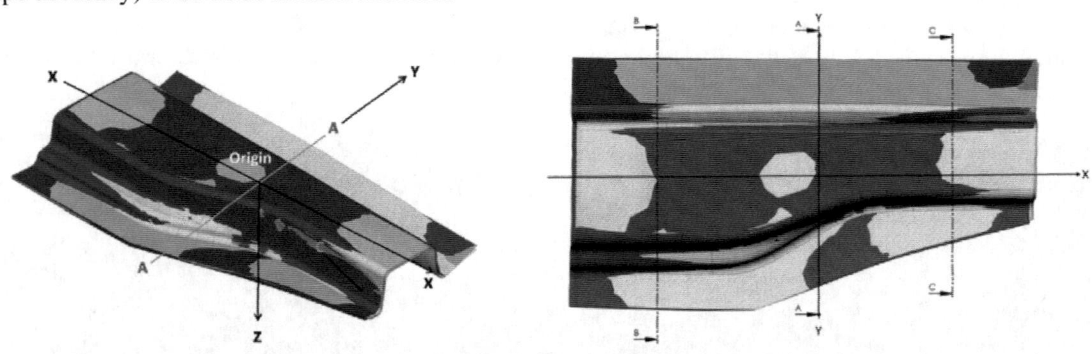

Figure 17. The flexrail shape accuracy/springback will be displayed in these three sections.

The surface hardness was measured (on the flexrails produced in accordance to the four tested routes in Figure 8) in the longitudinal and transverse directions and positions indicated in Figure 19. The results of these hardness measurements are depicted in Figure 20. As shown in this figure, the flexrail produced in accordance to the:

Reference route displays an even hardness of 200 HV in both the longitudinal and transverse directions.
GridBlank route exhibits the highest hardness, approximately 700 HV, in the laser heat treated tracks and in both longitudinal and transverse directions.

IOP Conf. Series: Materials Science and Engineering **159** (2016) 012023 doi:10.1088/1757-899X/159/1/012023

GridTube route shows a hardness of approximately 500-530 HV in the left hand track & somewhat lower hardness in the right hand track in both the longitudinal and transverse directions.
RapidLaser route displays a hardness of approximately 275-280 HV in the left hand track and somewhat higher hardness in the right hand track in both the longitudinal and transverse directions.

5. Discussion

The present paper is an account of the results obtained so far in an ongoing investigation. The following domains are scheduled for further studies.
Flexrails produced in accordance to the four production routes displayed in Figure 8 will be subject to further tests, such as 3- and 4-point flexure tests and crash tests.
The pattern generated by selective laser heat treatment will be predicted and optimized by simulations. Based on the intended component behaviour in use, the laser heat treatment pattern, including its exact shape & precise location, will be designed by reverse engineering and simulations and tested in practice.
Several business cases comprising different large production volume scenarios will be generated and evaluated.
The results of the above-mentioned studies will be accounted for, as soon as they are conducted/completed.

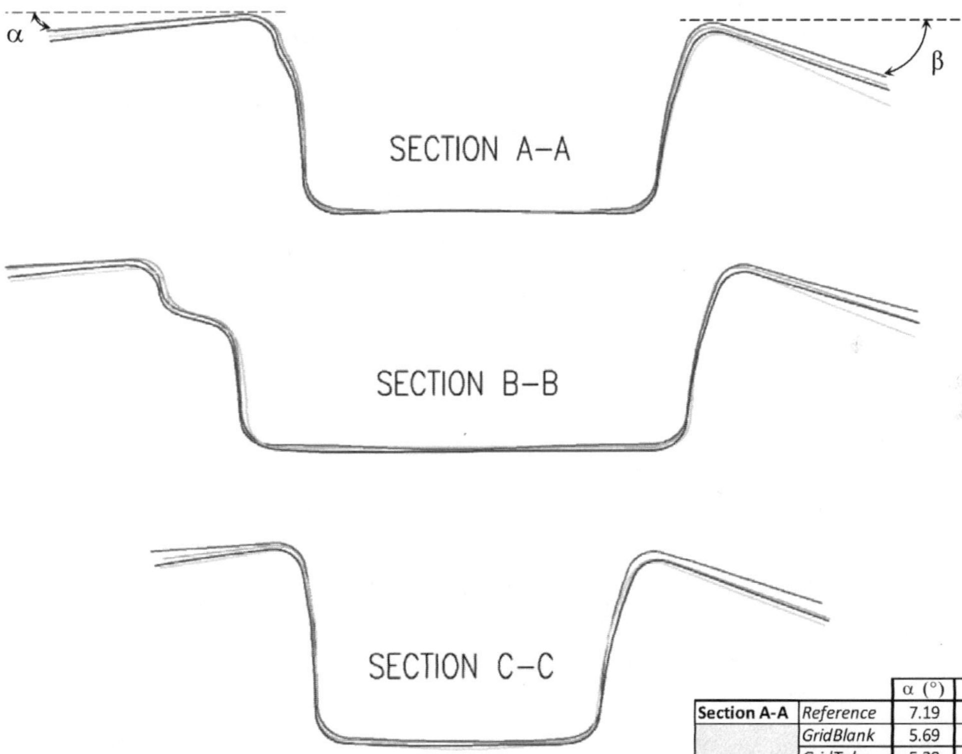

		α (°)	β (°)
Section A-A	*Reference*	7.19	16.95
	GridBlank	5.69	16.29
	GridTube	5.29	21.41
	RapidLaser	8.13	17.10
Section B-B	*Reference*	4.74	16.90
	GridBlank	3.25	15.40
	GridTube	5.76	20.60
	RapidLaser	6.63	15.47
Section C-C	*Reference*	9.18	17.45
	GridBlank	5.42	16.40
	GridTube	10.81	18.64
	RapidLaser	10.00	18.36

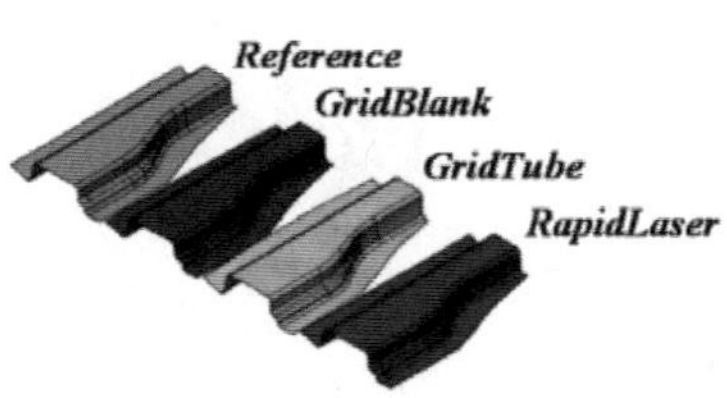

Figure 18. Flexrail shape accuracy/springback in three sections (see also Figure 17) for the four tested production routes (Figure 8).

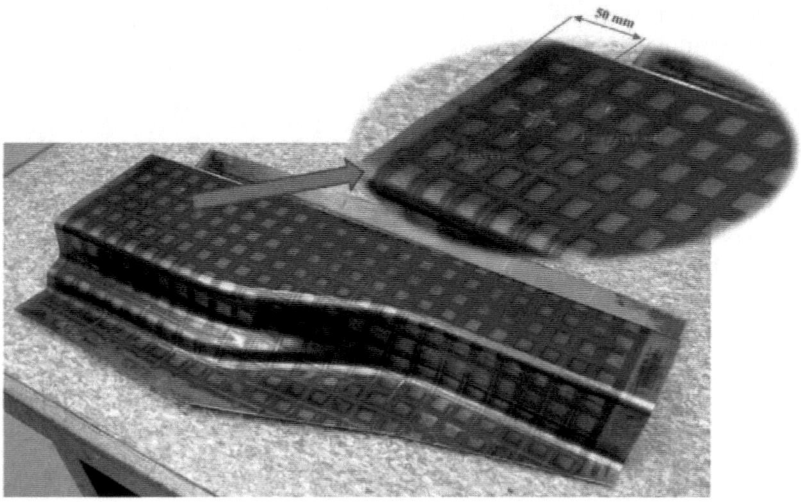

Figure 19. The surface hardness was measured in the indicated longitudinal and transverse positions.

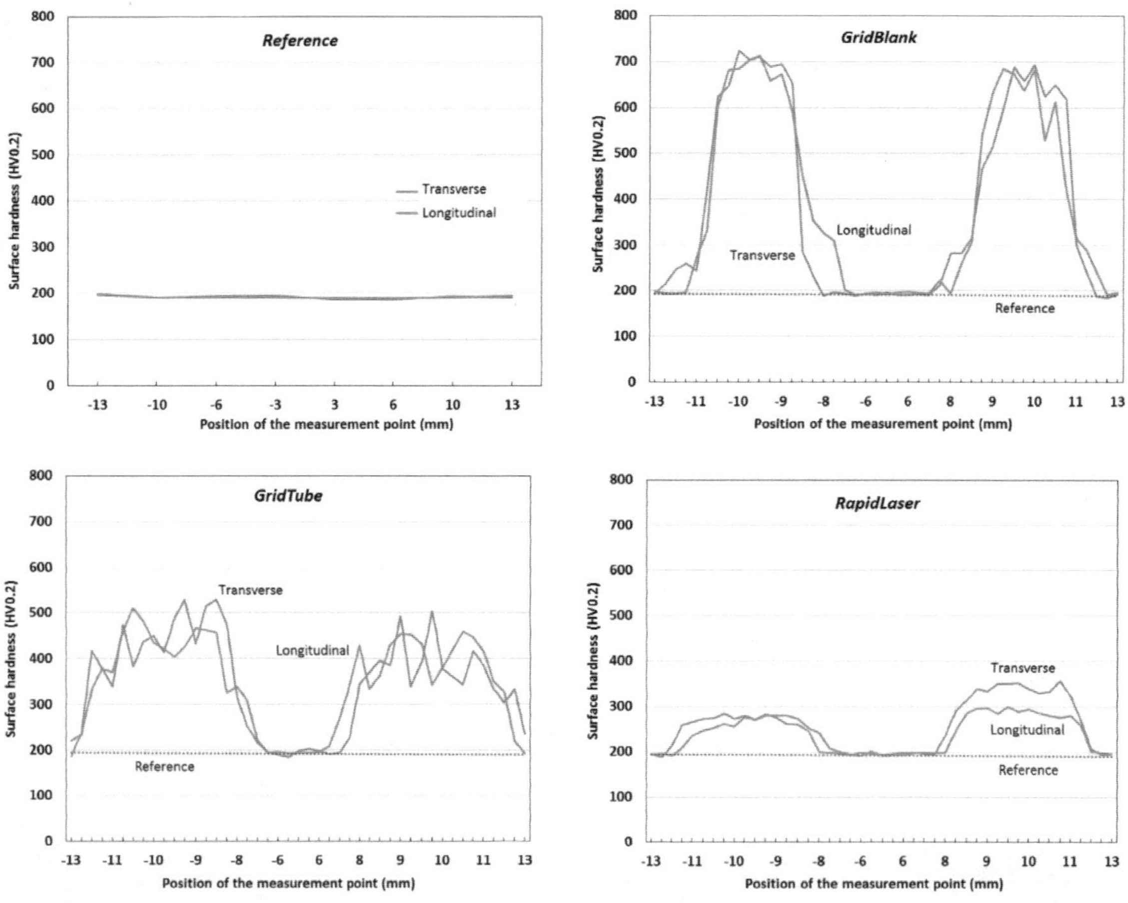

Figure 20. The results of the surface hardness measurements (Figure 19) on the flexrails produced according to the four tested routes (Figure 8).

However, the results accounted for in this paper indicate that the *GridBlank* and *RapidLaser* routes can be considered as the most efficient and flexible routes for large volume production. They are the most efficient and flexible routes for large volume production, since the selective laser heat treatment can be conducted at the material supplier or at a supplier between the material supplier and the stamping plant or at the stamping plant. The existing press/stamping lines can be used as usual. "No difference" needs to be taken into consideration, since the sheet material in combination with the selective heat treatment pattern and "normal" stamping generate the intended components with the required properties.

6. Conclusions

The chief conclusions drawn from this investigation are as follows:
It is possible to tailor boron steel sheet component properties by selective laser heat treatment. This selected laser heat treatment can be conducted on the blanks prior to stamping at ambient (room) temperature or on the component after stamping at ambient (room) temperature. The selective laser heat treatment accomplishes conversion into austenite and still air cooling results in the conversion of austenite into martensite.
The *GridBlank* production route, in which 1 mm thick blanks of Boloc 02 (Figure 9) was selectively laser heat treated, allowed to cool down (still air cooling), transported to the stamping plant where it was stamped at ambient (room) temperature, yields the best component (flexrail) shape accuracy and the best hardness (700 HV) but required 41% higher maximum press force than the none laser heat treated Boloc 02.
The selective laser heat treatment of the sheet material (the blank) before stamping yields the most efficient and flexible route for large volume production.

References

[1] Ocaña J L, Correa C, García-Beltrán A, Porrò J A, Díaz M, Ruiz-de-Lara L and Peral D 2015 Laser Shock Processing of thin Al2024-T351 plates for induction of through-thickness compressive residual stresses fields *J. Materials Processing Techn.* **223** 8-15
[2] Brecher C, Emonts M, Eckert M and Weinbach M 2014 Double Sided Irradiation for Laser-assisted Shearing of Ultra High Strength Steels with Process Integrated Hardening *Physics Procedia* **56** 1427-1435
[3] Che Jamil M S and Sheikh M A, Li L 2011 A study of the effect of laser beam geometries on laser bending of sheet metal by buckling mechanism *Optics & Laser Technology* **43** 183-193
[4] Guan Y, Sun S, Zhao G and Luan Y 2005 Influence of material properties on the laser-forming process of sheet metals *J. Materials Processing Techn.* **167** 124-131
[5] Sharma R S and Molian P 2009 Yb:YAG laser welding of TRIP780 steel with dual phase and mild steels for use in tailor welded blanks *Materials & Design* **30** 4146-4155
[6] Merklein M, Johannes M, Lechner M and Kuppert A 2014 A review on tailored blanks – production, applications and evaluation *J. Materials Processing Techn.* **214** 151-164
[7] Kleiner M, Geiger M and Klaus A 2003 Manufacturing of Lightweight Components by Metal Forming *CIRP Annals - Manufacturing Technology* **52** 521–542
[8] Merklein M, Geiger M and Vogt U 2007 Enhanced formability of aluminium blanks by local laser heat treatment *Proc. 5th Intl. Conf. Laser Assisted Net Shape Engineering (LANE)*, (Erlangen) pp 1279–1288
[9] Vollertsen F & Hofmann A 2000 Modelling the deep drawing of process optimized blanks *Proc. 20th Biennial IDDRG Congress* (Genval) pp 67–76
[10] Geiger M, Merklein M and Kerausch M 2004 Finite Element Simulation of Deep Drawing of Tailored Heat Treated Blanks *CIRP Annals - Manufacturing Technology* **53** pp 223–226
[11] Merklein M and Nguyen H 2010 Advanced laser heat treatment with respect for the application for Tailored Heat Treated Blanks *Physics Procedia* **5** 233-242
[12] Merklein M, Böhm W and Lechner M 2012 Tailoring Material Properties of Aluminium by Local Laser Heat Treatment *Physics Procedia* **39** 232-239

[13] Nguyen H and Merklein M 2012 Improved Formability of Aluminium Alloys using Laser Induced Hardening of Tailored Heat Treated Blanks *Physics Procedia* **39** 318-326
[14] Mohammadi A, Vanhove H, Van Bael A and Duflou J R 2012 Bending Properties of Locally Laser Heat Treated AA2024-T3 Aluminium Alloy *Physics Procedia* **39** 257-264
[15] Hino R, Kawabata K and Yoshida F 2014 Incremental Forming with Local Heating by Laser Irradiation for Magnesium Alloy Sheet *Procedia Engineering* **81** 2330-2335
[16] Neugebauer R, Scheffler S and Poprawe R and Weisheit A 2009 Local laser heat treatment of ultra high strength steels to improve formability *Production Engineering* **3** 347–351

IOP Conf. Series: Materials Science and Engineering **159** (2016) 012024 doi:10.1088/1757-899X/159/1/012024

Improving the crash behavior of structural components made of advanced high strength steel by local heat treatment

L Conrads[1], **M Daamen**[1], **G Hirt**[1] **and M Bambach**[2]

[1] Institute of Metal Forming, RWTH Aachen University, Intzestr. 10, 52056 Aachen, Germany
[2] Chair for Mechanical Design and Manufacturing, BTU Cottbus-Senftenberg, Konrad-Wachsmann-Allee 17, 03046 Cottbus, Germany

E-mail: conrads@ibf.rwth-aachen.de

Abstract. High manganese TWIP steel belongs to the second generation of advanced high strength steels. During the production of strip material, the microstructure and hence the mechanical properties of TWIP steel can be adapted to the specific needs of crash relevant structures. Whereas typically the whole steel strip is heat-treated after cold rolling, a local heat treatment can be applied to tailor the properties accordingly. In this work, a method is presented to identify a suitable process window for the local laser heat treatment of TWIP steel. The material is strain hardened and afterwards heat-treated at various temperatures for a short time. The influence of the respective heat treatment on microstructure and mechanical properties is evaluated and the most appropriate heat treatment is then reproduced using laser heating. To verify the effect of a local laser heat treatment at a structural component, crash boxes with different heat treatment patterns were produced and tested. The dynamic crash tests show that the local heat treatment can be used to improve the crash behavior of structural components. In addition, their deformation path can be influenced by using adapted heat treatment patterns and the crash behavior can be controlled.

1. Introduction

Advanced high strength steels (AHSS) have a great potential as lightweight material in the automotive industry compared to alternative materials as aluminum, plastics or carbon fiber reinforced plastics [1]. Strengths in the range of 1300-1500 MPa allow for the use of reduced sheet thicknesses and hence a weight saving of 30-40 % [2]. However, high strengths generally affect the formability. This complicates the processing of these steels and makes especially the cold forming difficult [3]. To still benefit from the high strength, the formability can be restored locally prior to the forming operation. The sheet metal is heat-treated within the areas prone to failure, while the rest remains at high strength. The subsequent forming can be performed at room temperature. The resulting 'Tailor Heat-Treated Blank' (THTB) shows a property distribution that is adjusted at best to the function of the final part [4].

Austenitic high manganese Twinning Induced Plasticity (TWIP) steel is part of the second generation of AHSS. It shows excellent crash behavior due to its high work hardening rate and hence is suitable for the use in structural components [5]. However, due to its low initial yield stress, TWIP steel cannot compete with industrially used dual phase steel DP800 in dynamic collapse tests [6]. Hence, the approach of a tailored heat treatment can be used to improve the crash behavior of structural components made of high manganese steel.

In this work, the high manganese TWIP steel Fe-28Mn-0.3C is investigated. A method is developed to identify suitable parameters for the local laser heat treatment: The material is strain hardened by cold rolling and the effect of different heat treatments on its microstructure and the resulting mechanical properties is evaluated via dilatometric experiments. The most promising heat treatment is then reproduced via laser heating. To investigate the effect of a local laser heat treatment on the crash behavior of a structural component, crash boxes with two different heat treatment patterns are manufactured. The patterns consist of alternating strain hardened and softened zones. Dynamic impact tests are performed to test the energy absorption capacity. The resulting energy-displacement curves are used to compare the different heat treatment patterns and assess their influence on the crashworthiness.

The goal is to show for a structural component that it is possible to improve and locally tailor the mechanical properties of AHSS based on a shortened process chain including local laser heat treatment, and thus, to reveal completely new design options. This yields the following research questions:

- How can the microstructure and mechanical properties of high manganese TWIP steel be influenced by a short-time (laser) heat treatment?
- Can the energy absorption capacity be increased by adapting the material properties in the different areas to the external loading condition during deformation?
- Does the deformation follow the pattern of softened and hardened zones?

2. Experimental work

2.1. Material and processing

The investigated Fe-28Mn-0.3C is a high manganese steel which forms twins during the deformation at room temperature. The TWIP behavior can be related to its stacking fault energy of 27 mJ/m^2 [7]. The material was produced at a lab-scale strip casting system of the Institute of Metal Forming (IBF) of RWTH Aachen University. The main advantages of twin-roll strip casting are the high cooling rates and the integrated forming process which enables a near-net-shape production of such high-alloyed steels. The applied process chain enables the production of high manganese TWIP steels with excellent mechanical properties compared to conventional strip production [8]. In table 1, the chemical composition of the cast strip is given.

Table 1. Chemical composition of as-cast Fe-28Mn-0.3C-alloy in wt% according to optical emission spectrometry.

Element (wt%)	Fe	C	Mn	Al	Si	N	S	P
Cast strip	Balance	0.284	27.84	0.003	0.107	0.004	0.002	0.01

The strip cast and inline hot rolled material with a thickness of 1.5 mm was subsequently cold rolled to 1.1 mm. The height reduction of 25 % strengthens the material effectively without eliminating its formability. Hence, the energy absorption capacity is increased while at the same time a sufficient formability of the strain hardened zones of the crash box is ensured. This facilitates a regular buckling during the crash test.

2.2. Dilatometric study

To identify a suitable temperature for the local heat treatment, different heat treatments with a maximum temperature ranging between 600 and 900 °C were applied using a dilatometer (DIL 805A/D, co. TA Instruments). Heating was carried out inductively under high vacuum. The heat treatment cycle is based on the general characteristics of a laser heat treatment, i.e. fast heating and exponential cooling rate. The used heating rate was 50 K/s, the holding time was set to 2 seconds and the cooling continues exponentially within two minutes down to 40 °C.

Then, the resulting microstructure was analyzed by means of electron backscatter diffraction (EBSD) at the Institute of Physical Metallurgy and Metals Physics (IMM) of RWTH Aachen University. The mechanical properties under quasi-static loading ($\dot{\varepsilon} = 0.0025\,s^{-1}$) were investigated using the deformation device of the dilatometer.

2.3. *Local heat treatment concepts*

To investigate the influence of a local heat treatment on the crashworthiness of the work hardened high manganese steel, two different heat treatment patterns were applied onto the crash boxes. The first pattern (see figure 1a) stems from the regular folding which was observed in the crash tests of the fully recrystallized high manganese TWIP steel [6]. It is assumed that the buckling of the folds, once started, is a purely geometrical effect. The measured distance of the tips of the folds was 20 mm. Hence, every 20 mm a heat-treated zone of 5 mm width was applied. The interaction of strengthened and softened areas is supposed to be beneficial for the crashworthiness. While the strengthened areas are supposed to increase the energy absorption capacity, the softened areas allow for a stable deformation process in the highly strained tips of the folds.

In the second heat treatment pattern (see figure 1b), the distance of the heat-treated zones was increased to 30 mm. Here, it is tried to influence the natural folding behavior and to control where the folding occurs. It is expected that the folds form preferably in the softened areas.

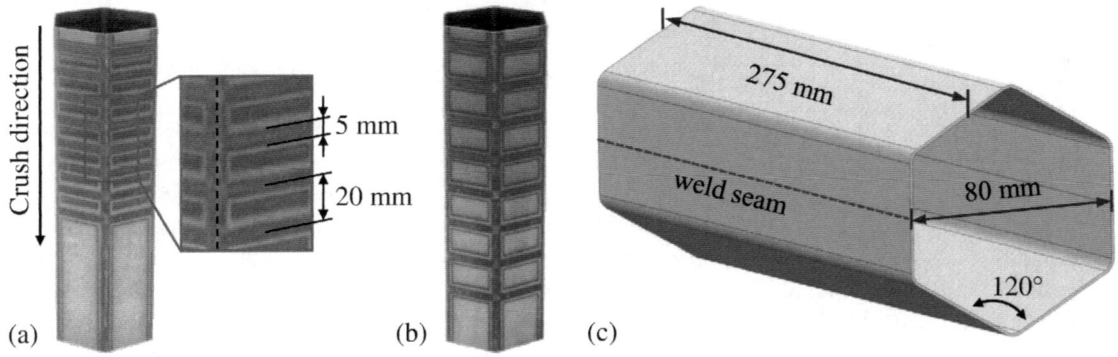

Figure 1. Locally heat-treated crash boxes made of Fe-28Mn-0.3C (a) with heat-treated zones in 20 mm distance (pattern A), (b) heat-treated zones in 30 mm distance (pattern B), (c) hexagonal crash box geometry.

As depicted in figure 1c, hexagonal crash boxes were manufactured. The plane, 25 % cold rolled sheets were heat-treated at the Fraunhofer Institute for Laser Technology (ILT) in Aachen using a fiber-coupled diode laser (Laserline LDF 12.000-100). In addition to the described patterns, the bending edges were locally heat-treated to avoid failure and to facilitate accurate bending. The laser beam width was 7 mm to make sure that the bent length of 5 mm is completely heat-treated. The feed rate for all heat-treated zones was 500 mm/min and the applied temperature was approx. 900 °C according to the results from the dilatometric study (see chapter 3.1). After the heat treatment, the sheets had to be roller levelled to eliminate the distortion resulting from the thermally introduced residual stresses. For each crash box, two square sheets were bent and then laser-welded at the ILT.

2.4. *Dynamic crash tests*

Crash tests of individual components are used to evaluate their deformation behavior under dynamic loading. The crash tests were performed at a test facility of the Forschungsgesellschaft Kraftfahrwesen mbH (fka). For pattern A, a test weight of 250 kg was dropped from a height of 2 m resulting in an impact energy of approx. 5 kJ. The drop height of 2 m was chosen to be able to compare the test results to the previously performed experiments. However, it is assumed that the strain hardened crash boxes are able to absorb even more energy. To use the full capacity of pattern B, which stretches over the whole crash box, a drop height of 3 m was applied in this case. This resulted in an impact energy of 8 kJ.

3. Results and discussion

3.1. Dilatometric heat treatment

Due to the high manganese content, the microstructure of Fe-28Mn-0.3C is austenitic at room temperature. Recovery and recrystallization are the only mechanisms that can restore the formability of the strain hardened strip during a heat treatment. The question is how the recrystallization proceeds at the investigated temperatures considering the degree of pre-deformation and the shortness of the heat treatment.

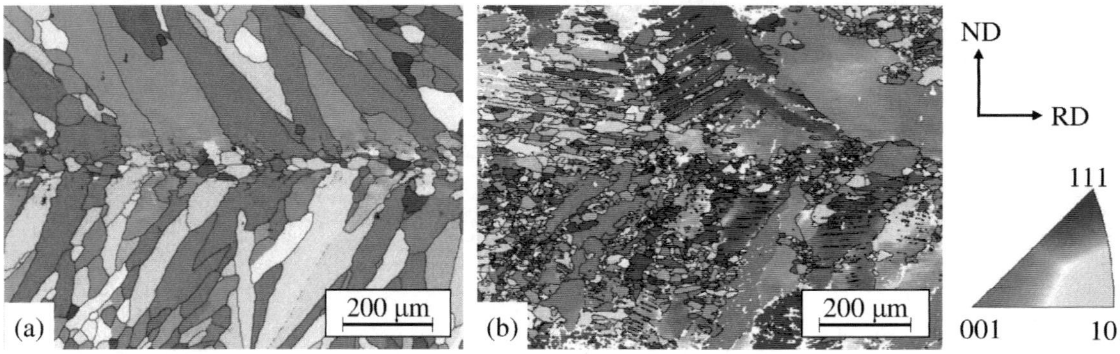

Figure 2. Microstructure of Fe-28Mn-0.3C. EBSD IPF mappings of the cross-section of (a) the as-cast and inline hot rolled strip and (b) the 25 % cold rolled sheet after the short-term heat treatment at 900 °C.

The EBSD mapping of the hot rolled Fe-28Mn-0.3C shows the typical solidification structure of strip-cast material (see figure 2a). The large grains are elongated and aligned according to the heat flow. On the other hand, the processed sample shows deformation twins and slip bands resulting from cold rolling (see figure 2b). However, the strain hardened structure has mostly dissolved after the short heat treatment at 900 °C and the microstructure is partially recrystallized. The cold rolling with 25 % height reduction is sufficient to trigger the recrystallization process, but non-recrystallized areas containing the initial hot strip microstructure are still apparent.

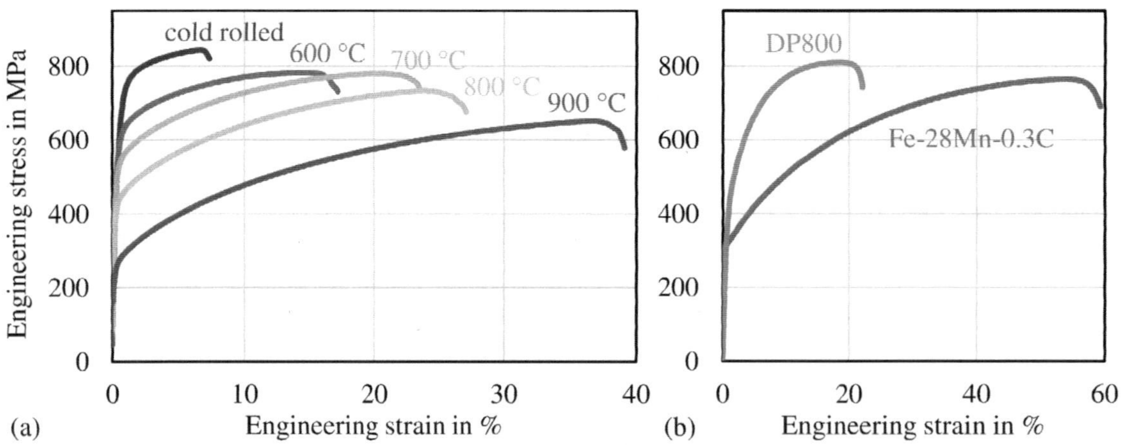

Figure 3. Tensile test results of (a) 25 % cold rolled Fe-28Mn-0.3C after different short-term heat treatments in the dilatometer and (b) fully recrystallized Fe-28Mn-0.3C compared to the dual phase steel DP800.

Figure 3a shows the mechanical properties under quasi-static loading after cold rolling and heat treatments at various temperatures. The cold rolled strip has an ultimate tensile strength of 844 MPa and a total elongation of 7 %. With advancing recovery and the onset of recrystallization, the strength decreases and the formability increases. After a heat treatment at 900 °C, the ultimate tensile strength equals the as-cast condition, i.e. 652 MPa [8]. The elongation, however, is with 39 % more than half restored as the EBSD mapping already suggested. Based on the results gained from the dilatometric study a local laser heat treatment of 900 °C is applied to restore a sufficient formability.

From other experiments it is known that a height reduction of 50 % in cold rolling is necessary to achieve a fully recrystallized microstructure during the short-time heat treatment at 900 °C. As aforementioned, the resulting formability of about 1 % is not high enough.

3.2. Crashworthiness

The mechanical properties shown in figure 3b help explain why the DP800 outperforms the high manganese steel in the dynamic collapse test. The fully recrystallized Fe-28Mn-0.3C cannot absorb enough energy within the crash relevant strain level compared to the industrially used dual phase steel. Using cold working, the initial yield stress and hence the energy absorption capacity of the high manganese steel can be enhanced. However, sufficient formability still needs to be provided for a stable deformation during crash. Consequently, the local heat treatment patterns were applied to soften the crash box in its highly strained areas.

Figure 4 shows the results of the dynamic crash test of the locally heat-treated crash box according to pattern A compared to the DP800 and a globally heat-treated crash box, also made of Fe-28Mn-0.3C. The latter was fully recrystallized after cold rolling. As intended, the strain hardening combined with the tailored local heat treatment improves the crashworthiness of the high manganese TWIP steel effectively. The energy of about 5 kJ was absorbed in a deformation path of only 85 mm. The fully recrystallized crash box needed almost twice the length, 166 mm, to absorb the same energy. Due to the local heat treatment, the high manganese steel outperforms the dual phase steel, which has a deformation path of 140 mm. The high strength of the strain hardened zones enables to consume more energy during the crash test, while the softened zones with its low yield stress allow for a stable crushing.

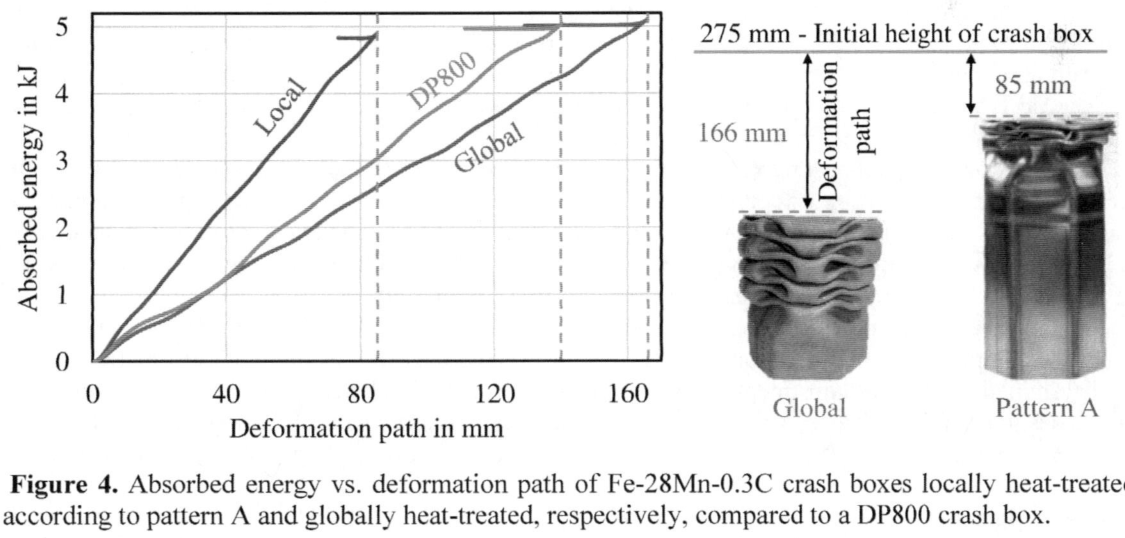

Figure 4. Absorbed energy vs. deformation path of Fe-28Mn-0.3C crash boxes locally heat-treated according to pattern A and globally heat-treated, respectively, compared to a DP800 crash box.

The influence of the softened zones become even more pronounced for the heat treatment pattern B. As visible in figure 5, a periodic folding according to the alternating material properties was realized. The softened zones serve as preferred location for the formation of folds in case of both heat treatment patterns. In contrast to pattern A, here, the "natural" distance of the folds, which is assumed to be determined by a purely geometrical effect, is shifted to higher values. While for the crash box in fully recrystallized condition a distance of 20 mm is observed, the distance of the folds could be successfully increased to 30 mm according to pattern B.

Both approaches reveal not only an enhanced crashworthiness but also completely new design possibilities for crash relevant structures based on local heat treatment. A locally heat-treated crash box with reduced sheet thickness absorbs the same energy compared to its fully recrystallized counterpart with thicker sheet. By decreasing the distance of the folds, more energy can be absorbed over the same length. Hence, lighter or smaller crash relevant structures can be applied to absorb the kinetic energy in case of a crash. Alternatively, the folding may be split in several predefined areas by introducing the softened zones discontinuously in the component.

Figure 5. Recordings of a high-speed camera during the crash test of the locally heat-treated Fe-28Mn-0.3C crash box with pattern B.

4. Conclusion

Due to its high work hardening rate, the investigated high manganese TWIP steel Fe-28Mn-0.3C promises a high energy absorption capacity in crash relevant components. The presented combination of strain hardening and local heat treatment makes use of the wide range of material properties of high manganese TWIP steel. In dilatometric experiments, it was shown that the recrystallization is triggered within a short-term heat treatment similar to laser heating.

By means of a tailored heat treatment the deformation behavior of crash boxes can be modified. On the one hand, the crashworthiness can be significantly improved by alternately introducing softened and hardened areas. Due to the higher formability within the softened areas, the crack formation is drastically reduced and a stable deformation progress ensured. As a result, the forming capacity of the material can be used to a larger extend and the energy absorption can be increased.

On the other hand, the local material softening serves to control and predict the crushing of a structural component. Both the distance of the folds as well as the preferred area of deformation can be predetermined by a local adjustment of the mechanical properties. Hence, crash relevant components can be designed and customized according to the constructive requirements. This demonstrates the great potential of a local laser heat treatment for the use of advanced high strength steels in lightweight design.

Acknowledgements

The authors gratefully acknowledge the financial support of the German Research Foundation DFG within the Collaborative Research Center (SFB) 761 "Steel – ab initio". Furthermore, the authors thank the Fraunhofer Institute for Laser Technology in Aachen for their support in the laser heat treatment and the Institute of Physical Metallurgy and Metals Physics of RWTH Aachen University for their support in EBSD mapping.

IOP Conf. Series: Materials Science and Engineering **159** (2016) 012024 doi:10.1088/1757-899X/159/1/012024

References

[1] Heuss R, Müller N, van Sintern W, Starke A, Tschiesner A and McKinsey&Company 2012 Lightweight, heavy impact - How carbon fiber and other lightweight materials will develop across industries and specifically in automotive *Advanced Industries*

[2] de Cooman B C, Kwon O and Chin K-G 2012 State-of-the-knowledge on TWIP steel *Mater. Sci. Tech.* **28** pp 513–27

[3] Kaupper M and Merklein M 2013 Bendability of advanced high strength steels - A new evaluation procedure *CIRP Ann. – Manuf. Tech.* **62** pp 247–50

[4] Merklein M, Johannes M, Lechner M and Kuppert A 2014 A review on tailored blanks - Production, applications and evaluation *J. Mater. Process. Tech.* **214** pp 151–64

[5] Frommeyer G, Brüx U and Neumann P 2003 Supra-ductile and high-strength manganese-TRIP/TWIP steels for high energy absorption purposes *Isij Int.* **43** pp 438–46

[6] Daamen M, Güvenç O, Bambach M and Hirt G 2014 Development of efficient production routes based on strip casting for advanced high strength steels for crash-relevant parts *CIRP Ann. – Manuf. Tech.* **63** pp 265–8

[7] Saeed-Akbari A, Mishra A K, Mayer J and Bleck W 2012 Characterization and prediction of flow behavior in high-manganese twinning induced plasticity steels: Part II. Jerky flow and instantaneous strain rate *Metall. Mater. Trans.* A **43** pp 1705–23

[8] Daamen M, Haase C, Dierdorf J, Molodov D A and Hirt G 2015 Twin-roll strip casting: A competitive alternative for the production of high-manganese steels with advanced mechanical properties *Mater. Sci. Eng.* A **627** pp 72–81

IOP Conf. Series: Materials Science and Engineering **159** (2016) 012025 doi:10.1088/1757-899X/159/1/012025

Press-hardening of zinc coated steel – characterization of a new material for a new process

T Kurz[1], P Larour[1], J Lackner[1], T Steck[1] and G Jesner[2]

[1] voestalpine Stahl GmbH, Research & Development – CTE, Business Unit Coil, voestalpine-Straße 3, 4020 Linz, Austria.
[2] BÖHLER Edelstahl GmbH & Co KG, Tooling – ETT, Mariazeller Strasse 25, 8605 Kapfenberg, Austria.

E-mail: thomas.kurz@voestalpine.com

Abstract. Press-hardening of zinc-coated PHS has been limited to the indirect process until a pre-cooling step was introduced before the hot forming to prevent liquid metal embrittlement. Even though that's only a minor change in the process itself it does not only eliminate LME, but increases also the demands on the base material especially in terms of hardenability or phase transformations at temperatures below 700 °C in general. This paper deals with the characterization of a modified zinc-coated material for press-hardening with pre-cooling that assures a robust process. The pre-cooling step itself and especially the transfer of the blank in the hot-forming die is more demanding than the standard 22MnB5 can stand to ensure full hardenability. Therefore the transformation behavior of the modified material is shown in CCT and TTT diagrams. Of the same importance are the changed hot forming temperature and flow curves for material at lower temperatures than typically used in direct hot forming. The resulting mechanical properties after hardening from tensile testing and bending tests are shown in detail. Finally some results from side impact crash tests and correlations of the findings with mechanical properties such as fracture elongation, tensile strength, VDA238 bending angle at maximum force as well as postuniform bending slope are given as well. Fracture elongation is shown to be of little help for damage prediction in side impact crash. Tensile strength and VDA bending properties enable however some accurate prediction of the PHS final damage behavior in bending dominated side impact load case.

1. Introduction

Direct hot forming is the dominating process for manufacturing the parts (Figure 1a). But as this process can, in case of zinc coated material that is more and more demanded for applications in the underbody, lead to liquid metal embrittlement (LME) induced cracking, zinc-coated PHS is primarily produced in the indirect process [1]. Nevertheless, an adaption of the conventional direct process with an integrated pre-cooling step between furnace and press (Figure 1b) can eliminate LME cracks [1]. Similar findings are reported in [2]. As this pre-cooling means lower cooling rates and/or longer times on the transfer the widely used 22MnB5 might be not sufficient to ensure hardenability or at least not the demanded robustness of the process regarding hardenability. Therefore an adapted steel grade is needed. This paper deals with the basic characterization of 20MnB8 that fulfills the needs for this changed process.

(a) Direct process for conventional hot stamping.

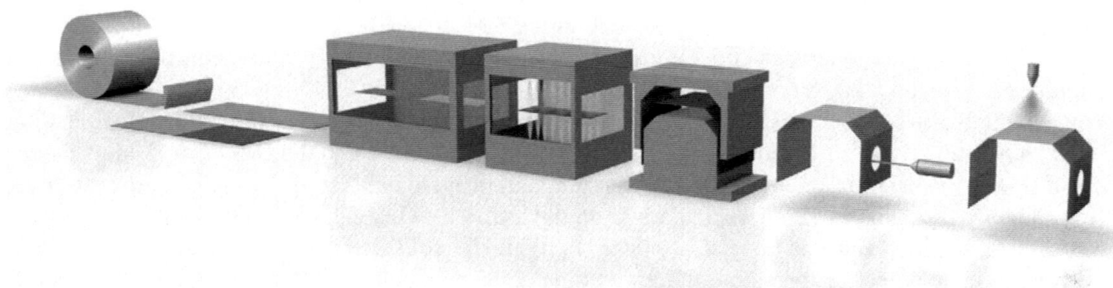

(b) Direct process with pre-cooling for hot stamping of zinc coated material.

Figure 1. Processes for press hardening: (a) Standard process for direct hot forming; (b) adapted process for zinc coated material with precooling before hot forming to prevent Liquid Metal Embrittlement (LME) micro-cracks [1].

Many efforts have been also put to correlate side impact behavior with small scale mechanical test results for press-hardening (PHS) sheet steel grades [3]. In particular 3-point-bending tests according to VDA238 standard [4], understanding all experimental artefacts influencing the bending angle determination [5], offer some more suitable evaluation criteria to assess the ability of press-hardened parts to deform plastically under crash conditions.

Side impact crash tests are carried out in the present investigation with new developed zinc-coated 20MnB8 press-hardening steels (phs-directform® 1500 GA) in comparison to zinc-coated 22MnB5 references (phs-ultraform® 1500 GI), in hardened condition with/without bake hardening heat treatment (170 °C, 20 min) applied to the material.

2. Material characterization

2.1. Chemical composition and hardenability

As can be seen in Table 1 the adapted material is still a manganese-boron type steel, but with increased amount of manganese (2 %) to ensure hardenability and slightly reduce the austenitization temperature needed. Additionally the carbon content has been slightly reduced to stay at the same strength level of around 1500 MPa that is typically specified by OEMs.

IOP Conf. Series: Materials Science and Engineering **159** (2016) 012025 doi:10.1088/1757-899X/159/1/012025

Table 1. Chemical composition (in weight percent) for the adapted steel composition 20MnB8 in comparison to a standard 22MnB5.

Steel grade	C	Si	Mn	P	S	Al	Cr	Mo	Ti	B	N
	%	%	%	%	%	%	%	%	%	%	%
22MnB5	0.220	0.20	1.22	0.010	0.0006	0.04	0.21	0.00	0.028	0.0027	0.0068
20MnB8	0.195	0.19	1.98	0.012	0.0006	0.05	0.03	0.00	0.030	0.0030	0.0054

The dilatometric studies have been carried out using a Bähr DIL 805A/D dilatometer. The samples had a sheet thickness of 1.5 mm and the size of 3.5 x 15 mm. Heating is done by inductive heating in this device with a heating rate of 25 K/s up to 900 °C for all experiments. After holding this temperature for 60 s the cooling cycle started with either 50 K/s for the isothermal TTT investigations to the isothermal holding temperature or the appropriate cooling rate for the continuous cooling investigations.

This adapted chemical composition significantly increases hardenability in terms of critical cooling rate (to reach almost 100 % martensite) from 40-50 K/s for the conventional 22MnB5 to 20 K/s for the 20MnB8 (Figure 2). But not only critical cooling rate is improved, but also the transformation at lower cooling rates is pushed to significantly lower temperatures. This significantly increases process robustness for a wide range of pre-cooling rates ranging from 20 K/s for free cooling to hundreds of K/s for contact cooling and pre-cooling temperatures of typically 450-650 °C. Both steel grades end up with about the same hardness level of approximately 465 HV1 for fully martensitic hardening.

The transfer from pre-cooling into press is even more critical in terms of phase transformation than cooling to pre-cooling temperature, as the transformation rate is significantly higher at pre-cooling temperatures of 450-650 °C as at 750 °C and above: This can easily be seen in Figure 3a for 22MnB5 where the isothermal transformation at e.g. 550 °C starts about 2 s after reaching the isothermal holding temperature.

Whereas for 20MnB8 the time to start the isothermal transformation at the same 550 °C is more than 20 s (Figure 3b) and therefore is long enough to transfer the sheet into the die and start hot forming without any transformation having started.

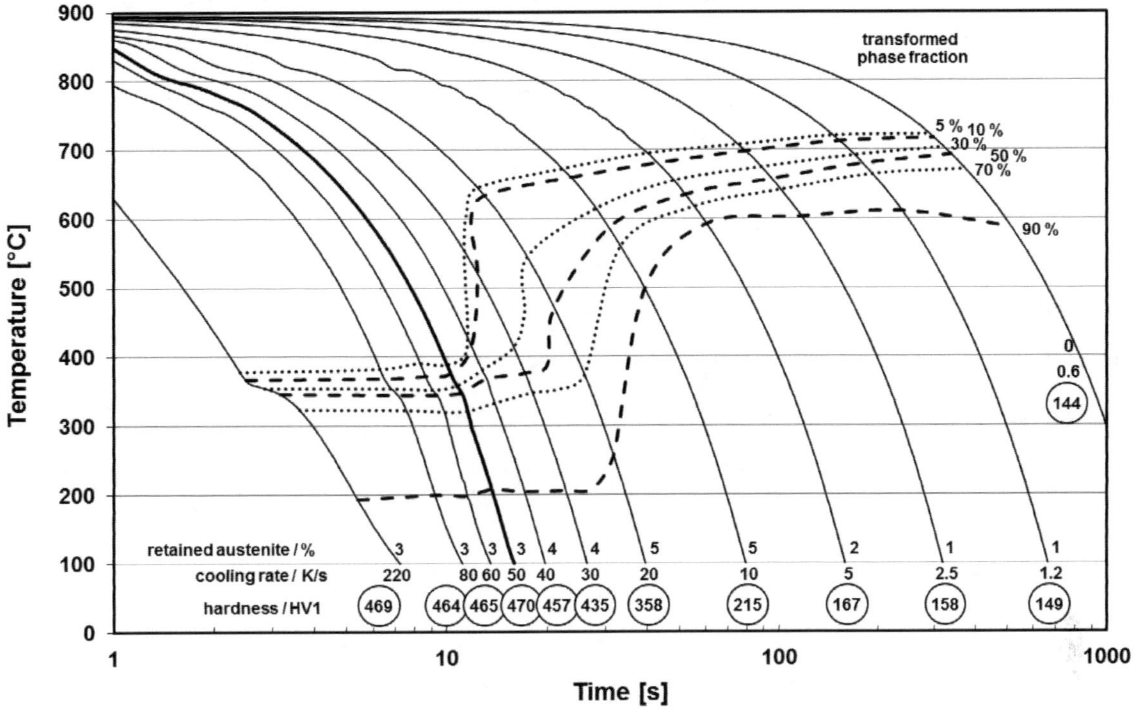

(a) CCT of 22MnB5

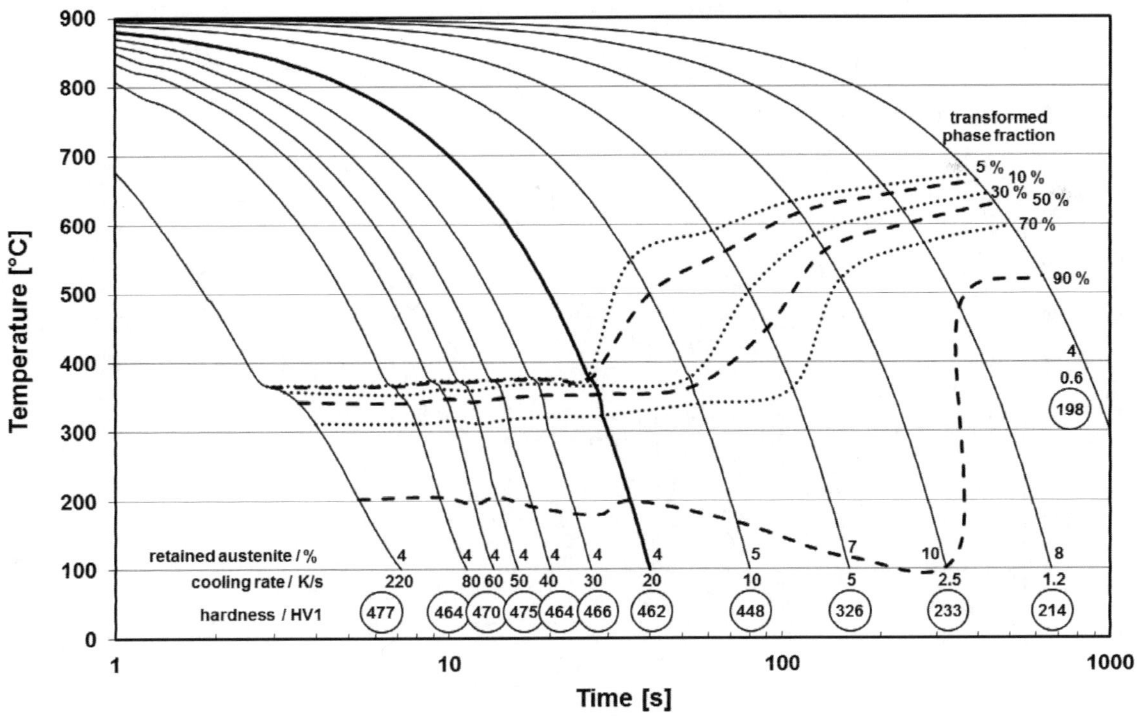

(b) CCT of 20MnB8

Figure 2. Continuous cooling transformation (CCT) diagram for a standard 22MnB5 (a) in comparison to the adapted steel composition 20MnB8 (b) for austenitization at 900 °C for 60 s.

IOP Conf. Series: Materials Science and Engineering **159** (2016) 012025 doi:10.1088/1757-899X/159/1/012025

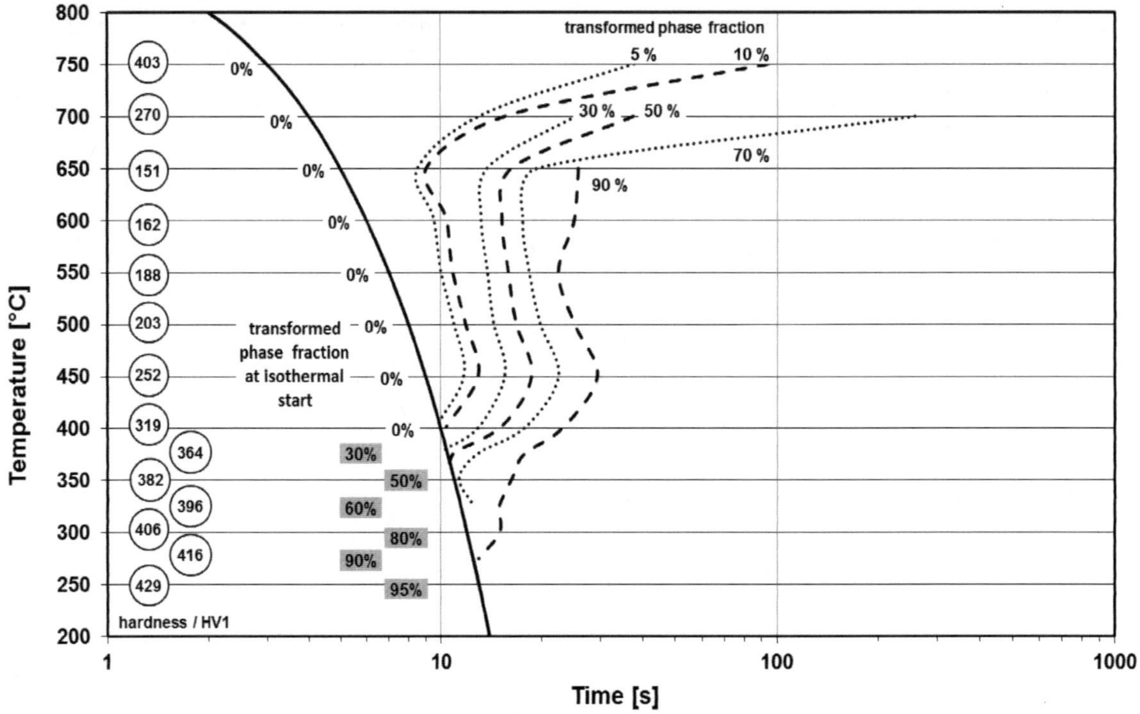

(a) TTT of 22MnB5

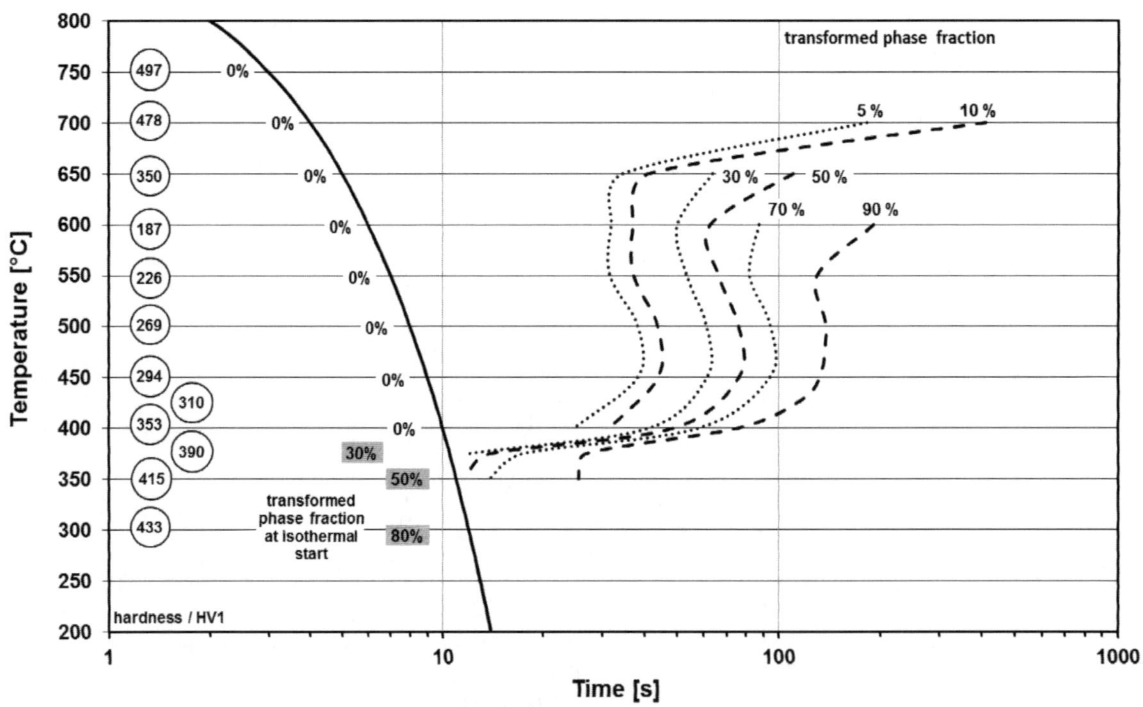

(b) TTT of 20MnB8

Figure 3. Isothermal transformation (TTT) diagram for a standard 22MnB5 (a) in comparison to the adapted steel composition 20MnB8 (b) with an initial cooling to isothermal holding temperature of 50 K/s.

2.2. Characterization of hot formability

For the simulation of the hot stamping process a Gleeble System 1500 was used. By a conductive heating in combination with compressed air nozzles it is possible to adjust the typical time temperature profile of the hot stamping process (Figure 4).

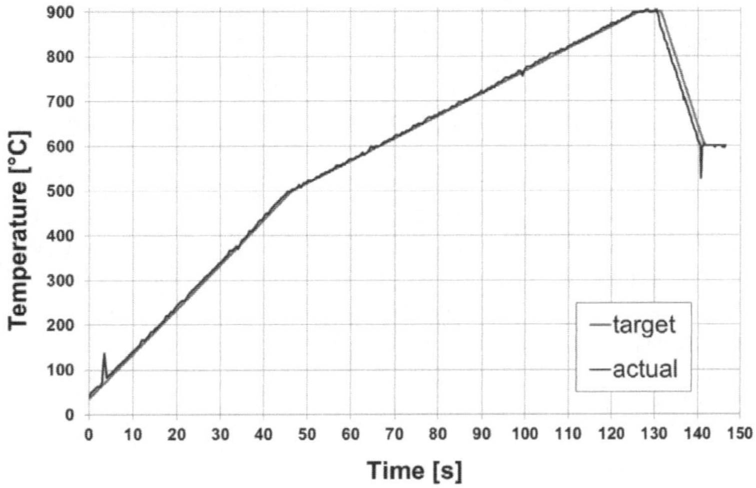

Figure 4. Typical time temperature profile used in the hot formability experiments for a deformation temperature of 600 °C, 20MnB8 GA.

Tests in a temperature range of 500 °C to 800 °C with steps of 50 °C and strain rates of 0.05, 0.5 and 5s^{-1} have been investigated. Each of the tests has been carried out at least three times. Temperature was controlled by Ni/Cr-Ni thermocouples, which were positioned by spot welding on the lower side of the tensile specimens. Based on the physical principle of conductive heating, a homogenous temperature range of about 15 mm can be realized in the middle of the specimen. However, the gradation of the temperature leads to an inhomogeneous material flow along the tensile specimens during the deformation [6].

In order to analyze the local strain distribution, the optical strain measurement system ARAMIS of the company GOM was used [7]. A stochastic pattern is applied on the samples before the process. Afterwards, the system calculates the deformation based on the displacement of single points. The geometry of the specimens for the test was designed according to the recommendation of EN482-2.

Figure 5 shows exemplarily the temperature dependency at 0.5 s^{-1} (Figure 5a) and the strain rate dependency at 600 °C (Figure 5b) for the 20MnB8 flow curves.

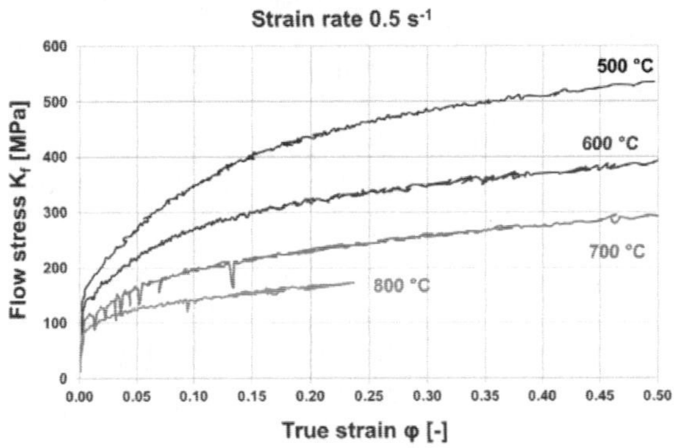

(a) temperature dependency

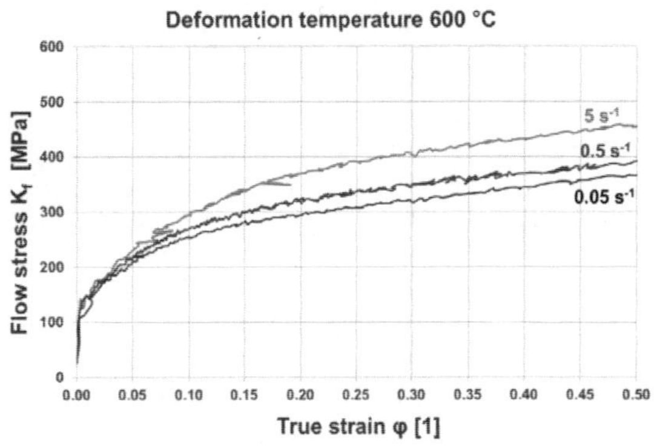

(b) strain rate dependency

Figure 5. Isothermal flow curves for 20MnB8 GA at 500 to 800 °C at a strain rate of 0.5 s^{-1} (a) and strain rate dependency at 600 °C (b).

2.3. Coefficient of friction and wear resistance

To obtain the optimum properties of the hot-formed parts, tool steels with improved physical properties are required. Especially high thermal conductivity should be achieved, but such steels are nearly not common used until now consequently. On the other hand a long tool life time is required and much more important than a further decrease of the cycle time.

To meet this request a sufficient high abrasive and adhesive wear resistance is demanded. Several hot work and cold work tool steels offer such properties and are available for hot stamping applications. In many cases neither only a high thermal conductivity nor only a high abrasive and adhesive wear resistance fulfills the requirements. Therefore, tool steels which possess both properties on a high level are essential.

Based on those demands several Böhler cold- and hot work-tool steels are tested in a tribological lab test to assess the wear resistance and the friction behavior for hot stamping of zinc-coated press-hardening steels. In this laboratory test the austenitization of the steel sheets and the transportation are set similar to a real hot stamping process. The determination of the coefficient of friction was performed in a strip drawing tensile test. The tested material had a thickness of 1.5 mm and was heated to 900 °C in a radiation furnace before being transferred to the friction test device in a fully automatic test facility.

In Figure 6a, the coefficient of friction is plotted vs. different tool steels for GA-coated steel sheets. Böhler K353 gives the lowest coefficient of friction for the investigated tool steels, followed by Böhler W360 ISOBLOC and Böhler W300 for GA 90/90 g/m^2 coated material. Furthermore the volume loss (abrasive wear resistance) of the different tool steel grades was determined by using the white light interferometer as shown in Figure 6b.

Considering the results of the coefficient of friction and the volume loss Böhler K353 and Böhler W360 ISOBLOC show the best performance for direct hot stamping of zinc coated press-hardening steel (e.g. phs-directform® 1500).

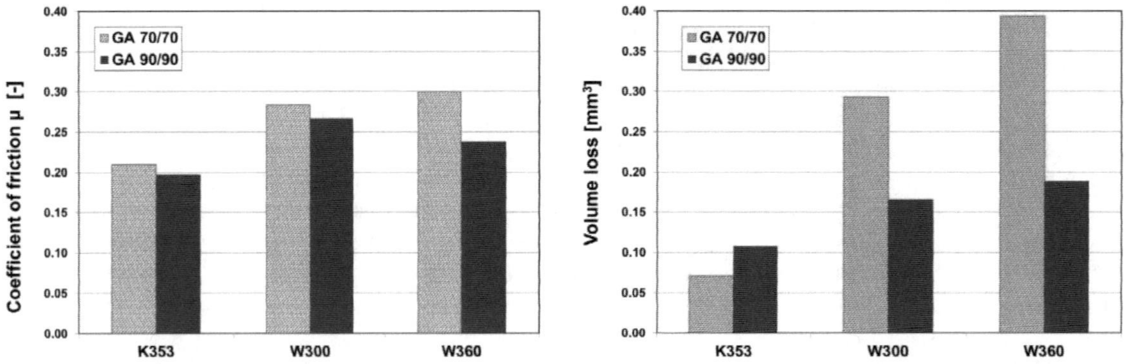

(a) coefficient of friction (b) volume loss

Figure 6. Coefficient of friction (a) and volume loss (b) of zinc coated press-hardening steels versus various available Böhler tool steels from strip drawing tests. Test conditions: normal force: 4 kN, surface pressure 8 MPa, test speed: 100 mm/s, 10 replicates.

2.4. Mechanical testing of investigated materials

The mechanical properties from ISO6892 tensile tests and VDA238-100 bending tests (Figure 7a) have been determined directly on untested crash samples in the bottom region in hardened as well as hardened and bake hardened condition (BH: 170 °C, 20 min). The crash sample 900 mm length is set transverse to rolling direction.

Tensile test are taken along crash sample axis transverse to rolling direction (T). The bending angle at maximum force (α_{Fmax}) and the post uniform bending slope (absolute value, |PUBS|) between maximum force and inflection point have been determined as illustrated in Figure 7b.

Crash fold axis and VDA bending line axis along rolling direction are most suitable for best crash-bending test correlation results (Figure 7c). Table 2 summarizes the mechanical properties of the investigated steel grades.

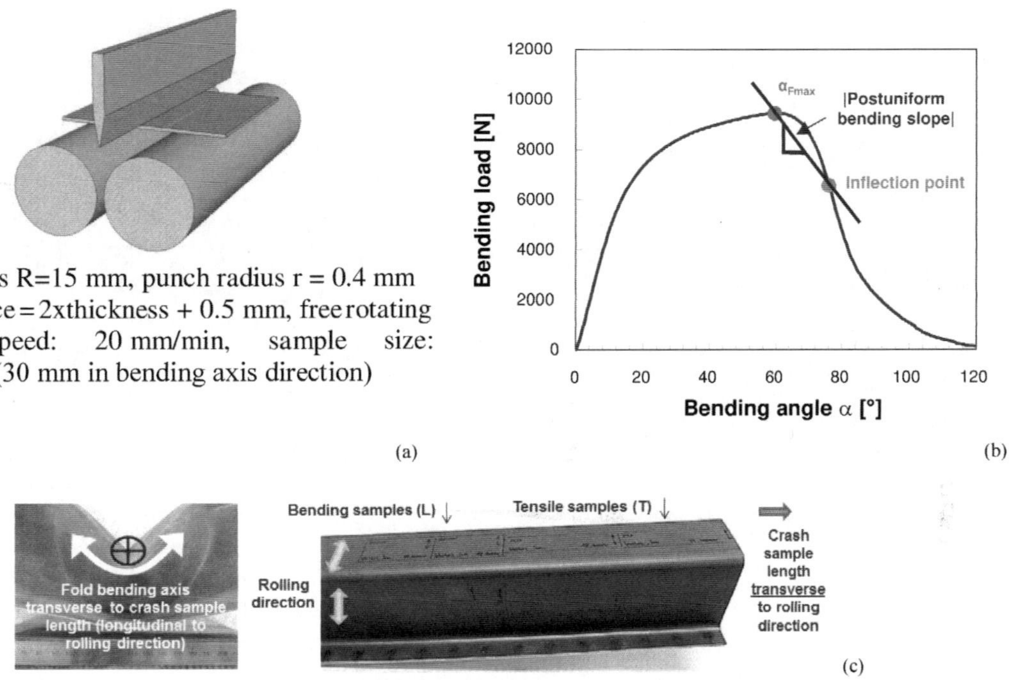

Figure 7. VDA238-100 bending test set-up: (a) Test set-up parameter; (b) Bending curve with α_{Fmax} and post uniform bending slope PUBS; (c) Sample location on (untested) crash samples bottom.

Table 2. Mechanical properties of investigated steel grades.

Material			ISO 6892 Tensile test					VDA238 Bending test		
Steel grade	Condition	Thick-ness mm	Direc tion	R_{p02} MPa	R_m MPa	A_g %	A_{25} %	Direc tion	α_{Fmax} °	|PUBS| N/°
20MnB8 GA direct	hardened	1.50	T	1039	1500	5	10	L	52	159
	hardened + BH	1.49	T	1165	1502	4	10	L	60	107
20MnB8 GA indirect	hardened	1.49	T	1029	1504	4	10	L	56	105
	hardened + BH	1.48	T	1152	1494	4	10	L	59	81
22MnB5 GI indirect-1	hardened	1.46	T	1185	1586	3	8	L	55	121
	hardened + BH	1.60	T	1180	1516	4	9	L	61	83
22MnB5 GI indirect-2	hardened	1.48	T	1028	1449	4	10	L	73	80
	hardened + BH	1.48	T	1132	1450	4	8	L	73	64

2.5. Side impact crash test setup & methodology

3-point side impact crash tests with 4 to 14 individual crash samples for each material condition are performed with an initial impactor speed in the range 20-30 km/h with an impactor mass of 86 kg, resulting in 40 mm up to 211 mm maximum bending deflection depth (Figure 8).

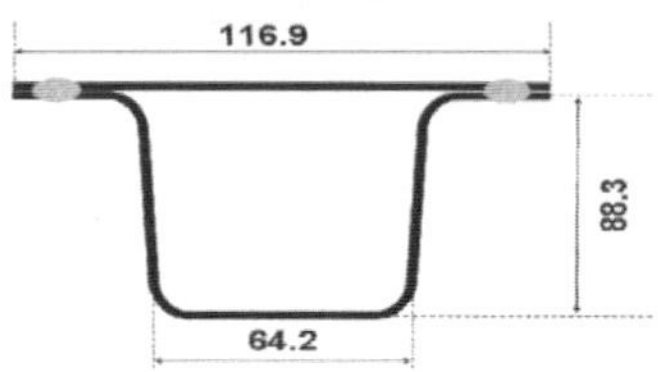

Load mass: 86 kg, span: 700 mm
Testing speed: 20~30 km/h
Sample length: 900 mm transverse to roll direction.
Impactor radius: 127 mm. Max. displacement: up to 211 mm (from video in loaded condition).
Spot welded bottom sheet out of HC340LAD, 1.5 mm.

Figure 8. Side impact testing set-up, test parameters.

As previously introduced in [3], a side impact crashindex (CI) as given in equation (1) is introduced for damage assessment after crash test, which is determined based on the normalized total crack length observed on tested crash boxes (Figure 9). The overall crack length is measured after crash test on side impact samples. The total small superficial surface crack length L_S is less weighted (20 %) in comparison to the total through-thickness big cracks length L_B ($L_S+L_B\leq 250$ mm):

$$\text{Crashindex}_{\text{sideimpact}}=\left(0.2\times\left(1-\frac{L_S}{250}\right)+1.0\times\left(1-\frac{L_B}{250}\right)-0.2\right)\times 100 \tag{1}$$

Best case: no crack ⇒ CI = 100 %
Worst case: crack length L_B= 250mm, L_S=0 ⇒ CI = 0 %

Crashindex: calculation example

Figure 9. Side impact crack length (L_B: big crack length, L_S: small superficial crack length).

Cracks start from bending fold sides in the flange area and propagate then in the bending fold bottom more or less depending on material bending damage resistance ability. The crack pattern is often not symmetrical. This does not matter for crashindex calculation, since only the overall crack length is taken into account. The crack length can be determined manually fast and easily with ± 5 mm precision on outer surface of crashed samples after testing. Material inherent stochastic crack behavior results finally results to a ± 5-10 scattering in crashindex values. This still allows some accurate quantitative assessment of crash damage amount with at least 5 crash replicates. Both through-thickness and small superficial cracks coexist and are included in equation (1) for crashindex calculation.

The maximum bending displacement before springback for each crash test is determined from high speed video recording (Figure 10). The side impact crash test is performed here in a 3-point-bending configuration with free ends and ~ 250 mm punch diameter. An early loss of contact between crash sample and impactor punch takes place (Figure 10). This gap occurs when the inner fold radius of the crash specimen is smaller than the punch radius, similar to VDA238 bending test at high angle (Figure 10). Figure 11 shows the fold bending angle of crash samples measured after crash test and springback. The fold bending angle is defined the same way as for VDA bending test (Figure 11).

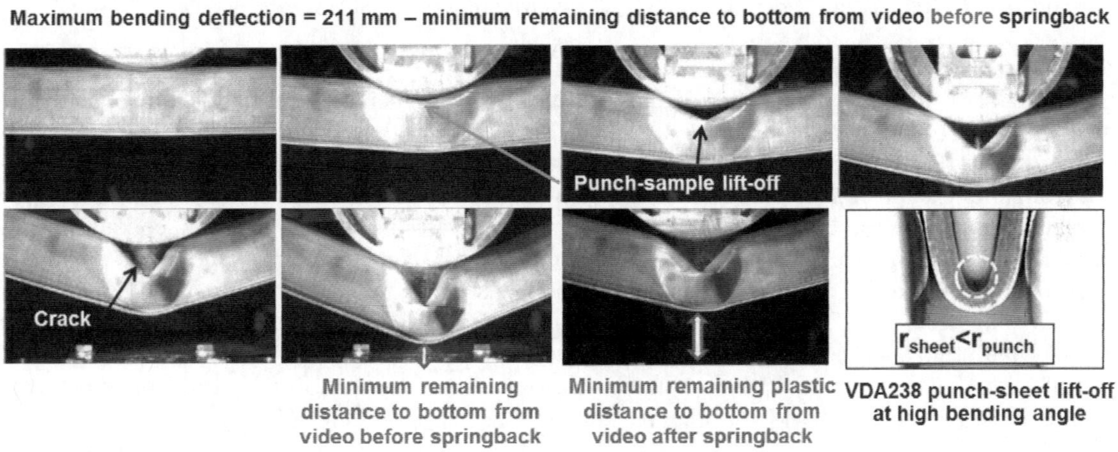

Figure 10. Crash impact video snapshot test development stages (lift off, failure, springback).

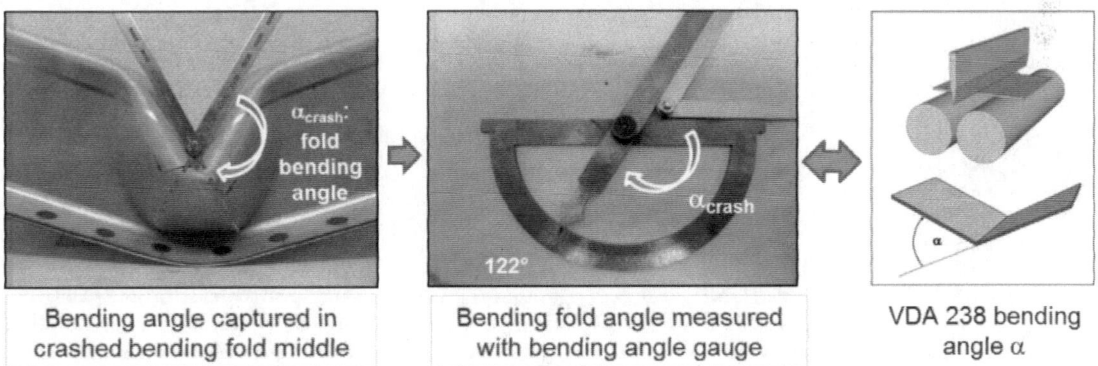

Figure 11. Bending fold angle measurement method on crashed sample after springback.

The crashindex is then plotted versus maximum bending deflection δ from video (Figure 12a). The resulting crashindex-displacement curves are interpolated at 150 mm maximum bending displacement ($CI_{\delta=150\,mm}$). The displacement at 100 % crashindex ($\delta_{Ci=100\,\%}$) is also determined. As shown in Figure 12b, the crashindex is plotted vs. side impact bending fold angle, as previously defined in Figure 11. The fold bending angle at first crack initiation (crashindex CI=100 %) is then interpolated.

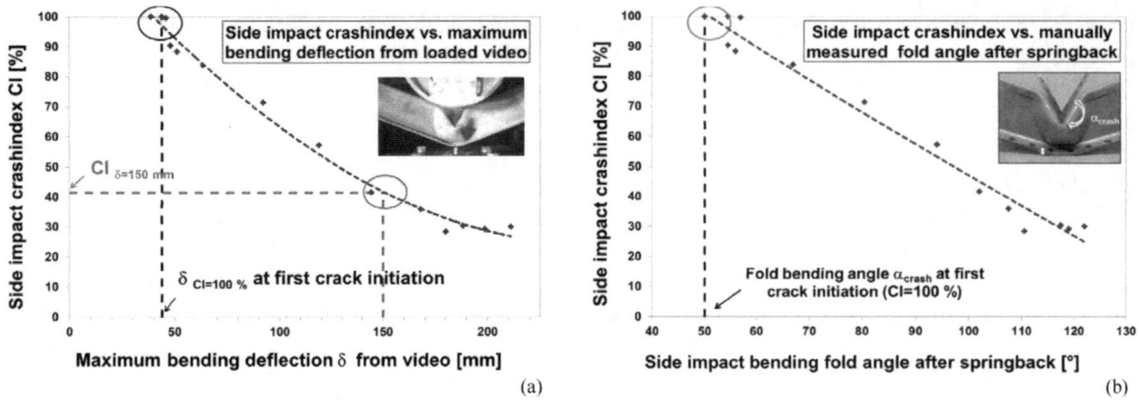

Figure 12. Crashindex test methods: a) Crashindex at 150 mm bending deflection $CI_{\delta=150\ mm}$ and deflection at 100 % crashindex $\delta_{Ci=100\ \%}$; b) Crash fold angle at 100 % crashindex $\alpha_{crash,CI=100\ \%}$.

2.6. Side impact crash test results

Figure 13 shows the side impact crashindex versus maximum bending deflection (40 to 210 mm deflection) before springback measured from video film in the hardened (Figure 13a) as well as hardened with 170 °C/20 min bake hardening conditions (Figure 13b). The bake hardening heat treatment shifts the crashindex curve up to a higher level. No statistically relevant difference can be observed between direct and indirect 20MnB8 materials crash behavior in any condition. The 22MnB5-2 material shows some much better crashindex level than 22MnB5-1 material, it should be however rather considered as an atypical positive outliner with an unusual bending angle in the upper range >70°. 22MnB5-1 indirect material shows some more typical bending angle (55°) and crashindex behavior. With this regard, no significant differences can be noticed between 22MnB5 indirect-1 and 20MnB8 direct/indirect material crash damage behavior. Some scattering in 22MnB5 material crash behavior should be expected, which is directly linked to the variation range of α_{Fmax}.

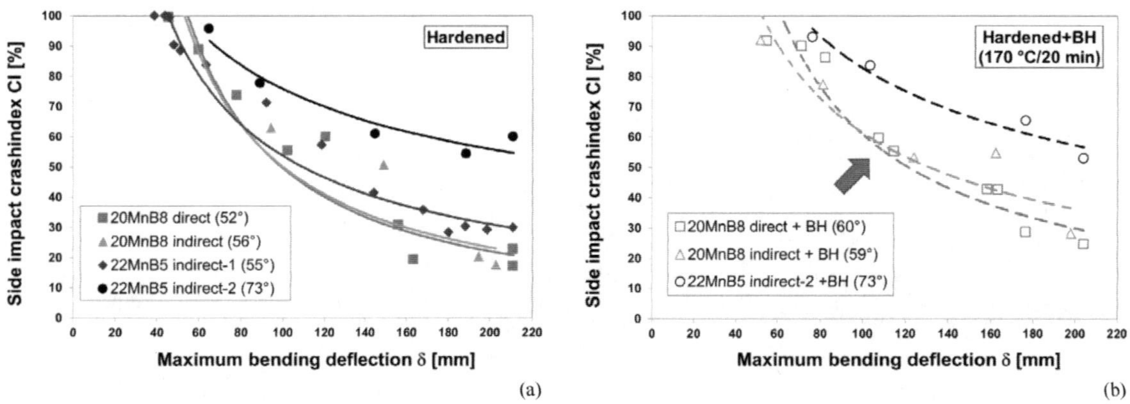

Figure 13. Crashindex vs. maximum bending deflection curves: a) hardened condition; b) hardened + BH (170 °C/20 min) condition.

Figure 14 shows the side impact crashindex versus crash bending fold angle after springback measured on crashed sample after testing according to procedure described in Figure 11, in both hardened (Figure 14a) as well as hardened & bake hardened conditions (Figure 14b). The crashindex behavior is quite similar when plotted versus maximum video bending deflection or versus bending fold angle after springback. The bending fold angle reaches up to 120-130° maximum after springback. The initial bending angle at fracture initiation (CI≤100 %) amounts to 50/70° in the hardened and 60/80° in the hardened & bake hardened conditions, which is exactly in the same range as the VDA238 bending angle α_{Fmax} according to Table 2.

IOP Conf. Series: Materials Science and Engineering **159** (2016) 012025 doi:10.1088/1757-899X/159/1/012025

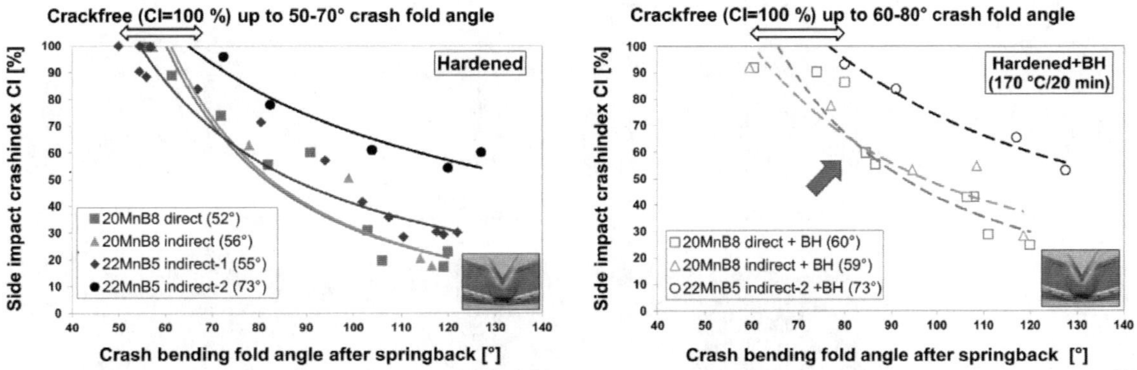

(a) (b)

Figure 14. Crashindex vs. crash bending fold angle after springback: a) hardened condition; b) hardened + BH (170 °C/20 min) condition.

Figure 15a shows the relation between VDA238 bending angle α_{Fmax} and the crash bending fold angle after springback α_{crash} at crack initiation. The correlation analysis is performed with present investigation results together with voestalpine PHS coated (GI, GA) 1.5 mm side impact database results. A direct correlation with almost the same absolute numerical angle values can be observed between VDA238 bending angle at maximum force α_{Fmax} and bending fold angle at crack initiation α_{crash}. The same type of bending loading with the same bending angle definition (Figure 11) is therefore observed between both testing. This explains the strong almost redundant correlation between bulk material bending properties and side impact crash behavior.

Figure 15b shows the evolution of maximum crash bending deflection from video with increasing crash fold bending angle after springback measured on crashed samples. The experimental results for all crash tests have been gathered from this investigation with the same initial 1.5 mm material thickness. The amount of springback between maximum bending deflection from video and remaining plastic bending deflection on crash samples after testing is also plotted versus crash fold bending angle. Some almost linear dependency of crash bending deflection and springback amount with crash fold bending angle can be observed. The amount of springback also increases up to 100 mm at 120-130° bending fold angle after springback (210 mm maximum video bending deflection). This indicates some considerable springback amount up to ≈50 % for press hardening steels with $R_{p0.2} > 1000$ MPa.

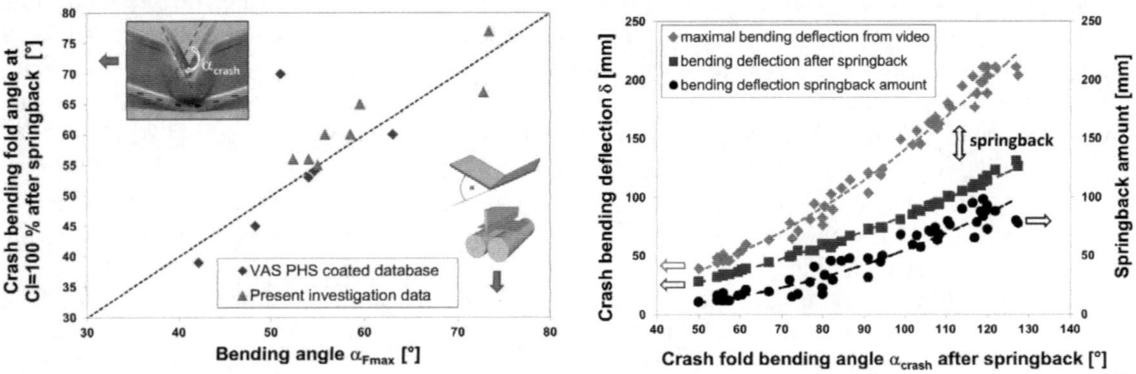

(a) (b)

Figure 15. a) Crash bending fold angle at crack initiation after springback vs. bending angle α_{Fmax}; b) Crash bending deflection (maximal from video/after springback) and deflection springback amount vs. crash fold bending angle α_{crash} after springback (present investigation data).

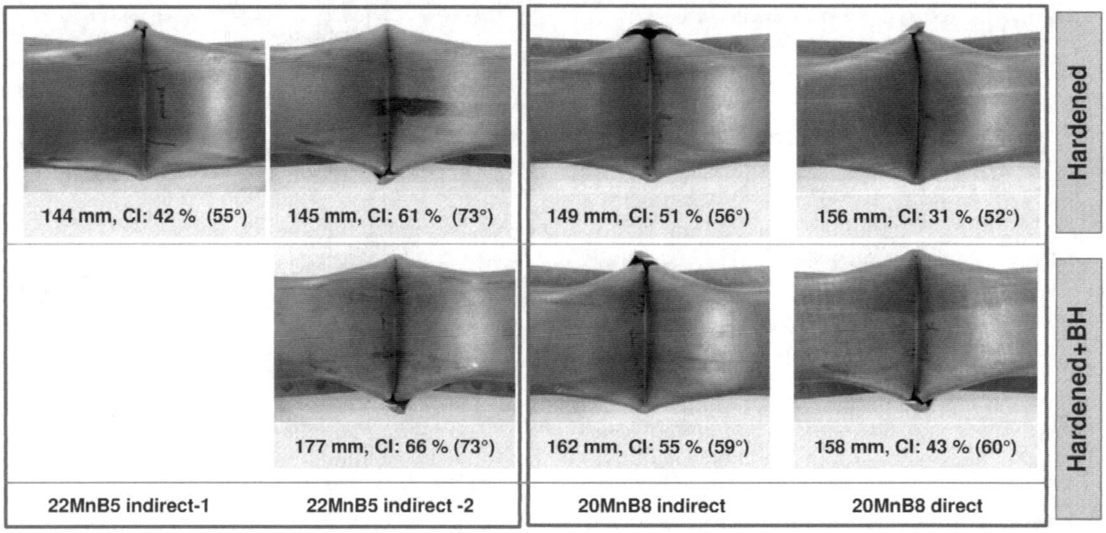

Figure 16. Crashindex around 150 mm bending deflection in hardened & hardened + BH conditions.

Figure 16 shows some representative crashed sample pictures from investigated materials for crash tests around 150 mm maximum bending video deflection. Major differences in crashindex values are obtained based on crack length in the bending fold bottom.

3. Discussion

The crashindex at δ=150 mm bending deflection ($CI_{\delta=150\,mm}$) as well as the bending deflection for CI=100 % ($\delta_{Ci100\,\%}$) at crack initiation are plotted for the present investigation results (a), as well as compared to voestalpine (VAS) side impact database results (b). The voestalpine internal side impact crash tests database relies on the same test set up and includes AHSS, UHSS as well as PHS materials in a broad range of tensile strength and tempering conditions for a better statistical validation.

$CI_{\delta=150\,mm}$ and $\delta_{Ci100\,\%}$ are plotted respectively versus tensile fracture elongation (Figure 17), tensile strength R_m (Figure 18), VDA238 bending angle α_{Fmax} (Figure 19) as well as bending post uniform slope |PUBS| (Figure 20). Both $\delta_{Ci100\,\%}$ at crack initiation and $CI_{\delta=150\,mm}$ follow the same trends.

No specific dependency can be observed for $\delta_{Ci100\,\%}$ or $CI_{\delta=150\,mm}$ with tensile fracture elongation (Figure 17a). No correlation can be recognized either for $CI_{\delta=150\,mm}$ based on the extended VAS database results (Figure 17b). The higher the tensile strength, the lower the $\delta_{Ci100\,\%}$, or $CI_{\delta=150\,mm}$ values (Figure 18a). $CI_{\delta=150\,mm}$ values decrease linearly from ~100 % around 1200 MPa down to ~0 % around 1900 MPa tensile strength (Figure 18b). $\delta_{Ci100\,\%}$ and $CI_{\delta=150\,mm}$ increase with increasing bending angle α_{Fmax} (Figure 19a). $CI_{\delta=150\,mm}$ increases linearly from ~0 % at 40° up to ~100 % around 80° α_{Fmax} bending angle (Figure 19b). $CI_{\delta=150\,mm}$ and $\delta_{Ci100\,\%}$ also decrease with increasing absolute bending postuniform slope |PUBS| (Figure 20a). $CI_{\delta=150\,mm}$ is following an inverse square root law vs. |PUBS| based on a broader statistic (Figure 20b).

The tensile fracture elongation A is therefore of little help for crashindex prediction. The knowledge of tensile strength R_m allows some rough positioning of the side impact crashindex level. Some more accurate prediction of crash damage behavior can be reached knowing the combination of bending angle α_{Fmax} and corresponding postuniform bending slope |PUBS|. This illustrates the need for VDA238 bending test, which is suitable for side impact crash damage prediction for sheet steels materials with $R_m \geq 1200$MPa and $\alpha_{Fmax} \leq 80°$.

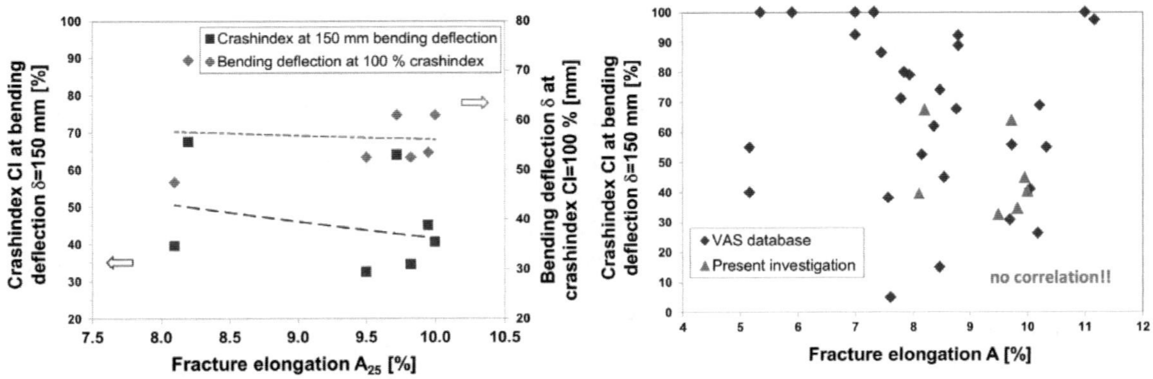

a) b)

Figure 17. a) $CI_{\delta=150\ mm}$ and $\delta_{Ci100\ \%}$ vs. tensile fracture elongation A_{25} (present investigation); b) $CI_{\delta=150\ mm}$ vs. tensile fracture elongation (VAS side impact database).

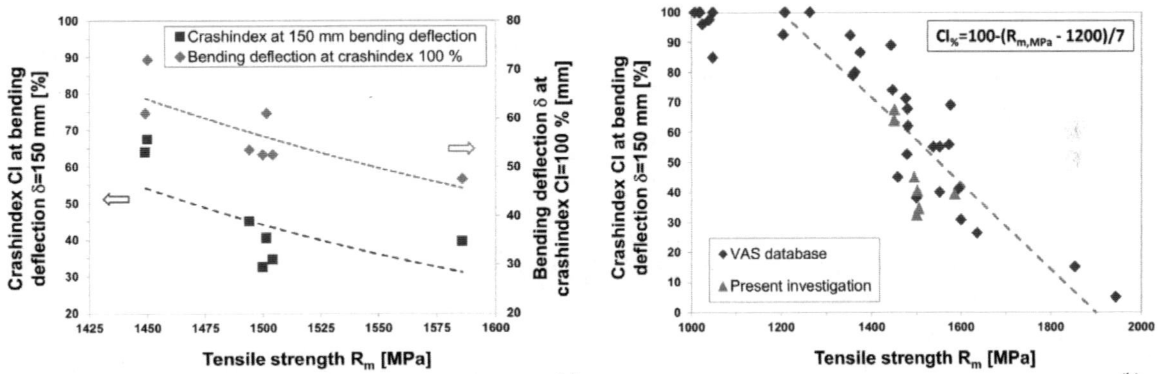

(a) (b)

Figure 18. a) $CI_{\delta=150\ mm}$ and $\delta_{Ci100\ \%}$ vs. tensile strength R_m (present investigation); b) $CI_{\delta=150\ mm}$ vs. tensile strength R_m (VAS side impact database).

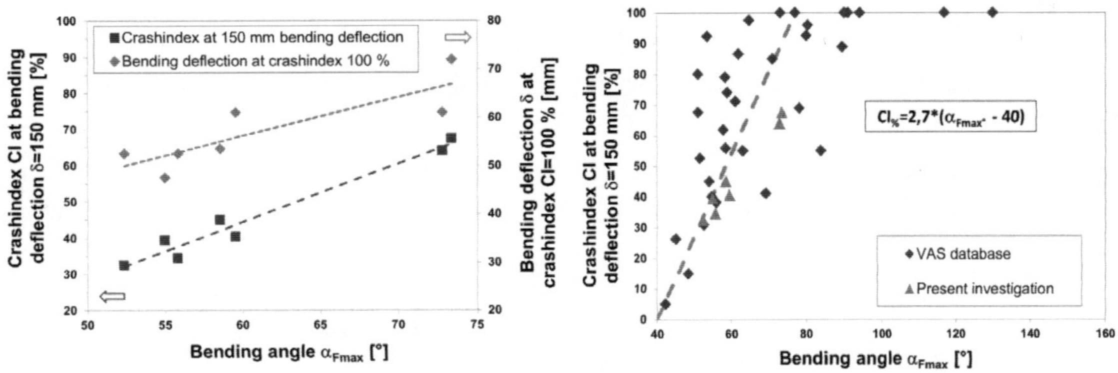

a) b)

Figure 19. a) $CI_{\delta=150\ mm}$ and $\delta_{Ci100\ \%}$ vs. VDA238 bending angle α_{Fmax} (present investigation); b) $CI_{\delta=150\ mm}$ vs. VDA238 bending angle α_{Fmax} (VAS side impact database).

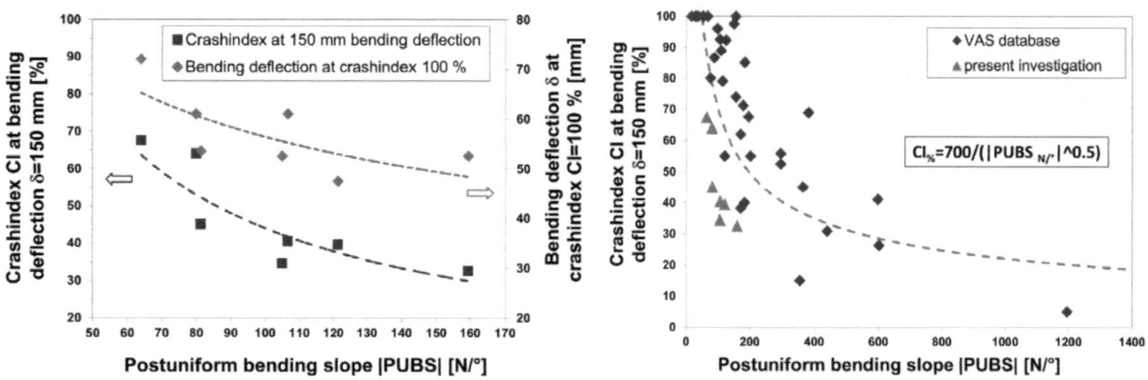

(a) (b)

Figure 20. a) $CI_{\delta=150\,mm}$ and $\delta_{Ci100\,\%}$ vs. postuniform bending slope |PUBS| (present investigation); b) $CI_{\delta=150\,mm}$ vs. postuniform bending slope |PUBS| (VAS side impact database).

4. Conclusions

As direct hot forming for zinc-coated press-hardening steel demands pre-cooling to deal with liquid metal embrittlement with higher demands regarding delayed transformation than that from 22MnB5, an adapted steel composition 20MnB8 was investigated.

CCT and TTT diagrams both show significantly increased stability of 20MnB8 to implement a wide range of possible pre-cooling methods and transfer technologies within a robust process.

Some information about the strain rate (0.05 to 5 s^{-1}) and temperature dependency (500 to 800 °C) of the 20MnB8 flow curves is also given.

Considering the results of the coefficient of friction and the volume loss Böhler K353 and Böhler W360 ISOBLOC show the best performance for direct hot stamping of zinc coated press-hardening steel (e.g. phs-directform® 1500).

Side impact crash tests performed on industrially produced crash members of 20MnB8 show a consistent crash behavior with 22MnB5. The strength level of all parts was at about 1500 MPa ultimate tensile strength. With the materials picked for the tests a wide range of bending angles – including ones with bake hardening applied to the crash members – has been observed to find the bending angle to be the best correlating laboratory test to predict crash performance for press-hardening steels.

This makes 3-point bending to an important influencing factor in predicting crash behavior and optimizing material properties. Increased bending angles do result directly in reduced cracking in crash tests or higher crash indices respectively.

Acknowledgements
Financial support from the Austrian Research Promotion Agency (FFG) is gratefully acknowledged.

References

[1] Kurz T, Schwinghammer H, Luckeneder G, Manzenreiter T and Sommer A 2015 Zinc Coated Press-hardening Steel – Challenges and Solutions *5th Int. Conf. Hot sheet metal forming of high performance steel CHS² (Toronto, Ont., Canada, May 31-June 03)* pp 345-354.

[2] Hensen G, Beentjes P, Abspoel M and Loiseaux J 2015 Unlocking the potential of zinc coated steel for hot forming by innovative process modifications *5th Int. Conf. Hot sheet metal forming of high performance steel CHS² (Toronto, Ont., Canada, May 31-June 03)* pp 85-92.

[3] Larour P, Naito J, Pichler A, Kurz T and Murakami T 2015 Side impact crash behavior of press-hardened steels-correlation with mechanical properties. *5th Int. Conf. Hot sheet metal forming of high performance steel CHS² (Toronto, Ont., Canada, May 31-June 3)* pp 281-289.

[4] VDA 238-100 test specification draft: Plate bending test for metallic materials. 12/2010

[5] Larour P, Hackl B and Leomann F 2013 Sensitivity analysis on the calculated bending angle in the instrumented bending test. *Proc. Int. Conf. IDDRG (Zurich, Switzerland, June 02-05)* pp 317-322.

[6] Lechler J 2008 Beschreibung und Modellierung des Werkstoffverhaltens von presshärtbaren Bor-Manganstählen. *Doctoral thesis, Fertigungstechnik-Erlangen, Germany, ISBN 978-3-87525-286-6.*

[7] Hagenah H, Merklein M, Lechner M, Schaub A and Lutz S 2015 Determination of the mechanical properties of hot stamped parts from numerical simulations. *9th CIRP Conf. on Intelligent Computation in Manufacturing Engineering, Procedia CIRP 33* pp 167-172.

Hot stamping of AA7075 aluminum sheets

J Mendiguren[1], E Saenz de Argandoña[1] and L Galdos[1]
[1]Mechanical and Manufacturing department, Mondragon University, Loramendi 4, 20500 Arrasate-Mondragon, Spain

Abstract. In this work the formability of a high strength aluminium alloy (AA7075-T6) for the stamping of an automotive component has been studied. Due to the low formability of the selected alloy, two different heat assisted forming strategies have been analysed. On the one hand, the W-temper process, where the thermal process is carried out prior to the forming operation. On the other hand, the hot stamping process, where the thermal process is carried out at the same time as the forming. The results showed that both technology were able to form the component avoiding any failure of the material. On the contrary, both processes reduced the final mechanical properties of the material compared to the as received material condition. However, the obtained mechanical properties doubled the strength of commonly used 5xxx and 6xxx aluminium alloys.

1. Introduction

In the last years, the automotive industry is facing an enormous challenge to lightweight the current cars mass and for complying with the new environmental and safety regulations. The weight of the car has to be reduced in order to reduce the fuel consumption and the CO_2 emissions while the car body has to be improved to overtake the new safety standards.

These objectives have been achieved by the introduction of new materials in the car body. New cold formable high strength steels are being developed with strength levels of around 1200 MPa. In the same context, carbon fiber is being used in high-class cars such as BMW 7 Series. Other OEMs, such as Audi and Jaguar Land Rover, have decided to push the use of aluminium alloys in order to make use of the good strength to weight ratio they present.

Medium strength 5xxx and 6xxx aluminium families are being already used in the body in white and skin parts of several commercial cars. However, the strength to weight ratio of these families is not yet the optimum one for security crash components and to get the desired accepted cost per kg saved. In this regard, the 7xxx family is the next potential aluminium series to be introduced on the market. The main disadvantage of this family is its reduced formability at room temperature.

One of the alternatives to increase the formability of the material is to increase the forming temperature. However, being an age-hardenable family, the increase of temperature during forming degrades the final mechanical properties of the material. Regarding the forming process, Lee et al. [1] studied the warm hydroformability of 7075 tubes between room temperature to 300ºC. The results showed that sufficient formability of high strength aluminium alloys could be achieved by the selection of the pertinent pre-treatment conditions and deformation temperatures. Wang et al. [2] presented a paper where material characterization and LDR tests were performed using the 7075-T6 material. They found that total elongation at fracture increased between 140 ºC and 220 ºC due to the increase in strain rate sensitivity, which controls diffuse necking and prevents plastic strain from concentrating in a localized neck.

Regarding the possible technologies to stamp the high strength aluminium alloys, three different technologies can be found: the warm forming, the hot stamping and the W-temper forming. On the warm forming, both blank and dies are heated during the process [2-3]. However, the heating of the dies critically increases the forming costs and set-up and maintenance aspects. The hot stamping process is being widely used on boron steels for safety components [4]. In this process the blank is heated in a furnace and then transferred to the press where the forming is performed with dies at room temperature, significantly reducing the cost of the process [5-6]. When using boron steels, the dies are kept closed after the forming for direct quenching of the material.

The W-temper process, on the other hand, is an uncoupled thermal-forming process [7]. First the blank is heated in the furnace and subsequently quenched for solution heat treating of the alloy reducing its strength and increasing the formability of the material. Before the natural ageing and strengthening of the material and taking advantage of that time window, the forming is performed to make use of the formability improvement.

2. Selected automotive component and aluminium alloy

In this work, a high strength aluminium automotive component has been formed with both hot stamping and W-temper methods. The selected component has been a structural component, namely the front side member of the Volkswagen Touran passenger car (see circle in red in 'figure 1'). In 'figure 2' the catastrophic result of the room temperature forming of the component using the 7075 aluminium alloy at T6 condition is shown.

Figure 1. Front side member of VW Touran.

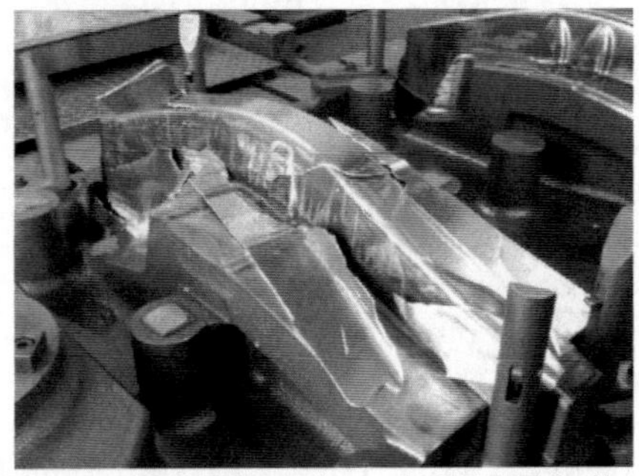

Figure 2. RT forming of 7075-T6 alloy.

For the study the aluminium alloy AA7075-T6 of 1.5 mm thickness has been used. In table 1 the as-received mechanical properties obtained by ASTM E8-04 standard tensile tests are summarized.

Table 1. As-received aluminum AA7075 mechanical properties.

Material	Thickness	Rp02	Rm	A%
AA7075-T6	1.5 mm	540 MPa	590 MPa	10 %

3. Temperature assisted forming

As mentioned before, two different manufacturing processes have been studied: a) the hot stamping process and b) the W-temper process. In the hot stamping process, the blank has been first heated up to 500 °C and maintained for 300s in a resistance furnace. Secondly, the blank has been transferred to the dies using a robot arm in approximately 7s. While the blank is still at high temperature, the forming process has been carried out and the dies have been kept closed during 10s to force the fast cooling of the material. ORAFOR WD from CONDAT Lubrifiants company, water diluted at 10% lubricant, was sprayed on the dies in order to avoid galling issues.

The second variant, the W-temper, is an indirect process where the microstructure of the material is modified prior to forming. Similarly to the hot stamping process, first the blank has been heated up to 480 °C and maintained for 300s in a resistance furnace. Then the blank has been transferred to a water bath using a robot arm and quenched for its solution heat treatment, modifying its initial microstructure. Finally, the forming has been conducted at room temperature without having to maintain the dies closed. In table 2 the advantages and disadvantages of each technology (compared to each other) are shown.

Table 2. Advantages and disadvantages of each technology.

Technology	Advantages	Disadvantages
Hot stamping	No critical time managing	Bigger process time (dies have to be closed after forming)
W-temper	Forming is performed at room temperature (less galling, tool maintenance, tool cost)	Critical time managing (forming before aging) Tool costs and maintenance are higher

A summary of the testing conditions used in this work are presented in table 3.

Table 3. Hot stamping and W temper manufacturing processes.

Manufac. Method	Heating time	Heating temperature	Pre-Forming Cooling	Press closing time
Hot stamping	300 s	500 °C	No	10 s
W-temper	300 s	480 °C	Yes	-

After forming, geometrical measurement of the components has been performed and the final mechanical properties have been evaluated from tensile samples cut from two different zones of the component.

4. Dimensional accuracy of components

A metrology inspection of the hot stamped and W-Temper formed components has been conducted using the GOM ATOS® system. 'Figure 4' shows the geometrical deviations of both components in comparison to the theoretical tool geometry definition.

Higher geometrical accuracy is observed for the hot stamping technology.

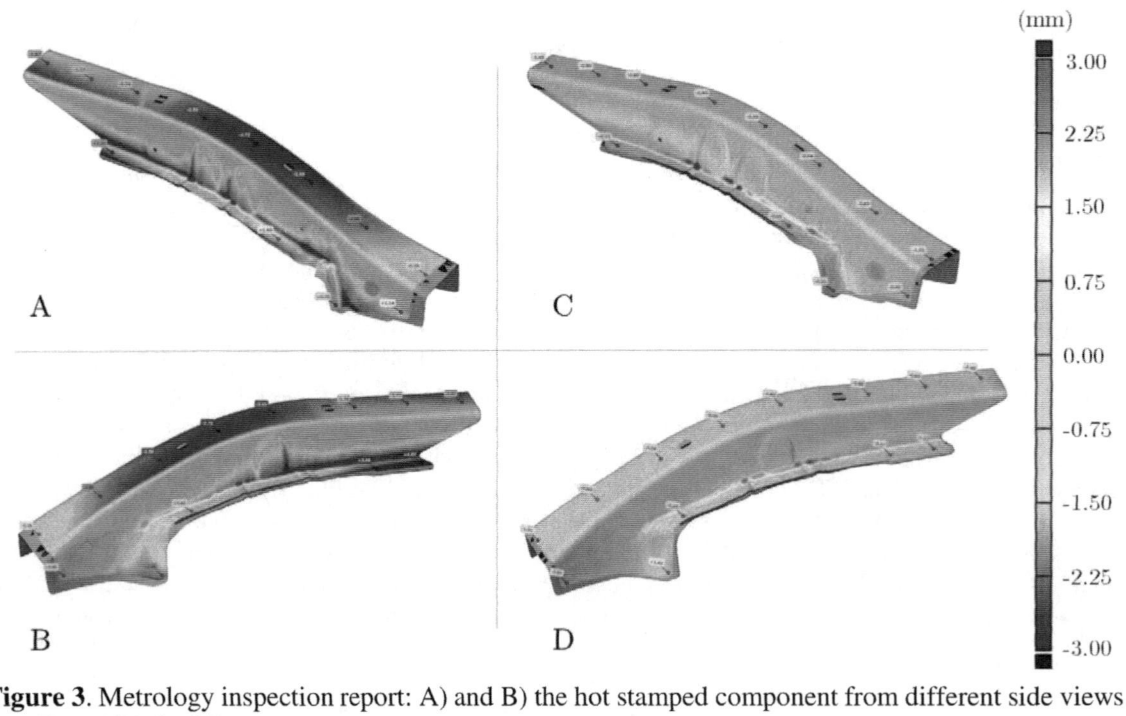

Figure 3. Metrology inspection report: A) and B) the hot stamped component from different side views and C) and D) the W-temper component.

5. Final material properties

As previously mentioned by Lee et al. [1], the thermal history of the blank during the process could degrade the final mechanical properties of the material. In order to analyse the mechanical properties of the final component, standard tensile samples after the ASTM E8-04 standard have been cut off using wire EDM technology from the component as shown in 'figure 5' for both process variants.

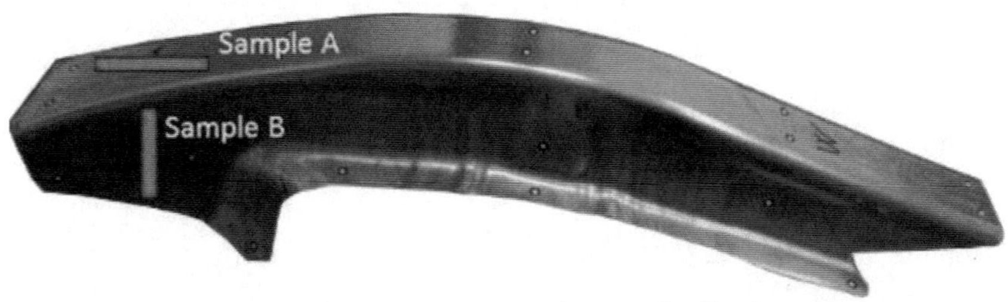

Figure 4. Tensile test samples location on the final component.

The 'figure 5' presents the comparative between the final mechanical properties and the as-received material for the both studied forming strategies. It can be shown that both technologies reduce the mechanical properties from a Yield Strength of about 540 MPa to a new one of around 280 MPa.

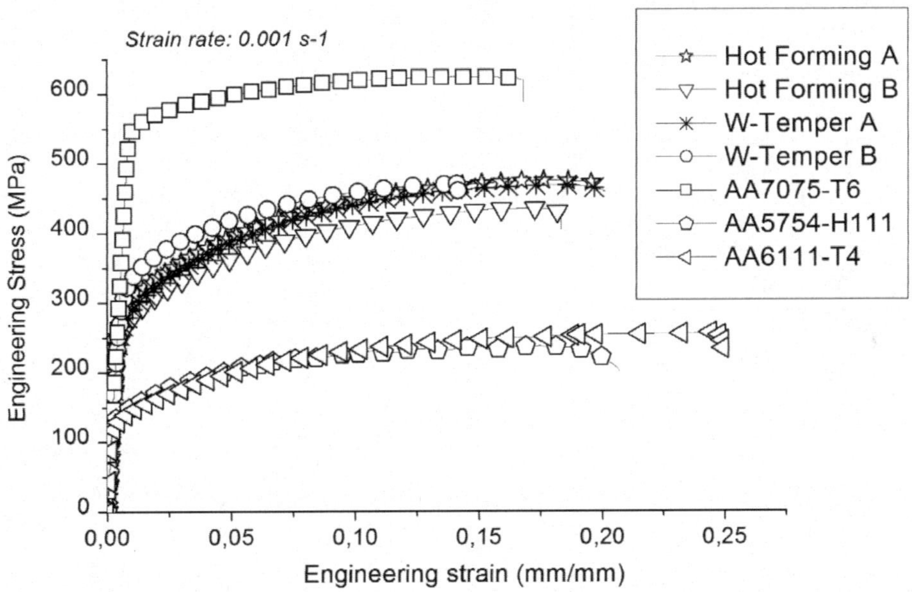

Figure 5. Comparison of tensile test results.

Square and triangle data markers represent the reference AA5754-H111 and the AA6111-T4 automotive aluminium materials' tensile properties. As it is clearly observed, the modified AA7075 aluminium components formed by hot forming and using the W-Temper process, even after losing mechanical properties from the as-received condition, still have approximately double strength compared to the well-known reference alloys. Comparison of changed strength levels due to bake hardening effects after a typical e-coating process cycles are not available.

6. Discussion and conclusions

In this work, two different technologies have been analysed in order to overtake the low ductility problem when forming the AA7075-T6 high strength aluminium alloy. The hot stamping and the W-temper technologies have been found to be valid to form the desired components without cracks.

Regarding the geometrical accuracy, the metrology inspections have shown that the hot stamping process gives higher accuracy compared to the W-temper process.

The main disadvantage of both processes is the degradation of the material mechanical properties during the forming process. However, the final properties obtained almost double the ones presented by 5xxx and 6xxx reference alloys used in the automotive industry. The post heat treating of formed components using the e-coat process used in serial production is still under study.

Acknowledgments
The authors would like to acknowledge the support of the industrial partner and automotive tool maker BATZ S. Coop. company.

References

[1] Lee M, Sohn S, Kang C, Suh D and Lee S Effects of pre-treatment conditionson warm hydroformability of 7075 aluminum tubes *J. Mater. Process. Technol.* 155–156 (2004) 1337–1343.

[2] Wang H, Yingbing L, Friedman P, Minghe C and Lin G Warm forming behaviour of high strength aluminium alloy AA7075 *Trans. Nonferrous Met.Soc. China* 22 (2012) 1–7.

[3] Kleiner M, Geiger M and Klaus A Manufacturing of Lightweight Components by Metal Forming, *CIRP Annals - Manufacturing Technology* 52-2 (2003) 521–542.

[4] Mendiguren J, Ortubay R, Saenz de Argandoña E and Galdos L Experimental characterization of the heat transfer coefficient under different close loop controlled pressures and die temperatures *Applied Thermal Engineering* 99 (2016) 813–824.

[5] Fan X, He Z, Yuan S and Lin P Investigation on strengthening of 6A02 aluminum alloy sheet in hot forming-quenching integrated process with warm forming-dies *Materials Science & Engineering A* 587 (2013) 221–227.

[6] Bariani PF, Bruschi S, Ghiotti A and Michieletto F Hot stamping of AA5083 aluminium alloy sheets, *CIRP Annals - Manufacturing Technology* 62- 1 (2013) 251–254.

[7] Zhao Z and Frankel GS The effect of temper on the first breakdown in AA7075 *Corrosion science* 49-7 (2007) 3089-3111.

IOP Conf. Series: Materials Science and Engineering **159** (2016) 012027 doi:10.1088/1757-899X/159/1/012027

A study on springback of bending linear flow split profiles

P Mahajan, C Taplick, M Özel and P Groche

Institute for Production Engineering and Forming Machines, Otto-Berndt-Straße 2, 64287 Darmstadt

E-mail: mahajan@ptu.tu-darmstadt.de

Abstract. The bending of linear flow split profiles made up of high strength materials involves high bending loads leading to high springback and geometrical defects. In addition, the linear flow split profiles are made stronger due to the high plastic deformation applied by the process itself. The bending method proposed in this paper combines the linear flow splitting process with a movable bending tool. The aim of the research was to investigate the effect of superimposed stresses exerted by the linear flow splitting process on bending load and springback of the profile by using a finite element model. The latter was validated by means of experimental results. The results show that the bending loads and the springback were reduced by increasing the superposition of stress applied by the linear flow splitting process. The reduction in the bending loads leads to a reduction in the cross-sectional distortion. Furthermore, the springback was compensated by controlling the amount of superimposed stress.

1. Introduction

Springback is a change in shape of a metal sheet after forming due to its elastic recovery. It is inevitable and causes shape discrepancies in the final product. In recent years, research on springback has grown considerably. Due to the demand on weight reduction mainly in the automotive sector, the application of high-strength materials has increased significantly. High springback, associated with high-strength materials, makes the study of prediction and compensation of springback a challenging task [1].

Several studies were undertaken to reduce springback in bending sheet metal. Bending high-strength steel under warm conditions reduces or even eliminates springback [2] [3]. One of the promising techniques to reduce springback is to superimpose stresses in the bending zone. The Stretch forming applies additional axial tension in the bending zone resulting in the reduction of springback. This axial tension can be controlled to reduce springback [4]. Another example of uniaxial superimposed stress is found in air bending. In this process, springback reduction of 60% was found when stresses were superimposed in the thickness direction [5]. The Torque Superposed Spatial (TSS) bending process superimposes torsional stresses to the bending moment. These stresses reduce bending forces, cross sectional distortion and springback [6]. The incremental tube forming process which combines tube spinning and bending operations applies multidimensional stresses resulting from the reduction in the diameter of the tube. This process shows a significant reduction in bending moment and springback [7].

The linear flow splitting process is a multi-station sheet-bulk metal forming process which allows continuous production of bifurcated profiles. This process is often carried out on high-strength

IOP Conf. Series: Materials Science and Engineering **159** (2016) 012027 doi:10.1088/1757-899X/159/1/012027

materials. Therefore, bending of linear flow split profiles is one of the most challenging tasks due to high springback.

Bending linear flow split profiles remains a challenge because of the requirement of mass production, flexibility and the use of high strength materials. In this paper, the process of bending linear flow split profiles is presented. A bending tool system is used to bend the profiles in a continuous manner. This tool system considers the aforementioned procedure of stress superposition in the bending zone. The aim of this paper is to analyze the effect of stress superposition on the reduction of bending forces and springback. A numerical model of the whole process is developed to investigate the effects of stress superposition on bending forces and springback. The model is verified with the help of experimental results for the bending force and the springback ratio. Furthermore, the model shows the effect of variation of superimposed stresses on the bending force and the springback ratio.

2. Linear Flow splitting

Linear flow splitting is a multi-station process which forms a metal sheet into a bifurcated profile at room temperature. The bifurcation of the metal sheet into flanges is made possible by a pair of obtuse angled supporting rolls and splitting rolls at each station, as shown in figure 1. The metal sheet is moved by means of rotating supporting rolls. At each station n, the splitting rolls are closed in by an incremental depth y_{inc}. The width of the supporting rolls is also reduced accordingly. Therefore, the surface area of the metal sheet edge is increased at each station. The process is continued until the desired geometry is achieved [8].

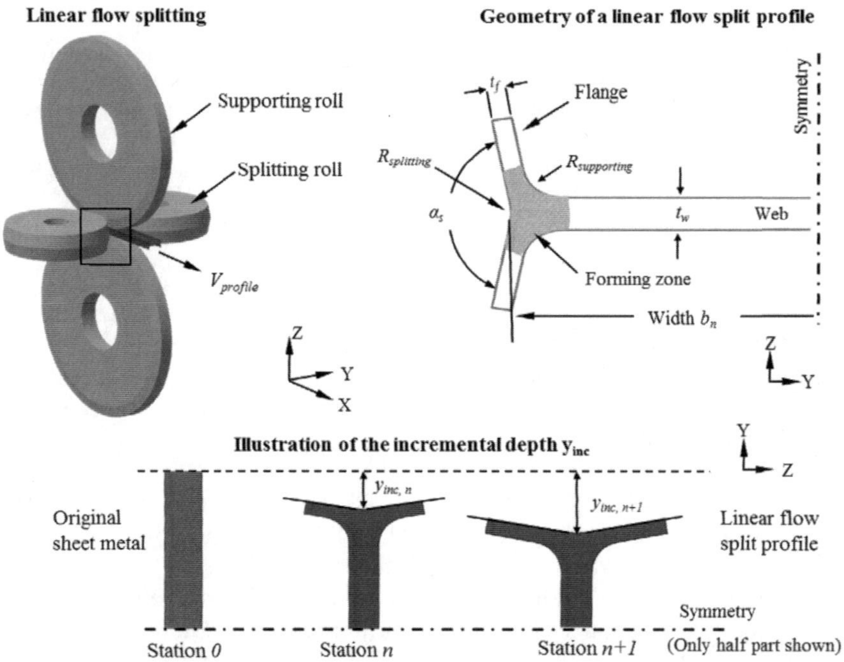

Figure 1. Principle of the linear flow splitting process

In the forming zone, splitting rolls exert high compressive stresses in transverse direction which allows the formation of flanges. However, some amount of the plastified material flows in the thickness direction. This thicker part of the web is rolled by the supporting rolls. The combination of these two deformation mechanisms allows high hydrostatic stresses in the forming zone. High hydrostatic stresses are beneficial in elevating formability of the sheet metal in the forming zone. Thus, the linear flow splitting process applies multi-axial stresses in the forming zone [9].

IOP Conf. Series: Materials Science and Engineering **159** (2016) 012027 doi:10.1088/1757-899X/159/1/012027

3. Experimental setup

The experimental setup consists of a decoiler, seven stations of the linear flow splitting process, a tool system for the bending operation and a profile cutting machine. A high-strength steel "RAWAEL 80" (Yield strength = 840 MPa) is used in all investigations. The tool system for the bending operation consists of a bending roll, a calibrated load cell and a blankholder mounted on a rotary table, as shown in figure 2 (a).

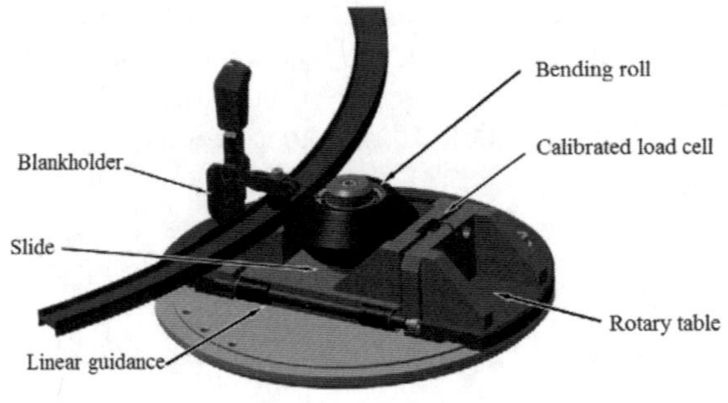

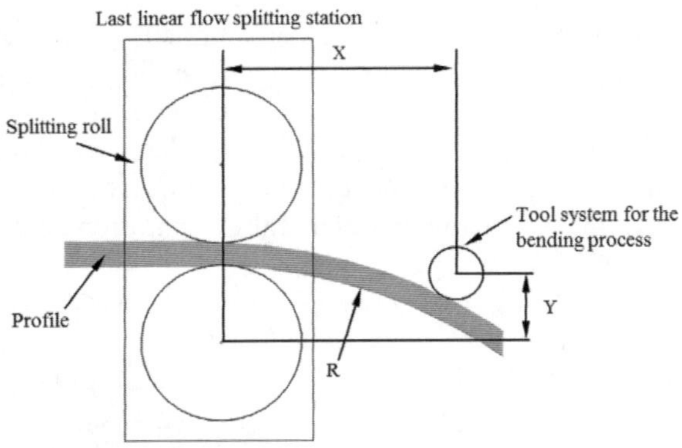

Figure 2. Bending process (a) Tool system, (b) Layout

In all investigations, the linear flow split profile is bent in transverse direction only. Due to the stress superposition by the linear flow splitting process in the forming zone, bending forces exerted on the tool system are expected to be lower than the forming forces. Therefore, the diameter of the bending roller is kept smaller than the splitting roller. The tool system for the bending operation is positioned directly after the last linear flow splitting station on the same height as the splitting rollers which allows a continuous production, as shown in figure 2 (b). The distances between the bending roller from the splitting rollers of the last station X and Y determine the bending radius R_1. The values of X and Y as well as the process parameters chosen for the experiments are given in table 1 (a) and (b).

Table 1 (a) Position of the bending tool, (b) Process parameters

(a)

X (mm)	Y (mm)	Bending radius R_1 (mm)
425.36	100	600
546.01	100	1000
644.46	100	1400
729.74	100	1800
806.06	100	2200

(b)

Process Parameter	Value
Velocity of the metal sheet, V_{sheet}	3 m/min
Initial sheet width, $b_{initial}$	50 mm
Width of the profile before bending, b_{n-1}	35.5 mm, 34.5 mm
Incremental depth at the last station, y_{inc}	1.5 mm, 2 mm
Total incremental depth, $y_{inc,\ total}$	9.5 mm

Therefore, widths of the profiles before the last linear flow splitting station are considered as 34.5 mm and 35.5 mm for incremental depths of 1.5 mm and 2 mm respectively. Therefore, the width of the

IOP Conf. Series: Materials Science and Engineering **159** (2016) 012027 doi:10.1088/1757-899X/159/1/012027

bent profile is constant irrespective of the changes in the incremental depth at last linear flow splitting station. In experimental investigations, the radius of the bent profile is measured by means of a coordinate measuring system. The received data is fitted in a curve to find out the radius of the profile.

4. Numerical model

The numerical model of the bending process is created by using the FE software package Marc Mentat 2012, as shown in figure 3. The model is created using PYTHON scripts to achieve consistency in the geometries and solved with an implicit quasistatic solver. In this model, the linear flow splitting operation on the last stand and the bending process are considered. Additional support roller sets are required to hold the profile in *Y*-direction. In experimental investigations, the geometry of the profile before the last stand is measured by a laser triangulation system. This geometry is considered to be an initial geometry for the profile. The degree of deformation accumulated due to previous stages of forming is implemented as an initial condition in the model.

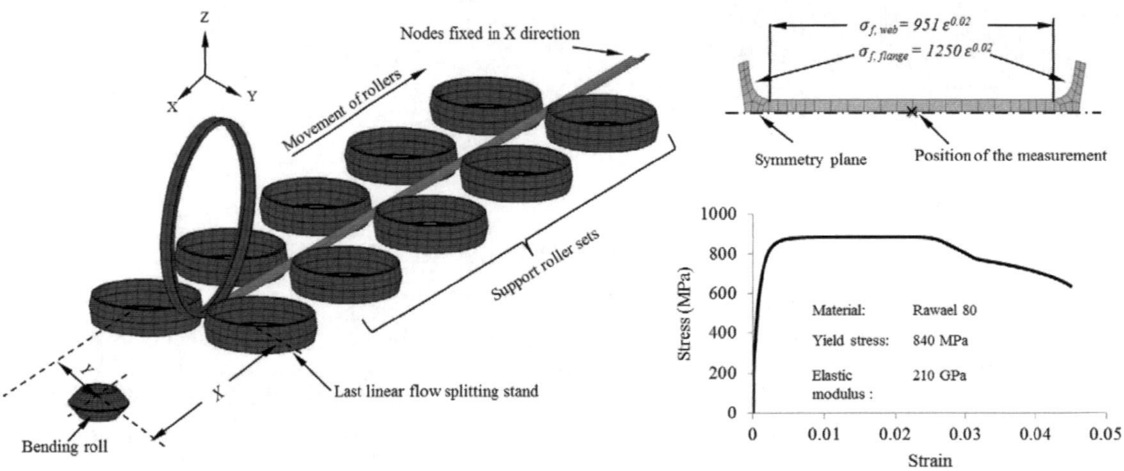

Figure 3. Numerical model of the process

The profile in the FE model is symmetric in thickness direction, as shown in figure 3. Initial material data was obtained from tensile tests to calculate the yield stress and the strain hardening coefficient. The material data for flanges and web is derived from experimental hardness measurements and fitted in Ludwik's equations $k_{f,flange} = 1250\varepsilon^{0.02}$ and $k_{f,web} = 951\varepsilon^{0.02}$. As a characteristic of the linear flow splitting process, flanges possess a higher strength than the material in the web because of high forming strains [9]. The Von mises yield criterion is considered with the isotropic hardening rule. All rollers are defined as analytical rigid bodies because of their high stiffness. The model is simulated using a Coulomb friction coefficient of 0.12 between rollers and the sheet. The model uses 8-noded hex elements with an assumed strain formulation to accurately simulate the behavior under bending process. Thermal effects are neglected.

The length of the sheet is taken as 2500 mm to achieve the steady state which is often assumed in continuous forming processes. To reduce the computational time, the nodes at the end of the sheet are fixed in all three directions and the rollers move across the sheet to retain the relative motion. At the start of the analysis, the bending roller is positioned at *X* distance from the last forming stand according to the desired bending radius R_l. Furthermore, the bending roller is moved in transverse and longitudinal direction until it reaches a designed position *Y*. After positioning the bending roller at *X* and *Y* distance from the splitting rollers, it is further moved only in longitudinal direction with the same velocity as that of the other forming rollers. All rollers are moved away from the profile at the end of the simulation. A python script measures the coordinate data from the nodes located at the middle of the bent profile. This data is curve fitted in Matlab to find the bending radius after the springback R_2.

5. Results and discussion

5.1. Verification of the numerical model

5.1.1. Bending force

The verification of the numerical model was carried out by comparing bending forces measured in the experiments and in the simulations for incremental depths of 1.5 mm and 2 mm, as shown in figure 4.

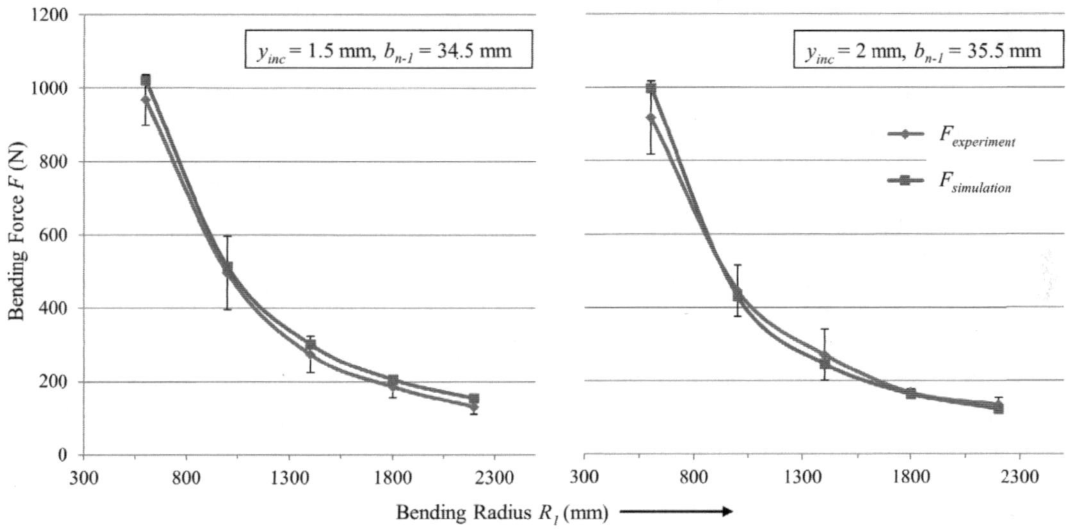

Figure 4. Verification of the numerical model

The bending forces predicted by the numerical model confirms the forces measured in the experiments for both incremental depths y_{inc} = 1.5 mm and y_{inc} = 2 mm. In addition, it can be observed that bending forces decrease with increasing the bending radii. The stress superimposed in the last forming stand increases with the increased incremental depth y_{inc}. Therefore, bending forces exerted for y_{inc}= 2 mm are less than for y_{inc}= 1.5 mm for all bending radii considered. It is interesting to note that the magnitude of the bending force is very small because the bending zone is already plastified in the last linear flow splitting station.

5.1.2. Springback ratio K

The springback ratio K is defined as the ratio of the radius before springback to the radius after springback R_2.

$$K = \frac{R_1}{R_2} \tag{1}$$

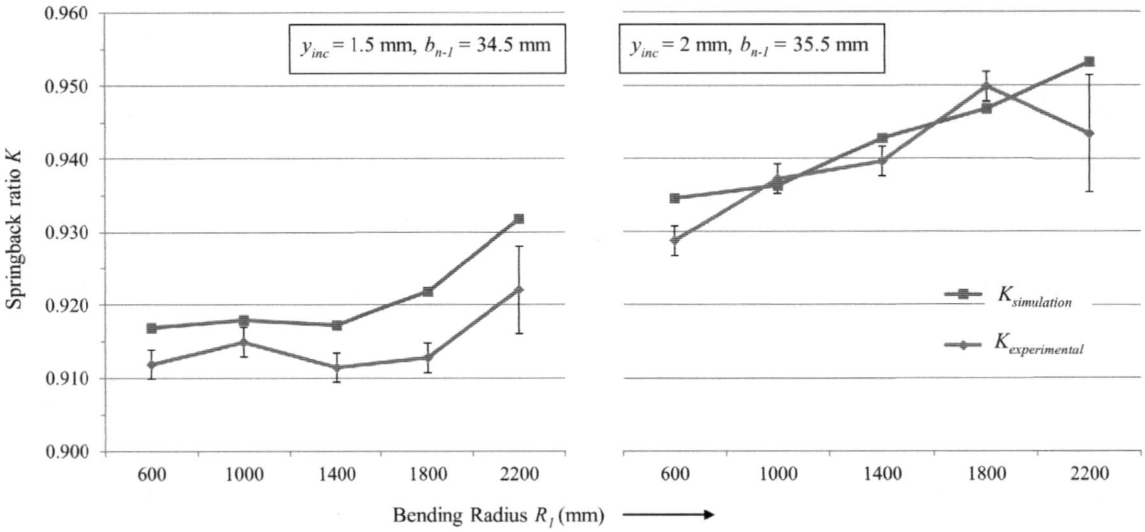

Figure 5. Variation of springback ratio K with bending radius R_1

Figure 5 shows the relationship between the bending radius R_1 and the springback ratio K for incremental depths of 1.5 mm and 2 mm. It was found that the springback ratio increases with increasing the bending radius for both incremental depths. Besides, at higher incremental depth of 2 mm, the springback ratio was higher for every bending radius compared to the incremental depth of 1.5 mm. This suggests that with higher incremental depths, the amount of springback decreases.

5.2. Effect of stress superposition

Stress superposition plays an important role in the reduction of the bending force and the springback ratio. To understand the amount of stress superimposed during the linear flow splitting process, the normal stresses along the width of the profile are plotted, as shown in figure 6 (a). The figure shows the influence of the incremental depth on the normal stress in the forming zone. This suggests that the stress superposition in this bending process can be controlled by adjusting the incremental depth in the last forming stand.

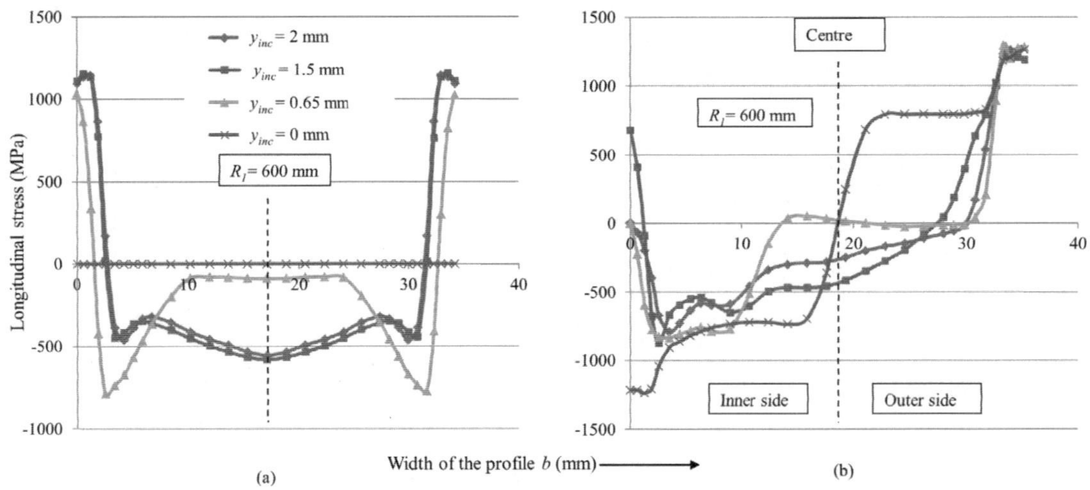

Figure 6. Normal stress in (a) linear flow splitting, (b) linear flow splitting and bending process

Figure 6 (b) presents the effect of the bending operation in the stress superimposed forming zone. The normal stresses were measured at the forming zone which also describes that bending takes place within the forming zone created by the linear flow splitting process.

Variation in stress superposition affects the bending force as well as the springback ratio, as shown in figure 7 (a), (b). It can be seen clearly that the increase in the stress superposition reduces the bending force. The springback ratio decreases with decreasing the incremental depth as well.

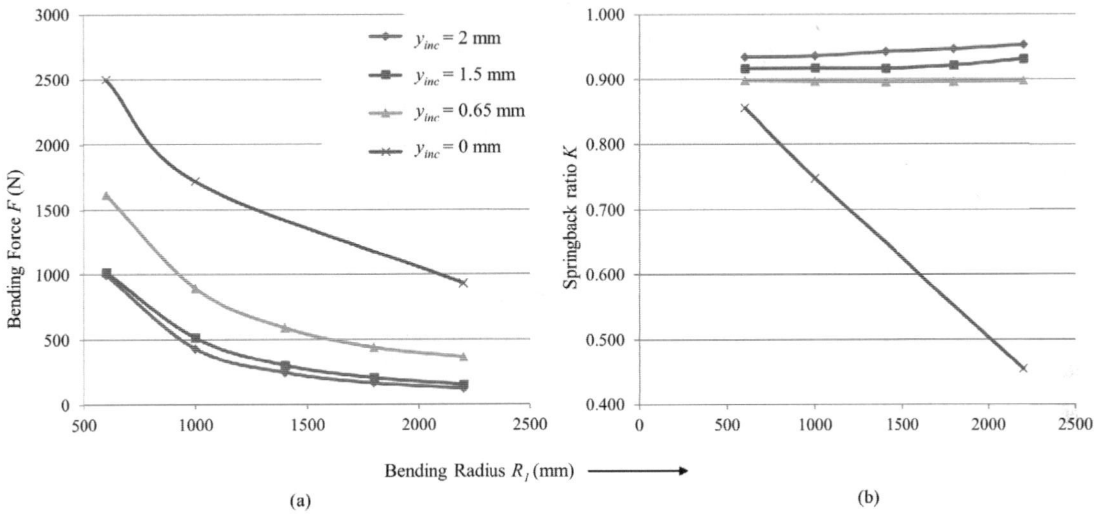

Figure 7. Effect of stress superposition on (a) bending force, (b) springback ratio

It is interesting to note that the springback ratio increases with increasing the bending radius when the stress is superimposed in the bending zone, whereas the springback ratio shows the opposite trend in the absence of superimposed stress, as shown in figure 7 (b). At incremental depth y_{inc}= 0 mm, bending forces were higher with lower springback ratios. At y_{inc}= 0.65 mm, it was found that the springback ratio remains constant irrespective of the bending radius considered for the investigations.

6. Conclusion

The numerical model for the process was developed. The model was successfully verified with the experimental results for bending forces as well as the springback ratio. The maximum amount of error in predicting bending forces and springback ratios was found to be 0.7% and 1% respectively.

The amount of stress superimposed in the bending zone plays an important role in the reduction of bending forces and the springback ratio. Bending forces were reduced at least by 86% whereas the springback ratio was increased at least by 9.4% in the presence of superimposed stress. Furthermore, the springback ratio with the stress superposition shows little variation for the considered bending radii. In current investigations, the springback ratio remained constant for an incremental depth of 0.65 mm irrespective of the bending radii.

The results from the numerical model as well as the experiments show the effectiveness of stress superposition in the bending zone to reduce higher bending forces and springback. With this bending process, it is beneficial to bend the linear flow split profiles in a continuous manner with less springback.

Acknowledgement
The investigations presented in this paper are supported by the German Research Foundation (DFG). The authors would like to thank DFG for funding the Collaborative Research Centre 666 "Integral sheet metal design with higher order bifurcations – Development, Production, Evaluation".

References
[1] Wagoner R, Lim H, Lee M 2013 Advanced issues in springback. *International Journal of Plasticity* **45** 3–20
[2] Yoshida T, Katayama T, Hashimoto K, Kuriyama Y July 2003 Shape Control Techniques for High Strength Steel.*Sheet Metal Forming Nippon steel technical report No.* **88**
[3] Yanagimoto J, Oyamada K 2007 Mechanism of springback-free bending of high strength steel sheets under warm forming conditions. *Annals of the CIRP* **56(1)**:265–268
[4] Hardt D, Norfleet W, Valentin V, Parris A 2000 In Process Control of Strain in a Stretch Forming Process. *J. Eng. Mater. Technology* **123(4)**, 496-503
[5] Chatti S, Hermes M, Weinrich A, Ben-Khalifa N, Tekkaya AE 2009 New incremental methods for springback compensation by stress superposition. *International Journal of Material Forming* Vol. **2** Suppl 1:817–820
[6] Chatti S, Hermes M, Tekkaya AE, Kleiner M 2010 The new TSS bending process: 3D bending of profiles with arbitary cross-sections. *Annals of the CIRP* **59(1)**:315–318
[7] Becker C, Tekkaya AE, Kleiner M 2014 Fundamentals of the incremental tube forming process. *CIRP Annals – Manufacturing Technology* **63** 253-256
[8] Groche P, Vucic D, Jöckel M 2006 Basics of linear flow splitting. *Journal of Materials Processing Technology* **183**: 249-255
[9] Muller C, Bohn T, Bruder E, Bruder T, Landersheim V, Dsoki C, Groche P, Veleva D 2007 Severe plastic deformation by linear flow splitting. *Mat.-wiss. u. Werkstofftech.* **38**, No. 10

Approaches for springback reduction when forming ultra high-strength sheet metals

R Radonjic and M Liewald

University of Stuttgart, Institute for Metal Forming Technology (IFU), Holzgartenstrasse 17, 70174 Stuttgart, Germany

ranko.radonjic@ifu.uni-stuttgart.de

Abstract. Nowadays, the automotive industry is challenged constantly by increasing environmental regulations and the continuous enhancement of standards with regard to passenger's safety (NCAP, Part 1). In order to fulfil the aforementioned requirements, the use of ultra high-strength steels in research and industrial applications is of high interest. When forming such materials, the main problem results from the large amount of springback which occurs after the release of the part. This paper shows the applicability of several approaches for the reduction of springback amount by forming of one hat channel shaped component. A novel approach for springack reduction which is based on forming with an alternating blank draw-in is presented as well. In this investigation an ultra high-strength steel of the grade DP 980 was used. The part's measurements were taken at significant cross-sections in order to provide a qualitative comparison between the reference geometry and the part's released shape. The obtained results were analysed and used in order to quantify the success of particular approaches for springback reduction. When taking a curved hat channel shaped component as an example, the results achieved in the investigations showed that it is possible to reduce part shape deviations significantly when using DP 980 as workpiece material.

1. Introduction

One of the most important tendencies in nowadays automotive industry is the development of lightweight car body structures due to increasing demands regarding fuel consumption and carbon dioxide emissions. Car body parts manufactured from ultra high-strength steels (UHSS) have an increased crash performance and are contributing to the improvement of the passenger's safety accompanied by noticeably reduction of vehicle weight [1]. It is necessary to emphasize that the steels mentioned before do have very high levels of tensile strength at relatively low values of elongation at fracture as well as relatively low anisotropy values. These mechanical properties mean that UHSS usually do have bad formability, a strong tendency towards wrinkling, and a tremendous elastic recovery or springback after forming. While such new steel alloys are becoming increasingly stronger, they simultaneously are more difficult to form into automotive parts. Larger amounts of springback are affected by many parameters such as sheet metal thickness, blank shape, level of ultimate tensile strength, hardening rate, Young's modulus, method plan, tool radii, tool clearance, and forming conditions as well. Springback in general can appear in several forms: angle change, sidewall curl, radii change, and twisting or torsion [2,7] (Figure 1). Angular change is the angle created between reference geometry and after forming obtained part sidewall shape. Sidewall curl is the curvature created in the sidewall of the part and usually occurs as a result of bending and unbending when a blank is deep drawn over die radius or through draw beads. Twist or torsion is caused by torsion moments in the longitudinal cross-sections of the part and it is more pronounced if the cross-section changes along the parts longitudinal direction. Besides these parameters inhomogeneous strain distribution along the formed part together with the elastic-plastic behavior of the workpiece material affects springback occurrence too. Due to that, the use of UHSS is actually increasing existing problems in manufacturing processes regarding quality control requirements and process robustness. In order to form these materials with a die, it is necessary to define the forming process with new boundary conditions, but keep the manufacturing costs as low as possible.

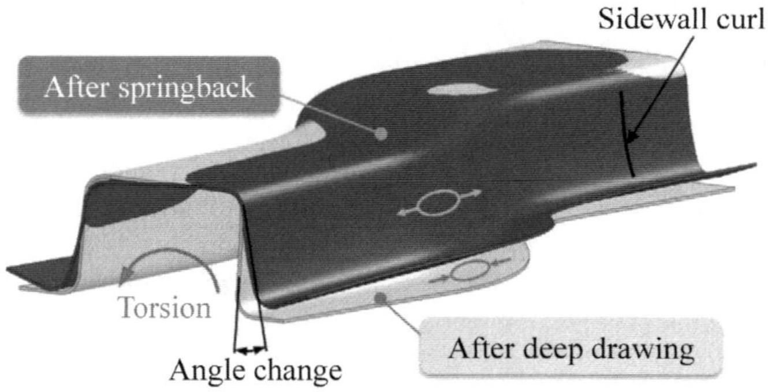

Figure 1. Different kinds of springback

There are different strategies which may contribute to reduce the springback amount after forming. Some of them are taking into account the application of geometrical methods such as the implementation of stiffening features into the part shape as well as addendum and binder surface modifications [3,6]. The other possibility which may contribute to achieve the final goal in terms of dimensional accuracy considers a range of methods based on modifying the stress-strain conditions in the part by appropriate arrangements such as additional restraining forces or even adjustable and controllable blank holding forces which are time dependent to punch stroke travel [2,4].

In this work, several, already known strategies for the reduction of springback were tested by forming one curved hat channel shaped component. First, the possibility to reduce springback was simulated using an additional operation (restrike) in the forming process. For this purpose, the vertical stiffening beads were embossed into the part sidewalls after the deep drawing process using appropriate tool shapes in one additional forming step.

The possibility to reduce the springback amount was investigated experimentally by applying different blank holder forces. The required experimental try-outs for this purpose were provided at the servo driven press AIDA.

At the end of this work, a new approach for springback reduction when forming ultra high-strength steel sheets is presented. The new approach implies the forming with the alternating blank draw-in. After such a forming process, the part geometry has an almost negligible deviation after the release with regard to the required part shape geometry. Forming with the alternating blank draw-in was tested experimentally as well.

2. Sheet material properties

HCT 980 X steel sheet (regarding to DIN EN 10338 standard) is selected as blank material for this investigation. The material mentioned belongs to a group of high strength dual phase steel grades. This material consists of a ferritic matrix including a hard martensitic second phase in the form of islands which generally cause relatively high strength level. Due to that, material HCT 980 X has very high values of ultimate tensile strength and yield strength, while the value for elongation is very small. HCT 980 X material has anisotropy parameters smaller than 1. Material properties for steel grade HCT 980 X are shown in Table 1.

IOP Conf. Series: Materials Science and Engineering **159** (2016) 012028 doi:10.1088/1757-899X/159/1/012028

Table 1. Material properties for steel grade HCT 980 X

Material type		HCT 980 X
Thickness [mm]		1.15
Poisson's ratio		0.28
Young's modulus [GPa]		210
Anisotropy parameters considering to rolling direction	r_0	0.735
	r_{45}	0.989
	r_{90}	0.64
Yield strength [MPa]		745
Ultimate tensile strength [MPa]		1070

3. Simulation of springback

Springback results from elastic recovery of the part and can be defined as a shape dimensional change which occurs during unloading when tools are removed in the forming process. In order to analyse springback behaviour of UHSS, various simulations for one curved hat channel shaped part geometry were provided (Figure 2a). The analysed part can be divided into two straight areas having a length of 125 mm and two curved areas with the radius of 325 mm. The drawing depth of the part amounts to 40 mm. The deep drawing process and springback were simulated using the FE-Code LS-Dyna. The modelling of the tool and the preparation of active tool surfaces (Figure 2b) for the simulation was performed using the software CATIA V5. The tool active surfaces were meshed with the shell elements type Belytschko-Tsay Shell (ELFORM 2). The blank was meshed with the fully integrated shell elements (ELFORM 16) taking into account nine integration points through the sheet thickness. During meshing it was paid attention to the existence of approximately one node per 5° of turn angle at the tool radius [5]. In the simulation, the material model *MAT_125 was used. The applied blank holder force was 1500 kN, while the friction coefficient was determined experimentally and amounts to 0.07.

Forming conditions	
Blank holder force [kN]	1500
Friction coeficient	0,07
Die radius [mm]	6
Punch radius [mm]	7

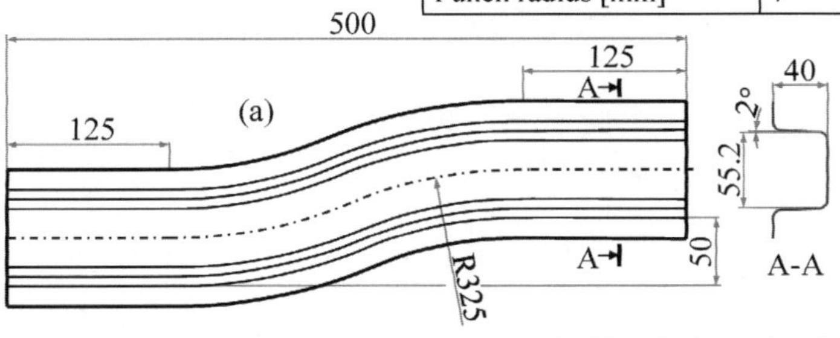

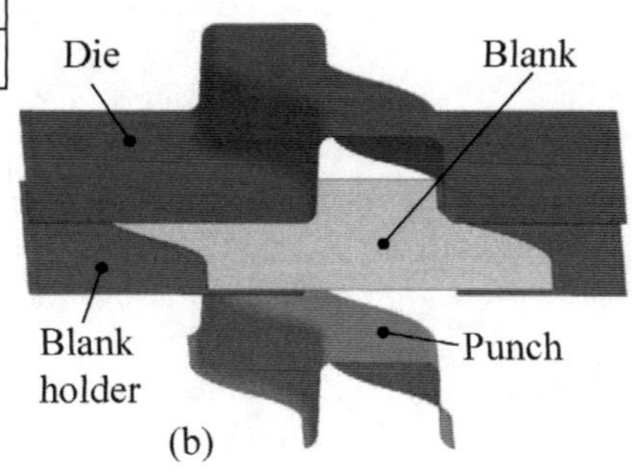

Figure 2. Simulation of springback

Figure 3 shows the calculated stress distribution for different integration points after the forming process. High tensile stresses dominate within integration point 1 (outer part fibre) in the part sidewall. Solely at the transition between punch radius and sidewall in convex part area compressive stresses are arising. Due to the superposition of bending stresses and the shifting of neutral fibre, integration point 5 in the part sidewall shows tensile stresses. In integration point 9 (inner layer) compressive stresses generally arise. Tensile stress was observed at the transition between punch radius and sidewall in the convex part area.

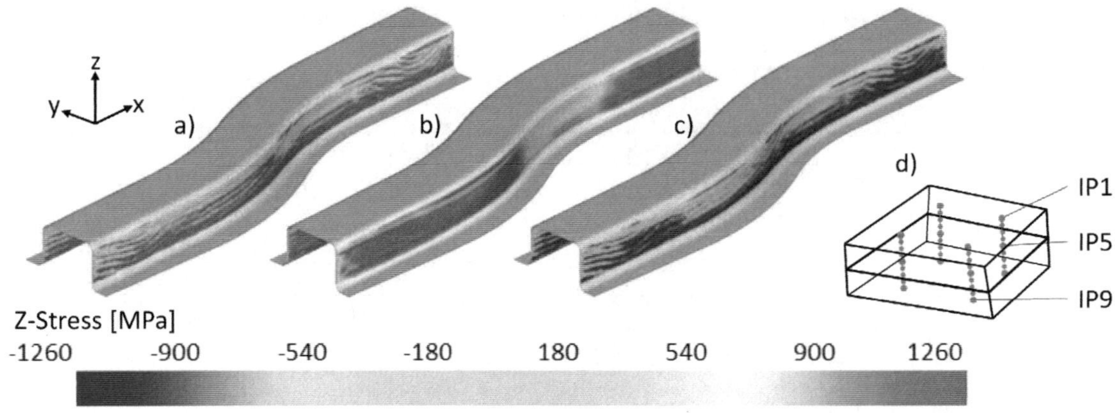

Figure 3. Stress distribution emerging in z-direction; a) for integration point (IP1), b) for IP5 and c) for IP9, d) fully-integrated shell formulation

After the part removal from the die cavity this kind of stress is released. The fibres which showed tensile stresses tended to be shorter, while fibres with the compressive stresses were elongated after the release. This will lead to the arising of the part sidewall curl in outer direction. The mentioned change of the part shape appears especially in the straight and concave part areas.

The calculated springback amount is shown in Figure 4a. It can be noticed that the very high part shape deviation in the lower sidewall area occurred (6 mm). This deviation represents the cumulation of the sidewall curl and angle change. Sidewall curl might be understood as the result of the bending stresses which occur due to bending and unbending at the die radius as well as additional tensile stresses in the part sidewall which result from stretching during deep drawing. Figure 4b shows the stress distribution in z direction for one selected finite element at the part sidewall during deep drawing. It can be seen that after the deep drawing process, a large difference in stress amount between outer (IP1) and inner part fibre (IP9) occurs (more than 2000 MPa). Due to this stress difference a large curl of the sidewall has occurred.

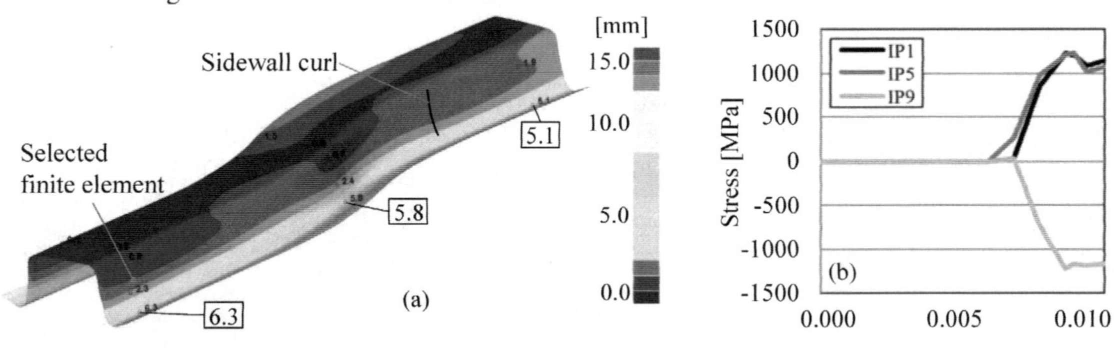

Figure 4. (a) Springback amount after deep drawing, (b) stress distribution in the part sidewall for selected finite element (IP1- outer part fibre, IP5- middle part fibre and IP9-inner part fibre)

3.1. Reduction of springback through a restrike operation

As aforementioned, the sidewall curl, indeed, has a tremendous influence on the total amount of springback. With the aim to reduce the occurred sidewall curl, a measure was simulated which implies an increase of the part stiffness through additional embossing of the corresponding features into the sidewall shape. The method plan used to achieve the complete part shape in simulation is defined as follows: deep drawing, restrike and springback (release of the part). Regarding to the defined method plan, the simulation of deep drawing was created (considering the forming conditions given in Figure 2) and calculated first. After the deep drawing simulation was finished, the released part was used as a workpiece for a second defined forming step – the restrike operation. As can be seen in Figure 5, the tool design for the restrike operation was based on a crash forming concept (without blank holder). Punch and die were modelled with the appropriate vertical stiffening features placed in the part sidewall. The width of the stiffening features amounts to 10 mm and the depth was modelled with 1 mm (Figure 5). The distance between particular stiffening beads is 40 mm. In the second forming step (restrike), first the part was positioned appropriately onto the punch bottom surface. Afterwards the die was moved down until the defined clearance with the punch was reached (sheet thickness + 10%). During this forming step, the corresponding stiffening beads or features were formed at the part shape. Resulting from this, the acting stresses in the part were changed. When comparing to the results presented in Figure 4a, this measure has contributed to springback reduction.

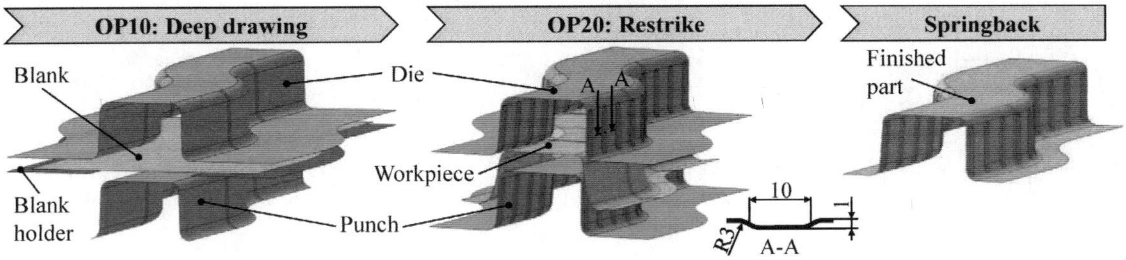

Figure 5. Simulation of restrike operation – method plan

Figure 6b shows the stress distribution during the restrike operation for the finite element which was defined at the same place as in the deep drawing operation. It is apparent that the large stress difference that occurred after the deep drawing between outer and inner part fibre will be reduced significantly during the restrike operation. This reduction of the stress difference has resulted in a smaller amount of springback (Figure 6a). The highest measured shape deviation in the lower part sidewall area now only amounts to 2 mm.

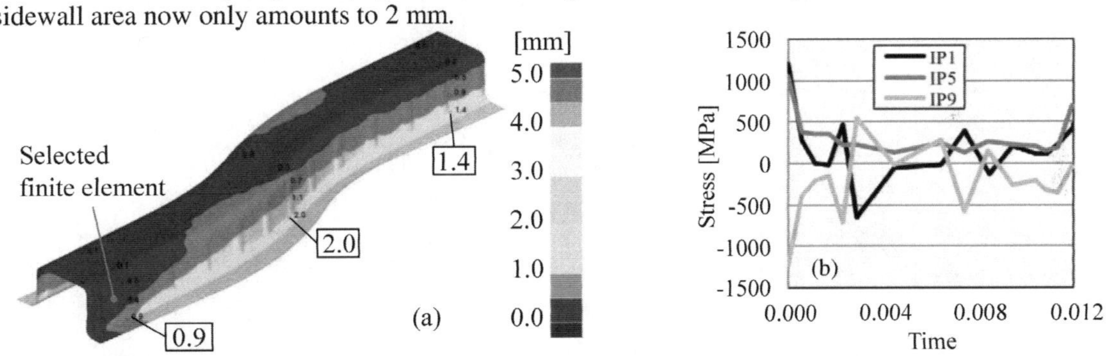

Figure 6. (a) Springback amount after restrike operation, (b) stress distribution in the part sidewall for selected finite element (IP1- outer part fibre, IP5- middle part fibre and IP9-inner part fibre)

4. Influence of blank holder force on springback

To investigate the influence of the blank holder force on the springback amount, numerous experimental tests were performed. For that purpose, three values of the blank holder force were analysed (300 kN, 900 kN and 1500 kN). For each applied blank holder force, five parts were formed at the press. Experiments have been performed at the servo driven press AIDA with a nominal force of 6300 kN and blank holder force of 1500 kN. The specimens were lubricated with 1.5 g/m² of the oil M100 produced by the company Georg Oest Mineralölwerk GmbH & Co. KG. After forming, part shape deflections were measured using the optical measurement system (GOM ATOS), and the results were compared with the reference geometry (CAD of the part). In order to compare the obtained results in an adequate manner, the part shape deviations were evaluated in three characteristic sections. Two cross sections (section 1 and 3) were defined near to the part end, 220 mm away from the middle of the part. Section 2 is located in the middle of the part and is oriented perpendicular onto the tangent of the part curvature. In these defined cross sections, angle change from both sides (left and right side) was evaluated. Figure 7a shows the deep drawing tool assembled to the servo press as well as the parts which were deep drawn in the experiment (Figure 7b). The defined measurement concept is shown in Figure 7c.

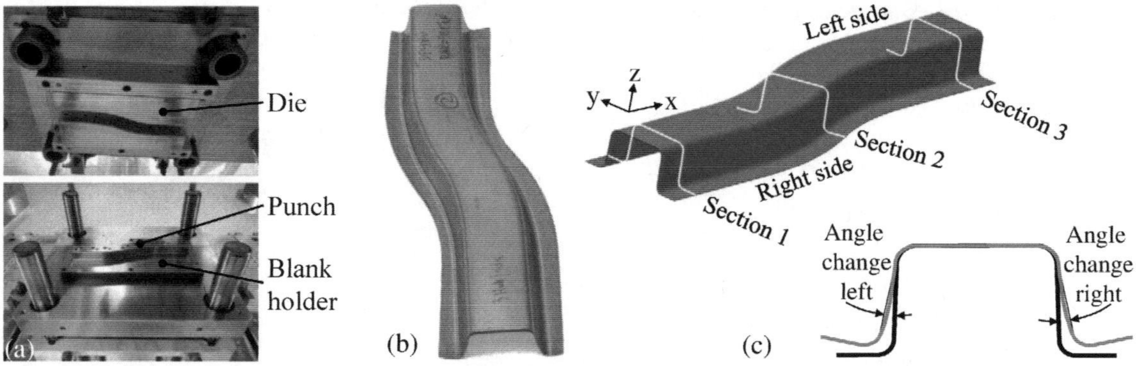

Figure 7. Experimental setup; (a) tool for deep drawing, (b) one of the parts formed in experiment, (c) measurement concept

Figure 8 shows measured angle change of the parts regarding different, applied blank holder forces. The highest angle change for a blank holder force of 300 kN is located at section 1 from the right side and reaches values bigger than 15°. By applying higher blank holder forces, the angle change can be reduced. The highest value of angle change for a blank holder force of 1500 kN is measured in the same section as described before and amounts to approximately 13.7°. When comparing obtained results being gained by different blank holder forces (300 kN and 1500 kN), it can be detected that in case of higher applied blank holder forces the angle change will be reduced by approximately 2°.

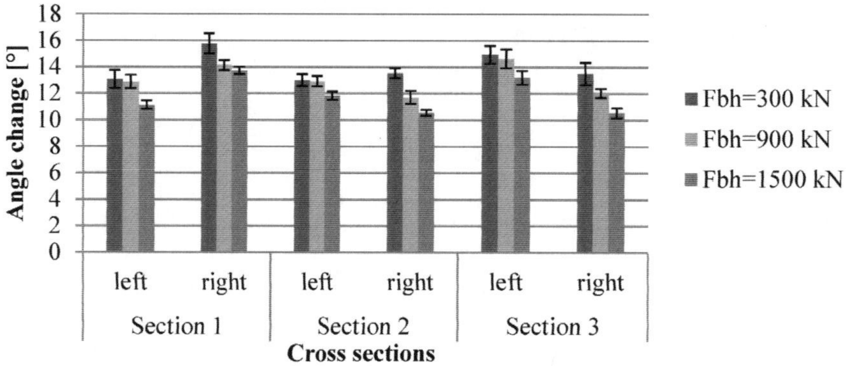

Figure 8. Influence of the blank holder force on springback

5. Pendular drawing - a novel approach for springback reduction

Forming with an alternating blank draw-in or pendular drawing shows a great potential for the successful reduction of springback. The uniqueness of this process lies in the fact that during forming the blank is drawn up to a specific drawing depth from only one side, which means that blank draw-in is only allowed from one side. After achieving the defined drawing depth (e.g for two-step forming, half of the total drawing depth), the blank will be drawn from the opposite side of the part. The workpiece area that is formed over the punch radius in the first step is located at the part bottom will be drawn back into the part sidewall in a second step. Resulting from that it comes to a repeated "bending and unbending" over the punch radius during this forming process. Forming with the alternating blank draw-in is shown schematically in Figure 9 (with two steps in Figure 9a and with three steps in Figure 9b).

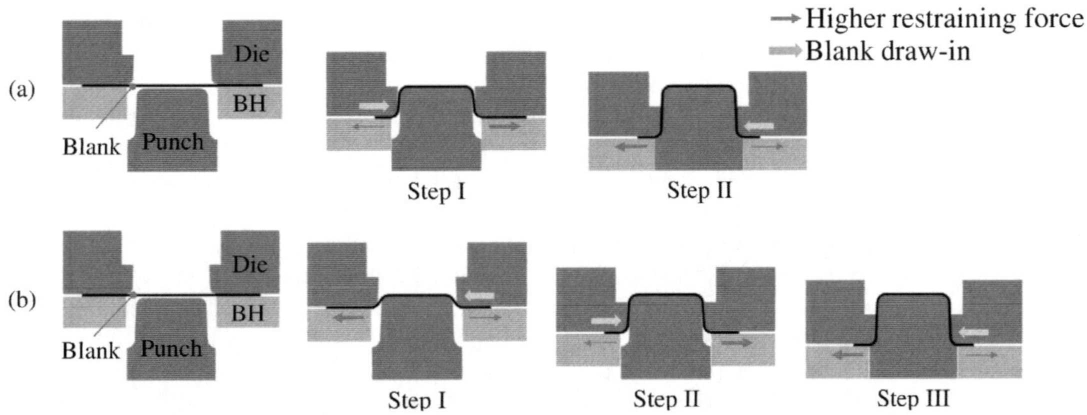

Figure 9. Forming with the two step (a) and three step (b) alternating blank draw-in

5.1. Analysis of the forming with the alternating blank draw-in

As mentioned in chapter 3, the bending stresses are one of the main reasons for the occurring of the sidewall curl as well as other kinds of springback. During conventional deep drawing, the corresponding workpiece area will be bent over the punch and the die radius only once. Bending over the die radius will result in a significant difference of stress amount between inner and outer side of the part wall. When forming with the alternating blank draw-in (pendular drawing) considering the phenomenon of the repeated bending and unbending over the punch radius, the difference in stress amount from inner and outer side of the part will be reduced. In order to better understand this phenomenon, forming with the alternating blank draw-in of the part shape geometry described before was simulated. For the first 20 mm of the drawing depth the blank is drawn-in just from the right side, and for the remaining 20 mm of the drawing depth the blank is drawn-in from the opposite (left) side. The alternation of the blank draw-in was influenced by applying the different friction conditions from the left and right side of the part taking into account the defined drawing depth and appropriate blank side. In the simulation, the friction coefficient between tool surfaces and blank for the blank side with the allowable draw-in, was defined with the value of 0.07. For the opposite blank side (not movable), the value of the friction coefficient was 0.17 and the applied blank holder force was 1500 kN. After the simulation was finished, the stress distribution (z-stress) was evaluated through the sheet thickness for one selected finite element placed in the middle of the part sidewall after first and after second forming step (Figure 10).

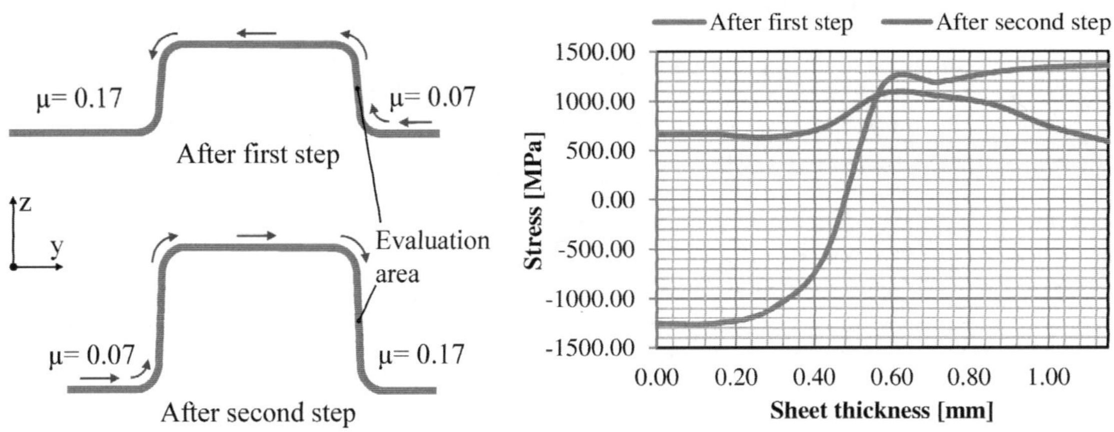

Figure 10. Stress distribution (z-stress) through the sheet thickness after first and second step in case of forming with the alternating draw-in

It can be seen that a significant difference in stress amount from the inner and outer part side for the chosen element occurs after the first forming step (from the inner part side the value of stress in z direction amounts to -1255,3 MPa while from the outer part side its value is 1357,9 MPa). This tendency of stress distribution in the part sidewall is comparable with the stress distribution, which normally occurs in conventional deep drawing. After second forming step, this difference in the middle of the part sidewall was reduced significantly (633,4 MPa from the inner part side and 588 MPa from the outer part side), see Figure 10.

It is necessary to say that the part sidewall undergoes a combined bending and drawing process during the classical deep drawing. After the classical drawing, tensile stresses at the outer part fibre and compressive stresses at the inner part fibre can occur, depending on the size of the die radius. During the release of the part, the point of time when the part is taken out of the die cavity, stress relaxation happens. Through the sheet thickness, the asymmetrical stress distribution generates one bending moment, which generally results in the curved part sidewall. During this action, the part fibres with the acting compressive stresses tend to elongate, while the part fibres with caused tensile stresses tend to be shorter.

It was mentioned that forming with an alternating blank draw-in results in a repeated bending and unbending of the workpiece over the punch radius. Due to this phenomenon, a stress superposition happens in the area, which is placed in the part sidewall at the end of forming process. The acting stresses from the outer and inner part fibre will have approximately the same value (Figure 10). Due to a significantly reduced difference in stress amount from the inner and outer part fibre, a smaller amount of springback can be expected.

5.2. Experimental try-out of forming with alternating blank draw-in

In order to verify the opportunities that were noticed in simulations, this kind of forming process was tested experimentally by using the tool shown in Figure 7a, and springback results were compared with the results obtained by conventional deep drawing. The mentioned tool was designed for conventional deep drawing performed on a single action press with a cushion. Because of that one of the possibilities to influence the blank draw-in was alternating lubrication of the appropriate blank side. This means that in case of two-step forming with alternating blank draw-in, just one side of the blank was lubricated (with 1.5 g/m² of the oil M100) while the second side was dry. The applied blank holder force for each step of this forming process was 1500 kN. For the first step, the forming depth on the press was defined to be 20 mm and due to differences in friction values only one side of the blank (the lubricated one) was drawn-in during the press stroke. After forming with the first defined depth, the part was prepared for the second step. Then, the second or opposite side of the blank was lubricated and the remaining oil from the first side was removed. In the second forming step, the opposite side of the blank (left in this case) was drawn-in. The obtained part shapes after the forming process were digitized using the optical measurement system GOM ATOS. An evaluation of the angle change at the cross sections that were defined earlier and part shape deviations in general was performed using the software GOM Inspect (Figure 7c).

Figure 11 shows the measured angle changes for the part which was formed using conventional deep drawing (applied blank holder force was 1500 kN) and for two parts which were formed using the two-step and three-step alternating blank draw-in. It becomes apparent that the part formed by conventional deep drawing has very high part shape deviations. The highest angle change was measured at the cross section 1 from the right side and amounts to 13.7°. Taking into account all angle changes at the defined cross sections for the formed part by conventional deep drawing, the average angle change is higher than 10°. The part formed using the two-step alternating blank draw-in shows significant smaller angle change for the observed cross sections. The highest measured angle change is located at the third cross section from the right side and amounts to 6.3°. It must be mentioned that the angle changes in every other cross sections are smaller than 4°. Furthermore, at the first cross section, the part sidewall from the right side has been deformed in opposite direction. Due to that, the angle change of -1.98° occurred.

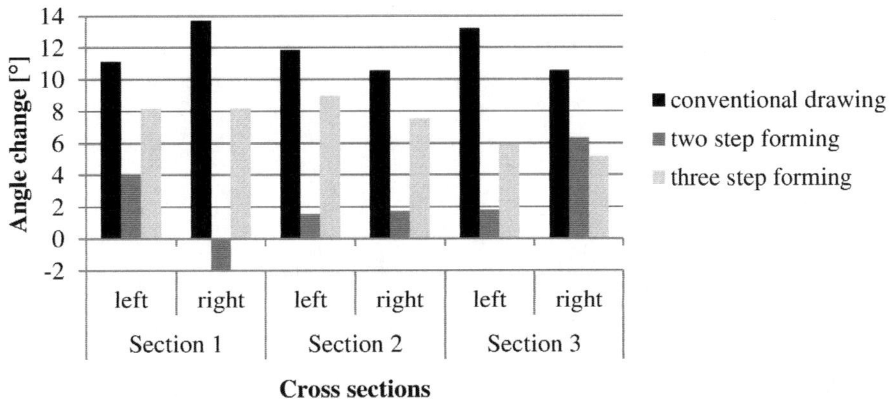

Figure 11. Influence of the blank draw-in on the angle change for applied blank holder force of 1500 kN (experimentally obtained results)

Figure 12 shows the deviation between the obtained part shapes according to the forming processes and reference geometries that were defined earlier (corresponds with the results shown in Figure 11). It is noticeable that the parts formed with the two-step alternating blank draw-in have obtained the curl in opposite direction (inwardly) at the part wall which is located from the side which was drawn-in first, (Figure 12b). This phenomenon can be explained due to fact that the part area which was bent over the punch radius in the first step was drawn in the second step back and is located approx. in the middle of the part sidewall. In this case, the stresses caused in the first forming step were superimposed in the second step. In order to minimize these failures, a forming strategy with three step alternating blank draw-in was tested. For this purpose, in the first 10 mm of the forming depth, the blank was drawn from the right side. During the next 20 mm of the press stroke, the blank was drawn from the opposite (left) side. And at the end, for the last 10 mm, the blank was drawn from the right side again. This measure has contributed to reducing of visible bulge or curl in the part bottom area (Figure 12c). When comparing the results obtained with the conventional deep drawing, it can be noticed that the angle change in all considered cross sections was reduced to approximately 3°. But for most of the observed cross sections, the obtained angle change is bigger than in case of forming with the two-step alternating blank draw-in.

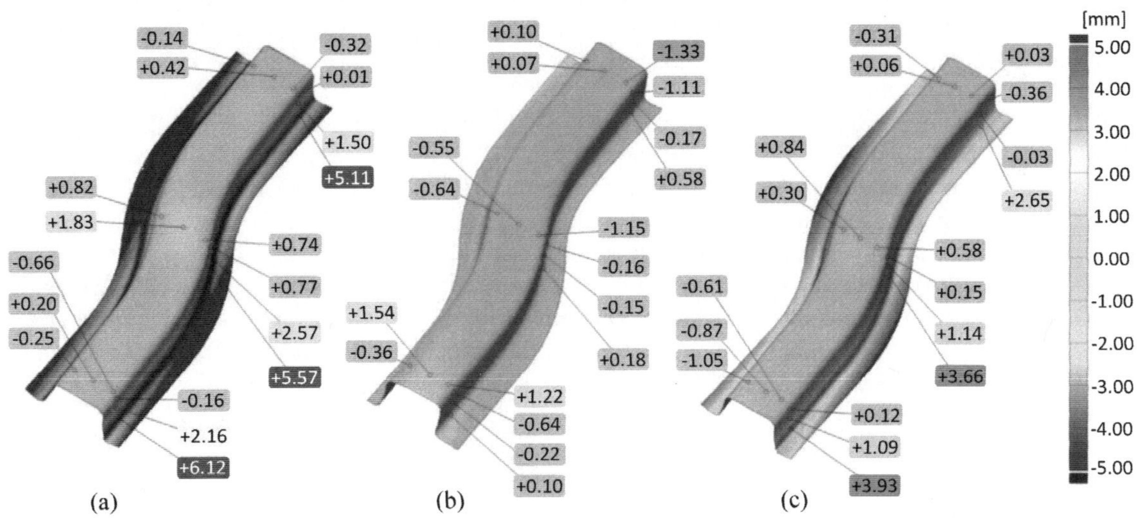

Figure 12. Part shape deviations measured by the optical measurement system GOM ATOS (a) after conventional deep drawing, (b) after two step forming, (c) after three step forming

6. Discussion

The results of this investigation present the applicability of different approaches for springback reduction. The simulation results for one curved hat channel shaped component showed a tremendous amount of springback, when using HCT 980 X as workpiece material. The significant difference in stress amount from inner and outer side of the part sidewall was identified as one reason for the mentioned amount of springback.

In order to reduce the stress difference and the springback amount, the tool geometry with numerous stiffening beads in the part sidewall was defined first (shown in Figure 5). This tool geometry was used to restrike the part shape that was obtained after the deep drawing process. After the restrike operation, the simulation results showed that the difference in stress amount from inner and outer part side in the observed area was reduced significantly. This appearance led to a reduced size of the sidewall curl and to a reduced total springback amount in general. In order to experimentally validate this kind of approach for springback reduction, the corresponding tool parts will be manufactured in this project.

Furthermore, numerous experiments were carried out with the available tool in order to investigate the influence of the blank holder force on the occurrence of springback. The obtained results showed that the level of applied blank holder force has a low influence on the occurrence of springback. In case of higher applied blank holder forces the springback amount will be slightly reduced. In order to reduce the angle change by 2° for the used part geometry it is necessary to increase the blank holder force by more than 1000 kN.

In this investigation, the forming with alternating blank draw-in (pendular drawing) was analysed as well. The simulation results showed that the difference in stress amount from inner and outer part side will be lower in case of forming with the alternating blank draw-in in comparison to conventional deep drawing, which means that this kind of forming process causes a superposition of the acting stresses in the part sidewall. This phenomenon led to a significant reduction of springback when forming steel grade HCT 980 X.

7. Conclusion

In this paper, the success as well as the applicability of several approaches to reduce springback when forming curved hat channel shaped components was analysed. Meanwhile, a new approach to reduce springback when forming curved hat channel shaped components was proposed. Forming with an alternating blank draw-in has shown a promising technique for springback reduction in forming of one hat channel shaped component out of steel grade HCT 980 X. The experimental results showed that after forming with the two-step alternating blank draw-in, the part shape deviations were reduced significantly (more than 50 %) in comparison to the obtained shape deviations that occurred after the conventional deep drawing.

The additional conclusions which can be drawn from this investigation are:

- Using an additional restrike operation during the forming process can be a helpful technique for springback reduction. Simulation results showed that embossing the part sidewall with the appropriate beads in the restrike operation minimizes the stress difference between inner and outer part side and leads to a reduced amount of springback.
- The blank holder force had a slight or almost negligible influence on springback occurrence when forming the hat channel shaped component. Experimental results showed that a higher blank holder force leads to a smaller amount of springback. The parts formed with a blank holder force of 1500 kN have shown an approximately 2° reduced angle change in comparison to the results obtained by forming with a blank holder force of 300 kN.
- When comparing all approaches analysed in this work, it was possible to achieve the most precise part shape geometry, which had the smallest amount of springback, by forming with the alternating blank draw-in.

Acknowledgments
The authors are grateful to the German Research Foundation (DFG) for its financial support.

References

[1] Hamad ul. H, Maqbool F, Güner A, Hartmaier A, Khalifa N.B and Tekkaya A.E 2015 Springback prediction and reduction in deep drawing under influence of unloading modulus degradation *Int. J. Mater. Form.* DOI 10.1007/s12289-015-1248-5

[2] Liewald M and Radonjic R 2014 *Proc. 4rd Int. Conf. on Steels in Cars and Trucks* (Brauschweig) p 38

[3] Liewald M, Wagner S and Radonjic R 2014 *Proc. 14th Stuttgart Int. Symposium Automotive and Engine Technology* (Stuttgart) p 496

[4] Liewald M and Radonjic R 2013 *Proc. Tools and Technologies for Processing Ultra High Strength Materials*-TTP2013 (Graz) p 323

[5] Peng C and Muammer K 2007 Simulation of springback variation in forming of advanced high strength steels *J. of Mater. Proc. Tech.* **190** 189

[6] Weinschenk A and Volk W 2014 Decrease of springback by geometrical modification of the sheet metal part *Adv. Mater. Research* **1018** 277

[7] Yoshida T, Sato K, Isogai E and Hashimoto K 2013 Springback problems in forming of high-strength steels sheets and countermeasures *Nip. St. Tech. Rep.* **103** 4

IOP Conf. Series: Materials Science and Engineering **159** (2016) 012029 doi:10.1088/1757-899X/159/1/012029

Considering the edge-crack sensitivity of a hot-rolled steel in forming simulation

T Gläsner[1], M Schneider[2], M Troitzsch[3] and S Westhäuser[2]

[1] Volkswagen AG, Brieffach 013/3370, 38112 Braunschweig, Germany
[2] Salzgitter Mannesmann Forschung GmbH, Eisenhüttenstr. 99, 38239 Salzgitter, Germany
[3] OSB AG for Volkswagen AG Group Research, 38436 Wolfsburg

E-mail: thorsten.glaesner@volkswagen.de

Abstract. The formability of sheet metal materials is locally reduced by shear cutting operations, and as a result the risk of a crack during further processing is increased at the edge. Materials particularly susceptible to this are described as sensitive to edge-cracking. A procedure for quantitatively determining edge-crack sensitivity and for applying corresponding characteristic values has not been previously established. Below, two test methods and an approach for using the results in an extended forming limit diagram are presented. The producibility of a collar drawn test component as well as a chassis component is reevaluated using this extended forming limit diagram.

1. Introduction

Because of economic goals and legal requirements, the automotive industry faces the challenge of reducing the CO_2 emissions of vehicles. A very promising approach to this challenge is lightweight steel construction. A reduction in the vehicle mass directly influences the CO_2 emissions. For every 100 kilograms of mass spared, the CO_2 emissions are reduced by 8.5 g/km [1]. The steel industry has recognized this trend and developed new products and application possibilities. Steel grades are becoming stronger and stronger, sheet metals used are becoming thinner and thinner, and component geometries are becoming more and more complex. Therefore, the requirements for the mechanical properties of the initial material increase with each new component generation. At the same time, forming technology faces appropriate challenges.

An example of such a challenge is the edge-crack sensitivity of shear-cut materials. A shear cutting process can significantly reduce the formability of the newly produced edge in comparison with the formability of the base material. In subsequent forming operations, crack formation starting from the edge can occur because of the reduced formability. Up to now it was not possible to predict this crack formation with sufficient accuracy using existing experimental and simulation methods. In the field of forming simulation, the evaluation of the formability of a sheet metal by using the forming limit curve (FLC) in the corresponding forming limit diagram (FLD) is the state of the art. However, a universally applicable approach to simulating edge-crack sensitivity that has been tested and verified on a meaningful number of practical components has not been previously established. However, approaches that add limits for the cut edge region to the forming limit diagram have been published [1a].

In cooperation with ESI GmbH, Salzgitter Mannesmann Forschung GmbH, Volkswagen Component Toolmaking, and Volkswagen Group Research, an industrially usable procedure has now been defined and has been applied to a chassis component. The "hole expansion with Nakajima punch" and "Kobe hole tensile test" testing methods used for this purpose are presented below.

2. Technical status

The hole expanding test as per ISO 16630 is currently the only globally standardized test for determining edge-crack sensitivity. In this test, a hole with a diameter of 10 mm is made in the sheet metal specimen by shear cutting (die clearance: 12%) and then expanded with a conical punch (head cone angle: 60°). The operator stops the expansion as soon as the operator visually detects a crack extending through the entire thickness of the sheet metal. Consequently, the results of this test greatly depend on the perception and reaction speed of the operator. The test result is given by the hole expansion ratio, which is defined as the ratio of the hole diameter increase to the initial hole diameter [2].

Because of the described test method, which allows, among other things, a strong influence on the measurement results by the operator, the hole expansion ratios determined from at least three individual specimens generally have very high scattering. This high scattering can be observed both within a test location and across test locations [3]. Therefore, an increase in the number of specimens, among other things, is called for in [3] with the goal of providing better statistical assurance.

In addition to the standardized hole expanding test, there are numerous other test methods that have been developed in order to quantify edge-crack sensitivity [3a].

3. Characterisation methods, testing procedure und material describtion

3.1. Determination of edge forming limit

The test setup for determining a forming limit curve is used for the "hole expansion with Nakajima punch." As in the case of the ISO 16630 hole expanding test, the test consists of two steps. First, a hole (diameter: 20 mm) is made in a square specimen (edge length: 200 mm) by punching (die clearance: 12%); see figure 1. In the second step, the specimen prepared in such a way is expanded with a hemispherical punch (diameter: 100 mm). On the basis of the ISO 16630 hole expanding test, the selected punch speed must be less than or equal to 1.0 mm/s. The specimen must be placed in such a way that the punching direction corresponds to the forming direction. The specimen must be precisely centered to achieve reliable results. The test is immediately stopped as soon as a crack extending through the entire thickness of the sheet metal can be detected. As in the case of the ISO 16630 hole expanding test, at least three specimens are tested per setting. The crack initiation occurs more abruptly in the "hole expansion with Nakajima punch" than in the ISO 16630 hole expanding test [5]. For this reason, a crack width correction, as presented in [6], should be performed in the evaluation. The result is, again, the hole expansion ratio described above. However, in contrast to the ISO 16630 hole expanding test, a stochastic pattern can be applied to the surface of the sheet metal before the forming and a detailed elongation analysis for the region of the specimen near the edge can be performed with the ARAMIS optical measurement system from the company GOM. A comparable analysis is not possible with a standard ARAMIS system for the hole expansion as per ISO 16630, because the specimen surface moves out of the initial sheet metal plane to a high degree during the test. The crack initiation and the hole expansion ratio can be automatically detected and determined by means of an evaluation macro based on Visual Basic, which contains defined crack criteria [7].

For the "Kobe hole tensile test," a rectangular tensile specimen is used (250 x 40 mm^2) as per [8], in which a hole having a diameter of 10 mm is punched in a centered manner (die clearance: 14%); see figure 1. Then the specimen is drawn using a tensile test machine at a speed of 10 mm/min until a crack is initiated at the hole edge. In contrast to the procedure as per [8], the characteristic value is determined virtually. As in the case of the "hole expansion with Nakajima punch," a stochastic pattern is used so that the displacements and therefore the elongations in the hole region can be detected by using the ARAMIS measurement system. The specimen is evaluated by means of a virtual measurement length. In the initial state, this measurement length has a length of 2 mm and a distance from the edge of 1 mm and is oriented in the direction of the longitudinal axis. The characteristic value of this test method is given by the ratio of the extension of this measurement length at the start of cracking to the initial measurement length.

IOP Conf. Series: Materials Science and Engineering **159** (2016) 012029 doi:10.1088/1757-899X/159/1/012029

However, the gauge length value of 2 mm specified in [8] lies within the order of magnitude of the element size used in the finite element simulation. Because an optical strain measurement system (ARAMIS) is used in both the "hole expansion with Nakajima punch" and the "Kobe hole tensile test," the gauge length can be variably adapted. In this way, an elongation limit value depending on the element size can be determined in accordance with requirements.

The two testing methods, "hole expansion with Nakajima punch" and "Kobe hole tensile test," are presented below. These testing methods allow optical measurement systems to be used, which can significantly reduce the operator influence and therefore the scattering of the test results [4].

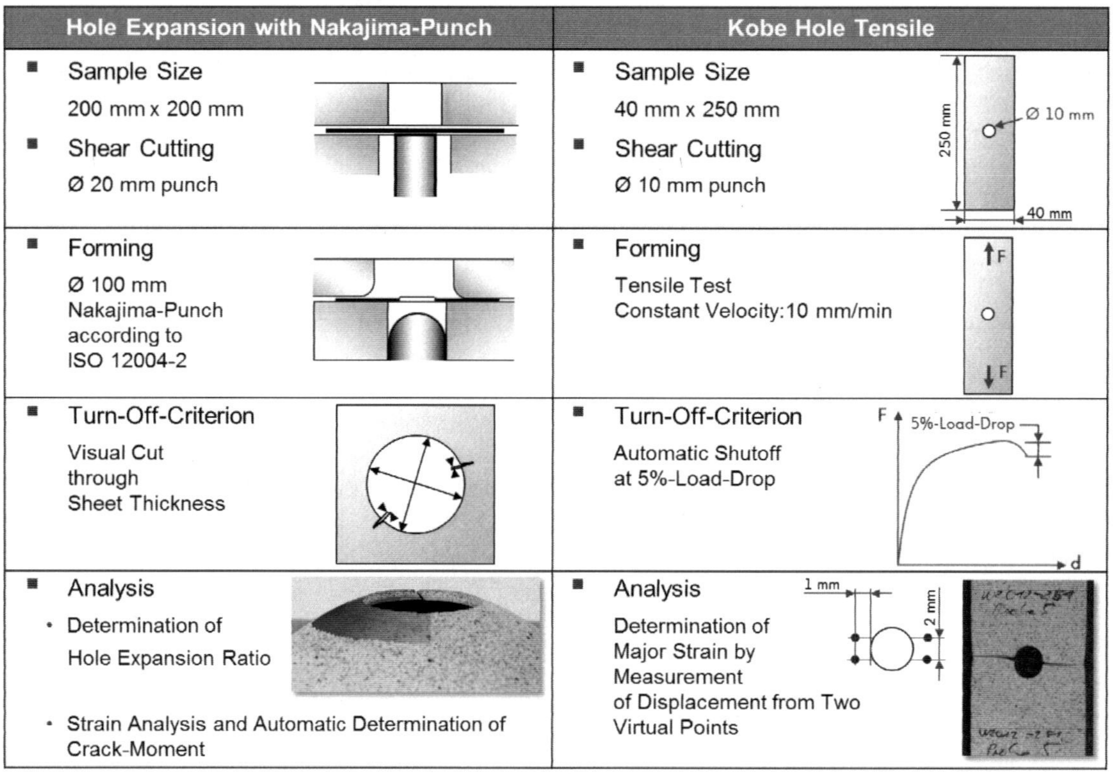

Figure 1. Test procedure of the "hole expansion with Nakajima punch" and "Kobe hole tensile test" edge-crack tests

The determined "edge forming limit" characteristic value is then integrated into an FLD as additional information. Figure 2 shows the traditional FLC (blue) and the addition of the edge forming limit (red) as a horizontal line. In the example below, an edge forming limit of about $\varphi = 0.43$ has been plotted.

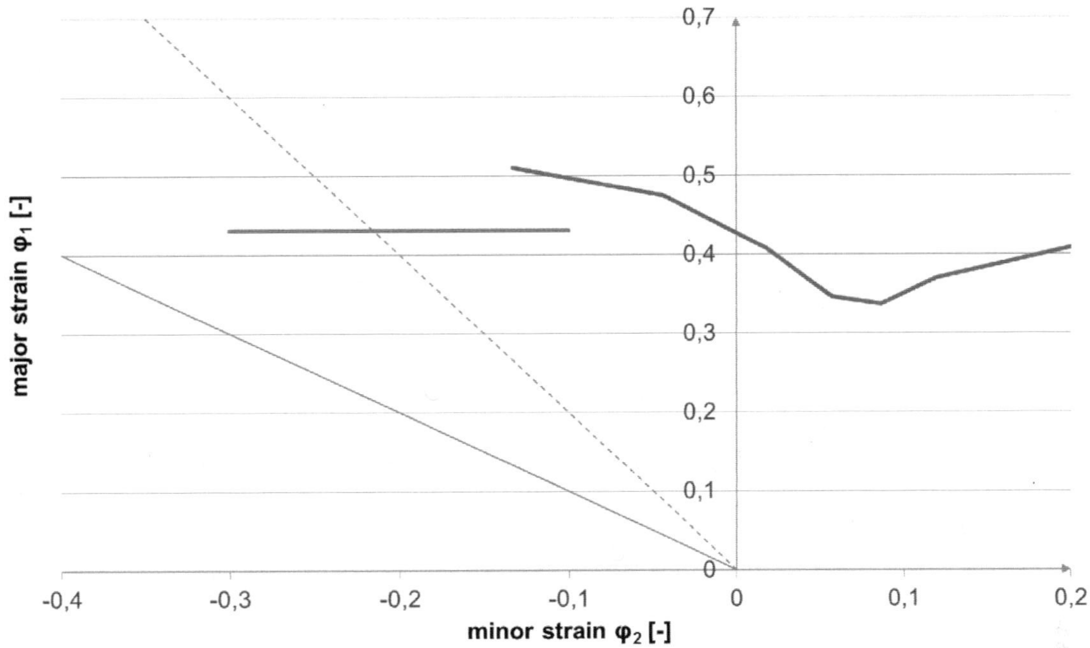

Figure 2. Forming limit diagram with forming limit curve and edge forming limit [1a]

3.2. Test geometry

Modularly constructed test tools are used for the shear cutting of the test blanks and the collar drawing on simple test geometries. These test tools were designed and developed at the Institute of Metal Forming and Casting at Technische Universität München. Both tools can be converted for the required parameters in a few steps (see figure 3).

Figure 3. Test tool for shear cutting and collar drawing

Thus, different pre-holes and die clearances can be realized with the cutting tool. In the collar-drawing configuration, the expansion ratio can be varied. For the various test series, a hole with Ø 50 mm is made in each blank (200 mm x 160 mm x 3.5 mm) by shear cutting. Then this hole is expanded by a conical punch. A total of three different expansion ratios are tested (see table 1). The expansion ratio λ (see formula 1) is set by varying the drawing punch diameter. The pre-hole was held constant at Ø 50 mm.

$$\lambda = \frac{dz - 50\,mm}{50\,mm} \circ 100\,\% \qquad (1)$$

IOP Conf. Series: Materials Science and Engineering **159** (2016) 012029 doi:10.1088/1757-899X/159/1/012029

Table 1. Tool parameters and resulting expansion ratios

Expansion ratio λ [%]	Pre-hole diameter [mm]	Collar-drawing punch diameter [mm]
30	50	65
50	50	75
70	50	85

3.3. Test material

SZBS800 belongs to the group of complex phase steels and is distinguished by a very high yield point and tensile strength because of its largely bainitic microstructure. The most important mechanical properties are listed in table 2 below on the basis of the material data sheet of Salzgitter Flachstahl GmbH [9].

Table 2. Mechanical properties of the test material SZBS800 [9]

Characteristic	Value
Test material	SZBS800
Yield strength	$\geq$ 680 MPa
Tensile strength	800 – 980 MPa
Elongation at break	$\geq$ 12 %

This complex phase steel is well suited particularly for forming processes. Particularly in the case of form rolling, bending, and collar drawing, SZBS800 can outperform other steel grades [9]. This is due to the low edge-crack sensitivity of SZBS800. The elongation progression along the cutting edge of the hole can be shown by using polar diagrams generated from results of the "hole expansion with Nakajima punch" (see figure 4). The elongation progressions shown indicate that the greatest elongations and contractions, which describe the anisotropy of the material, occur perpendicularly and parallel to the rolling direction. Therefore, it can be assumed that the crack nucleation likewise occurs in accordance with this orientation [6].

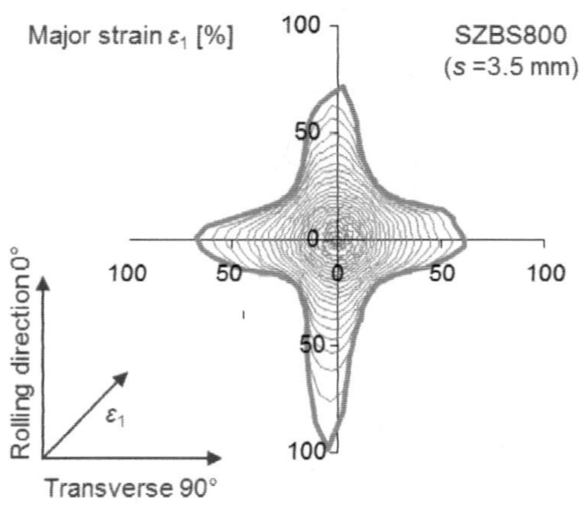

Figure 4. Polar diagram of the elongation distribution in the hole expansion [6]

4. Results

4.1. Results for determining the edge forming limit

The results of the "hole expansion with Nakajima punch" and "Kobe hole tensile test" testing methods are shown in figure 5. In the case of the "hole expansion with Nakajima punch," the results are from the macro evaluation. In the case of the "Kobe hole tensile test" testing method, the results were determined in accordance with the rolling direction (longitudinal, diagonal, and transverse) with five individual specimens in each case. The minimum and maximum individual results in each case were not taken into account in the calculation of the mean or the standard deviation. Therefore, each mean and standard deviation given in figure 5 represents the average from three specimens.

The characteristic values of the "hole expansion with Nakajima punch" and of the "Kobe hole tensile test" with an orientation transverse to the rolling direction are at the same level. Because the specimen of the "hole expansion with Nakajima punch," which is loaded uniformly over the contour of the punched edge, tends to crack in the rolling direction (0° or 180° direction), these two test variants have the same crack propagation direction. Also striking is that the specimen of the "Kobe hole tensile test" with an orientation transverse to the rolling direction represents the case in which the material can withstand the smallest strains. The characteristic values for the "Kobe hole tensile test" with orientations longitudinal and diagonal to the rolling direction are higher on average. At this point, it has to be mentioned, that the influences of tool contact, gauge length and strain gradients are ignored. The focus of this work is on the use of the test result in forming simulation.

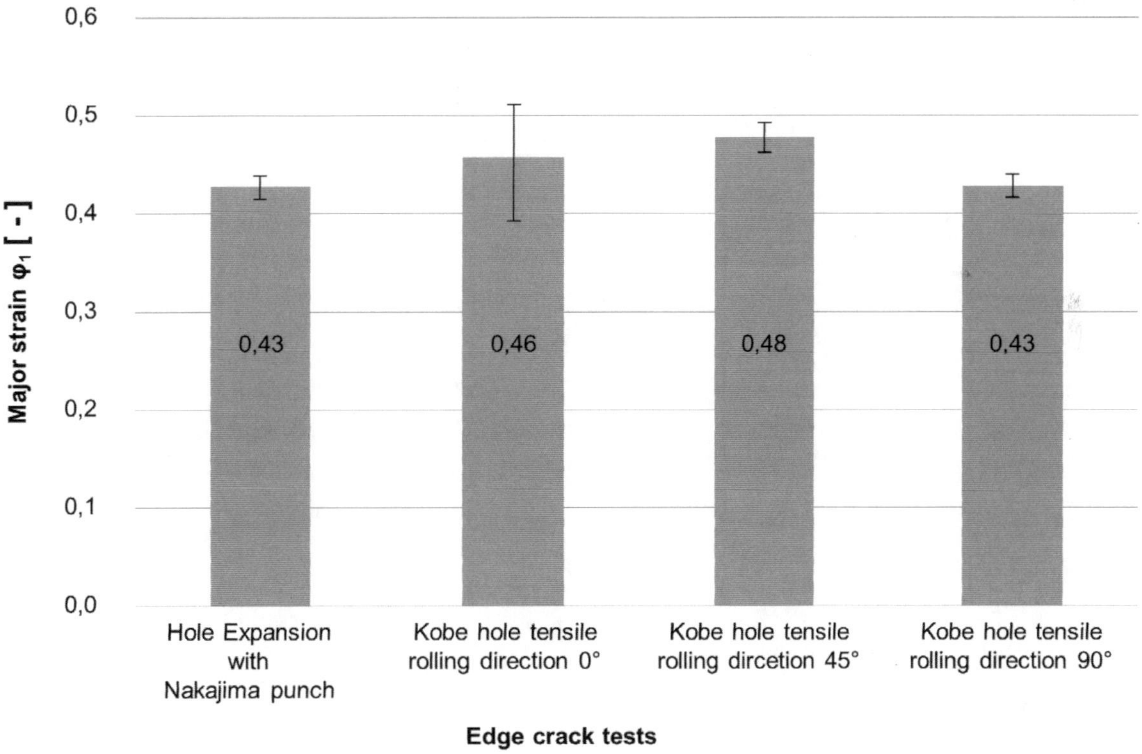

Figure 5. Results of the "hole expansion with Nakajima punch" and "Kobe hole tensile test" edge-crack tests

The simulation results were compared with the real forming tests by using the simulation software PAM-STAMP from ESI GmbH. The comparison of the achieved results with those of the ARAMIS measurement shows that there is close qualitative and quantitative correspondence. This comparison is shown in figure 6 for the first major strain in the "hole expansion with Nakajima punch" as an example.

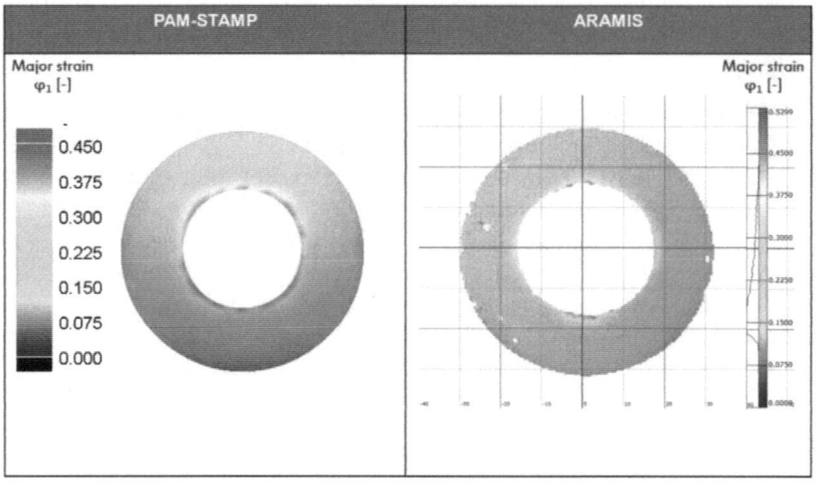

Figure 6. Comparison of the major strains in the "hole expansion with Nakajima punch" on the top side between the simulation and the laboratory specimen

Figure 7 shows the forming limit diagram extended with the edge forming limit.

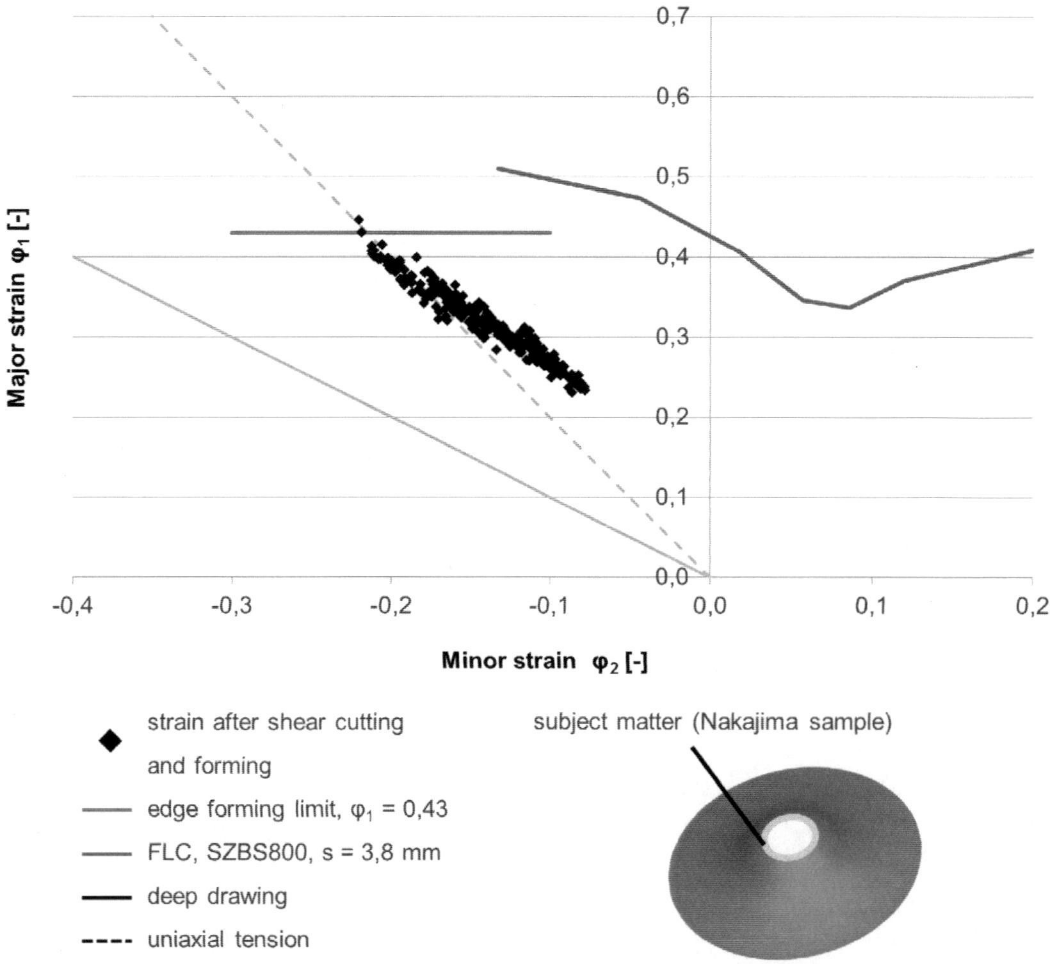

Figure 7. Forming limit diagram with strain limits for the edge-crack sensitivity, and edge elements from the numerical simulation of the "hole expansion with Nakajima punch"

IOP Conf. Series: Materials Science and Engineering **159** (2016) 012029 doi:10.1088/1757-899X/159/1/012029

If the elongations from the simulation of the Nakajima specimen with corresponding punch travel are plotted in this diagram, a risk of cracking due to the elongation limit being exceeded is indicated.

Likewise for the Kobe hole tensile test, the results from the numerical simulation with the PAM-STAMP software correspond closely to the experimental results with regard to the elongation distribution and localization. In contrast to the "hole expansion with Nakajima punch", the region of maximum strains is already predetermined because of the specimen geometry. Figure 8 shows the addition of the elongation limit arising from the edge tests to the forming limit diagram. Here again, the point of failure is correctly predicted.

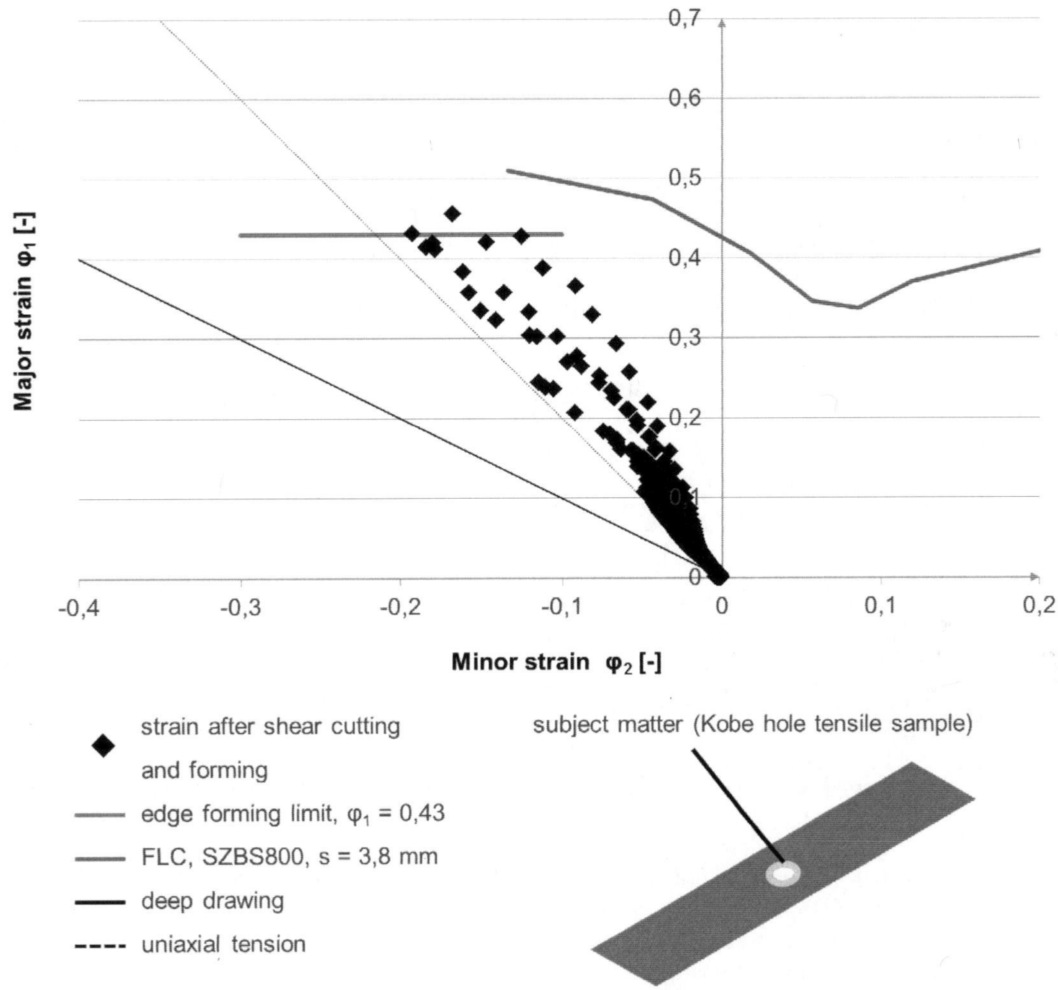

Figure 8. Forming limit diagram with elongation limit for the edge-crack sensitivity, and edge element elongation from the numerical simulation of the "Kobe hole tensile test"

4.2. Application of the results to a test geometry

The "edge forming limit" characteristic value is validated using a simple test geometry in a first step. For this purpose, both simulation calculations (including of the expanded FLD) and experimental series of tests were performed and compared with each other. The simulations were performed with the simulation software AutoForm (version 6) from AutoForm Engineering GmbH and were checked with the experimental results (see test tool). The software offers the advantage that the "edge forming limit" can be stored as a parameter by using an edge-crack criterion. The failure criterion as per McEwan-Underhill-Langerack is used here [1a]. For the shear cutting, the previously determined value of

IOP Conf. Series: Materials Science and Engineering **159** (2016) 012029 doi:10.1088/1757-899X/159/1/012029

$\varphi_1 = 0.43$ was entered for this purpose. An optimized cut edge can be achieved by two-stage shear cutting (φ_2) [11, 12, 14]; in this case, the possible edge forming limit is increased ($\varphi_{1\,(\text{Two-Stage Cutting})} > \varphi_1$). Thus, it is possible to analyze the influence of different qualities of the cut edge in the simulation.

First, the traditional evaluation methods of thickness, formability, and forming limit diagram (without the addition of the edge forming limit characteristic value) are shown and compared with the real test in figure 9. The simulation results show that the strain in the edge region increases significantly with increasing expansion ratio. Satisfactory predictions with regard to edge-crack sensitivity are possible only for expansion ratios of λ = 30% and 70%. The predictions in the limit range (λ = 50%) are insufficiently accurate. If only the FLD is used as a criterion, the strain of the edge elements approaches the forming limit curve in the diagram region of uniaxial tension but does not exceed the forming limit curve. In reality, however, cracks over the thickness of the sheet metal can already be identified.

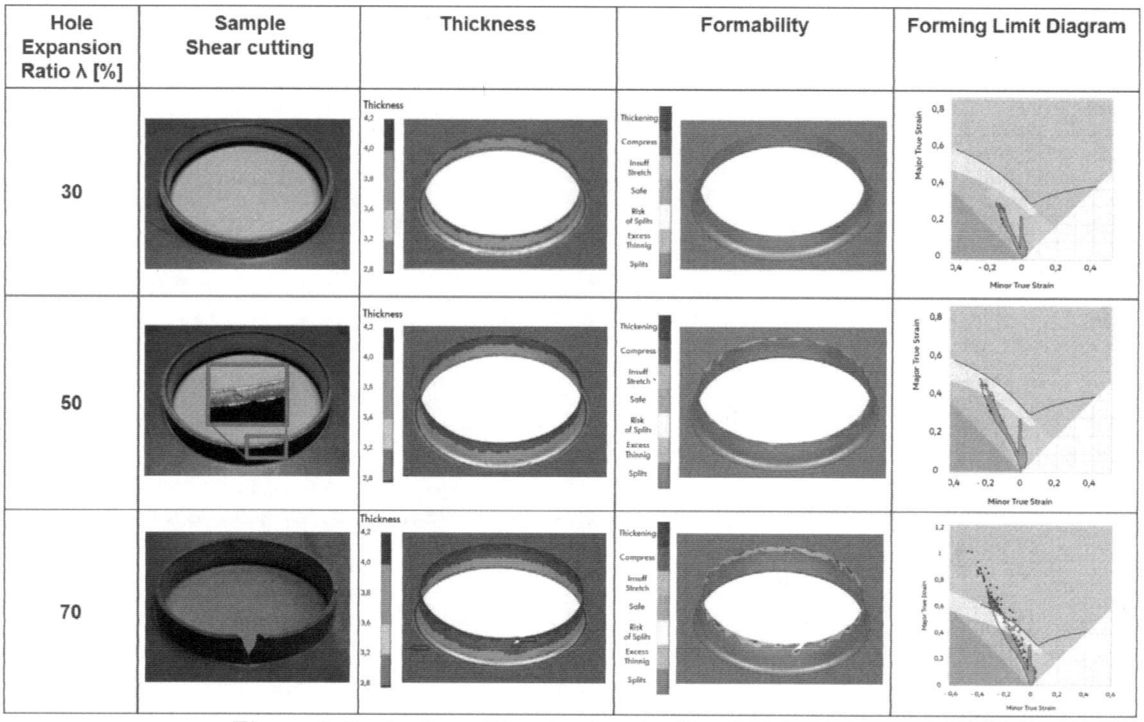

Figure 9. Evaluation of the simulated collar-drawing process

When the edge-crack criterion (including the previously determined elongation limits) is used, the edge-crack sensitivity of SZBS800 is found to be clearly predictable in the evaluation of the forming process (see figure 10). Both in the case of normal cutting and in the case of the optimized shear cutting method, the two-stage shear cutting, the simulation prediction and the experimental tests correspond very closely to each other. If the starting hole is expanded into a collar at an expansion ratio of 30%, it is found that there is no risk of edge cracking in the case of normal cutting or two-stage cutting. The experimental series of tests for these two cases confirm this. In contrast, first cracks occur at an expansion ratio of λ = 50% after normal cutting, whereas a pre-hole cut in two stages shows no tendency toward edge-crack sensitivity [11, 12]. If the expansion ratio is increased further, the risk of edge cracks also increases in the case of two-stage cutting, but edge-cracks still cannot be visually detected at λ = 70% on the specimen.

Hole Expansion Ratio λ [%]	Sample Shear Cutting	Cut-Edge-Criterion Shear Cutting $\varphi_1 = 0{,}43$	Sample Two-Stage Shear Cutting	Cut-Edge-Criterion Two-Stage Shear Cutting $\varphi_{1\,(Two\text{-}Stage\ Cutting)} > \varphi_1$
30		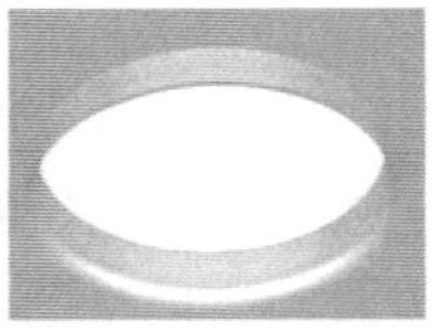		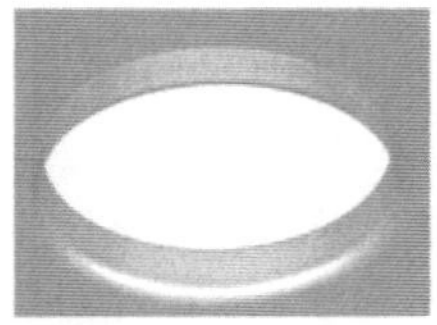
50	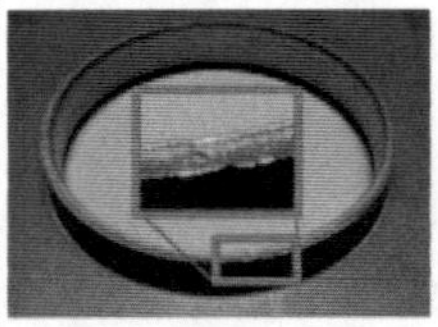	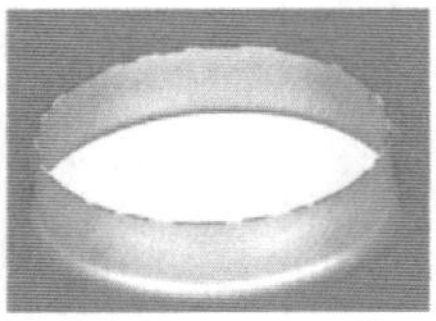		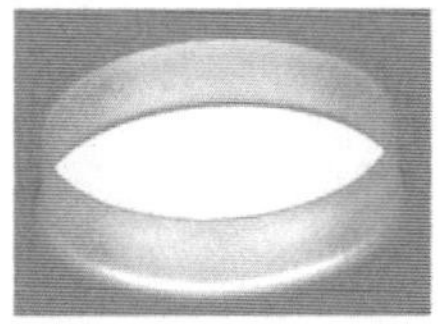
70	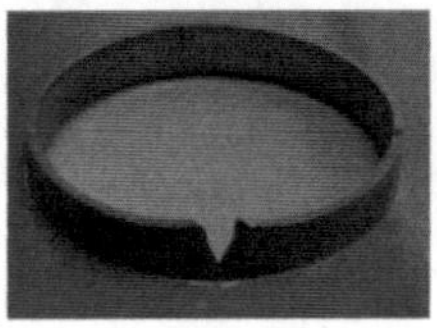	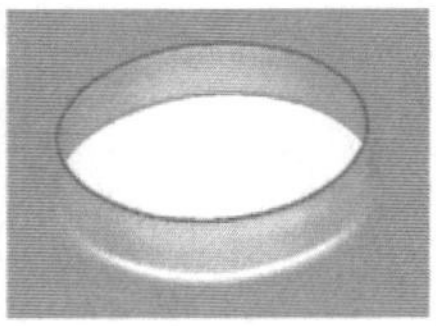		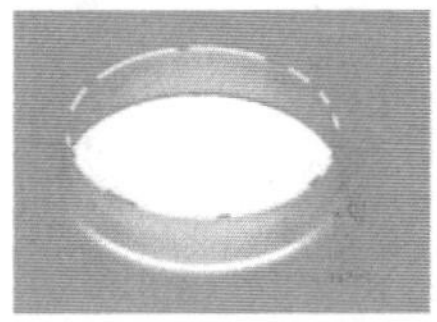

Figure 10. Application of the cut-edge criterion to the test geometry

4.3. Application of the findings to a chassis component

Below, the modular transverse matrix MQB-A track control arm for the Golf platform of the Volkswagen Group is used as an example of how the procedure, which has previously been applied only to geometries on a laboratory scale, can be put into practice. In addition, the simulation calculations were performed in this case with the PAM-STAMP software from the company ESI GmbH, again in order to establish broad acceptance of the "edge forming limit" characteristic value. On the basis of experience from the series production process, it is also assumed that the standardized simple cutting process used to produce the laboratory specimens would result in a failure of the component edges in the production of the rear track control arm.

To ensure in advance that the elongations in the forming simulation correspond to the conditions of the component, a forming analysis of the series production process is performed (see figure 11).

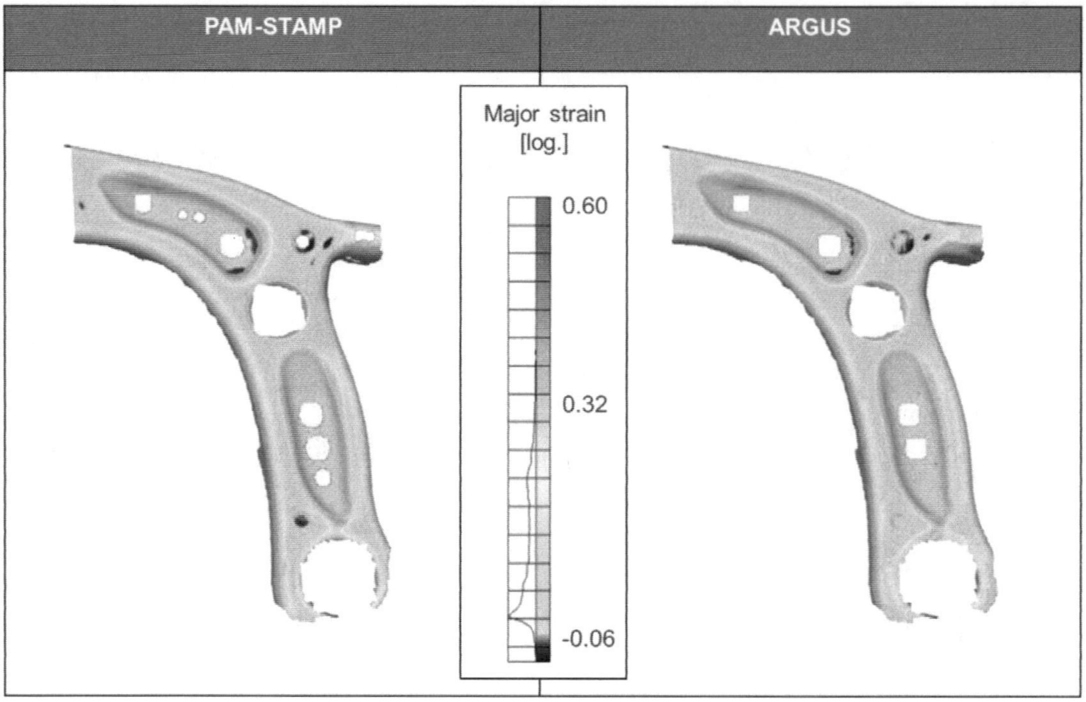

Figure 11. Comparison of the major strains from the forming simulation (PAM-STAMP) and the forming analysis (ARGUS)

For this purpose, equidistant points are electrolytically applied to the flat sheet metal before forming. The sheet metal is fed to the series production process and is removed again after forming. Three-dimensional coordinates and elongation distributions are calculated from digital photos of the formed sheet metal by using the optical elongation measurement system ARGUS from the company GOM. These correspond to the surface of the sheet metal [13]. Figure 11 shows the distribution of the major strain from PAM-STAMP on the left and from ARGUS on the right. The correspondence is high enough for the further examination.

Below, various regions of the MQB-A track control arm are checked to determine whether the edge forming limit ($\varphi_1 = 0.43$) determined on the laboratory scale can also be applied to a production component. The simulation examinations are performed at three points in total. The first two examples concern regions that are not significantly formed before the cutting. In contrast, the third region is an example of a component zone that is pre-elongated, cut, and further formed.

In figure 12, the element elongations of the PAM-STAMP simulation for an outer edge of the MQB-A track control arm are shown in the expanded forming limit diagram, which also contains the edge forming limit in addition to the traditional forming limit curve.

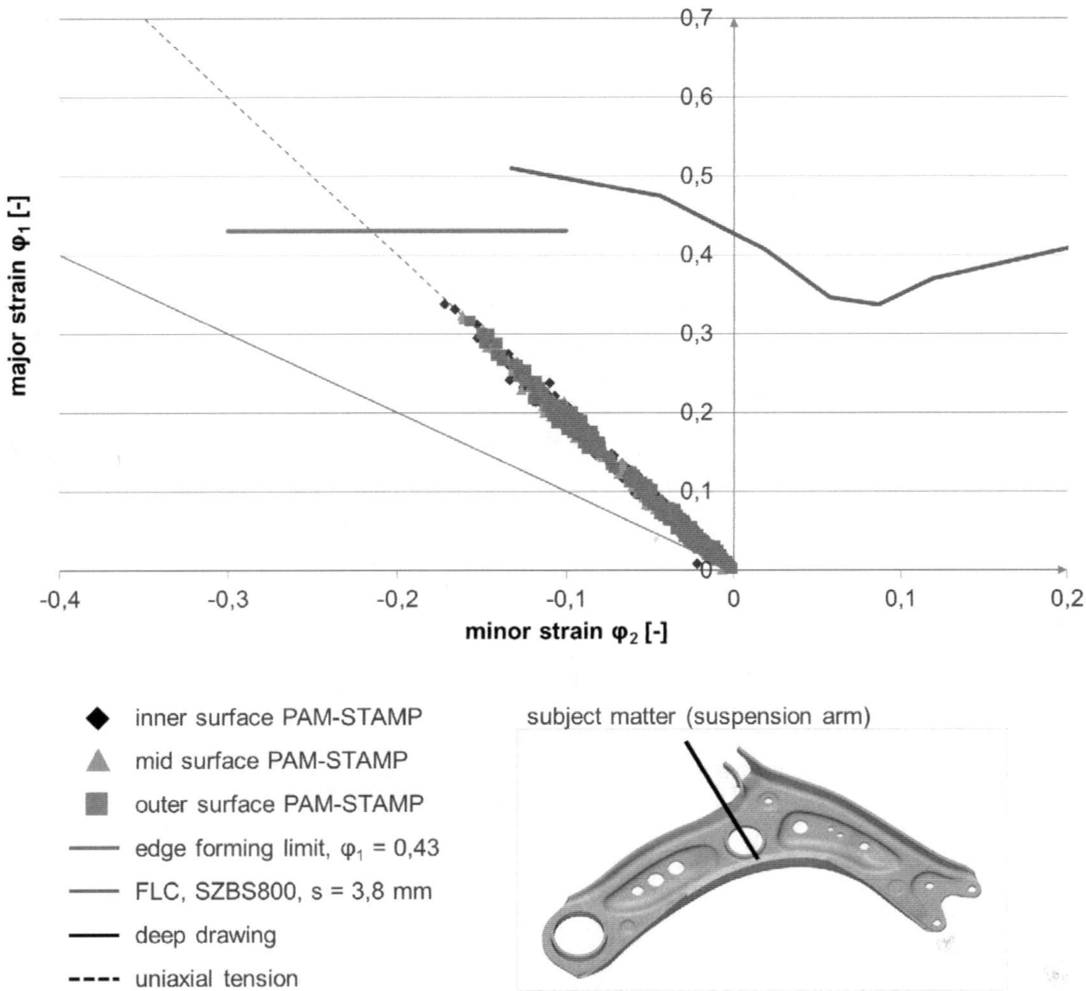

Figure 12. Evaluation of the producibility of the outer edge of the track control arm

As in the edge-crack tests, it is found that the examined region close to the edge is loaded in accordance with uniaxial tension. All elements are clearly below the FLC as well as the edge forming limit and therefore can be assessed as not at risk of edge-cracking. In addition, no influence of the sheet metal plane to be evaluated can be found. The inner, middle, and outer surfaces are only exposed to an elongation that can be withstood.

Rim holes, produced by shear cutting of a hole which is expanded by collar drawing, traditionally have proven to be particularly sensitive to edge-cracking. In collar drawing in flat sheet metal, the sheet metal segment lying over the drawing die opening is bent around the edges of the drawing die and the drawing punch. During this forming, the diameter of the pre-hole is increased while the thickness of the sheet metal is simultaneously reduced. The subject of the examinations is the collar in the center of the component (see figure 13). Because there is an elongation gradient over the thickness of the sheet metal here due to the forming, a different pattern results for each evaluation plane. Therefore, the evaluation plane having the greatest elongation must be selected for the evaluation of producibility. In the present case, this means that the rim hole is identified as not producible in the case of a standardized simple punching process. This corresponds to the empirical values. For this reason, a two-stage shear cutting process is used in practice. The two-stage shear cutting process damages the material to a lesser degree in comparison with the one-stage shear cutting process and thereby enables significantly greater formability of the edge [11].

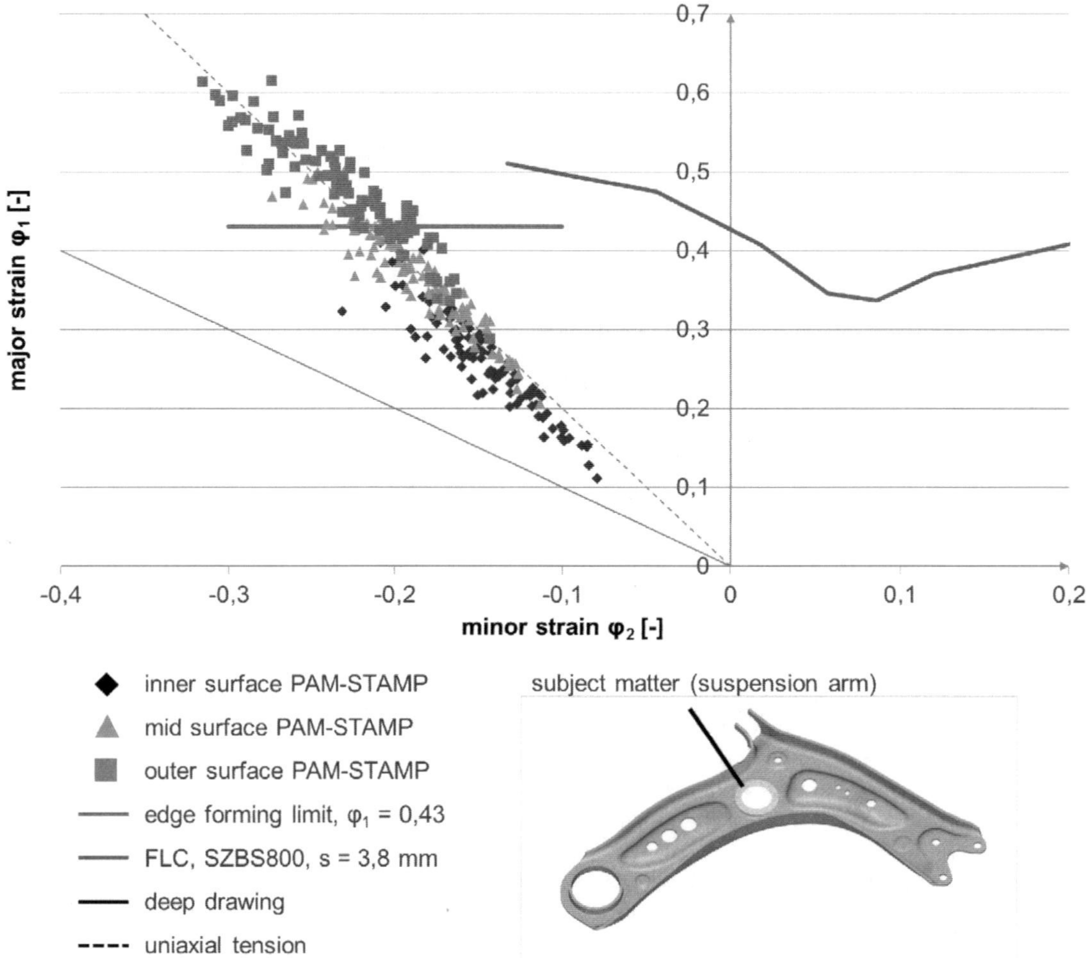

Figure 13. Evaluation of a rim hole with regard to producibility

In contrast to the previous examinations, the region examined below (see figure 14) was already formed before the cutting. In accordance with the production method, the region is formed, cut, and finally formed again. If the evaluation described above is performed here and the elongations accumulated from the two forming stages are plotted, a failure of the edge is predicted. This contradicts experience from the real production process.

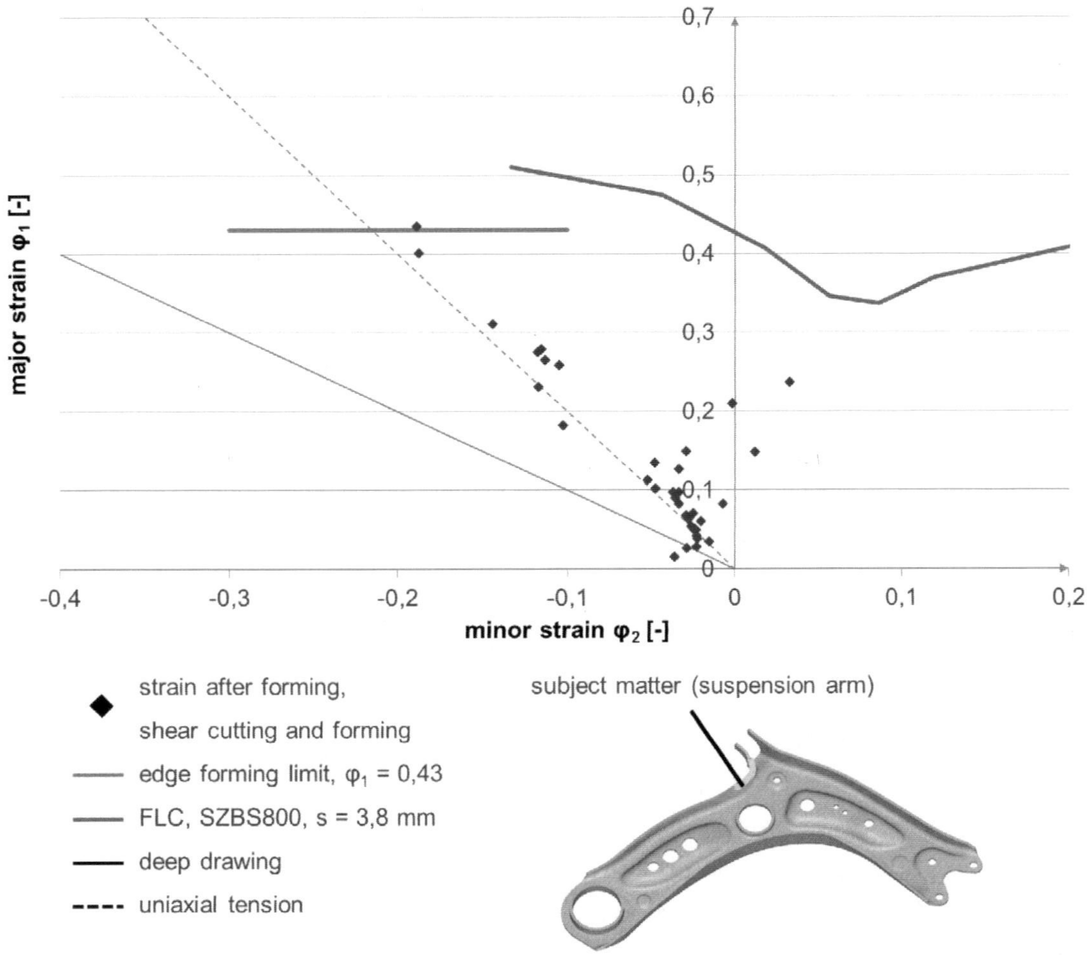

Figure 14. Evaluation of the producibility of a region that is formed, cut, and formed again (prediction 1)

This apparent misjudgment from prediction 1 could be avoided if merely the elongation path since the cutting operation were indicated in the post-processing. Figure 15 shows the corresponding illustration of the second prediction. However, this procedure is based on the assumption that the pre-strains has no influence on the formability of a shear-cut edge. This optimistic assumption appears to be just as improbable as the previous prediction. How the pre-damage (before the cutting process) affects the formability of a cut edge must be determined in the future. A test variant of the two presented edge-crack tests, with pre-strained specimens, would be conceivable. Currently, the two assumptions and predictions made are the possible limits for this.

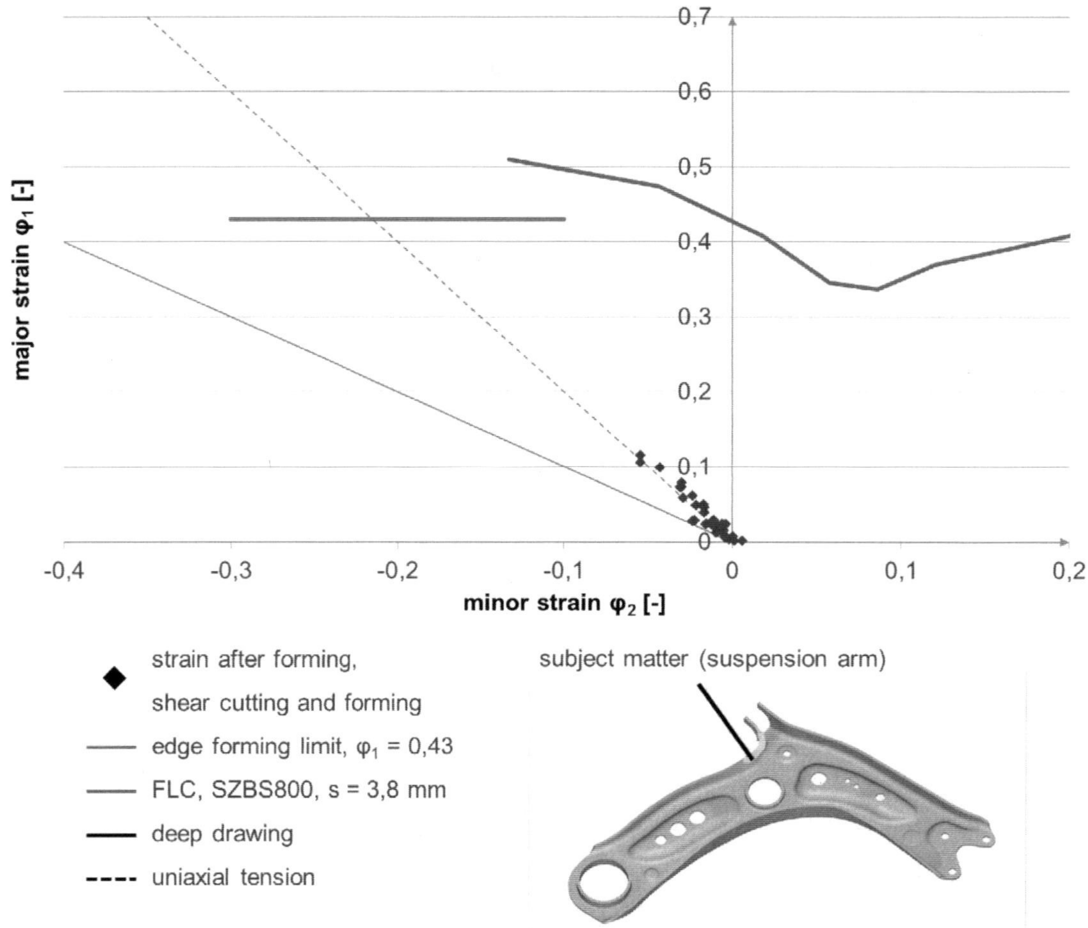

Figure 15. Evaluation of the producibility of a region that is formed, cut, and formed again by examining the elongation since the cutting (prediction 2)

4.4. Deriving a work instruction

The following instructions can be derived from the above examples of how to implement the described approach for taking the reduced formability of a shear-cut sheet metal edge into account in a forming simulation.

- Because this is pure post-processing, the approach can be performed regardless of the material model used.

- The blank edge to be analyzed must have a fine mesh consisting of approximately square quadrilateral shell elements (standard value: 1.0 mm edge length).

    - If complex cutting operations are part of a multi-stage forming process, an attempt must be made to achieve the aforementioned mesh quality in the edge region by suitable software settings.

    - In PAM-STAMP, for example, this can be approximately achieved by using the function "Optimize for flanging" and the sub-function "Force orthogonal edges."

- In the post-processing, the elements which, by means of two nodes, are part of the outer edge to be analyzed are selected.

- It must be ensured that not the membrane plane but rather the outer fiber having the highest major strain/tangential elongation is selected.

- If material that was previously in the as-delivered condition is cut and formed, the state of uniaxial tension is predominant at the edge.

    - In order to also take the anisotropy into account here, a *tolerance range on each side* must be provided.

- In order to still evaluate a slight pre-deformation before the cutting or a slightly nonlinear elongation path and yet detect excessive deviations, an analysis of the elongation path is recommended.
    - In the PAM-STAMP software, for example, the post-processing function "Zones by linearity" is suitable for this purpose.
    - In general, a comparison of the accumulated comparison strain with the comparison strain (based on the current major and minor strain) should enable this.

- It is recommended that characterization tests be performed with different edge qualities so that a working range can be defined.

These recommendations are intended as a basis of discussion that can be refined or corrected as experience is gathered.

5. Summary

In the production of complex chassis components, it has been found in daily practice that sheet metal edges produced by shear cutting have significantly reduced formability in comparison to the starting material, and that these regions tend to exhibit increased crack formation in subsequent forming operations. A concrete prediction of edge-crack sensitivity was previously not possible.

Therefore, the goal was to identify a testing method that can evaluate edge-crack sensitivity and at the same time provide input variables having low for the forming simulation. The "hole expansion with Nakajima punch" and the "Kobe hole tensile test" meet these requirements. Common to these two methods is that the evaluation can be performed by optical measurement and therefore a detailed strain analysis is possible for the region near the edge. Major strains shortly before the onset of cracking are determined and are added to the existing forming limit diagram as additional information.

In the experimental tests, it was found that the two methods lead to comparable and repeatable major strain values. Corresponding simulation models of the two edge-crack tests were constructed and were compared with the elongation measurements from the experiment with regard to the strain distribution.
The producibility of a collar drawn component as well as track control arm was examined on the basis of the determined characteristic values and the defined procedure. In both cases the described procedure provides reliable predictions. Concerning the track control arm this procedure is limited to all component edges that were not significantly formed before the cutting process and that move in the range of uniaxial tension. However, further research is necessary if there is deviation from this described elongation path or if multi-stage forming processes lead to nonlinear elongation paths.

References

[1] Buchmayer, B.: *Innovative Beiträge der Umformtechnik zum Leichtbau von Kraftfahrzeuge (Innovative Contributions of Forming Technology to the Lightweight Construction of Motor Vehicles),* BHM, volume 152, 2007

[1a] McEwan, C., Underhill, R., Langerak, N., Botman, G., de Bruine, M.: *A New Approach to Predicting Edge Splits – The Combined FLC/HEC Diagram.* International Deep Drawing Research Group 2009, Golden, USA

[2] N.N.: International standard ISO 16630, *Metallic Materials – Sheet and Strip – Hole Expanding Test,* 2009

[3] Braun, M., Borsutzki, M., Bülter, M., Carlsson, B., Larour, P., Richter, A., Atzema, E., Geisler, S.: *Ergebnisse eines europäischen Ringversuchs zum Lochaufweitungsversuch nach ISO 16630 (Results of a European Inter-Laboratory Test for the Hole Expanding Test as per ISO 16630),* Tagungsband Werkstoffprüfung (Materials Testing Conference Volume), Berlin, Germany, 2011

[3a] Schneider, M., Geffert, A., Peshekhodov, I., Bouguecha, A., Behrens, B.-A.: *Overview and Comparison of Various Test Methods to Determine Formability of a Sheet Metal Cut-Edge and Approaches to the Test Results Application in Forming Analysis*, 2015, Material Science and Engineering Technology. 46: 12: 1 145 – 1 225

[4] Braun, M.: *Determination of Edge Cracking Sensitivity – Estimation of Reduced Formability of Sheet Steel Edges Produced by Shearing Using Different Test Methods*, Umformen im Karosseriebau (Forming in Car Body Engineering), Bad Nauheim, Germany, 2013

[5] Schneider, M., Eggers, U.: *Investigation on Punched Edge Formability*, in: International Deep Drawing Research Group 2011, Bilbao, Spain, 2011

[6] Schneider, M.: *Bewertungsmöglichkeiten für das Restumformvermögen einer durch Scherschneiden vorgeschädigten Blechkante (Evaluation Options for the Residual Formability of a Sheet Metal Edge Pre-Damaged by Shear Cutting)*, EFB 2012, Bad Boll, Germany, 2012

[7] Behrens, B., Vucetic, M., Peshekhodov, I., Schneider, M.: *Numerische Beschreibung der Schädigung beim Scherschneiden und des verbleibenden Restumformvermögens für Dualphasenstähle mithilfe der Multiskalensimulation (Numerical Description of the Damage During Shear Cutting and of the Remaining Residual Formability for Dual-Phase Steels by Using Multi-Scale Simulation*, final report, Forschungsvereinigung Stahlanwendung e.V. (FOSTA), 2015

[8] Watanabe, K.; et al.: *Simple Prediction Method for the Edge Fracture of Steel Sheet During Vehicle Collision – Evaluation of Fracture Limit from the Edge Using Small-Sized Test Pieces*, 5th LS-Dyna user forum, Ulm, Germany, 2006

[9] N.N.: *Datasheet SZBS800, Salzgitter Flachstahl,* Salzgitter, Germany, 2010

[10] Gläsner, T., Imat, V., Theiß, M., Volk, W., Hoffmann, H., Golle, R.: *Erhöhung der Vorhersagegenauigkeit in der Umformsimulation durch eine optimierte Vernetzungsstrategie (Increasing the Prediction Accuracy in Forming Simulation by Using an Optimized Mesh Strategy),* http://www.umformtechnik.net/binary_data/3175583_2015-02-23-radiale-vernetzung.pdf, 2015

[11] Gläsner, T., Sunderkötter, C., Plath, A., Bednarz, M., Volk, W., Hoffmann, H., Golle, R.: *Methods to Decrease Cut Edge Sensitivity of High Strength Steels.* Key Engineering Materials 611 – 612, 2014, pp 1 294 – 1 307

[12] Gläsner, T., Schaffner, T., Sunderkötter, C., Volk, W., Hoffmann, H., Golle, R.: *Studies on the Effect of Tool Wear on the Production Process in Two Stage Shear Cutting*, Key Engineering Materials 651-653, 2015, pp 1 261 – 1 270

[13] Friebe, H., Galanulis, K., Schneider, M.: *Validation and Optimisation of Numerical Simulation by Optical Measurement of Tools and Parts*, Conference, International Deep Drawing Research Group (publisher): Olofström, Sweden, 2008

[14] Westhäuser, S., Schneider, M.: *Edge cracking sensitivity – advanced experimental and numerical methods of valuation*, International Conference on Steels in Cars and Trucks, Braunschweig, 2014

Modelling of fracture effects in the sheet metal forming based on an extended FLC evaluation method in combination with fracture criterions

P Hora[1], M Gorji[1], and B Berisha[1]

[1] ETH Zurich, Institute of Virtual Manufacturing, Tannenstr. 3, 8092 Zurich, Switzerland

E-mail: hora@ivp.mavt.ethz.ch

Abstract. The industrial based prediction in sheet metal forming bases still on the Forming Limit Diagrams (FLD) as formally proposed by Goodwin [1]. The FLD are commonly specified by the Nakajima tests and evaluated with the so called cross section method. Although widely used, the FLC concept has numerous serious limitations. In the paper the possibilities for a specific prediction of crack limits based on an extended FLC concept (X-FLC) will be discussed. The new concept demonstrates that the Nakajima tests are not only appropriate for the evaluation of the necking instability but for the detection of the real crack strains too. For the evaluation of the crack strains a local thinning method as proposed by Gorji et al. [3] is applied and tested for special 6xxx and 5xxx Al-alloys as well as for the corresponding multilayer FUSION material.

1. Introduction

In the sheet metal forming beside the limits induced by flow instability which results in localized necking also crack phenomena occurs.

Various crack types can occur e.g. cracks induced by hemming, by small die curvature radii, edge cracks as well as shear cracks (s. Figure 1). The corresponding strains are usually considerably higher than the instability limits predicted by FLC's and need another theoretical approach.

In the sheet metal forming there have been different approaches e.g. [7] and [8] to specify the corresponding strain limits by additional curves in the FLC diagrams.

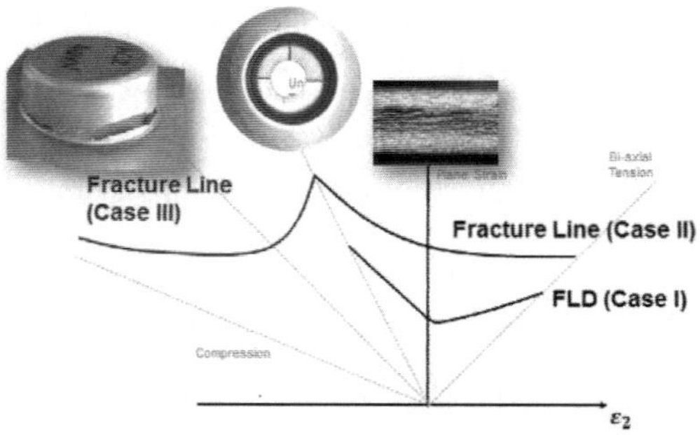

Figure 1. Different type of cracks in sheet metal forming: **(left)** Die bending crack, **(middle)** Hole expansion edge crack; **(right)** Hemming surface crack, Gorji [4].

Another approach was initiated by Wierzbicki [2] and his co-authors Bai and Bao who introduced extended Johnson-Cook models with the additional influence of the Lode-Parameter L (s. Figure 2). In this way the strain limits will be defined in dependency of stress parameters $\eta = \sigma_H/\sigma_{eq}$ and $\xi = L$ instead in dependency of strains as used in the FLC-concept.

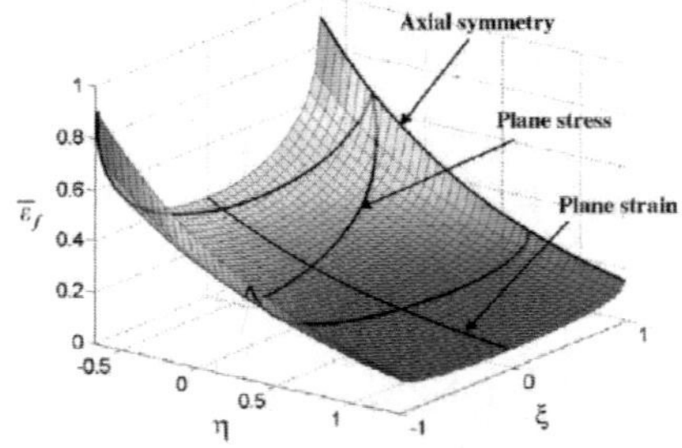

Figure 2. Characteristic shape of a triaxiality diagram [2].

Concerning the difficulties in the application of the triaxiality method two aspects have to be mentioned: the evaluation of different stress stats requires very complex specimens (s. Figure 3), which cannot be fabricated of thin sheet materials and due to the fact, that the stress-ratio changes during the experiment the parameters are path dependent and will be used in the diagram only as average values. This aspect was more deeply discussed by Gorji [4].

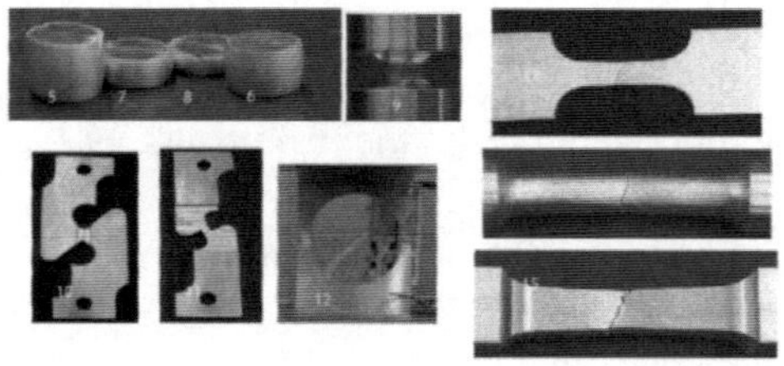

Figure 3. Proposed specimens for the evaluation of the triaxiality diagrams, [2].

Because of the mentioned weaknesses of the triaxility experiments the industrial oriented fracture tests are based typically on a bending test and on a hole expansion test, see Figure 4.

IOP Conf. Series: Materials Science and Engineering **159** (2016) 012030 doi:10.1088/1757-899X/159/1/012030

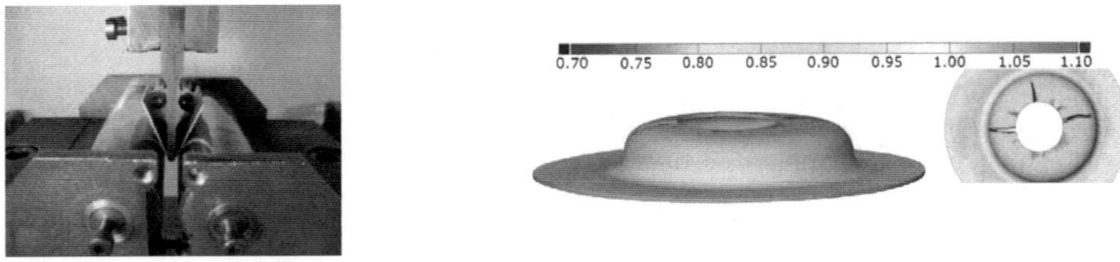

Figure 4. Sheet test for evaluation of crack behaviour: Bending test and hole expansion test.

The presented contribution shows a generalized crack prediction concept based only on Nakajima tests in combination with a special deep drawing test. The so detected experimental data will be approximated based on 4 different crack criterions, highlighting the newly introduced linear fracture line by Gorji et al. [5].

In addition to mono-layer materials also a multi-layer material, FUSION^{TM} [Novelis], has been investigated, see Figure 5. It is composed of a soft AA5005 alloy outside (clad) and a hard AA6016 alloy inside (core). The layer specific failure prediction is indispensable for multi-layer materials with strongly different behavior of the layers, as will be shown in chapter 5.

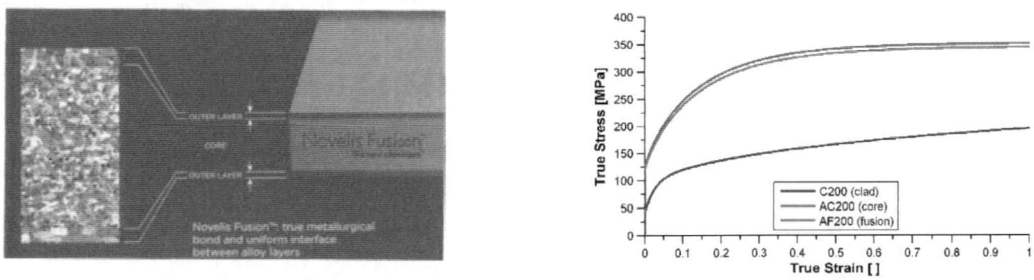

Figure 5. (left) Multilayer FUSION material composed of core and a clad material – [source Novelis]; **(right)** Yield curves of the single core-AA6016, clad-AA5005 and of the combined FUSION material. Material data are given in chapter 2, Table 1 and Table 2.

Figure 5 right shows that both materials, core and FUSION, have practically the same yield curve, whereas the yield curve of the clad material is much lower. The crack-prediction method – linear fracture line - will be demonstrated in an application on a part called "triangle test" with a very small curvature radius of 3 mm for two different materials: in Figure 18 for a mono-layer AA6016 material and in Figure 21 for the multi-layer Al sheet FUSION.

2. Material data

In the framework of this investigation both materials, AA6016 and in combination with the FUSION AA5005 as clad, have been used. The specific material models are given below.

Uniaxial tensile tests and the hydraulic bulge tests were performed to determine the yield curves. The well-known Hockett-Sherby approximation has been used for the description of the core material AA6016 and FUSION AF200

$$\sigma_y = B - (B - A) \exp[-m\varepsilon_{eq}^n] \quad (1)$$

whereas for the clad material AA5005 a combined approach of Ghosh and Hockett-Sherby showed better approximation

$$\sigma_y = k(C_1 + C_2(C_3 + \varepsilon_{eq})^{C_4}) + (1-k)(B - (B - A)\exp[-m\varepsilon_{eq}^n]) \quad (2)$$

The parameters of the models are given in Table 1.

Table 1. Parameters of the yield curves

	A	B	m	n	C_1	C_2	C_3	C_4	k
AA6016	123.7	352.4	5.62	0.87					
AA5005	39.7	123	88.72	1.23	8.9	234.8	0.003	0.34	0.6
FUSION	119.2	345.4	5.44	0.862					

Further, parameters of the YLD-2000 yield loci are given in Table 2:

Table 2. Material properties and YLD-2000 parameters determined at 4 % equivalent strain

	σ_0	σ_{45}	σ_{90}	σ_b	R_0	R_{45}	R_{90}	R_b
AA6016	189.2	182.6	184	186.7	0.67	0.5	0.67	1
AA5005	95.8	90.8	94.1	91.3	0.68	0.33	1.33	1
FUSION	184.8	176.1	176.7	176.5	0.69	0.43	0.73	1
	α_1	α_2	α_3	α_4	α_5	α_6	α_7	α_8
AA6016	0.947	1.017	0.961	1.032	1.021	1.013	0.967	1.153
AA5005	0.8757	1.1421	1.1462	1.0152	1.033	1.0469	0.938	1.1879
FUSION	0.9201	1.0806	1.0183	1.0508	1.0377	1.0842	0.9686	1.1564

The X-FLC applies a FLC as well as the crack failure data, as shown in Figure 6 and Figure 7:

AA6016	
ε_2	ε_1
-0.215	0.43
-0.086	0.288
-0.036	0.238
-0.023	0.228
0	0.22
0.011	0.218
0.022	0.22
0.045	0.225
0.11	0.245
0.186	0.265
0.28	0.28

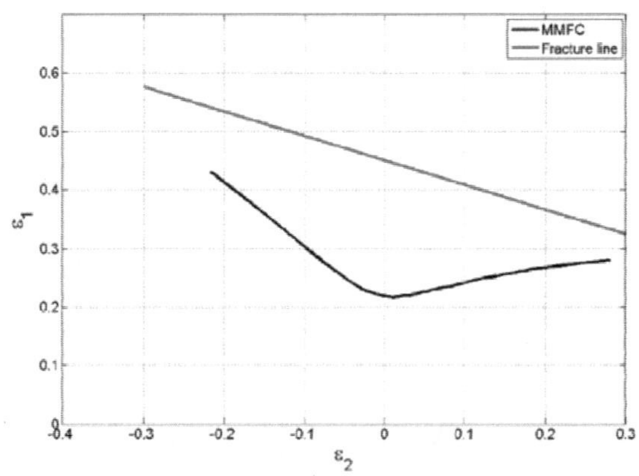

Figure 6. X-FLC data for AA6016: **(left)** Data based on MMFC approach (FLC-3 as shown in section 5.3); **(right)** The corresponding plot of the FLC and the fracture line.

The fracture strains for the core material (AA6016) are approximated with the following linear model

$$\varepsilon^f = 0.45 - 0.42 * \varepsilon_2 \quad (3)$$

The same procedure is applied also for the clad material AA5005 shown in Figure 7.

AA5005	
ε_2	ε_1
-0.308	0.615
-0.108	0.36
-0.039	0.26
-0.025	0.245
0	0.23
0.012	0.233
0.024	0.238
0.051	0.255
0.14	0.31
0.249	0.355
0.388	0.388

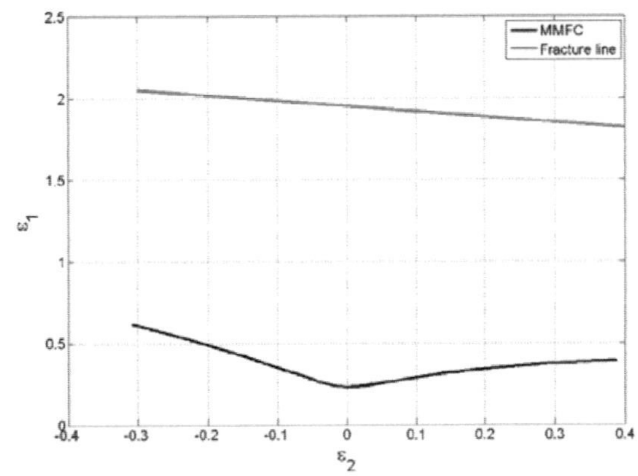

Figure 7. X-FLC data for AA5005: **(left)** Data based on MMFC approach (FLC-3 as shown in section 5.3); **(right)** The corresponding plot of the FLC and the fracture line.

The fracture strains for the clad material (AA5005) are approximated with the following linear model

$$\varepsilon^f = 1.95 - 0.33 * \varepsilon_2 \tag{4}$$

The fracture strains of the clad material AA5005 are much higher than those of the core AA6016, as shown in Figure 7 and Figure 6, respectively.

3. Limitations of standard FLC prediction methods

The standard FLC evaluation methods bases on Nakajima tests which use spherical punches with diameter of 100.0 mm. For sheet thicknesses of 1.0 mm the bending influence is then negligible. For those reasons the FLC describes only the necking behaviour for practically flat sheets. If the t/r ratio (sheet thickness to radius ratio) becomes significant, the FLC will be transformed to higher values. A more detailed discussion about the influence of the curvature on the FLC can be found in Hora et al. [6].

3.1. Influence of bending effects on deep drawing behaviour

How significant the bending influence is, can be demonstrated by following example. Figure 8 shows a deep drawing example with a die radius of r=3.0 mm for the material AA6016.

Figure 8. FLC based prediction of the critical depth H. Die radius 3.0 mm, material AA6016. **(left)** Experiment H ≈ 29 mm. **(right)** theoretical prediction H ≈14.5 mm Gorji [3],[4].

IOP Conf. Series: Materials Science and Engineering **159** (2016) 012030 doi:10.1088/1757-899X/159/1/012030

The cracks in the region of the die curvature will be often identified as "shear cracks". An investigation done by Gorji [4] with different square blanks and different blank holder forces demonstrated, that dependent on the selected parameter combination the localized necking occurs as classical bottom neck or as a upper radius crack, see Figure 9. By increasing the die radius to r=5.0 mm the upper crack disappears. In Figure 9 the parameter β represents the ratio between edge length L of the squared blank and the punch diameter.

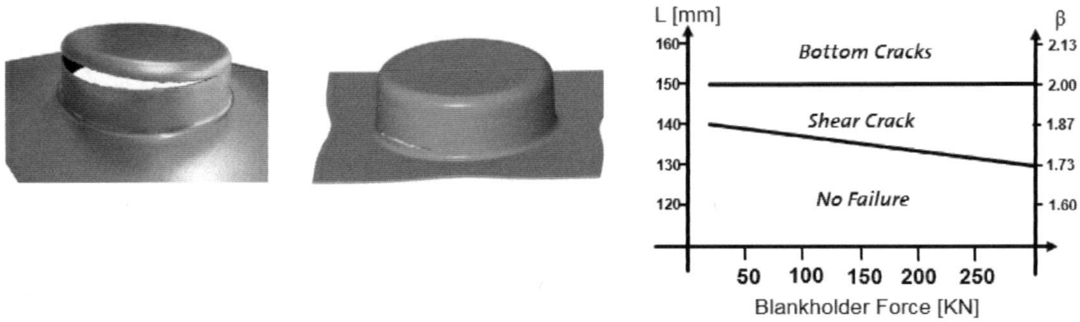

Figure 9. Formability diagram of AA6016 sheet sample, Gorji [4].

Figure 8 demonstrates that a FEM simulation based on shell elements and classical FLC is not able to predict this influence correctly.

4. Experimental detection of crack limits for sheets

For the evaluation of such shear cracks the physical strain crack limits $\varepsilon_{maj}^{crack}(\beta)$ and not the classical $\varepsilon_{maj}^{FLC}(\beta)$ are needed, where beta is the strain ratio $\beta = \varepsilon_{maj}/\varepsilon_{min}$.

In contrast to the specimens shown in Figure 3 the goal was to use the established "standard" sheet tests. For this reason, the authors proposed the combination of two methods – the Nakajima tests combined with an additional cup drawing test for detecting the behaviour for $\beta < -0.5$.

4.1. Experimental detection of crack limits based on Nakajima tests

The first method evaluates the crack strain based on the thinning strains ("Thinning Method") measured on the fractured Nakajima specimens, see Figure 10. The detailed evaluation procedure is given in [4] and [5].

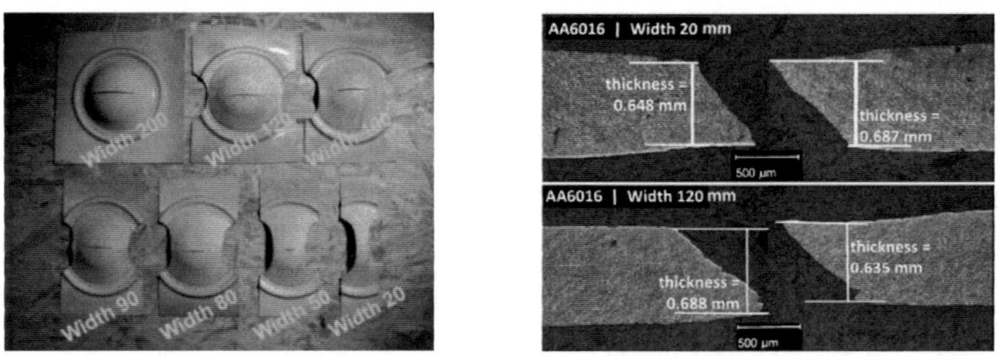

Figure 10. Evaluation of the fracture strain by the local detection of fracture thinning on the Nakajima specimens. Gorji [3],[4].

Figure 11 shows the so detected crack strains $\varepsilon_{maj}^{crack}(\beta)$ for the material AA6016 in relation to the classical FLC limits evaluated by the cross section as well as the time dependent method. The parameter B represents the width of the plastic zone in the so called LL-FLD (Localized Level - FLD) as proposed by Hora et al. [9]

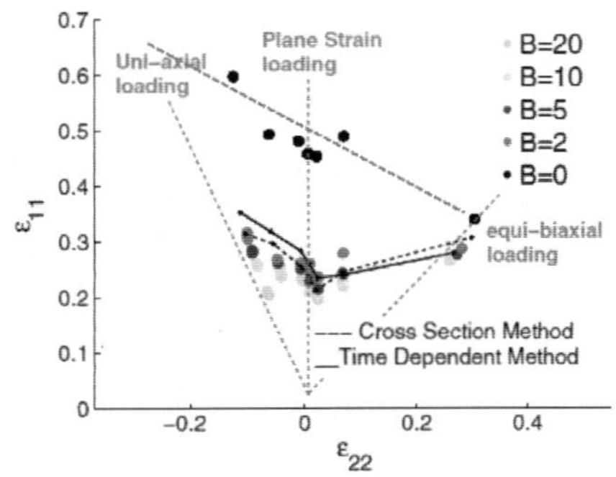

Figure 11. Evaluation of the fracture strain by the local detection of fracture thinning on the Nakajima specimens and comparison with the FLC. Gorji [3],[4].

4.2. Experimental detection of crack limits based on Deep Drawing tests

The Nakajima based test are restricted to the stress range

$$0 \leq \alpha = \frac{\sigma_2}{\sigma_1} \leq 1.0$$

In many deep drawing applications, the largest strains occur on the left side of the FLC. For those reasons a special DD test with a quadratic blank and a relatively small die curvature of r=3.0 mm was applied to get an additional "point" specifically in the deep drawing (compression-tension combination) range. The evaluation of the strains is based on the comparison of the real part (height and draw in at fracture) with the FEM simulation, Figure 12. As critical fracture strain, the strain on the surface of the sheet (top layer of the shell), was defined.

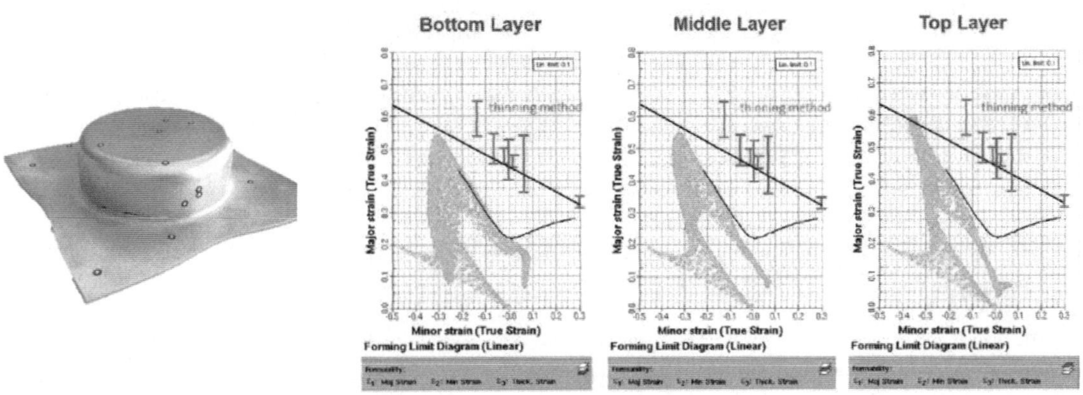

Figure 12. Calibration of the fracture line based on a deep drawing experiment and the corresponding strain distribution at the crack time step, Gorji [4].

The combination of the Nakajima fracture strains with the additional DD fracture strains can be used as data base for the determination of a generalized fracture line.

4.3. Extrapolation and interpolation of the fracture points based on different failure criteria

The experimentally predefined strain has been compared with the theoretical limits of different failure criteria. The check was especially done applying following 4 different criterions:

- Maximum shear stress criterion
- Equivalent strain criterion
- Johnson-Cook criterion and
- Linear fracture line criterion

Figure 13 shows the results of the theoretical fitted failure curves with the experimental points, which were determined from the Nakajima experiments shown in Figure 10. Given that the fracture strains in the principal strain space show practically a linear distribution (Figure 13 left), it is reasonable to use a linear model for the description of the fracture behaviour. Moreover, a stable behaviour of the model is ensured, as there is no rapid change in the description of the fracture behaviour, see (Figure 13 left).

Remarkable is the fact, that all criterions fit the position of the measures points quite well, but that the criterions deviate significantly in the left "deep drawing" range. Especially this range influences strongly the virtual results for deep drawing operations.

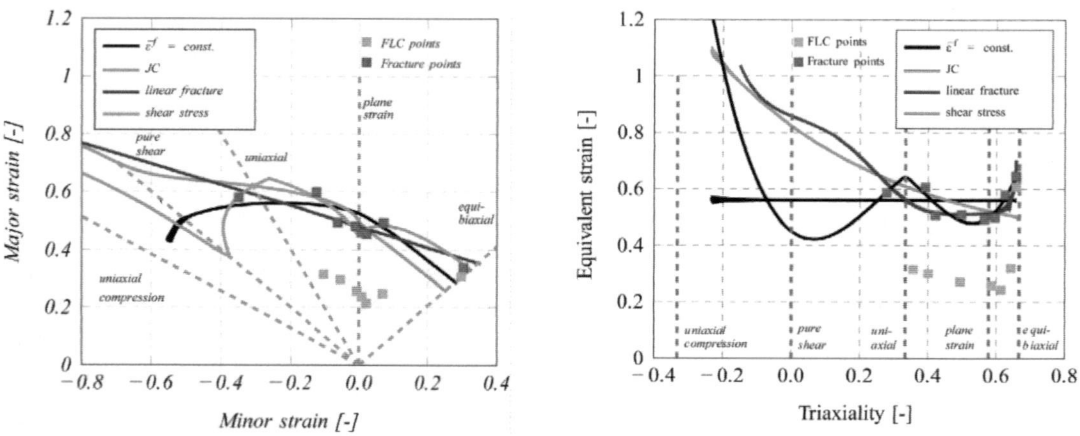

Figure 13. Comparison of different fracture criteria in principal strain space and in triaxiality-equivalent strain space, [4].

As will be demonstrated in chapter 5 the "linear fracture line" describes the behaviour in the most accurate way and delivers for real applications the best fits with the real behaviour. The JC-criterion will deliver only slightly different results.

5. FE-Implementation of the X-FLC concept for monolayer and multilayer materials

5.1. Layer based failure prediction with shell elements

The classical failure predictions bases on a mono-layer FLC prediction. If the fracture is initiated by a surface crack, as it is the case by small bending radii, an extended X-FLC method has to be applied, see Figure 14. The crack develops and initiates only when the critical layer reached the crack limit. Points above the FLC will then be interpreted as conditional stable (this corresponds to the real physical behaviour!) and not as usually assumed as failed by cracks.

IOP Conf. Series: Materials Science and Engineering **159** (2016) 012030 doi:10.1088/1757-899X/159/1/012030

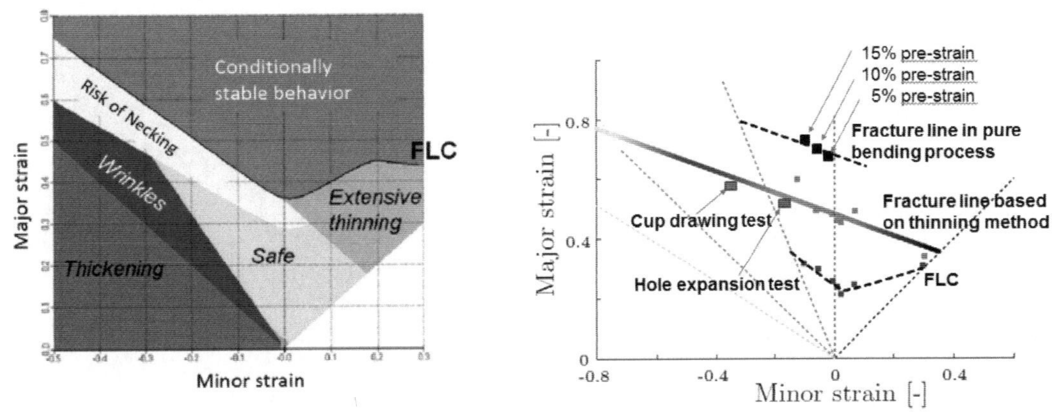

Figure 14. Extended X-FLC description considering the fracture ("crack") limits by the bending strains.

The FEM-implementation was done in that way, that the crack failure was checked specifically for each layer. If a critical strain was detected, the specific shell layer was deactivated by setting the stresses to zero. In the LS-Dyna code the implementation was done by the subroutine UMAT41. The *PART COMPOSITE functionality of FE-code LS-Dyna has been employed instead of the regular shell element. Based on this element formulation the mechanical properties and thickness distribution of each layer can be described separately. The implementation contains following main steps:

- computation of the principal strains from the strain tensor
- comparison of the computed principal strains with the fracture line
- if the computed strains are above the fracture line, then all stress components are set to zero

5.2. Validation tests

For the validation of the above implementation a new "triangle" test was designed. The tests have been done with the monolayer material AA6016 as well as with the FUSION material. Figure 15 demonstrates the significant change of the deep drawing behaviour for the both materials.

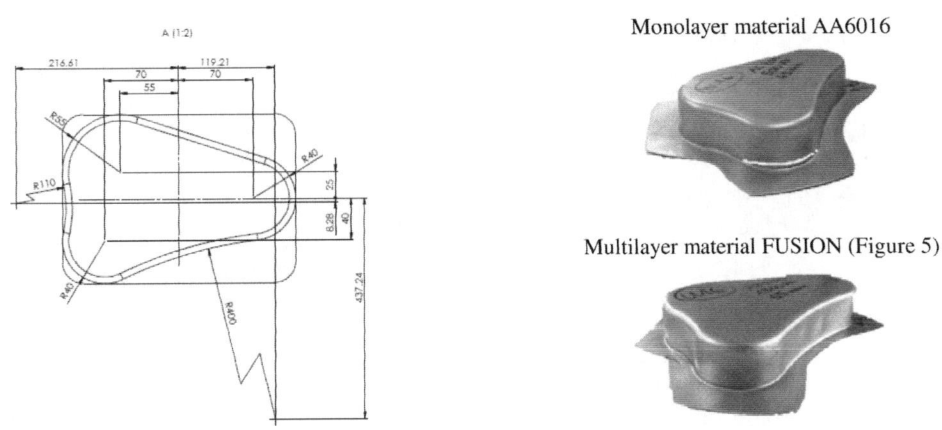

Figure 15. Triangle part - **(top)** Monolayer material AA6016. Critical depth H = 43 mm; **(bottom)** FUSION material still no failure at H=55 mm.

IOP Conf. Series: Materials Science and Engineering **159** (2016) 012030 doi:10.1088/1757-899X/159/1/012030

The crack of the AA6016 material, in the die region, is induced by the small die radius of 3.0 mm. Due to the better bending behaviour of the FUSION material, the multilayer material does not fail. This special crack behaviour can be only understood on the base of an additional fracture consideration as introduced by the X-FLC concept.

5.3. Prediction of crack limits for parts with small die radii and a monolayer structure

The experimental behaviour of the monolayer material was demonstrated in Figure 15. The rupture occurs at a depth of H ≈ 43 mm.

Before the application of the extended X-FLC method the classical FLC approach shall be discussed. For the determination of the FLC three different methods have been compared:

- FLC (1) Evaluation of the Nakajima test – Cross Section evaluation method. Most left point B20
- FLC (2) Evaluation of the Nakajima test – Cross Section evaluation method. Most left point B50
- FLC (3) Numerically evaluated FLC – MMFC criterion [6], adjusted with FLC0 (plane strain point)

For the numerical failure prediction a relative failure value of $\varepsilon_{maj}/\varepsilon_{FLC}$=1.0 will be used as critical measure. The specific results are plotted in Figure 16: method FLC (1) delivers a critical depth of H=22 mm, with method FLC (2) the depth increases to 41 mm. Method FLC(3) shows approximately the same behaviour as (2) with H=45 mm.

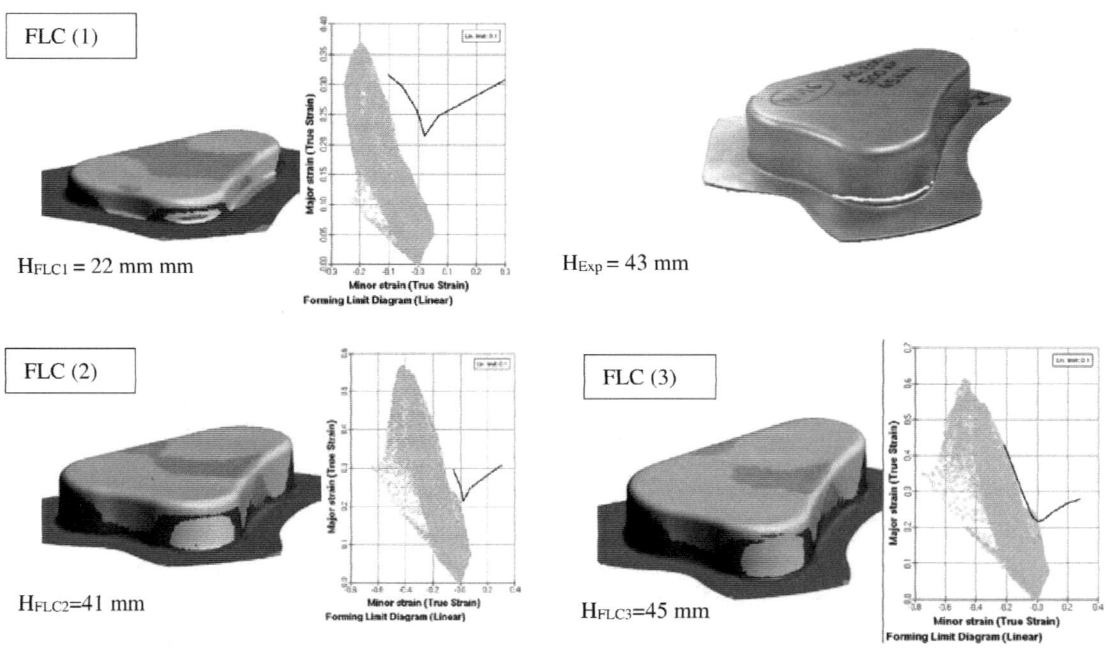

Figure 16. Triangle part. Monolayer material AA6016. **FLC 1:** Prediction of critical depth based on a monolayer measured FLC (1): height H = 22 mm; **FLC 2:** the same FLC without the "tensile" point FLC (2), height H = 41mm. **FLC 3:** MMFC-criterion adjusted with FLC0, height H = 45mm.

The differences of the 3 methods are obviously caused by the changes of the FLC in the extrapolated left range, see Figure 17. Only a slight difference in the slope, based on the left two FLC points, influences strongly the "classical" FLC-prediction of the critical state. In this sense the use of the Nakajima B20 specimen (method 1) is misleading, if extrapolation on the left side of the FLC is needed. The reason may be the stress boundary condition in the case of the B20 specimen instead of the strain BC in the real DD case. To avoid this uncertainty, the shape of the FLC on the left side should be checked by additional cup drawing test as showed for example in Figure 12. But even then the prediction of the critical depth is still not "robust" in a process control sense.

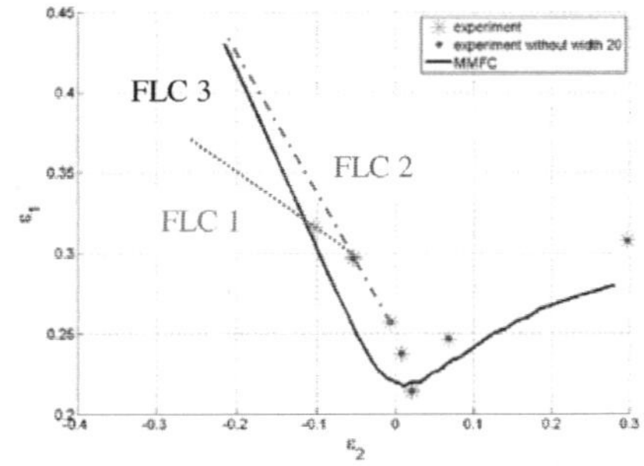

Figure 17. Comparison of the different FLCs in the extrapolated left range (Material: AA6016).

A "robust", reliable prediction in the case of the very small die radii can be only achieved if the extended X-FLC concept is used. The extended concept is based on:

- FLC determination with method (FLC 3) and validated with an additional cup drawing test
- determination of the fracture line based on the thinning method
- Layer based evaluation of the strains

Figure 18 demonstrates the correct prediction of the depth as well as of the crack position for the triangle test.

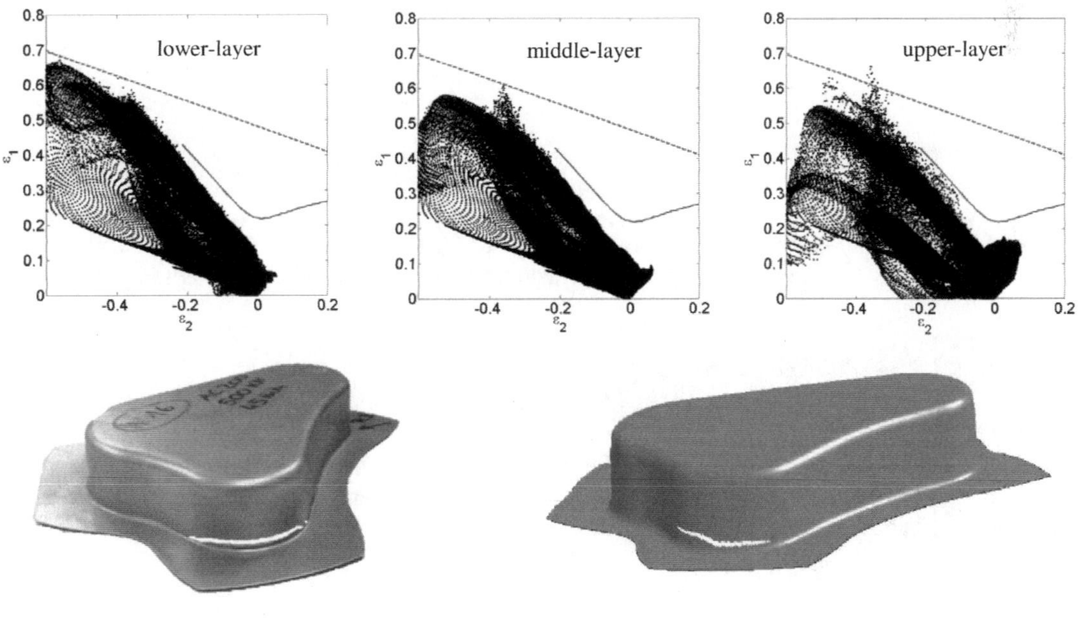

Figure 18. Material AA6016. FE Simulation of triangular experiment with linear fracture criterion – dashed line (strain distribution of lower, middle and upper-layer); solid line: FLC based on MMFC-criterion, Gorji [3].

5.4. Prediction of crack limits for multilayer FUSION materials

Beside Figure 15 right for the triangle part also Figure 19 for a simple cup drawing part demonstrates again clearly how significantly the DD behaviour for the multi-layer material changes.

Monolayer Material AA6016 **FUSION Material**

Figure 19. Comparison of the forming behaviour: **(left)** monolayer material AA6016; **(right)** FUSION material, Gorji [4].

The reason for the significant change of the forming behavior is not a shift of the FLC – which is more or less the same for the monolayer and FUSION-material – but the significant shift of the fracture strain for the clad material, as proven with the "thinning method" and showed in Figure 20.

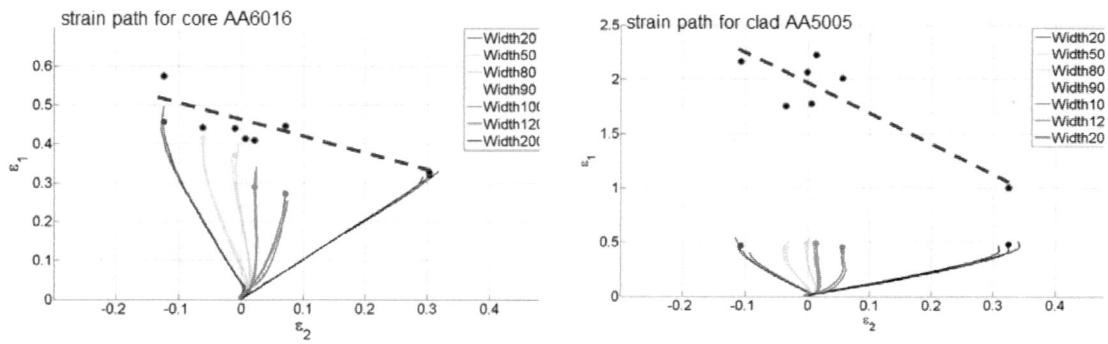

Figure 20. Evaluation of the fracture strains for the core (AA6016) and the clad (AA5005) materials, where dots represent measured data based on thinning-method.

5.5. FE-model for multilayer composite materials

For the modelling of the multi-layer FUSION material, Figure 5, eleven integration points (IPs) through the thickness of the composite shell elements have been employed. One IP for each clad outer layer with the thickness of 0.06 mm and nine IPs for the core with thickness of 0.0978 mm for each layer with a total core thickness of 0.88 mm. The integration points of the core and clad have their own material properties i.e., hardening curve, standard or modified Yld2000-2d yield function and fracture limit as specified in Figure 14 and Figure 20.

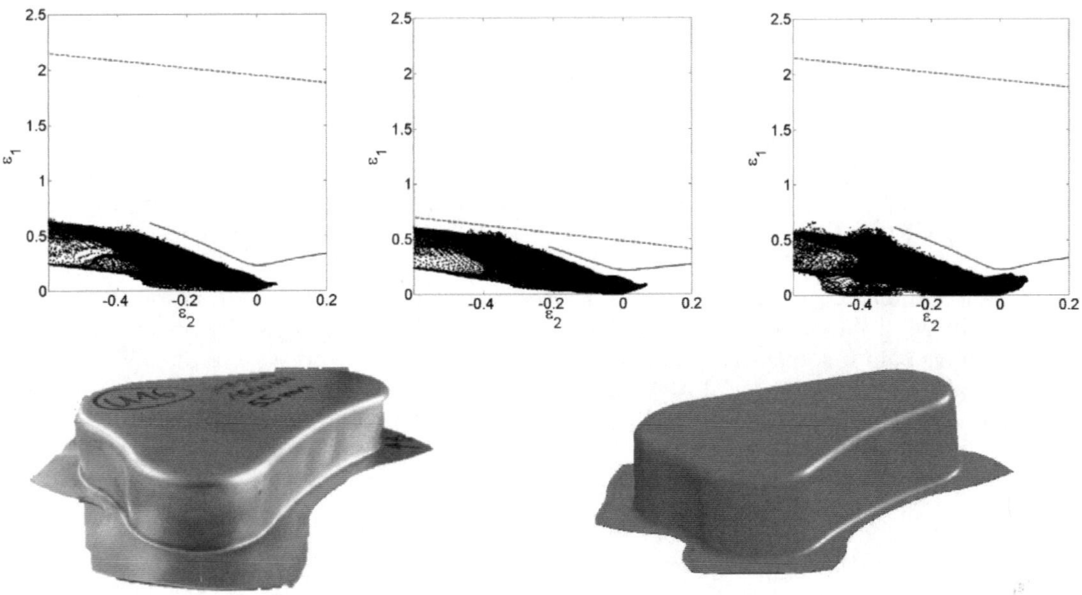

Figure 21. Forming behaviour of FUSION material: FE simulation of triangular experiment with linear fracture criterion (dashed line) specifically defined for the layers (strain distribution of lower, middle and upper-layer); solid line: FLC based on MMFC-criterion, Gorji [3].

Figure 21 demonstrates, that based on the X-FLC concept, the "not failed" behavior of the FUSION material was again correctly predicted.

6. Conclusions

The contribution demonstrates on selected examples that the classical FLC prediction is not applicable if the parts have either small die radii or are composed by layers with different properties.

In this case - beside the FLC - a crack limit curve has to be specified too. For the detection of such critical strain the thinning method, evaluating the fracture strains on Nakajima specimens, have been used. The multi-layer failure identification was implemented into the explicit FEM code LS-Dyna. It was also shown that, in contrast to the monolayer material 6xxx, the presented FUSION material allows forming operations without failure even for small die radii.

Acknowledgements

The authors are very grateful to the CTI (The Swiss Innovation Promotion Agency) for the financial support of part of this work within the project 13082.1 PFIW-IW. Also thanks to J. Timm and E. Combaz (Novelis Switzerland SA) for providing the material and performing a part of the experiments. K. Wiegand (Daimler AG), M. Selig (AutoForm Development GmbH) and E. Müller (GOM International AG) are also gratefully acknowledged for their contribution on the project results.

References

[1] G. M. Goodwin, Application of strain analysis to sheet metal forming problems in the press shop, Soc. Of Automotive Engineering, Nr. 680093, pp. 380-387, 1968

[2] Y. Bai, T. Wierzbicki: Application of extended Mohr-Coulumb criterion to ductile fracture, Int J Fract (2010) 161:1–20

[3] M. Gorji, B. Berisha, P. Hora, F. Barlat: Modeling of localization and fracture phenomena in stress and strain space for sheet metal forming, Int J of Mater Form, DOI 10.1007/s12289-015-1242-y, 2015.

[4] M. Gorji, Instability and fracture models to optimize the metal forming and bending crack behavior of Al-Alloy composites, Diss ETH 23066, 2015

[5] M. Gorji, B. Berisha, N. Manopulo, P. Hora. Effect of through thickness strain distribution on shear fractur hazard and its mitigation by using multilayer aluminum sheets, J Materials Processing Technology, 232: 19-33, 2016

[6] P. Hora, L. Tong, M. Gorji, N. Manopulo, B. Berisha: Significance of the local sheet curvature in the prediction of sheet metal forming limits by necking instabilities and cracks. Proceedings of NUMIFORM -2016.

[7] R. Schleich, M. Sindel, M. Liewald, Investigation on the effect of curvature on forming limit prediction for aluminium sheet alloys, International Journal of Material Forming, 2:69-74, 2009

[8] K. Isik, M. B. Silva, A.E. Tekkaya, P.A.F. Martins, Formability limits by fracture in sheet metal forming, J of Processing Technology, 214:1557-1565, 2014

[9] P. Hora, B. Berisha, M. Gorji, N. Manopulo, A generalized approach for the prediction of necking and rupture phenomena in the sheet metal forming, IDDRG12, 79-93, 2012